POLLUTION PREVENTION HANDBOOK

Edited by
Thomas E. Higgins

LEWIS PUBLISHERS

Boca Raton London Tokyo

Library of Congress Cataloging-in-Publication Data

Higgins, Thomas E., 1948–
Pollution prevention handbook / Thomas E. Higgins.
p. cm.
Includes bibliographical references and index.
1. Factory and trade waste—Management. I. Title.
TD897.H53 1995
628.4—dc20 94-44430
ISBN 1-56670-145-7

Lewis Publishers is an imprint of CRC Press

International Standard Book Number 1-56670-145-7

Library of Congress Card Number 94-44430
Printed in the United States of America
1 2 3 4 5 6 7 8 9 0

Printed on acid-free paper

This book is dedicated to our families. You endured our absences while we were working on the projects described in this book and were putting the lessons learned from these projects down on paper for the book. Thank you for your patience and support.

We also thank the staff who worked on the projects used to develop materials for the book. Your efforts were critical to the level of detail and preparation of the book and the high quality of the examples and case studies that we included.

Last, but not least, we wish to thank our clients. We thank you for inviting us to work on your projects. They challenged our creativity and were professionally rewarding. And we thank you for turning pollution prevention from a regulatory theory into a cost-effective practice by "giving it a try" at your facilities.

About the Author

Thomas Higgins is a Vice President and Director of Industrial Water and Wastewater Engineering, and Director of the Office of Innovation for CH2M HILL. During his 10 years at CH2M HILL, Dr. Higgins has been project manager and technical adviser on pollution prevention projects for numerous industries, including aluminum, aerospace, defense, chemicals, computer, electric utilities, electronics, explosives, household appliance, iron and steel, leather tanning, metal plating and surface finishing, photoprocessing, optical and transportation companies in the United States as well as in Australia, Canada, Hungary, Poland, Russia and Taiwan. He has written over 60 papers and has given numerous lectures and classes on waste management and pollution prevention. Before joining CH2M HILL, Dr. Higgins was an Associate Professor of Civil Engineering at Arizona State University, where he consulted, performed research and taught courses in environmental engineering. He holds a B.S. in civil engineering and an M.S. and a Ph.D. in environmental engineering from the University of Notre Dame. He is a registered professional engineer in eight states and is a Fellow of the American Society of Civil Engineers, a Diplomate of the American Academy of Environmental Engineers and a member for the Water Environment Federation.

About CH2M HILL

In the mid-1930s, three Oregon State College engineering students—Holly Cornell, Jim Howland, and Burke Hayes—spent time after class talking with one of their professors, Fred Merryfield, about starting up an engineering firm some day. Following graduation, these students went on to graduate school, and into industry. With the outbreak of World War II, all four entered the military services, where they served as military engineers. Though separated, the four continued to correspond and plan. In January, 1946 in Corvallis, Oregon, they formed the partnership of Cornell, Howland, Hayes, and Merryfield, specializing in providing sanitary engineering design services in the Pacific Northwest.

Later, the name CH2M came about. Some say a client suggested it; others remember it as resulting from a word game, derived from the chemical name for water,

H2O. However the name evolved, the firm earned a reputation for developing innovative water treatment technologies, including multimedia filtration, which led to their design of the world's first major advanced wastewater treatment plant at Lake Tahoe. The Tahoe project led to a merger with Clair A. Hill and Associates of Redding, California, in 1971, and the addition of HILL to the firm's name.

Throughout its history, the firm has enjoyed steady growth geographically, expanding to more than 70 offices with 6,000 employees providing environmental consulting and engineering services to industrial and governmental clients worldwide. An important part of many projects has been providing pollution prevention services, such as baseline surveys, pollution prevention opportunity assessments and plans, technology demonstrations, new facility and facility modification designs, startup and staff training.

From its founding in two small rented rooms above a hardware store in Corvallis Oregon, CH2M HILL has grown to be among the largest architectural/engineering firms in the world and a leader in providing environmental engineering services. Based on the long-range vision of its early principals, the company is entirely employee-owned.

Preface

Waste and the pollution that inevitably results from it are a part of our everyday lives. We experience it when we eat a candy bar and are left holding a sticky wrapper. We are faced with a significantly larger disposal problem when an automobile reaches the end of its useful life. When painting around the house, we expose ourselves and the environment to solvent vapors. It has been reported that each American generates over twice his or her weight in waste each day, when municipal solid waste, industrial solid waste and air emissions are included.

What can be done personally to reduce this waste generation and to prevent the resulting pollution? We can buy products that have minimal or reusable packaging. We can buy automobiles with above-average repair records and keep them for a longer time. We can paint with water-based paints to reduce air emissions and eliminate the need to use organic solvents for cleanup.

Preventing pollution at work is an extension of these common-sense practices. McDonald's found that the bulk of their solid wastes was generated behind the counter—principally packaging for the materials coming in through the back door. They instituted a program in which their suppliers used returnable containers. This significantly reduced their waste generation and costs. They also reduced the volume of waste generated by their customers by switching from styrofoam containers to paper wrappers for their sandwiches.

This book is written for people interested in preventing pollution in the workplace. It is written for the manager of a government facility tasked with preparing a pollution prevention plan; for the foreman at an automobile assembly line looking for ideas on how to improve productivity; for the environmental manager of a multinational company looking for a less expensive method to achieve compliance with environmental regulations.

The initial basis for this book was a series of CH2M HILL reports for the Department of Defense evaluating the effectiveness of waste minimization projects at government-owned manufacturing and maintenance facilities. The findings and case studies in those reports were combined with lessons learned from numerous other CH2M HILL projects in preparing *The Hazardous Waste Minimization Handbook*, published in 1989. This current book, *Pollution Prevention Handbook*, updates the earlier book and substantially expands it to include pollution prevention techniques applicable to air emissions, wastewater and solid wastes in addition to hazardous wastes. Each of the chapters was written by present or former employees of CH2M HILL who wrote from their experiences on projects with the firm.

The book is divided into three major sections. The first section deals with setting up a pollution prevention program. There are chapters on laws and regulations, performing pollution prevention assessments, setting up recycling programs, using a waste exchange, treatment to reduce waste disposal and performing economic evaluations of projects.

The second section deals with operations common to a broad range of industries or maintenance facilities: machining and metalworking, solvent cleaning, CFC use for refrigeration and fire protection, metal plating and surface finishing, painting, paint removal; and motor oil and antifreeze changeout. The third section deals with specific industries: aluminum, construction and demolition, electric utilities, food processing, iron and steel, petroleum, pharmaceuticals, and pulp and paper. Each chapter provides data on waste generation, followed by proven pollution prevention techniques illustrated with examples and case studies.

Contents

List of Figures

Figure

List of Tables

Table

List of Case Studies

Case Study

POLLUTION PREVENTION HANDBOOK

CHAPTER 1

Introduction and Purpose

*Thomas Higgins**

WHAT'S IT CALLED THIS WEEK— OR CAN'T ANYBODY MAKE UP HIS MIND?

Waste minimization? Waste reduction? Pollution prevention? Source reduction? Toxics use reduction? Responsible care? Why have so many terms been developed, which is correct, and why do they keep changing? The U.S. Environmental Protection Agency (U.S. EPA), Congress, and environmental and industrial groups keep coining new terms, and the result is a confusing collection of remarkably similar terms having politically motivated, subtle differences.

Regardless of which name it is given, pollution prevention has been a long-term congressional goal, as is evident in most environmental legislation. The Federal Water Pollution Control Act of 1972 set a goal of "zero discharge" of wastes to the nation's waters by 1985. The original hazardous waste act enacted in 1976 was titled the Resource Conservation and Recovery Act (RCRA), not the Hazardous Waste Disposal Act, and the Hazardous and Solid Waste Amendments (HSWA) to RCRA enacted in 1984 included:

> The Congress hereby declares that it is to be a national policy of the United States that, where feasible, the generation of hazardous waste is to be reduced or eliminated as expeditiously as possible. Waste nonetheless generated should be treated, stored, or disposed of so as to minimize the present and future threat to human health and the environment.[1]

All the names for pollution prevention have similar meanings and are frequently used interchangeably. Toxics use/reduction and source reduction are the most narrow terms, being restricted to raw material or production process changes. Minimization incorporates out-of-process recycling and treatment to reduce the quantity of material disposed of. Pollution prevention promotes a hierarchy that favors source reduction, but also promotes other minimization techniques or treatment where

*Dr. Higgins is a vice president in CH2M HILL's Reston, Virginia, office and may be reached at (703) 471–1441.

production changes are infeasible. These distinctions are more political than practical.

From a private-sector perspective, the best approach is one that cost-effectively reduces waste, minimizes worker exposure to toxic materials, optimizes use of raw materials, improves a product's competitiveness in local and world markets, and enhances a company's image of responsibility to the community. Compared to meeting these goals, the argument over terminology is trivial. To describe the broad range of alternatives available to achieve these goals, I will use the term pollution prevention throughout this book.

SUCCESSFUL WASTE REDUCTION—A CURE FOR WHAT AILS US

The significant issues of the day—global competitiveness, the consumer movement, environmental issues, work force issues, local community issues, and the end of the Cold War—all have a common focus and potentially, a common solution. These issues involve the public, as well as academic, regulatory, and business communities. The responsibility for resolving problems associated with these issues also rests with these communities. Proposed methods for dealing with these problems constitute the glue that binds this book together.

Global Competitiveness Issues

Competing in the world market involves a number of seemingly unrelated issues. As a result of the continuously improving *productivity* of our international competitors, it has become commonplace for the cost of electronics goods to stay the same or decrease while the features of the goods are improved. At the same time, *resource scarcity* is causing the price of raw materials used in production to increase. Some argue that U.S. manufacturers face a *regulatory burden* that many of our competitors do not, making it difficult for American products to compete in the world market. Finally, improving *product quality* is critical in any market.

Example. Honda and Toyota have been able to charge a premium for their cars, which are perceived by their fiercely loyal customers to be of high quality, whereas other car companies have had to use rebates, low-interest financing, and high-pressure sales tactics to sell their cars.

Consumer Issues

The "*green movement*" affects individuals' buying decisions. Products that are perceived to be environmentally friendly (that is, "green") command premium prices. Our biggest global market is our domestic market, and *product quality* is just as critical at home as it is abroad.

Environmental Issues

Businesses face significant liabilities for the environmental impacts of their operations. *Hazardous waste liability* cannot be eliminated, not even by contracting with a responsible disposal firm. Increasingly, corporate officers risk *criminal liability* for violations of environmental permits. *Permit standards* are becoming increasingly stringent, thus requiring increasingly sophisticated technologies that increase the cost of treatment. Businesses are also affected by global environmental issues such as *ozone depletion, global warming*, and *nuclear waste disposal*.

As part of the HSWA, every generator of hazardous waste was required as of September 1, 1985, to certify on manifests that:

> A program is in place to reduce the volume or quantity and toxicity of hazardous waste determined to be economically practicable.

and that:

> The proposed method of treatment, storage, or disposal will minimize the present and future threat to human health and the environment.

The Office of Technology Assessment (a research arm of Congress) argued in a report to Congress that treatment is not a desirable practice, that only *reduction* of waste at the source or during the production process is acceptable, and that treatment or even recycling, if it is not directly connected to the production process, is unacceptable.[2]

The U.S. EPA, in attempting to come up with a term acceptable to others, borrowed from the name of a waste reduction program developed by the 3M Company, Pollution Prevention Pays. Abbreviated as 3P, the name was a play on the company's name. The 3M Company emphasized that reducing discharges through production changes could pay for itself by reducing the cost of waste treatment and disposal and optimizing the use of raw materials. The EPA adopted two of the three Ps—"pollution prevention"—to replace the term "waste minimization."

Environmental groups favor *source reduction*, a term that stresses changes in production processes to reduce generation of wastes. They also favor *toxics use reduction*, a term used to denote reduction or elimination of toxic raw materials used in production.

The chemical industry has promoted *responsible care*, their term for a conscientious approach to the production and use of chemicals throughout their life cycles.

Work Force Issues

Workers are exposed to toxic chemicals in the workplace. Illnesses that can result from this kind of exposure cost employers in terms of lost time, increased health insurance costs, and disability payments. Reducing the use of toxic or hazardous materials can result in *a safe and healthy work environment* that is as important to workers as skyrocketing *health insurance* premiums are to management.

Community Issues

Businesses have a significant responsibility to their local communities. *Community right-to-know* laws require reporting of hazardous material releases to the local environment, including those that are legally permitted. Many local *landfills* are filling up, and new ones are difficult to site and increasingly expensive.

An Outbreak of Peace

Who would think that winning the Cold War would result in environmental problems such as those associated with disposing of *nuclear disarmament wastes*, *cleaning up Russian and American military installations*, and *reducing the emissions* from government-run state enterprises.

The Common Thread—Pollution Prevention

All of these issues are interrelated. If each is handled separately, conflicts can result that interfere with specific solutions to individual problems. When environmental problems are handled as production improvement opportunities, a common solution can be developed. A successful corporate pollution prevention program can:

- improve global competitiveness
- enhance consumer acceptance of products
- reduce environmental impacts
- improve working conditions
- enhance community relations

IMPROVING GLOBAL COMPETITIVENESS

Lower Costs

It has been said that waste is really a resource out of place. It is true that most manufacturing wastes consist either of inefficiently used raw materials or potential products that are flushed down the drain, shipped to a landfill, or discharged into the air. Throwing potential resources away costs money, increasing product costs and hurting our competitiveness. Conversely, reducing waste can reduce operating costs and improve corporate profitability.

Example. The single-largest discharger of metals to a Silicon Valley wastewater treatment plant was a circuit board manufacturer. This manufacturer had undergone repeated bankruptcies and restructuring. The high amount of waste was just one sign of inefficiencies in the manufacturer's mismanagement of resources.

Example. An ice cream manufacturer was faced with a large surcharge for discharging wastewater having high biochemical oxygen demand (BOD) levels to the local wastewater treatment plant. He was making multiple flavors of ice cream each day and needed to wash down his equipment between flavors. The cleaning process resulted in a large loss of product and heavy BOD loads on the treatment plant. He solved his problem by increasing his freezer capacity so that he only had to produce one or two flavors of ice cream a day to maintain a sufficient inventory to meet his customers' needs. This solution increased his production efficiency while decreasing his sewer bill.

Example. A study performed by CH2M HILL for the Department of Defense (DOD) found that conversion to a dry-media paint-stripping technique at Hill Air Force Base (AFB) resulted in an annual savings of $5 million.[3] If the method was adopted by all three branches of the military, DOD has estimated that the annual savings would exceed $100 million.[4]

Similar results have been achieved by both large and small industries.[5-7] The common finding is that pollution prevention can (and should) be justified on the basis of cost.

Example. The 3M Company has had its Pollution Prevention Pays program since 1975.[8] In the first 10 years, the 1,500 projects supported under this program resulted in a savings to the company of more than $235 million ". . . for pollution control facilities that did not have to be constructed, for reduced pollution control operating costs, and for retained sales of products. . . ."

Higher Quality

Waste reduction and product quality are not mutually exclusive. The most successful waste reduction projects have resulted in improved product quality. Conversely, improving product quality can reduce waste.

Example. At a large auto manufacturer, the greatest environmental problem was paint-stripping waste; that is, waste produced in removing paint from finished cars that didn't meet quality goals and waste paint from paint booths that was inefficiently applied to the cars. Improving the efficiency and quality of paint application would help to solve both problems.

Example. Plastic-media blasting (PMB), a paint-stripping method developed at Hill AFB in Ogden, Utah, eliminates the use of solvents in stripping aircraft for remanufacture. Not only does PMB reduce paint-stripping waste, it also produces a superior surface for repainting.

Example. DuPont's Tyvek manufacturing plant in Richmond, Virginia, measures quality by tracking uniformity of product thickness and quantity of product rejected by quality control at the plant or by customers. Maintaining thickness and reducing rejects not only promotes customers' perception of quality but also reduces raw material usage and waste generation.

ENHANCING CONSUMER ACCEPTANCE

Consumers have become increasingly aware of the environmental ramifications of the products they buy. They are concerned about potential exposure to hazardous components in products, as well as the costs of disposing of waste and running out of landfill space. Sometimes a solution has several benefits; for example, reducing the overpackaging of consumer products can reduce the cost of raw materials and at the same time provide a marketing edge for a product.

Example. When detergent was originally developed, it was so concentrated that only a quarter cup was required for a load of laundry. Consumers initially rejected it because the boxes were so much smaller than an equivalent amount of soap powder, which required at least a cup per load. As a marketing ploy, detergent was made bulkier by adding phosphate fillers, with a resulting environmental impact. Detergent manufacturers are now eliminating the phosphate and promoting their concentrated formulas as a means of reducing both harmful chemicals and packaging waste.

REDUCING ENVIRONMENTAL IMPACTS

Early hazardous waste legislation (RCRA and the Comprehensive Environmental Response, Compensation, and Liability Act [CERCLA]) imposed "cradle to grave" liability on waste generators for the waste that they produced and disposed of. Additional legislation has had a tendency to extend this liability, as one environmental engineer put it, from "sperm to worm." This expanded view can be interpreted as a call for hazardous waste birth control.

Example. A major aerospace manufacturing facility used a caustic etching process to perform final machining on aluminum structural panels. Approximately 40,000 gallons of depleted caustic waste was disposed of by deep-well injection. The facility is working to implement a process change to regenerate the caustic and remove the aluminum as aluminum hydroxide, which will be sold to a chemical company.

Water Quality

Increasingly stringent wastewater discharge standards are being based on water quality effects rather than being based on treatment technology. These standards can lead to a requirement that discharge sewers be "fishable and swimmable." Treatment costs to meet these standards are expensive, and in some cases, technically unachievable, thus forcing companies to go to zero discharge.

Example. An aerospace facility in Florida installed a zero-discharge rinse system on its circuit board line. The recycle system consisted of ion exchange demineralization with metals recovery by electrowinning and sale of the recovered scrap metal. Brines were treated by vapor recompression evaporation to recover more water for process use. These high-tech processes achieved zero discharge, but at a high price.

Ozone Is Never Where You Need It

In the stratosphere, ozone protects us from ultraviolet radiation. Chlorofluorocarbons (CFCs) react with ozone and break it down, causing depletion of ozone. Other volatile organic compounds (VOCs), such as auto exhaust, react with oxides of nitrogen (NO_x) and sunlight to produce ozone in health-threatening concentrations at ground level. Reducing use and emissions of these compounds will lessen these impacts.

Example. Bell Laboratories has switched from CFCs to a citrus-based cleaner for cleaning electronics components.

Example. Hughes Missile Division in Tucson switched from solvent paints to powder coating for part of their missiles, thus reducing emissions of VOCs and producing parts with a superior finish and more effective corrosion protection.

IMPROVING WORKING CONDITIONS

Overall improvements in product quality and productivity have been effected by a blurring of responsibility and authority between labor and management. Involving workers in pollution prevention programs has had similar benefits, and really is an extension of the quality-circle approach to manufacturing.

Example. A golf ball manufacturer's biggest waste problem was stripping paint from the tripods that hold balls while they are being painted. When this problem was discussed with one of the painters, he excused himself for a moment and returned with the plastic cover to a coffee cup. He placed the cover on top of the tripod and proceeded to paint golf balls on the plastic cover. When the cover got so clogged with paint that it was no longer functional, he replaced it with a new one, rather than having to strip paint from the tripod using phenolic or methylene chloride-based chemicals. When asked why he had not done this earlier, his reply was that no one had ever indicated that there was a problem or asked his help.

Reducing worker exposure to toxic materials will reduce illness among workers. This translates into better working conditions, reduced sick leave, and lower costs for health and disability insurance.

Example. Conversion to PMB paint stripping by Hill AFB eliminated worker exposure to toxic paint-stripping compounds and reduced the volume of toxic waste.

ENHANCING COMMUNITY RELATIONS

Under the Emergency Planning and Community-Right-To-Know (EPCRA) section of the Superfund Amendments and Reauthorization Act (SARA), companies are required to report releases of listed toxic compounds to the environment, even if those releases are permitted. Keeping these emissions to a minimum makes for good community relations.

Example. Management at an oil refinery was surprised to find that leaking valves and fittings were the main source of VOC emissions. Implementing a preventive maintenance program reduced losses of product and improved the environmental standing of the refinery relative to other dischargers in the region.

Landfills and trash disposal are community issues. Reducing the quantity of waste produced by companies reduces disposal costs and extends the life of community landfills. Setting up in-plant recycling efforts for paper, cardboard, and other materials or buying materials in returnable containers makes good business sense, and enhances a company's image in the community.

Example. Frequently, the product of CH2M HILL's engineering consultation is paper—reports, plans, and specifications. Many of the company's offices have set up recycling programs for office paper and other recyclables, such as aluminum, and recycled paper is used for reports.

METHODS FOR REDUCING WASTE

Several methods have been successful in reducing waste. They fall into one of the following categories:

- Change materials purchasing and control methods.
- Improve housekeeping.
- Change production methods.
- Substitute less-toxic materials.
- Reduce wastewater flows.
- Segregate wastes.
- Recycle or reclaim wastes.
- Treat waste to reduce volume and/or toxicity.

A short description of each of these methods follows, as well as illustrations of how these changes have resulted in reduced waste and cost.

Change Materials Purchasing and Control Methods

Purchasers of raw materials and supplies frequently focus on the unit cost of new material and do not consider the cost of disposing of unused material or waste generated in use. Means for minimizing disposal of new or unused material include:

- Reducing to a minimum the number of different products that serve the same function (i.e., cleaning fluids, cutting oils, etc.). This streamlining mitigates expired shelf-life problems and reduces the number of empty or partially used containers to dispose of.

- Buying products in container sizes appropriate to the actual use. Buying quart containers of a perishable product can ultimately be less expensive than purchasing gallons or drums of that product at a lower unit cost and later having to dispose of the unused portion.

- Reducing the inventory of hazardous materials to a minimum and ensuring that old containers are rotated from the back of shelves to the front when new material is purchased. This will reduce the volume of new materials that needs to be disposed of because their shelf-life has expired.

These measures will reduce the cost of raw materials and waste disposal, as well as the investment tied up in inventory.

Example. At a large repair and rework facility, the majority of hazardous waste collected was either partially used material or unopened containers of new material that had exceeded its shelf-life. At the same facility, a solvent-recycle program was rendered useless by a purchasing agent who continued to order the same quantity of new solvent, regardless of the need for new material. Excess purchasing resulted in increased amounts of expired unused solvent that had to be disposed of to make room for subsequent shipments.[3]

Improve Housekeeping

Excessive waste production often results from sloppy housekeeping. Leaking tanks, valves, or pumps can cause process chemicals to spill onto the floor, requiring cleanup and disposal of wastes. Poor cleaning of parts before placing them in process tanks can greatly reduce the useful lives of process chemicals and can increase both the volume of waste requiring disposal and the cost of chemical replacement.

Example. At two similar plating shops, hard-to-replace parts were hard-chrome plated in preparation for machining and reassembly. At one shop, the plating baths were dumped an average of once a year at a cost of $25,000 per year. At the other shop, only one plating bath had been dumped in 27 years of operation. The principal reason given for the difference was careful cleaning and masking of parts before plating.[3]

Key advantages of housekeeping changes are that they can usually be implemented quickly, they require little if any capital investment, and they are likely to result in substantial reduction in production of waste.

Change Production Methods

There is a tendency to continue using the same manufacturing process even after improved methods have been developed. The old saying, "If it ain't broke, don't fix it," comes to mind. However, sometimes the process can be broken in other than a production-related sense if it produces too much waste. Frequently, changes that reduce waste can also provide production improvements.

Example. For more than 50 years, the Navy had used the same methods for hard-chrome plating. However, when improved methods were developed in the Cleveland area, the Navy created a program, using these methods, to reduce the waste from chrome-plating operations. The result was a zero-discharge plating process that produced a more uniform plate while using considerably fewer tanks.[3]

Substitute Less-Toxic Materials

Frequently, a nontoxic material can be substituted for one that is toxic or that causes a special waste treatment problem. The reduced cost of disposal or the reduced exposure of workers to a toxic material can justify the change.

Example. Cadmium was traditionally plated from a bath containing cyanide. Cyanide wastes require separate treatment from other plating wastes, thereby increasing the complexity and cost of waste treatment. When Lockheed successfully switched from an alkaline cyanide cadmium bath to an acidic noncyanide cadmium bath, plating quality was equivalent and overall costs were reduced. The reduced costs resulted from simplified treatment, which counteracted a higher cost for plating chemicals.[3]

Reduce Wastewater Flows

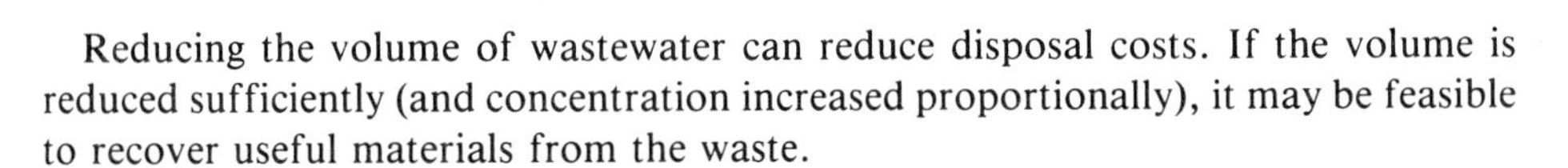

Reducing the volume of wastewater can reduce disposal costs. If the volume is reduced sufficiently (and concentration increased proportionally), it may be feasible to recover useful materials from the waste.

Example. At the Pensacola Naval Air Rework Facility's plating shop, a recirculating spray-rinse system was installed on the hard-chrome plating line, which reduced rinse volumes to less than the evaporation rate from the plating bath. The rinsewater was returned to the plating bath, resulting in recovery of plating chemicals and elimination of rinsewater discharges.[3]

Segregate Wastes

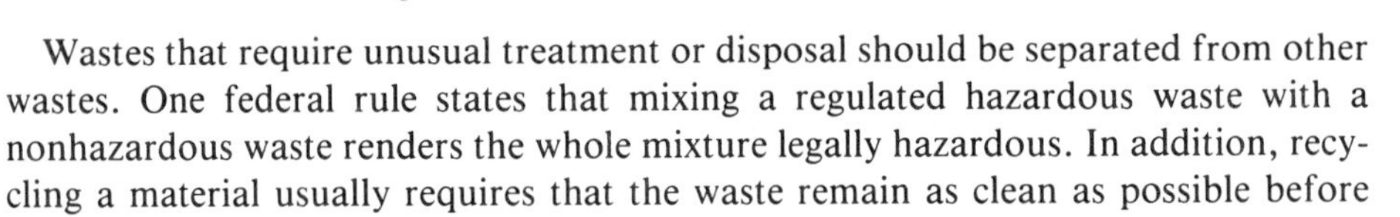

Wastes that require unusual treatment or disposal should be separated from other wastes. One federal rule states that mixing a regulated hazardous waste with a nonhazardous waste renders the whole mixture legally hazardous. In addition, recycling a material usually requires that the waste remain as clean as possible before reuse.

Example. If solvent recycling using simple stills is to be feasible, individual solvents must be segregated. At one facility, an attempt to recover heptane from a used storage tank containing waste heptane was abandoned when it was found that the reclaimed heptane failed to meet specifications. This failure resulted from different waste solvents being discharged to the storage tank.[3]

Recycle/Reclaim/Reuse

Recycling a chemical is often less expensive than purchasing new material and paying disposal costs. Sometimes a material may no longer meet the specifications for a process for which it was originally purchased, but it may still be suitable for other more tolerant uses.

Example. A high-purity solvent is used for cleaning liquefied gas lines at a federal research facility. When the solvent no longer meets the stringent purity requirements for this application, the facility collects and uses the solvent for general cleaning at the facility, rather than disposing of it off the site.

Reduce Volume and/or Toxicity

Disposal costs are based on the type and volume of waste. If a waste has hazardous characteristics, then it must be disposed of in a hazardous waste disposal facility. If the waste is treated to eliminate these hazardous characteristics, it can be disposed of in an industrial waste landfill at a lower cost. Transportation and disposal costs can be decreased by reducing the volume of the waste.

Example. A plant producing a waste acid stream purchased caustic to neutralize the acid for disposal with the plant's wastewater. This plant also produced alkaline sludge that was disposed of in a hazardous waste landfill. By neutralizing the waste alkaline sludge with their waste acid, the sludge could be disposed of in a commercial (nonhazardous waste) landfill. The result was reduced raw-material costs and reduced disposal costs.

Incineration is one means of reducing a waste's volume and toxicity. Incineration can accomplish the following tasks:

- Reduce the volume of waste, producing inert ash.
- Change the character of waste by combusting organic constituents.
- Change the state of the waste by evaporating and combusting liquids, leaving only solid residues.
- Treat gases and vapors.

By burning the organic or combustible fraction of wastes, the volume of waste for ultimate disposal can be reduced to the minimum inorganic content, or ash. A volume reduction of 70% to 95% is readily achievable for industrial or office rubbish. This combustion can reduce the apparent toxicity of the waste caused by organic constituents or can change the character of the waste. Evaporating liquids and burning the combustible fraction of the liquids will leave behind only inorganic ash for disposal.

In addition, an incinerator can provide direct treatment of gases or vapors in air. Contaminated air can be ducted directly to a secondary combuster, mixed with fuel, and burned, thus destroying organic constituents. This technique precludes having to treat the water discharged from a wet scrubber or dispose of solids from carbon adsorbers.

Besides reducing the volume or toxicity of wastes, incineration can provide a means for recovering energy or minerals. By using energy recovery techniques, a combustion system can provide steam or electricity.

Mineral acids can be recovered from the combustion of waste materials. Burning chlorinated solvents generates hydrochloric acid (HCl), which can be recovered in an absorption tower as an 18% to 20% solution in water. Wastes with a high concentration of sulfur can be burned and the sulfur recovered as sulfuric acid. This latter process requires catalyst beds and absorbers in series.

Other minerals can be recovered by burning solid waste. Gold can be recovered by burning scrap circuit boards. Sandpaper grit can be recovered and recycled by burning scrap sandpaper.

IF IT AIN'T BROKE, DON'T BREAK IT

Methods for waste management are listed here in order of preference:

- Use smaller quantities of toxic materials or substitute less-toxic materials.
- Change processes to reduce waste.
- Recycle wastes in the process.
- Recycle wastes off the site.
- Recycle wastes on the site.
- Treat waste to reduce volume or toxicity.
- Dispose of wastes in a manner that is protective of human health and the environment.

Legislation and the resulting regulations can promote pollution prevention or, just as easily, inhibit it. It is good public policy to favor moving wastes as far up on the above hierarchy as is feasible. However, strict enforcement of the hierarchy beyond that which makes economic sense is counterproductive. Micromanagement of manufacturing methods leads to a breakdown in production similar to that experienced in Russia and Eastern Europe under communism.

Legislators and regulators can become preoccupied with the process of implementing pollution prevention-planning, making organizational changes, setting goals, and collecting data on the effectiveness of a program. Effective pollution prevention is not achieved through any of these, but rather is accomplished through specific projects. The most effective of these projects are usually initiated and implemented by a "champion" on the production floor. Planning works to the extent that it improves the ability to select the optimal mix of projects for funding, which can be very important when funds are limited.

REGULATORY REQUIREMENTS AND THE FREE MARKET

Passing legislation and implementing regulations take time, typically 5 years for federal legislation. This is usually too slow to be effective. Changes in the free market are not instantaneous either, but they are more responsive. The high cost of waste disposal and long-term liability for waste consigned to landfills have helped reduce waste among well-managed industries.

The resources used in responding to unreasonable regulatory requirements could be better spent on practical solutions. Regulators should avoid sending out questionnaires without regard to the amount of effort required to gather and present the data requested, a clear idea of what the data will be used for, and the means to process it when it arrives. Only Columbo can get away with "just one more question," and he is usually dealing with criminals.

Example. A recent pollution prevention questionnaire sent to an airline by regulators asked for detailed mass balances on all of its operations that used toxic materials or produced hazardous waste. Rigorous completion of the questionnaire would have required the time of the entire environmental staff of the airline. The airline contracted the work to CH2M HILL.

Recent state laws have required companies to perform pollution prevention audits and to prepare plans for preventing pollution. These requirements can be effective in focusing an industry's attention on effective pollution prevention projects; however, states need to avoid placing planning requirements on the regulated community that consume resources (time and money) that would be more effectively invested in projects. Companies that have already implemented pollution prevention programs should not be burdened by unreasonable planning requirements.

As a rough rule of thumb, planning and monitoring costs should normally consume less than 15% of a pollution prevention budget, and the pollution prevention budget should be somewhat related to the cost of waste disposal and liability.

Example. At a pollution prevention class, a plant manager asked what kind of a pollution prevention plan he needed to prepare and how much effort should be put into it. His plant produced a barrel of solvent every 2 or 3 years. He indicated that he had replaced chlorinated solvent in all but one process for which he could find no acceptable substitute. He was sending the remaining waste to a permitted treatment, storage, and disposal (TSD) facility where it was blended as a fuel. The level of effort that he had put into planning was proportionate to his waste problem; therefore, further analysis was not needed.

A FREE MARKET APPROACH

Legislators and regulators tend to distrust corporate claims of pollution prevention. The most common concern is that reported reduction of waste could be due to reduced production. This concern has led to a demand that companies relate waste generation to production volume. Production rates, however, are difficult to measure in units that are relevant to waste production. Few industries produce a single product on a single unchanging production line. Most produce a mix of products on a variety of production lines, each with its own inherent efficiency. If an efficient production process increased production and a less efficient operation reduced production, the result would be reduced waste even though neither process was changed. Relating waste production to gross national product is an interesting statistic, but requiring individual companies to prove that waste reduction is not simply caused by a downturn in production is a burden and not particularly meaningful.

Instead of micromanaging companies, a simple, focused regulatory effort that uses market forces to promote waste reduction is recommended. This effort should concentrate on five areas.

- Enforce waste disposal regulations. Companies that compete unfairly should be penalized because of improper or illegal waste disposal practices.

- Make waste disposal costs realistic. The incentive of reduced costs will motivate well-managed companies to reduce waste.

- Allow mismanaged companies to go out of business. Companies that use resources inefficiently will not survive.

- Require waste reduction plans proportional to waste generation.

- Set focused, clear goals, and leave it to individuals to achieve them.

Effective pollution prevention is possible if a cooperative effort is made by legislators, regulators, the business community, and the public. Working together will make it happen.

REFERENCES

1. Resource and Conservation and Recovery Act, Section 1003(b), as amended by the Hazardous and Solid Waste Amendments, Public Law 98-616 (November 1984).
2. Serious Reduction of Hazardous Waste, U.S. Congressional Office of Technology Assessment, Washington, D.C., OTA-ITE-317 (September 1986).
3. Higgins, T.E. "Industrial Processes to Reduce the Generation of Hazardous Waste at DOD Facilities," Phase 2 Report, NTIS No. AD-A159-239 (July 1985).
4. The Assistant Secretary of Defense for Manpower, Installations, and Logistics. "The 1984 Annual Environmental Protection Summary," 1984.
5. Huisingh, D., L. Martin, H. Hilger, and N. Seldman. "Proven Profits from Pollution Prevention: Case Studies in Resource Conservation and Waste Reduction," Institute for Local Self-Reliance, Washington, DC, 1986.
6. Campbell, M.E., and W.M. Glenn. "Profit from Pollution Prevention: A Guide to Industrial Waste Reduction and Recycling," Pollution Probe Foundation, Toronto, Ontario, 1982.
7. Overcash, M.R. *Techniques for Industrial Pollution Prevention* (Chelsea, MI: Lewis Publishers, 1986).
8. Susag, R.H. "Waste Minimization: The Economic Imperative," *Waste Minimization Manual*. Government Institutes, Inc., Washington, DC, 1987, pp. 2-7.

CHAPTER 2

Government and Industry Programs for Reducing Pollution

*Kathi Futornick**

INTRODUCTION

During the 1970s, with the enactment of the first environmental statutes and the establishment of the U.S. Environmental Protection Agency (U.S. EPA), we had a vision of a quick resolution of our environmental problems. Although we have made some progress, our most significant achievement over the years is the recognition that environmental problems are complex, and that long-term solutions are evasive. Unlike other government programs, environmental programs do not have a commonly recognized bottom-line; there are no absolutes and few precedents. Therefore, the search for solutions has evolved into an interactive process, requiring much learning, relearning, and consensus building.

Under the first U.S. environmental laws, we called on the new U.S. EPA to identify all dangerous pollutants, set standards for safe exposure, recycle wastes where feasible, and manage the remaining wastes for the protection of public health and the environment. Early environmental regulations were established in a command-and-control mode. We were convinced that all our environmental problems could be solved—that every citizen would be breathing clean air by the mid-1970s and that no pollutants would be discharged to waterways by the late 1970s.

The "end-of-the-pipe" technical fixes that we were counting on have not proved to be a magic cure. The levels of safety of pollutants have been difficult to assess, and end-of-the-pipe solutions having distinctly different approaches for various media—air, water, solid waste, and hazardous waste—have resulted in an environmental shell game, moving pollutants from one media to another. The annual cost for compliance with environmental regulations has increased from $26 billion in 1972 to an estimated $115 billion in 1993 without effectively eliminating pollution.

Today we are in a transition period, questioning established practices and learning to cooperate across all boundaries: agency, industry, environmental and public involvement groups, consumers, and members of the general public. Meanwhile,

*Ms. Futornick is Manager, Environmental Affairs for the Port of Portland, Oregon, and may be reached at (503) 731-7236.

industries both here in the U.S. and abroad continue to grow. If we continue to deal with environmental issues as we have in the past, we can expect more releases of pollutants and much greater costs to clean them up.

Traditional approaches, such as treatment and disposal, will not succeed as the primary methods of managing our waste. To many in the environmental field, our environmental programs have undergone a paradigm shift. Today's new challenges are being addressed through a wide range of prevention methods. Some of these methods move up the pipe, to the source, to reduce generation of wastes rather than rely on the end-of-the-pipe solutions that have been the benchmark of our environmental programs. Some methods find innovative ways of using the wastes as ingredients in industrial processes or as substitutes for commercial products. To further encourage these innovations, a wide range of regulatory and nonregulatory programs are being developed.

Several thousand pounds of hazardous waste are generated annually for each person in the United States. In 1990 in the U.S., some 4.8 billion pounds of 320 toxic chemicals were released into the air, water, or land or transferred for treatment or land disposal. Federal and state governments and private industries agree that in order to effectively reduce the generation of hazardous waste, facility-wide programs need to be established. Reducing generation of waste is a practical way of curtailing the amount of waste being treated and disposed of. Government agencies and private industries have institutionalized pollution prevention programs to lower waste management costs and liabilities. For many good reasons, pollution prevention is here to stay.

This chapter looks at the historical development of federal and state programs and at current laws and regulations governing pollution prevention. Also included is a discussion of industrial pollution prevention programs and international initiatives for reducing the generation of waste.

FEDERAL PROGRAMS

Currently, U.S. environmental protection programs emphasize management and cleanup of hazardous wastes after they are generated. Many billions of dollars are spent annually for managing and cleaning up waste. Toxic and hazardous wastes are often treated before disposal; however, treatment seldom renders the waste completely nontoxic. After treatment, the waste is usually discharged to the land, air, or water according to permit requirements.

Many of the pollution-control laws enacted by Congress, such as the 1977 Clean Air Act, the Clean Water Act of 1977, and the Emergency Planning and Community Right-to-Know Act (Title III of SARA 1986), have indirectly promoted pollution prevention by requiring increased waste management practices and increased liabilities. Congress also gave the U.S. EPA regulatory authority for toxic chemical use by passing the Federal Insecticide, Fungicide, and Rodenticide Act (FIFRA) in 1972 and the Toxic Substances Control Act (TSCA) in 1976. The Resource Conservation and Recovery Act (RCRA) has encouraged reduction of waste through a certification program, and the Pollution Prevention Act of 1990 sets national policy for a hierarchy of pollution prevention with source reduction first, recycling and reuse next, then treatment, and as a last resort, land disposal.

Hazardous and Solid Waste Amendments (1984)

Congress developed the first national policy of industrial pollutant prevention when it passed the Hazardous and Solid Waste Amendments (HSWA) of 1984. The HSWA officially amends the Solid Waste Disposal Act (SWDA), which is commonly referred to as the amendments (passed in 1976) to RCRA. These amendments set out national policy for waste reduction in Section 1003 of RCRA.

> *Objectives and National Policy of RCRA*
>
> *The Congress hereby declares it to be the national policy of the United States that, wherever feasible, the generation of hazardous waste is to be reduced or eliminated as expeditiously as possible. Waste that is nevertheless generated should be treated, stored, or disposed of so as to minimize the present and future threat to human health and the environment.*

With this, Congress defined a two-tier national waste reduction policy. First, the generation of hazardous waste is to be reduced or eliminated. Second, waste management practices should minimize risks to human health and the environment. It also recognized that by not producing hazardous waste to begin with, there is no risk to human health and the environment. In addition, it promotes good management practices that lower the risks from wastes and identifies land disposal as the least-preferred management practice. The objectives of RCRA further state that one way to protect health and the environment is by ". . . minimizing the generation of hazardous waste and land disposal of hazardous waste by encouraging process substitution, materials recovery, properly conducting recycling and reuse, and treatment . . ."[1]

Although Congress clearly mandated a two-tier approach to waste management in its national policy statement, specific provisions of the HSWA focus on waste management practices and the transfer of pollution to one medium: the land (Table 2.1). The HSWA does not directly require industry to minimize waste. It only requires companies to report on their current waste generation and what they are planning to do to reduce the volume or quantity and toxicity of wastes they are generating.

The U.S. EPA, in the preamble to rules and regulations pertaining to waste reduction, did not address the RCRA national policy statement pertaining to waste minimization, and set forth a voluntary and nonintrusive program.[2] The language and positioning of the regulations appeared to shift the emphasis from reducing waste generation to land disposal as a hazardous waste management practice. The U.S. EPA did not assume a leadership role for waste reduction and it continued to be a low-priority for some time. In the U.S. EPA's Operating Guidance FY 1987, waste reduction was not one of the program priorities.[3]

Although early environmental statutes presented some opportunities for achieving environmental protection through pollution prevention, the end-of-the-pipe approach to pollution control continued to dominate. The ineffectiveness of these historical attempts at waste management are reflected in the overall expenditures for pollution control. Although reducing waste generation presents many environmental and economic benefits, in 1986 more than 99% of federal and state environmental spending was applied to management of waste after it was generated; less than

Table 2.1. Key Provisions for HSWA Waste Reduction

Section of the Act	Congressional Action	Requirement
Section 3002(a)(6)[b]	Reporting Procedures	Generators subject to reporting requirements were to summarize ". . . efforts undertaken during the year to reduce the volume of toxicity of waste generated; and changes in volume and toxicity of waste actually achieved during the year in question in comparison with previous years prior to [November 8, 1984]."
	Manifest System	Generators were now required to certify on their manifest that is prepared for all regulated shipments of hazardous waste that a program was in place "to reduce the volume or quantity and toxicity of such waste to the degree determined by the generator to be economically practicable; and . . . the proposed method of treatment, storage, or disposal, is that practicable method currently available to the generator which minimizes the present and future threat to human health and the environment."[a]
Section 3005(h)	Permits	Conditions for issuance of any permit for treatment, storage, or disposal of hazardous waste, that a statement be made no less than annually, that "the generator of the hazardous waste has a program in place to reduce the volume or toxicity and quantity of such waste to the degree determined by the generator which minimizes the present and future threat to human health and the environment."
Section 8002(r)	EPA Study	Congress required EPA to prepare a Congressional report on "the feasibility and desirability of establishing standards of performance or of taking other additional actions under this Act to require the generators of hazardous waste to reduce the volume or quantity and toxicity of the hazardous waste they generate, and of establishing with respect to hazardous wastes required management practices or other requirements to assure such wastes are managed in ways that minimize present and future risks to human health and the environment. Such report shall include any provision for legislative changes which . . . are feasible and desirable to implement the national policy established in Section 1003."

[a]Generators are defined in 40 CFR 260.10 as "any person, by site, whose act or process produces hazardous waste identified or listed in Part 261 of this chapter or whose act first causes a hazardous waste to become subject to regulation.

1% was spent to reduce the generation of waste. The level of national spending in 1986 for pollution control was about $70 billion.[4]

Emergency Planning and Community Right-to-Know Act

The Emergency Planning and Community Right-to-Know Act (EPCRA) was written in response to the 1984 chemical releases at Bhopal, as Title III of the SARA of 1986. EPCRA requires owners and operators of facilities to report information on chemicals stored and on releases to the environment. EPCRA has become the government's greatest tool for gathering and disseminating information on potential or actual releases of chemicals. Public interest groups have used EPCRA information to publish lists such as the "Toxic 500." In turn, targeted companies have worked to reduce the amount of toxics released. Between 1990 and 1991, releases reported under EPCRA were reduced by 9%. So indirectly, EPCRA has become a regulator of toxic chemical release.

Pollution Prevention Act of 1990

Not until 1990 did Congress expand the scope of its pollution prevention policy beyond the land to cover releases of pollutants to all media. In the Pollution Prevention Act, Congress set a new national policy for waste reduction. It declared that "source reduction is fundamentally different and more desirable than waste management and pollution control."[5] The Act established a hierarchy for pollution prevention: source reduction first, recycling and reuse next, then treatment, and as a last resort, land disposal. The new national strategy of source reduction first moves away from a command-and-control regulatory approach and toward a voluntary cross-media approach.

The Pollution Prevention Act was enacted because Congress concluded that even where feasible, opportunities for pollution prevention often are not realized because "existing regulations, and the industrial resources they require for compliance, focus upon treatment and disposal, rather than source reduction; existing regulations do not emphasize multimedia management of pollution; and businesses need information and technical assistance to overcome institutional barriers to the adoption of source reduction practices."[6] The Act mandates that the U.S. EPA promote government involvement in order to overcome institutional barriers by providing grants to state technical assistance programs under a new federal pollution prevention program. The Pollution Prevention Act also amends EPCRA to require certain industries to *explain* what steps are being taken to reduce pollution; the Act does not *require* industries to reduce their pollution. The additional requirements of the Act are listed in Table 2.2.

Clean Air Act Amendments

The Clean Air Act Amendments (CAAA) were passed with provisions that also support pollution prevention, particularly reduction of pollution at its source. These amendments provided the U.S. EPA with the opportunity to promote waste reduction and pollution prevention and offer a change in emphasis from traditional end-of-pipe waste management programs.

Federal responsibilities for waste reduction expanded with the passage of the CAAA. Federal responsibilities concerning industrial sources of air pollution have

Table 2.2. Key Provisions of the Pollution Prevention Act of 1990

Provision	Mandate
Pollution Prevention Clearinghouse	Establish a clearinghouse to compile and disseminate pollution prevention information to businesses.
Regulatory Review	Review regulations and programs to assess their effect on pollution prevention efforts.
Coordination	Coordinate pollution prevention activities in each EPA office and promote pollution prevention practices in other federal agencies.
Procurement	Congress authorized $8 million to support state grant programs from FY 1991 through 1993. EPA was further required to investigate opportunities to use federal procurement to encourage pollution prevention.
Auditing	Establish standard methods to measure pollution prevention progress.
Award Program	Establish an annual award program to recognize industrial pollution prevention activity.
TRI Reporting	Expand the Toxics Release Inventory (TRI) reporting requirements to include questions about pollution prevention and recycling activities at individual facilities during the reporting year.

Source: Summarized from the U.S. Environmental Protection Agency Office of Pollution Prevention, Pollution Prevention Division. In conjunction with the Office of Solid Waste, Office of Toxic Substances, Pollution Prevention 1991, Progress on Reducing Industrial Pollutants. Washington, DC, October 1991, pp. 120–121.

included setting National Ambient Air Quality Standards (NAAQS) for total suspended particulates, sulfur dioxides, carbon monoxide, nitrogen dioxides, ozone, and lead; setting National Emission Standards for Hazardous Air Pollutants (NESHAP); and imposing new source performance standards (NSPS) on emissions from new stationary sources of pollution. The CAAA also established federal responsibilities with regard to oversight of state programs that set up permitting systems to control actual emissions.

The findings and purposes of Title 1, Air Pollution Prevention and Control, of the CAAA are:

> ". . . that the prevention and control of air pollution at its source is the primary responsibility of states and local governments;" and
>
> ". . . that Federal financial assistance and leadership is essential for the development of cooperative Federal, State, regional, and local programs to prevent and control air pollution."[7]

Unlike the original Clean Air Act and previous amendments, the 1990 Amendments added a clear mandate for pollution prevention. The amendments also required that the EPA "conduct a basic engineering research and technology program to develop, evaluate, and demonstrate nonregulatory technologies for air pollution prevention."[8] Other provisions of the CAAA that are relevant to pollution prevention are listed in Table 2.3.

Clean Water Act

The Clean Water Act (CWA) established in 1972 was the first major amendment to the Federal Water Pollution Control Act to establish a policy for eliminating the discharge of pollutants to navigable waters. Since 1972, technology-based regulations have been imposed on industrial plants. The CWA splits dischargers into two groups: those emitting pollutants directly into surface waters and those discharging indirectly through sewers to publicly owned treatment works (POTWs). Indirect users of POTWs that have pretreatment programs must comply with pretreatment standards for toxic pollutants.

The most significant pollution prevention program under the CWA is in the provisions for a general permit for stormwater discharges at industrial facilities and construction sites. The CWA requires each permittee to develop a stormwater pollution prevention plan.

Pollution Prevention Strategy

In February 1991, the U.S. EPA issued a Pollution Prevention Strategy, which clarifies the pollution prevention position and objectives.[9] This strategy is designed

Table 2.3. Other Provisions for Pollution Prevention in the Clean Air Act Amendments 1990

Provision	Discussion
Reduced emissions of hazardous air pollutants	In establishing emission standards for hazardous air pollutants, EPA is required to take into consideration: • Cross-media impacts • Substitution of materials or other modifications to reduce the volume of hazardous air pollutants • potential for process changes • Eliminate emissions entirely
Reduced sulfur dioxide emissions	Title IV is devoted to acid rain control; among its provisions is the allocation of annual allowances for sulfur dioxide emissions for fossil-fuel-fired facilities.
Improved fuel quality	Title II addresses provisions relating to mobile sources, including sale of cleaner-burning reformulated gasoline in the most smog-ridden cities.
Phase-out of substances	Chlorofluorocarbons and halons are to be phased out beginning in 1992 and EPA is required to ban the use of unsafe substitutes and to further regulate gasoline additives.

to provide guidance and direction for efforts to incorporate pollution prevention within the U.S. EPA's existing programs, and to establish a program that will achieve these objectives within a reasonable time frame. In addition, the U.S. EPA is attempting to confront its internal institutional barriers that make it difficult for separate environmental programs to achieve common goals. Established programs and their goals are listed in Table 2.4. The U.S. EPA, in order to achieve these goals, has begun to incorporate pollution prevention into every EPA program and activity.

As recently as 1993, U.S. EPA Administrator Carol Browner established pollution prevention as the central ethic of the agency's efforts to protect human health and the environment. At the national level, the U.S. EPA is beginning to raise the awareness of the benefits of pollution prevention. With few exceptions, the programs sponsored and advocated by the U.S. EPA are voluntary. This trend toward voluntary programs is supported by recent programs developed by the U.S. EPA and other federal agencies.

- **Source Reduction Review Project (SRRP).** The short-term goal of the Source Reduction Review Project is to ensure that source reduction measures and multimedia issues are considered when air, water, and hazardous waste standards affecting 17 industrial categories are developed. For the long term, the project tests different approaches to provide a model for the regulatory development process throughout the U.S. EPA. For example, the U.S. EPA is developing a regulation affecting the pulp and paper industry that will promote process changes to reduce the quantity of pollutants released to air, water, and land.

- **Pollution Prevention in Enforcement Settlement Policy.** U.S. EPA's negotiators are strongly encouraged to incorporate pollution prevention conditions into settlements—*both criminal and civil—involving private entities, federal facilities, and municipalities. The conditions can either correct an existing violation ("injunctive relief") or constitute a "supplemental environmental project" that the party performs.*

- **33/50 Program.** This is a voluntary initiative to reduce toxic-waste generation from industrial sources. The U.S. EPA targeted 17 chemicals for reductions of 33% by the end of 1992 and 50% by the end of 1995. To date, more than 1,150 companies have signed up to participate, committing to more than 354 million pounds of reductions in toxic chemical emissions.

- **Green Lights Program.** Pollution prevention and energy conservation are complementary goals of a new policy directive endorsed by U.S. President Clinton. In October 1993, Clinton announced the Climate Change Action Plan, an initiative encompassing more than 30 programs administered by the U.S. EPA, U.S. Department of Energy, and other federal agencies. These programs, which include the U.S. EPA's Green Lights Program, are market-driven and voluntary. By emphasizing the economic benefits of conserving energy and reducing waste, these programs are expected to sell themselves.

  The first of the U.S. EPA's market-driven, nonregulatory "green" programs, Green Lights encourages voluntary reductions in energy use through more efficient lighting technologies. More than 700 participants have agreed to survey their facilities and, where cost-effective, upgrade lighting efficiency in 90% of their square footage within 5 years. Green Lights partici-

Table 2.4. EPA Developments to Implement Pollution Prevention

Program	Goal
Waste Minimization Branch	Established in the Office of Solid Wastes and Emergency Response for coordinating the development of waste minimization and pollution prevention policy.
Office of Pollution Prevention	Established in the Office of Policy, Planning and Evaluation for coordinating agency-wide development of pollution prevention policy.
Risk Reduction and Engineering Laboratory	Charged with conducting research on industrial pollution prevention and waste minimization technologies.
Office of Pesticides and Toxic Substances	Declared pollution prevention to be a primary goal. This office administers TSCA and FIFRA.
Pollution Prevention Advisory Committee	Composed of directors and several regional managers. Established to ensure that pollution prevention is incorporated throughout EPA's programs.
American Institute for Pollution Prevention	Jointly established by EPA and the University of Cincinnati in June 1989 to support pollution prevention through four following priority areas: economics, education, implementation, and technology.
Pollution Prevention Information Clearinghouse	A joint project of the Office of Research and Development and the Office of Pollution Prevention. The PPIC disseminates pollution prevention information.

pants have reduced electrical usage by 35,000 kilowatts, for an annual savings of $6.9 million in electricity costs.

- **Energy Star Computers.** Energy Star is a voluntary partnership between the U.S. EPA and manufacturers that sell 60% of all desktop computers and 80% to 90% of all laser printers in the United States. These companies are now introducing products that automatically "power down" to save energy when not in use. Consumers will easily recognize the more efficient systems, because they will be labeled with the *Energy Star* logo.

- **Design for the Environment (DfE).** DfE is a cooperative effort between the U.S. EPA and industry to promote consideration of environmental impacts at the earliest stages of product design. Initial projects include evaluating alternative dry-cleaning processes and more environmentally preferable substitutes for toxic chemicals used in printing processes. Other projects include designing a more environmentally conscious computer workstation and funding studies of alternative means of synthesizing important industrial chemicals. A new focus of the DfE program is a joint effort with the accounting and insurance professions to integrate environmental considerations into capital budgeting and cost accounting systems.

- **EPA-GSA Cleaners Project.** Through this joint effort, the U.S. EPA and the U.S. General Services Administration (GSA) are developing procurement criteria for cleaning products based on considerations of efficacy, human health, and environmental safety. The ultimate objective is to advance the pollution prevention ethic throughout the federal supply system, and then among other public and private sector purchasers.
- **Water Alliances for Voluntary Efficiency (WAVE).** WAVE encourages hotels and motels to install water-saving techniques and equipment. Hotel chains such as Marriott, Sheraton, and Hilton have signed partnership agreements with the U.S. EPA to retrofit their facilities with water-efficient bathroom fixtures, dishwashing equipment, cooling towers, landscaping, and irrigation. WAVE intends to expand the program to other commercial buildings and institutions, including office buildings and schools.

STATE PROGRAMS

Unlike most environmental programs in which the federal government has led the way in developing laws and regulations, states have led in developing and implementing pollution prevention programs. Several state programs have served as the models for federal programs. For example, the Toxics Release Inventory provisions of the federal EPCRA were modeled after the New Jersey Industrial Survey of 1979.[10] During the 1980s, early state programs helped promote pollution prevention by disseminating information and providing technological assistance and in the late 1980s by requiring pollution prevention plans. This section looks at the contributions of state programs to pollution prevention.

State pollution prevention programs have evolved for several reasons. Two of those reasons were precipitated in response to federally mandated programs. First, in the early 1980s, programs focused on waste minimization. This was in response to federal requirements that generators certify that they have a waste minimization program in place (Section 3002, HSWA Amendments, 1984). Second, in 1986, SARA required states to have a capacity assurance plan that demonstrated how hazardous wastes generated within the state are to be managed. This prompted many states to develop more comprehensive waste reduction programs in order to meet this federal requirement. State programs began to focus on facility-wide management of wastes from all media. Recently, pollution prevention has become the focus of many state programs.

Before 1985, few state laws dealt with any aspect of pollution prevention. As of 1993, more than 30 states had passed or were debating pollution prevention laws. The provisions encompassed by these laws vary, but generally include facility-wide plans and technical assistance. In July 1989, Massachusetts and Oregon enacted new legislation emphasizing reduction in the use of toxic chemicals and, for the first time, requiring industries to prepare comprehensive pollution prevention plans. The statutes identified in Table 2.5 were enacted to cover pollution prevention.

State pollution prevention programs try to meet a variety of objectives. Some of the most common ones include dissemination of technical assistance and technical

Table 2.5. State Laws for Pollution Prevention Planning

State	Statutory Authorization	Date	Implementing Agency	Program Name
Alabama		1990	Alabama Department of Environmental Management	Alabama Waste Reduction and Technology Transfer Program
		1985		Hazardous Material Management & Resource Recovery Program
Alaska	Alaska Solid and Hazardous Waste Management Act HB 478	1990	Pollution Prevention Office Alaska Health Project	Waste Reduction Assistance Program
Arizona	State Omnibus Hazardous Waste Act	1991	Pollution Prevention Unit, Arizona EPA	Arizona Waste Minimization Program
Arkansas			Hazardous Waste Division–Department of Pollution Prevention and Ecology	Arkansas Pollution Prevention Program
California	Hazardous Waste Source Reduction Management and Review Act	1989	Department of Toxic Substance Control Office of Pollution Prevention and Technology Development	
		1984		Toxic Substance Control Program
Colorado	State Pollution Prevention Act	1992	Colorado Department of Health	Pollution Prevention and Waste Reduction Program
		1988		Waste Minimization Assessment Center
Connecticut			Connecticut Hazardous Waste Management Services	Connecticut Technical Assistance Program
	P.A. 91–376		Connecticut Department of Environmental Protection, Waste Management Bureau	
Delaware	Delaware HB 585, Waste Minimization/Pollution Prevention Act	1990	Department of Natural Resources and Environmental Control	Delaware Pollution Prevention Program
Florida		1988	Florida Department of Environmental Regulation	Waste Reduction Assistance Program
Georgia	Georgia Amendments to the Hazardous Waste Management Act	1990	Environmental Protection Division, Georgia Department of Natural Resources	Georgia Multimedia Source Reduction and Recycling Program
		1983		Hazardous Waste Technical Assistance Program

Table 2.5. State Laws for Pollution Prevention Planning (Continued)

State	Statutory Authorization	Date	Implementing Agency	Program Name
Hawaii		1984	Department of Health	Hazardous Waste Minimization Program
Idaho			Division of Environmental Quality, Idaho Department of Health and Welfare	
Illinois	Toxic Pollution Prevention Act (TPPA)	1989	Illinois EPA Office of Pollution Prevention	Voluntary Toxic Pollution Prevention Innovation Plan Program
	Illinois Pollution Prevention Act (IPPA)	1992		
Indiana	House Enrolled Act No. 1412	1993	Office of Pollution Prevention and Technical Assistance	Indiana Pollution Prevention Program
	Indiana Industrial Pollution Prevention and Safe Material Act	1990	Indiana Department of Environmental Management	
Iowa	Iowa Senate File No. 2153, Section 29	1990	Waste Management Authority Division, Department of Natural Resources	
Kansas			Kansas Department of Health and Environment	State Technical Action Plan
Kentucky	Laws of Kentucky Codified Sections as 224.980–224.986		Kentucky Department for Environmental Protection	
		1988		Kentucky PARTNERS
Louisiana	Louisiana Ènvironmental Quality Act	1987	Louisiana Department of Environmental Quality	
Maine	Amendment to the Reduction of Toxic Use, Waste and Release Act	1992	Department of Environmental Protection	
	Toxics Use & Hazardous Waste Reduction Act	1989	Department of Oil & Hazardous Materials Control	
Maryland			Maryland Department of the Environment Technical Extension Service	
		1990		Hazardous Waste Program
Massachusetts	H. 6161 Toxic Use Reduction Act	1989	Bureau of Waste Prevention Massachusetts Department of Environmental Protection Office of Safe Waste Management	

Michigan			Michigan Office of Waste Reduction Services Environmental Technology Board	
Minnesota	Toxic Pollution Prevention Act	1990	Minnesota Pollution Control Agency Minnesota Office of Waste Management	Minnesota Technical Assistance Program
Mississippi	S.B. 2568 Comprehensive Multimedia Waste Minimization Act	1990	Mississippi Department of Environmental Quality	Mississippi Waste Reduction/Waste Minimization Program Mississippi Technical Assistance Program
Missouri			Division of Environmental Quality Missouri Department of Natural Resources	Waste Management Program
		1972	Environmental Improvement and Energy Resource Authority	
Montana			Department of Health and Environmental Services	
Nebraska			Hazardous Waste Section Nebraska Department of Environmental Quality Pollution Prevention Office	
Nevada			Bureau of Waste Management Division of Environmental Protection	Business Environmental Program
New Hampshire	New Hampshire Bill 5835B	1990	New Hampshire Department of Environmental Services/Waste Management Division	
New Jersey	Pollution Prevention Act	1991	New Jersey Department of Environmental Protection and Energy Office of Pollution Prevention	
New Mexico			New Mexico Environmental Department Hazardous Waste & Radiation Bureau	Municipal Water Pollution Prevention Program

Table 2.5. State Laws for Pollution Prevention Planning (Continued)

State	Statutory Authorization	Date	Implementing Agency	Program Name
New York	S. 5276-B Hazardous Waste Reduction and RCRA Conformity Act	1990	New York Department of Environmental Conservation	
	Facility Planning for Air and Water Releases under Environmental Conservation Law	Draft 1993		
North Carolina	North Carolina SB324 Hazardous Waste Management Act	1989	North Carolina Department of Environment, Health and Natural Resources	Pollution Prevention Program
		1987	Bureau of Pollution Prevention	
		1983	Office of Waste Reduction	
North Dakota			North Dakota Department of Health and Consolidated Laboratories	No formal state program
Ohio	Requires underground injection facilities to prepare waste minimization plans for industrial wastes generated	1992	Ohio Environmental Protection Agency Pollution Prevention Section Ohio Technology Transfer Organization	
Oklahoma			Oklahoma State Department of Health	Pollution Prevention Technical Assistance Program
Oregon	H.B. 3515 Hazardous Waste Management Commission Act	1989	Oregon Department of Environmental Quality	Waste Reduction Assistance Program
Pennsylvania			Pennsylvania State University	Pennsylvania Technical Assistance Program
			Pennsylvania Department of Environmental Resources	
Rhode Island		1987	Rhode Island Department of Environmental Management	Hazardous Waste Reduction Program
South Carolina		1990	South Carolina Department of Health and Environmental Control	Center for Waste Minimization

State	Legislation	Year	Agency	Program
South Dakota			Department of Environmental and Natural Resources	Waste Management Program
Tennessee	H.B. 2217 Hazardous Waste Reduction	1990	Department of Health and Environment	
			Tennessee Valley Authority	Waste Reduction Assessment and Technology Transfer Training Program
Texas	Waste Reduction Policy Act	1991	Texas Water Commission	
	Texas SB 1521	1989	Waste Minimization Unit	
Utah			Utah Department of Environmental Quality	
Vermont	H. 733 An Act Relating to the Management of Hazardous Waste	1990	Agency of Natural Resources	
			Vermont Department of Environmental Conservation	Pollution Prevention Program
		1988		Vermont Waste Minimization Program
Virginia	Bill establishes pollution prevention as the environmental strategy of choice for the commonwealth	1993	Department of Environmental Quality	
			Virginia Department of Waste Management	Waste Minimization Program
Washington	Facility pollution prevention planning required HB 2390	1990	Washington Department of Ecology	Waste Reduction, Recycling and Litter Control Program
West Virginia			West Virginia Division of Natural Resources	Pollution Prevention and Open Dump Program
		1989		Generator Assistance Program
Wisconsin	Wisconsin Act 325 Pollution Prevention Law	1989	Wisconsin Department of Natural Resources	Hazardous Pollution Prevention Audit Grant Program
			Wisconsin Department of Development Bureau of Solid & Hazardous Waste Management	
Wyoming			Wyoming Department of Environmental Quality	Solid Waste Management Program

information, fostering incentives for pollution prevention, and increasing awareness of pollution prevention opportunities. These elements of state programs are typically voluntary. Mandatory elements, such as facility planning and performance standards, are found in some state programs. In 1990, eight states—Connecticut, Iowa, Maine, New Hampshire, New York, Rhode Island, Vermont, and Wisconsin—passed laws to eliminate heavy metals (mercury, lead, cadmium, and hexavalent chromium) from packaging materials.

Several states are also developing programs that authorize regulators to certify companies as environmentally conscientious based in part on their pollution prevention programs. In turn, these companies can use the certification in "Green" marketing programs. Oregon offers tax credits for the construction of facilities that prevent, control, or reduce air, water, solid, or hazardous waste pollution. In Illinois, facilities that volunteer "toxic pollution prevention innovation" plans receive expedited processing of their permit applications and support for variance requests. The Massachusetts Department of Environmental Protection has been developing a program that consolidates air, water, and waste requirements. The program, known as the Blackstone project, involves inspectors who look at a whole plant for violations of all media permits, and if found, they recommend that companies seek assistance from the state in applying source reduction technologies rather than media-specific solutions. More than 80% of the companies in the project said they preferred Blackstone inspections to media-specific ones.

The varying state approaches have posed problems for industry, particularly those with multi-state facilities. Complying with different state pollution prevention programs has prevented companies from adopting cost-efficient and uniform approaches to reducing waste. A uniform state policy for pollution prevention is being encouraged by some.[11] For many industries with multi-state facilities, a uniform state policy might include common goals from the many state approaches, such as common performance standards, documentation of progress from the same baseline year, and technical assistance and training programs.

During 1993, the U.S. EPA published three approaches for consideration of a uniform national approach to waste recycling. Source reduction is the preferred approach, waste recycling is a second choice, and treatment and disposal are less desirable options. Recycling waste by using waste materials as feedstocks in manufacturing operations deserves special mention. Waste recycling has been used with much hesitancy because of inconsistent federal and state regulatory interpretations and the potential encumbrances of RCRA requirements. A consistent national approach separate from RCRA requirements has been encouraged by both the U.S. EPA and industry. Hazardous waste recycling moves hazardous waste management from a "cradle-to-grave" philosophy to a "cradle-to-cradle" approach.

INDUSTRIAL PROGRAMS

Economic incentives to develop pollution prevention programs have always been there. American industry has been engaged in pollution prevention activities individually, through company initiatives, and collectively, as members of trade associations. As attention paid to pollution prevention increases and regulatory agencies develop award programs, many companies are beginning to publicize their waste reduction programs and successes. This section focuses on four trade associations that have set forth waste reduction programs for their members, and examines waste reduction programs in several individual industries.

The Chemical Manufacturers Association (CMA), the American Petroleum Institute (API), the National Paint and Coatings Association, and the National Electrical Manufacturers Association are industrial associations that have comprehensive programs for pollution prevention. The CMA program is one of the most comprehensive of the industrial programs, requiring its members to adopt specific codes of management practice, including pollution prevention principles. API has similar principles for its members.

The Chemical Manufacturers Association

The CMA program, known as Responsible Care, began in 1988 to improve the chemical industry's management of chemicals. All CMA members are required to participate in this program, which is based on 10 guiding principles. The principles address protection of health, safety, and the environment. The Responsible Care program is also based on Codes of Management Practice. One of the codes, known as the Waste and Release Reduction Code, mandates:

- a clear commitment by senior management through policy, communications, and resources, to ongoing reductions in each of the company's facilities in release to the air, water, and land, in the generation of wastes
- a quantitative inventory at each facility of wastes generated and releases to the air, water, and land, measured or estimated at the point of generation or release
- evaluation, sufficient to assist in establishing reduction priorities, of the potential impact of releases on the environment and the health and safety of the employees and the public
- education of, and dialogue with, employees and members of the public about the inventory, impact evaluation, and risks to the community
- establishment of priorities, goals, and plans for waste and release reduction, taking into account both community concerns and potential health, safety, and environmental impacts
- ongoing reduction of wastes and releases, giving preference first to source reduction, second to recycle/reuse, third to treatment
- measurement of progress at each facility in reducing the generation of wastes and in reducing releases to the air, water, and land, by updating the quantitative inventory at least annually
- ongoing dialogue with employees and members of the public regarding waste and release information, progress in achieving reductions, and future plans
- inclusion of waste and release prevention objectives in research and design of new or modified facilities, processes, and products
- an ongoing program for promotion and support of waste and release reduction by others[12]

The American Petroleum Institute

In 1989, the API Board of Directors identified seven key public policy trends of importance to the petroleum industry. From these trends, the board selected the environment as its highest priority, and used it as the focal point of a new strategic planning effort. The petroleum industry's commitment to protect the environment is embodied in its program called STEP—Strategies for Today's Environmental Partnership. The foundation of the industry's efforts to improve environmental, health, and safety performance is API's environmental mission statement and guiding environmental principles. Acceptance of the principles is a condition of membership in API. API's Pollution Prevention Task Force is spearheading the industry's commitment to identifying and implementing pollution prevention opportunities.

Others

The National Paint and Coating Association (NPCA) began its Pollution Prevention Program in 1990. According to the NPCA policy statement, the goal of the program is "the promotion of pollution prevention in our environment through effective material utilization, toxics use, and emission reduction and product stewardship in the paint industry."[13] The policy statement also expresses the board's recommendation that "each NPCA member company establish an ongoing waste and chemical release reduction program" including "priorities, goals, and plans for waste and release reduction giving preference first to source reduction, second to recycling/reuse, and third to treatment."[14]

The National Electrical Manufacturers Association (NEMA) provides pollution prevention assistance to its members through education and technical assistance. NEMA has sponsored waste reduction workshops for member companies.[15]

Individual companies have also established pollution prevention programs resulting in significant decreases in hazardous waste generation. Table 2.6 summarizes pollution prevention programs of 24 different companies.[16]

INTERNATIONAL PROGRAMS

Have other governments come to the conclusion that pollution prevention is important? The answer varies. Some governments, such as the United Kingdom, have decided to focus efforts on waste management and have developed end-of-the-pipe policies similar to those of the United States. Japan, on the other hand, has focused on reuse and recycling technologies. European programs for pollution prevention have been in place since the 1970s. The governments of France, Germany, Sweden, Norway, Denmark, the Netherlands, and Austria have been leaders in pollution prevention, at times expending more dollars than the United States.[17] Table 2.7 provides a summary of selected international programs.

Because of a lack of data and differences in definitions for hazardous wastes, comparisons between pollution prevention programs of other governments and those of the United States are difficult. European countries emphasized pollution control to facilitate interest and investment in waste reduction. European governments have taken a nonregulatory approach and have focused on cooperating with and assisting industries. Waste reduction in Europe has been used as a tool to enhance industrial efficiency, growth, and international competitiveness. Some of

Table 2.6. Company-Wide Pollution Prevention Programs and Goals[a]

Company/Program	Scope	Goal	Accomplishments
Allied Signal Waste Reduction Program	Includes waste minimization under RCRA, as well as nonhazardous waste, and evaluates various disposal alternatives and methods for detoxification.	Reduce the quantity and toxicity of hazardous waste that must be stored, treated, or disposed of as much as is economically practicable.	The amount of cyclohexylamine waste produced in 1987 was only 15% of the volume of the same waste produced in 1984. The amount of waste finish oil was reduced nearly 90% from 1984 to 1987.
Amoco Waste Minimization Program	Primary focus on minimizing hazardous waste disposal, also minimizing and tracking nonhazardous wastes.	Eliminate the generation and disposal of hazardous wastes.	Between 1983 and 1988, Amoco reduced its hazardous waste by 86%, saving the company about $50 million.
AT&T Environmental Program	Industrial source reduction and toxic-chemical-use substitution are priorities.	Achieve a 50% reduction of CFCs by 1991, and 100% by 1994. Eliminate toxic air emissions of all types by the year 2000, with a 50% reduction by 1993 and a 95% reduction by 1995. Decrease disposal of total manufacturing process wastes by 25% by 1994.	Substituted a citrus-based cleaner and other organic compounds for solvents that are used to clean electronic equipment. Eliminated CFC use in circuitboard manufacturing processes through use of the AT&T Low Solid Fluxer.
BASF	Toxic air emissions reduction.	Decrease toxic air emissions (lbs/yr) by 89% by December 1992 (base month and year: July 1989).	
Boeing Waste Minimization Program	Focus on process changes that reduce the volume and/or toxicity of hazardous materials used in operations.	Reduce use of hazardous materials Minimize the generation of hazardous waste. Ensure proper handling and disposal of all waste.	Case study; a chemical substitution in one photoresist stripping operation has increased stripping speeds by 50%, and, because of its longer useful life, should reduce annual hazardous waste generation by 50%.
BP America Waste Minimization Program (1989)	Adoption of EPA's environmental management hierarchy, with source reduction preferred.	All facilities are to have annual waste minimization goals.	One refinery processes spent caustics into acids, which are then sold. The unit paid for itself in less than 8 months.

Table 2.6. Company-Wide Pollution Prevention Programs and Goals[a] (Continued)

Company/Program	Scope	Goal	Accomplishments
3M Pollution Prevention Pays (3P, 1975) and 3P Plus (1989)	Elimination of pollution sources through product reformulation, process modification, equipment redesign, recycling, and recovery of waste materials for resale.	By 2000, cut all hazardous and nonhazardous releases to air, land, and water by 90%, and reduce the generation of hazardous waste by 50% (base year: 1987).	From 1975 to 1989, the 3P program has cut 3M pollution in half, per unit of production, with implementation of 2,511 recognized 3P projects throughout the company from 1975 to 1990.
Monsanto Priority One (TRI wastes)	Source reduction, reengineering, process changes, reuse, and recycling to reduce hazardous air emissions and TRI solid, liquid, and hazardous wastes.	A 90% reduction in hazardous air emissions from 1987 to 1992. A 70% reduction in TRI solid, liquid, and gaseous wastes from 1987 to 1995.	From 1987 to 1990, Monsanto achieved a 39% reduction in hazardous air emissions.
Northrup, B-2 Division Zero Discharge Pollution Prevention	Primary focus on hazardous waste minimization; goals have been set for other solid waste reduction; program includes projects to reduce water usage, stationary air pollutants, and water pollution; targeted areas include elimination of ozone-depleting chemicals (ODC) and toxic chemicals.	Eliminate the generation and disposal of hazardous waste by 1995; reduce solid waste disposal by 25% by 1995 and by 50% by 2000; reduce water usage by 20%; reduce mobile air pollution by 25% by 1992; eliminate ODC use by 1993; eliminate toxic chemical use/risk by 1995.	More than 50% reduction in hazardous waste generation in last 18 months; reduced ODC emissions by 50% in the last 12 months through substitution; reduced mobile air pollution by approximately 280 tons/year; reduced water usage by an average of 28% in each of the last 6 months; reduced solid waste disposal by 70% in 30 months.
Occidental	Reduction of toxic air emissions.	Decrease toxic air emissions (lbs/yr) by 78% by December 1992 (base month and year: December 1988).	
Polaroid Toxic Use and Waste Reduction Program (TUWR, 1987)	Industrial source reduction and toxic chemical use substitution are priorities, followed by recycling and reuse.	Reduce toxic use at the source and waste per unit of production by 10 percent per year in each of the 5 years ending in 1993 and, as a corollary, emphasize increased recycling of waste materials within the company.	Using 1988 as the base year, Polaroid's Environmental Accounting and Reporting System (EARS) reported an 11% reduction in toxic use and waste during 1989.

Raytheon	Toxic chemical use substitution.	Pledged to eliminate CFC–113 and methyl chloroform from its printed circuit board operations in five states by 1992.	
Scott Paper	Integrated and multifaceted approach, including source reduction, recycling and reuse of materials, and landfilling of unusable residual waste.	Design products and packaging to reduce volume of waste material, which Scott terms "source reduction." Decrease dioxin levels at paper mills by reducing chlorine usage or altering its method of application, or by adopting new technologies or replacements for chlorine bleaching.	By the end of 1989, about 20% of the pulp used for sanitary tissue products was made from recycled fiber, and Scott plans to approximately double its recycled capacity. The Duffel, Belgium, mill uses a process that uses less water and less fiber. Developed a system for source reduction known as "precycling" in which paper products are packaged in larger quantities, thus saving materials that otherwise would have been wasted.
Sheldahl (Northfield, Minnesota)	Industrial source reduction and toxic chemical use substitution for hazardous air pollutants.	For methylene chloride, pledged 90% emissions reduction by 1993 and 64% use reduction by 1992.	
Texaco	Reduction in toxic air emissions.	Decrease toxic air emissions (lbs/yr) by 92% by February 1991 (base month and year: July 1990).	
Xerox	Toxic chemical use substitution, materials recovery, and recycling.	Reduce hazardous waste generation by 50% from 1990 to 1995.	Substituting d-Limonene for chlorinated solvents allowed Xerox to reduce the amount of solvents emitted to the atmosphere from about 200,000 pounds in 1982, to an estimated 17,000 pounds in 1990. A high-pressure water-strip operation has enabled Xerox to recycle 800,000 pounds of nickel and 2 million pounds of aluminum tubes per year, and to return 160,000 pounds of selenium to suppliers for reuse.

Table 2.6. Company-Wide Pollution Prevention Programs and Goals[a] (Continued)

Company/Program	Scope	Goal	Accomplishments
Chevron Save Money and Reduce Toxics Program (SMART, 1987)	SMART adopts EPA's hierarchy, with an emphasis on industrial source reduction, toxic chemical use substitution, and recycling for hazardous and nonhazardous solid wastes.	Reduce hazardous waste generation by 65% by 1992 and recycle what is left. Find nontoxic alternatives to toxic materials and processes. Devise safer operating procedures to reduce accidental releases. Ensure that pollution reductions in one area don't transfer pollution to another.	From 1987 to 1990, Chevron reduced hazardous waste by 60% and saved more than $10 million in disposal costs. Case study: Chevron used to dispose of tank bottoms in landfills. It now uses a centrifuge to separate oil from water; it reuses the oil and treats the water, leaving only a small amount of solid waste to be landfilled (less than 5% of the original sludge).
Dow Waste Reduction Always Pays (WRAP, 1986)	Industrial source reduction and onsite recycling.	Increase management support for waste reduction activities, establish a recognition and reward system for individual plants, compile waste reduction data, and communicate information on waste reduction activities. Decrease SARA 313 air emissions by 50% by 1995 (base year: 1988). Decrease toxic air emissions (lbs/yr) 71% by December 1992 (base month and year: December 1988).	SARA 313 overall releases are down from 12,252 tons in 1987 to 9,659 tons in 1989, a 21 percent reduction. Offsite transfers are down from 2,855 tons in 1987 to 2,422 tons in 1989, a reduction of 15 percent. Air emissions for 1989 showed a 54 percent decrease from 1984.
Exxon	Toxic air emissions reduction.	Decrease toxic air emissions (lbs/yr) by 14% by January 1991 (base month and year: December 1988).	
General Dynamics Zero Discharge (1985)	Industrial source reduction, toxic chemical use substitution, recycling, treatment, and incineration.	Have no RCRA manifested wastes leaving company facilities.	Nearly 40 million lbs. of hazardous waste discharge was eliminated from 1984 to 1988 (approximately 72%), while sales increased from $7.3 billion to $9.35 billion over same period.

General Electric Pollution, Waste and Emissions Reduction Program (POWER, 1989)	Program encompasses all waste streams (e.g., hazardous, nonhazardous, packaging, and ultimate disposal of product) and adopts EPA's hierarchy, which places source reduction first.	Prevent or minimize "the generation or release of wastes and pollutants, to the extent technically and economically feasible, throughout the life cycle of the product, including its design, production, packaging, and ultimate fate in the environment." Decrease toxic air emissions (lbs/yr) by 90% by December 1993 (base month and year: December 1988).	GE Appliances' Louisville plant has reduced its production of hazardous wastewater treatment sludge by 95%; GE Plastics' Ottawa plant has reduced its butadiene emissions by more than 90%; GE Medical Systems' E. Dale Trout plant has reduced its generation of hazardous waste by 74%; business-wide, GE Power Delivery has reduced its CFC usage by 72%; and, company-wide, GE reduced its SARA 313 reported releases by 11% from 1987 to 1988.
Goodyear Toxic air emissions reduction	Industrial source reduction.	Decrease toxic air emissions (lbs/yr) by 71% by January 1991 (base month and year: December 1988).	Decreased air emissions from operations through improved maintenance and monitoring of equipment and through decreased use of acrylonitrile, butadiene, and styrene.
Hoechst Celanese Toxic chemical emission reductions	Adopts EPA's hierarchy, which makes source reduction a priority, with a main focus on substituting cleaner production processes for those generating high volumes of wastes.	An average 70% cut in emissions of TRI chemicals (with 80% reductions at 9 plants with the highest emissions) by 1996 (base year 1988).	
IBM	Industrial source reduction and toxic-chemical-use substitution are priorities, followed by recycling/reuse/reclamation, incineration, detoxification, and disposal in a secure or sanitary landfill, in that order.	Pledged to eliminate ozone depleting chemicals from IBM products and processes by end of 1993 and to recycle 50% of solid waste by 1992.	Hazardous waste generation was reduced 38% from 1984 to 1988; 84% of IBM's hazardous waste was recycled in 1988; 28 percent of all solid waste from IBM United States operations was recycled in 1988; IBM U.S. emissions were reduced 20% from 1987 to 1988; and IBM U.S. CFC emissions were reduced by 25% between 1987 and 1988.

Source: The information was obtained from EPA's Pollution Prevention 1991 report.

[a]A number of companies were reviewed by the U.S. EPA for the 1991 report but not included in the information provided.

Table 2.7. International Programs in Waste Reduction

Country	Discussion
Austria	All new industrial facilities must demonstrate that they employ state-of-the-art low-waste technology before receiving an operating permit. The Environment Fund provides loans and grants for waste reduction and recycling projects.
Denmark	Recycling, reuse, and waste reduction projects are funded through grant money provided under the Act on Recycling, Reuse, and Reduction of Waste.
France	Clean technologies have been pursued to revive productivity and creativity in industry, thereby increasing international competitiveness. The Mission for Clean Technologies provides funding for waste reduction projects. ANRED and the National Agency for Encouragement of Research (ANVARD) fund a wider variety of waste-related projects. France also provides rapid depreciation allowances for pollution prevention investments.
Germany	The principal federal environmental agency, Umweltbundesamt (UBA), has no regulatory authority, but acts as a facilitator for waste reduction. Recent amendments require that generation of pollution should be avoided and low-waste technologies used. The UBA funds both waste reduction and recycling projects.
Japan	The Waste Disposal and Public Cleansing Law of 1970 has developed and promoted waste recycling and reuse options to address environmental concerns. The Toxic Substances Control Law is used to place controls on toxic substances, such as a tax on air pollutants. The National Institute of Hygiene has been engaged in research to reduce specific toxins such as dioxins.
The Netherlands	Research and development of clean technologies is promoted to alleviate waste problems in Holland and to market these technologies elsewhere. The Netherlands Committee on Environment and Industry provides R&D grants for clean technologies; at the same time, waste generation, treatment, storage, and disposal are taxed. The government is using voluntary and command-and-control programs to reduce pollution to between 70 and 90% of 1985 levels by 2010.
Norway	Environmental regulations cover all media by regulating the entire industrial sector. Norway's Pollution Control Authority provides grants and loans for both clean technologies and pollution control.
Sweden	Waste reduction projects are funded through its Environmental Protection Act.
United Kingdom	The major emphasis of programs is on waste management and not waste reduction.

the earliest initiatives in pollution prevention came from international organizations. The United Nations Economic Commission for Europe (ECE) sponsored the first International Conference on Non-Waste Technology in Paris in 1976. In 1979 the ECE adopted a detailed "Declaration on Low- and Non-Waste Technology and Reutilization and Recycling of Wastes."[18] The ECE recommended action on both the national and international levels to develop and promote low- and non-waste technologies. In 1985, the ECE established an environmental fund for demonstrating innovative technologies for reducing pollution.

Western European governments have initiated pollution prevention programs that are broader in scope than those of the United States. In European programs, pollution prevention applies to hazardous as well as nonhazardous wastes and to products as well as process wastes. Initiatives include offsite recycling, incineration, and other clean technologies.

European governments have relied heavily on economic measures, in the form of grants or loans, to fund research on new low-waste technologies, and on tax incentives and disincentives to influence the action of generators. Nearly every Western European country sponsors a grant or loan program, or both.

The Role of the European Community

The 1957 Treaty of Rome that established the European Community (EC) specified that an objective of the EC (Article 130r) is to "preserve, protect, and improve the quality of the environment, and to contribute towards protecting human health and to ensure a prudent and rational utilization of natural resources."[19] However, even though the EC adopted several environmental directives, it was not until 1973 that the EC Council of Ministers specified objectives for reducing and preventing pollution by "developing protective measures" and by requiring that the "polluter pay." Between the early 1970s and the mid-1980s, the scope of the EC environmental regulations steadily expanded to air, water, noise pollution, and waste disposal.

The EC's expansion of environmental regulations and directives during this period was the result of political forces and public pressures. However, because the Treaty of Rome provided little framework for balancing environmental protection with other EC goals, the most important being the establishment of a common market, the resulting environmental regulations were a compromise that reflected competing economic and political interests. Not until the Single European Act (SEA) went into effect in July 1987 was environmental quality considered an EC objective in its own right, not requiring economic justification.[20]

The EC has been aggressive in addressing environmental issues and has produced large volumes of regulations and directives that address product labeling, marketing, and distribution; ultimate disposal of residuals and packaging; cradle-to-grave environmental monitoring; oversight of the manufacturing process; land-use issues; environmental management systems and audits; employee working conditions; waste and source reduction, and pollution prevention. Currently being considered are additional environmental regulations on labeling (Ecolabeling) and environmental auditing (Audit Scheme).

In addition to voluntary and nonvoluntary government programs, there are several independent nongovernmental actions that have had a widespread and significant effect on industry. These actions include adoption of the Montreal Protocol and Standardization of Environmental Management.

The Montreal Protocol

In 1987, 31 nations signed the Montreal Protocol, in which they pledged to reduce the production of chlorofluorocarbons by 50% by the end of the century. In March 1989, the EC announced that its 12 member states agreed to cut production of chlorofluorocarbons by 85% as soon as possible and to eliminate production completely by 2000. In addition, after 5 years of negotiations, the EC approved a directive that set nitrogen and sulfur emission standards for large power plants, thus controlling a major source of acid rain in much of northern Europe.

Total Quality Environmental Management and Sustainable Development

"Total quality environmental management" and "sustainable development" are familiar phrases in the environmental policy arena and raise the underlying issues of continuous improvement and pollution prevention to a global level. In the past 5 years national and international conferences, programs, and appointed commissions all have demonstrated universal support for economic development, and simultaneously, wise use of natural and environmental resources. This section briefly chronicles the development of these phrases and their meanings.

Sustainable development has been a much-used phrase since 1987, when it was popularized by the report of the World Commission on Environment and Development. The report, *Our Common Future*, notes that sustainable development allows the present generation to enjoy economic development and improved standards of living without compromising the ability of future generations to fulfill their economic needs and aspirations.[21]

Total quality environmental management (TQEM) takes total quality management (TQM) one step further by incorporating environmental programs into a TQM program. TQEM provides a framework for an interactive approach to managing the growing number of environmental requirements and potential liabilities.

TQEM and sustainable development are distinct American and European initiatives that appear to be converging in both priorities and tools. Sustainable development has its roots in European, South American, and Asian ecological and economic concern over industrial processes that are not sustainable because of high consumption of energy, forests, and agricultural lands, and contamination of the air, land, and water.

Sustainable development and TQEM programs share a holistic approach to dealing with environmental issues. Pollution prevention programs requiring reductions in the use of toxic substances and in waste at the source, and requiring facility planning, initially introduced the basic concept of reevaluating the manufacturing process from a holistic perspective. **Pollution prevention planning** provides the framework for the interaction between management and workers on all environmental issues. **TQM** provides the framework for continuous improvement and customer satisfaction. **Sustainable development** uses these concepts as a framework for achieving a global balance between economic development and use of natural and environmental resources. The process has been evolutionary, and companies with established pollution prevention programs can continue to evolve and incorporate the more recent concepts of TQEM and sustainable development.

International Chamber of Commerce

In 1991, the International Chamber of Commerce (ICC) drafted the Business Charter for Sustainable Development, 16 principles for business practices to reverse nonsustainable consumption trends. In one year, more than 600 companies signed a pledge of commitment to ICC sustainable development principles.

The Global Environmental Management Initiative

The Global Environmental Management Initiative (GEMI)—to some extent, the U.S. counterpart to the ICC—is supported by 24 member companies and over 300 participating companies committed to the sustainable development principles and to TQEM. GEMI has helped merge sustainable development and TQEM initiatives through development of the Environmental Self-Assessment Program (ESAP), a detailed self-audit protocol and scoring system designed to help companies monitor their progress toward the ICC charter goals.

Agenda 21

In June 1992, the United Nations hosted an "Earth Summit" in Rio de Janeiro, Brazil. The summit brought together representatives of 178 nations and more than 115 heads of state. The agenda for the UN Conference on Environment and Development (UNCED) was to address diverse global issues such as global climate change, deforestation, and conservation of biological diversity, as well as to produce an "earth charter" for the next century, appropriately titled Agenda 21. The role for industry in pursuing Agenda 21 is focused on sustainable development concepts and is laid out in the Business Charter for Sustainable Development developed by the ICC.

President's Council on Sustainable Development

In June 1993, President Clinton appointed 25 representatives from industry, government, and organizations to the new President's Council on Sustainable Development. He charged the council with helping to craft U.S. policies that will encourage economic growth, create jobs, and protect the environment. David Buzzelli, vice-president and corporate director of Environment, Health and Safety, and Public Affairs for the Dow Chemical Co. and co-director of the new council, stated that the mission of the council can be realized by answering one question: "How can we move U.S. business to a pattern that prevents pollution?"

From early concepts of pollution control to a global vision for sustainable development, industry, governments, and the public are striving to get ahead of their environmental problems and make improving the environment a compatible goal with industries. Pollution prevention is here to stay, and applying its concepts effectively will help lead us into the next century.

REFERENCES

1. Solid Waste Disposal Act, Section 1003(a)(6).
2. *Federal Register*. Vol. 50. July 15, 1985.
3. U.S. Environmental Protection Agency, Operating Guidance FY 1987. Washington, DC: Office of the Administrator, March 1986.
4. U.S. Congress, Office of Technology Assessment. Serious Reduction of Hazardous Waste: For Pollution Prevention and Industrial Efficiency, OTA-ITE-317. Washington, DC: U.S. Government Printing Office, September 1986.
5. Pollution Prevention Act of 1990. HR 5931, Section 6602(4).
6. Pollution Prevention Act of 1990, Section 6602(3).
7. The Clean Air Act, Section 102(a) and 103(a).

8. Clean Air Act as amended in 1990, Section 901(c).
9. U.S. Environmental Protection Agency, Pollution Prevention Strategy, *Federal Register* 56:7849–7864.
10. U.S. Environmental Protection Agency Office of Pollution Prevention, Pollution Prevention Division, and the Office of Solid Waste, Office of Toxic Substances, *Pollution Prevention 1991-Progress on Reducing Industrial Pollutants.* Washington, DC, October 1991, p. 71.
11. Scagnelli, J.M., "Uniform State Policy Would Aid Pollution Prevention," *Pollution Engineering*, June 1991, pp. 66–70.
12. CMA, Improving Performance in the Chemical Industry. September 1990, pp. 9–15.
13. NPCA, "Paint Pollution Prevention Program Policy Statement," *Pollution Prevention Bulletin*, April 1990.
14. U.S. Environmental Protection Agency, Office of Pollution Prevention, Pollution Prevention Division and Office of Solid Waste, Office of Toxic Substances, Pollution Prevention 1991-Progress on Reducing Industrial Pollutants. Washington DC, October 1991, pp. 41–42.
15. *Ibid*, page 42.
16. U.S. Environmental Protection Agency Office of Pollution Prevention, Pollution Prevention Division, and the Office of Solid Waste, Office of Toxic Substances, Pollution Prevention 1991-Progress on Reducing Industrial Pollutants. Washington, DC, October 1991, pp. 45–49.
17. U.S. Congress, Office of Technology Assessment, Serious Reduction of Hazardous Waste: For Pollution Prevention and Industrial Efficiency, OTA-ITE-317. Washington, DC. U.S. Government Printing Office, September 1986.
18. United Nations Economic Commission for Europe, Declaration on Low- and Non-Waste Technology and Reutilization and Recycling of Wastes. Geneva, Switzerland, November 1979.
19. Kirchenstein, J.J., and R. A. Jump. "The European Ecolabels and Audits Scheme: New Environmental Standards for Competing Abroad," *Total Quality Environmental Management*, Volume 3, Number 1. Executive Enterprises Publications, New York, Autumn 1993.
20. Vogel, D. *Environmental Policy in the European Community. Environmental Politics in the International Arena.* Edited by S. Kamieniecki. State University of New York, 1993.
21. World Commission on Environment and Development (WCED), *Our Common Future* (Oxford University Press, 1987), p. 8.

CHAPTER 3

Setting Up a Successful Pollution Prevention Program

*Thomas Higgins**

INTRODUCTION

Why Plan for Pollution Prevention?

Successful pollution prevention programs result from the selection and implementation of cost-effective projects that optimize the use of raw materials, reduce waste, and minimize worker exposure to toxic materials, community impacts, and long-term corporate liability for discharges to the environment.

It has been said that planning does not prevent pollution—only successful projects prevent pollution. Too much effort placed on pollution prevention programs can steal resources better invested in projects. Planning probably should account for less than 15% of investment planned for a pollution prevention program. That 15% can be a good investment, however, if it focuses resources on projects that have the greatest impacts on waste generation.

Example. A $100,000+ vapor compression evaporator was installed on a rinse tank and recovered $100 worth of chromium per year. This expensive device was installed as a result of a poor understanding of the amount of waste produced in the process.

Planning is cost-effective to the extent that it optimizes investments in projects and results in net savings. A pollution prevention plan is usually prepared after a pollution prevention assessment or audit.

Profit from Success; Learn from Failure

Successful pollution prevention projects usually result from a trial, error, and correction process rather than a single insightful jump to perfection. Success stories are useful for a company that has the same technical problems and similar production facilities and personnel as those cited in the story. However, information about

*Dr. Higgins is a vice president in CH2M HILL's Reston, Virginia, office and may be reached at (703) 471-1441.

the ultimate waste reduction technology or production change may not show why a particular project succeeded, or provide the information necessary to develop a successful pollution prevention program. Therefore, this chapter presents a framework for setting up a successful corporate pollution prevention program and illustrates the points with examples of successful and not-so-successful projects.

Essential Elements of Successful Projects

Although specific circumstances and reasons lay behind the success or failure of each pollution prevention project, after dozens of projects were evaluated, successful projects were determined to have two characteristics in common:

- Production personnel were strongly motivated to implement and maintain the necessary changes.
- Technologies used were "elegant in their simplicity."

These two characteristics were integral parts of each successful project; moreover, at least one of these elements was missing in each failure.

MOTIVATING AND TECHNICAL ELEMENTS

Top Management Effectively Supports the Project

For successful programs, effective support is provided at a management level that controls both production and environmental policy decisions. Frequently, waste disposal and environmental protection are viewed as service functions (overhead costs) that are subservient to production (profits). However, successful pollution prevention usually requires production changes that support environmental protection. In a cost-cutting and staff-reduction environment, it is usually difficult to justify adding personnel for environmental protection. The following example shows how one individual coped with this personnel problem.

Example. At the aircraft repair facility at Robins AFB, an individual set up a cost-effective solvent recycling program. Management allowed him to consolidate partial manpower slots from the individual shops that were served to receive, process, and supply recycled solvent. The program more than paid for itself in reduced costs for purchase of virgin solvent and disposal of waste solvent.

Rewards and Recognition Are Provided for Successful Efforts

An incentive program for submitting ideas is useful only if the "good ideas" for potential pollution prevention projects are successfully implemented and maintained. In addition to rewarding those who originate the ideas, an effective program also provides rewards and recognition for individuals who do the hard work of implementing projects and making them succeed. Well-publicized rewards and recognition provide incentives to other workers. However, elaborate programs in which panels waste time evaluating unfocused ideas can steal resources from a pollution prevention program.

Operating Personnel Are Involved in Planning

In successful programs, operating personnel are involved in the planning, design, and installation of the modification. Participation promotes a "buy-in" to the change, ensuring motivated operation and maintenance of the modified process.

A "Champion" Implements the Project

Many successful projects are led by a champion who strongly believes in the modification, ramrods the project, and overcomes development and startup problems. A champion has to overcome the inertia protecting an existing process that "works" but produces an excessive quantity of waste. Even a difficult-to-operate process can be made to work by an individual who is too stubborn to believe that it cannot be done.

Example. A highly successful solvent recycling program was set up by a former Kentucky "moonshiner" who would not accept the "truism" that central recycling programs do not work. He was able to convince management to buy stills and assign him manpower slots to recycle solvents. He also built a solar still to recover contaminated jet fuel for use in operating ground equipment.

Implementation of the Program Reduces Production Costs

Production managers are evaluated on the basis of cost control, production yields, and product quality. Modifications that reduce production costs are more likely to gain support than projects that simply reduce waste disposal costs, which are usually shared companywide—if at all—through reduction in overhead charges.

Implementation of the Program Increases Productivity

Reducing personnel requirements, thus increasing productivity, is a strong incentive for change.

Example. The plastic-media blasting process for stripping paint was originally developed by a technician at Hill AFB. The technician wanted to eliminate the use of methylene chloride and phenolic paint strippers in order to improve working conditions for his coworkers. In developing the process, he stressed that the new process reduced manpower requirements from 240 work-hours to 24 work-hours per aircraft. That productivity improvement (and cost savings) is the driving force behind the rapid adoption of the process.

Waste Disposal Costs Are Charged to Production Units

At facilities where modifications were successful, production people were aware of the true costs of hazardous waste disposal and considered those costs when making decisions to implement pollution prevention programs.

Production People Appreciate Environmental Impacts of Waste Disposal

Most people are motivated by a desire not to pollute the areas where they work and their families live. Letting workers know the effects of the waste they discard can motivate them to be more careful.

Example. One of CH2M HILL's engineers previously worked as an environmental coordinator for a major aircraft manufacturer. Solvents were found in the storm drains. She traced the source to shop floor drains. She had signs made that read "NEXT STOP IS YOUR HOME FAUCET," and the discharges stopped.

Implementation of the Program Improves Product Quality

The quality of American products is an important basis for selection by the end user or consumer. When production changes adopted to reduce waste also improve product quality, they tend to be successfully implemented. Conversely, changes that decrease product quality are soon abandoned.

Example. A zero-discharge hard-chrome plating process was implemented at a Navy aircraft maintenance shop. The overall package of improvements included increased uniformity of the plate. As a result, subsequent machining and rejection of parts was considerably reduced, and the successful process has been implemented Navy-wide.

Change Is "Evolutionary" Rather Than "Revolutionary"

On successful projects, off-the-shelf equipment can be adapted to a new application; special or complex equipment is avoided. The greater the number of modifications attempted at the same time and the more experimental or less developed the equipment is, the greater the likelihood that the process will prove unreliable and be abandoned in favor of an existing method that has all of the "bugs" worked out of it.

Example. PMB paint stripping was successful partly because the modification used existing sand-blasting equipment, a technology that is well developed. The plastic media used for the new method causes less wear and tear on the equipment than sand; hence, little modification was required. The modifications that were adopted made the equipment lighter and easier to use, taking advantage of the media's reduced aggressiveness.

Equipment Is Adapted to Local Conditions

Care is taken to tailor the modifications (even for off-the-shelf equipment) before transferring it to facilities where it has not been tested. Soliciting (and using) suggestions from the local operators helps to ensure that they will "buy into" the change. If a suggestion is not used, care is taken to explain that decision to the individual who made the suggestion.

Example. Zero-discharge chrome plating was adopted at a number of Navy plating shops, on the basis of the general principle of reduced rinsewater use and return of concentrated rinsewater to the plating tank to make up for evaporation. In one shop, platers used a recirculating spray rinse to reduce flow. At another, platers used countercurrent multiple rinses. Both met the goal of reducing rinsewater and recovering metal to the plating tank, but each was adapted to the individual users.

New Equipment Is Simple to Operate, Reliable, and Easy to Maintain

Successful modifications are straightforward and simple to operate, thus requiring minimal training of personnel. If a new process is unreliable and reduces production rates, the old methods tend to be reinstated quickly. In successful programs, maintenance requirements for the new equipment are minimal.

Example. An unsuccessful system for recovery of chrome from plating rinsewater employed both ion exchange and vapor recompression evaporation. The resulting equipment was complex, difficult to operate, and frequently shut down for maintenance. This system was soon replaced by the simpler, more successful zero-discharge plating system of the previous example.

Extensive Training Is Provided for Operating Personnel

Reduction projects usually require personnel to learn to operate new equipment or to operate existing equipment using new methods. Successful projects include training more operators than are required, thereby ensuring that backups are available when needed. Sometimes projects are unsuccessful because they are abandoned when key personnel leave the facility without training replacements.

Example. Initially, at one plating facility, only a few platers were trained to use the new zero-discharge chrome plating system, and management counted on them to train others. Unfortunately, the trained platers moved on to other jobs before they were able to train others. The project was salvaged by setting up a new training program in which all platers are trained in the new operating methods. At a successful plating shop at Cherry Point Naval Aircraft Depot, time is provided for the best plater to run a training program for less experienced platers.

TIPS FOR SETTING UP A SUCCESSFUL POLLUTION PREVENTION PROGRAM

On the basis of experience with successful and unsuccessful pollution prevention projects, the following suggestions are made to assist in setting up and running a corporate pollution prevention program.

Make Pollution Prevention Part of the Company's "Corporate Culture"

Production people are evaluated on their success in meeting production goals, usually for product quality or quantity. For workers to institute and sustain pollution prevention changes as part of their daily duties, they need to be convinced that it is an important goal of the company. The commitment has to be real, has to affect production decisions, and has to be a part of job performance evaluations for the individual workers. The message needs to come from the top, not just from the appointed environmental "nag."

In addition, incorporating processes that produce small amounts of waste into a new production facility is easier than retrofitting them into an existing facility.

Example. A research chemist at Dow Chemical was getting ready to pilot test and scale up production of a new chemical. In analyzing the reaction kinetics, he noted that a side reaction formed an intermediate compound that combined with the desired product and reduced the yield by over 20%, as well as producing a waste by-product. He noted that the boiling point of this undesirable intermediate was much lower than the other reactants and could be distilled off as it formed, before it could react with the other intermediate reactants. He incorporated distillation into the pilot and full-scale production plant and improved yield of his primary product to close to 100%. The quality of the product was superior because of removal of the now-pure by-product, for which he then found a market.

Corporate Roles in a Pollution Prevention Program

The *Board of Directors* has the responsibility of setting the policy direction and committing resources (manpower and capital) for a pollution prevention program. Senior management has the responsibility of implementing that policy. As a start, the company should publish a pollution prevention policy (Figure 3.1). An important sign of board and management commitment is establishing the position of Waste Reduction Director for the company.

The *Waste Reduction Director* should report to someone who has responsibility for both production and environmental compliance. When conflicts arise between the two, the company should ensure that environmental impacts are deemed as important as production.

The *Production Department* has the primary responsibility for implementing a waste reduction program. This group interacts with the other departments to determine what realistic waste reduction goals can be achieved, to set specifications for raw materials in conjunction with the purchasing department, to test changes to ensure that they do not adversely affect production costs or product quality, to provide marketing with an understanding of the impacts of changes on products, and to participate with marketing and R&D in developing new environmentally friendly products or manufacturing methods.

The *Purchasing Department*'s decisions have a profound effect on waste production. Limitation of the quantity and number of toxic raw materials is critical. In addition, considering disposal costs while making purchase decisions is essential. Materials should be purchased in container sizes and lots that will encourage their complete use. Getting a good deal on bulk containers doesn't help when only small quantities are used and the remaining unused material becomes a waste. Setting up a stock rotation program is imperative so that the first stock in is the first stock used, to limit disposal of materials due to expiration of shelf life. Materials should be purchased in returnable containers whenever feasible.

The *Accounting Department* can assist by measuring the total costs of waste disposal, including handling, treatment, and long-term liability, and assigning those costs to the individual departments responsible for the waste generation.

The *Marketing Department* can respond to low waste-producing requirements of

[Company Name]
Pollution Prevention Policy Statement

[Company Name] is committed to a proactive policy of protecting the environment in all of our activities. This pollution prevention statement is based on our commitment to providing:

- A clean and safe environment in our community
- A safe and healthy workplace for our employees
- Compliance with all applicable laws and regulations
- An equitably priced product to our customers
- Prudent stewardship of our stockholder's equity

To accomplish these objectives, we will implement programs to reduce or eliminate generation of waste through source reduction and other pollution prevention methodologies. This policy extends to air, wastewater, and solid and hazardous wastes. In addition to meeting the above objectives, there are other important benefits related to pollution prevention. One of these is to reduce future liability for waste disposal, since **[Company Name]** realizes that compliance with current waste management standards does not relieve future liability for disposal site cleanup. Another benefit of pollution prevention is to reduce waste management costs, which comprise an increasing percentage of **[Company Name]** operating costs.

[Company Name] is committed to reducing the weight and/or toxicity of any wastes generated. As part of this commitment, **[Company Name]** gives priority to waste reduction at the source. Where source reduction is infeasible, other pollution prevention methods, such as recycling, will be implemented where feasible. Those wastes that cannot be avoided will be converted to useful products or beneficially used, where feasible. Remaining wastes for which no pollution prevention option is warranted will be effectively treated (to decrease volume or toxicity) and responsibly managed. **[Company Name]** will select waste management methods that minimize their present and future effect on human health and the environment.

Pollution prevention is the responsibility of *all* of our employees. **[Company Name]** is committed to identifying and implementing pollution prevention opportunities through solicitation, encouragement, and involvement of all employees.

Figure 3.1. Sample corporate pollution prevention policy statement.

customers and promote environmentally friendly products to provide a marketing edge. Marketing also can help find a market for recycled by-products generated in a waste reduction project.

The *Research and Development Department* can help develop new products that reduce or eliminate the use of toxic raw materials, reduce the generation of waste during production, and reduce the environmental impacts of these products when used by customers.

The *Engineering Department* is responsible for implementing plant improvements that will reduce waste generation in the facility operations.

The *Training Department* is responsible for training all staff in waste minimization techniques, as well as providing specific training to production staff in implementing production changes or operating new equipment. Training is especially critical in implementing low-cost housekeeping changes to reduce waste, such as stressing the need not to mix wastes destined for recycling.

The *Environmental Department* is responsible for ensuring that management considers present and future regulatory requirements when making production decisions.

Set Goals

The first step in establishing a waste reduction program is to define the goals for the effort. Trying to reduce all wastes, maximize use of all raw materials, and reduce all worker exposure to everything can be too broad a direction to take, at least initially. Typical goals have included:

- Reduce the quantity of RCRA wastes that are shipped off the site to reduce disposal costs and long-term liability.
- Reduce water usage in a process because of a community-wide shortage.
- Find a replacement solvent for CFCs because of the planned phaseout of their production.
- Reduce solid waste generation because of decreasing landfill space.
- Reduce permitted discharges to the local community environment as reported under EPCRA (Emergency Planning and Community Right-to-Know Act).

These goals are then used to direct data-gathering in the inventory or assessment phase of the program.

Inventory Waste Generation by Individual Production Units

After the specific goals of an assessment are agreed on, an audit or inventory of wastes (or raw materials usage, if reduction of usage is the goal) is necessary to be able to focus on the rest of the program. The audit is used to determine the quantity and characteristics of wastes, the locations of waste-generating processes, and the specific activities or production units that generate the wastes. This inventory need not be exact, but rather should concentrate on order-of-magnitude accuracy, since the inventory is to be used to focus efforts on the largest waste streams, because focusing on them usually provides the greatest potential for reduction.

Creative methods can be used to generate the inventory. Existing data should be used whenever available rather than setting up a separate data collection system.

Example. A synthetic fibers plant used nontraditional sources of data for one of their fibers facilities. These included analytical data from waste facilities, as well as accounting records for materials purchasing, product specifications, and batch cards from production units.

To the extent feasible, waste production should be measured and listed by individual production units. These details allow the company to evaluate individual processes to determine where waste reduction efforts should be concentrated. Waste production data are necessary to evaluate the benefits of making any changes.

Target Waste Sources

On the basis of the goals of the project, efforts should be targeted on the major waste generation activities. Wastes can be ranked by volume or cost of handling and disposal. It has been proposed that waste toxicity or degree of hazard to human health or the environment should be a factor in ranking wastes for reduction efforts. However, adding risk to the equation is difficult and subjective and is of little practical use to a production manager. Because waste disposal costs are often associated with the relative risk of the waste, simply ranking the wastes by cost of treatment and disposal is a reasonable approximation for including relative risk in targeting wastes.

Example. Ryan Delcambre, Dow's first waste reduction coordinator, found that approximately 80% of his company's waste was produced by 20% of its production facilities. He found a similar relationship within each plant. He used this information to target his waste reduction resources on these potential high-payback processes and facilities.

Publicize Waste Disposal Costs

Frequently, production personnel do not realize the true cost of waste disposal. Knowing the costs can sufficiently motivate them to make changes that will eliminate or reduce waste disposal costs.

Charge Production Units for Waste Disposal Costs

Production budgets should include the costs of hazardous waste disposal so that the costs can be used in production decisions. The budget should also include a contingency fund for potential liabilities.

Study Production Units

The facilities that generate the most waste should be visited to determine what the specific sources of waste are and to gain an understanding of how raw materials are used and how wastes are produced. Simply focusing attention on the major sources often is sufficient to initiate a change.

Example. At one aerospace facility, 10 barrels of solvent-contaminated rags were produced each month. These barrels were disposed of by central maintenance, and the $600 per barrel disposal cost was paid from a general overhead budget. However, after production people learned of the cost of disposal, housekeeping changes were implemented that reduced the waste by 90%.

Prepare Process Flow Diagrams and Materials Balances

A process flow diagram with a materials balance can be useful in identifying the individual production units or steps in a production process in which the bulk of the waste is generated or in which the raw materials are consumed. It is not necessary to include all of the unit processes in the process flow diagram but rather to concentrate on the process steps that have the greatest effects on materials usage or waste generation. Example process flow diagrams for a chemical plant and the individual manufacturing units are shown as Figures 3.2 and 3.3.

The goal in preparing a material balance is not to achieve complete closure but rather to highlight the principal sources of waste generation and to better understand the particular processes that contribute to them.

Generate Pollution Prevention Alternatives

Ideas are generated for pollution prevention on the basis of the mass balance and ranking of wastes by generation rates and sources. Pollution prevention methods can range from low-cost housekeeping improvements or adoption of best management practices through production changes or recovery of a useful product from a waste. The general categories of pollution prevention are as follows and are described in greater detail in later sections of the book:

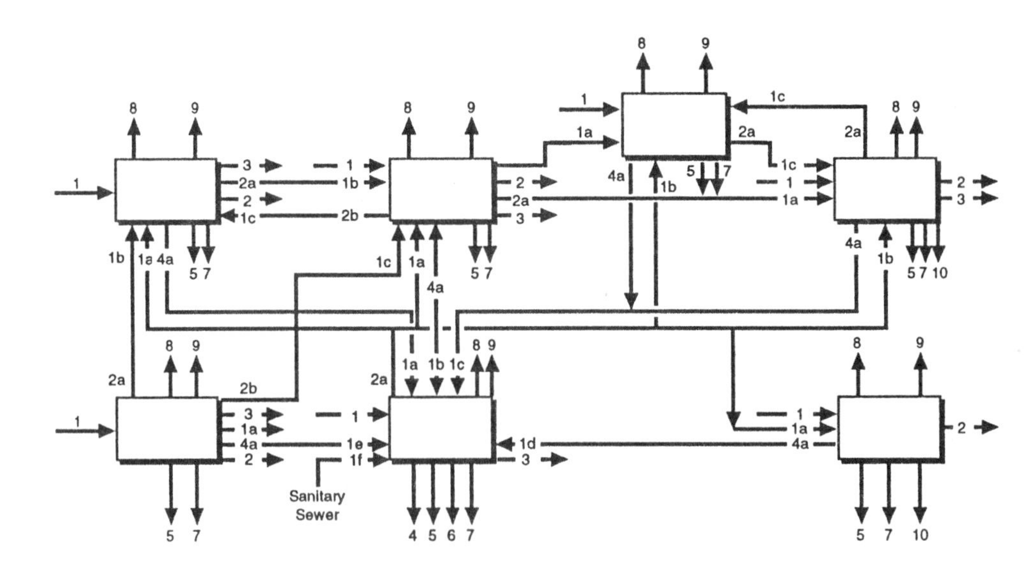

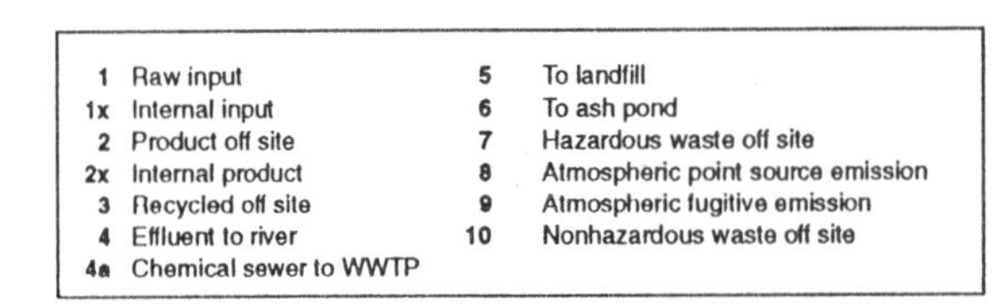

Figure 3.2. Process flow diagram for a chemical plant.

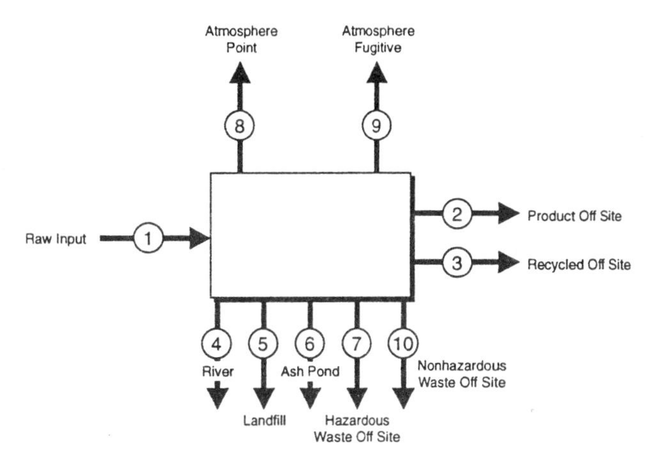

Figure 3.3. Process flow diagram for individual manufacturing unit.

- housekeeping improvements
- process modifications
- material substitution
- waste segregation
- material recycling and reuse
- treatment to reduce discharge

Pollution prevention alternatives can best be generated in a brainstorming workshop in which production and environmental personnel participate. Production people are closest to the waste generation processes and can screen processes for feasibility. They also may want to implement process changes for other than environmental reasons. Moreover, production people need to buy into any change if it is to be maintained.

Example. A client's waste sources were ranked by cost of disposal (see Table 3.1). Ammoniated citric acid and phosphoric acid wastes were generated when liquid oxygen pipelines were cleaned. The wastes were classified as hazardous because of corrosivity (pH). By simple mixing and neutralization, these wastes could be con-

Table 3.1. Annual Waste Generation Rates for Example Laboratory

Description	ID No.	Annual Generation Rate	Units	Unit Cost	Total Disposal Cost
Ammoniated citric acid	AB22A03	80,000	Gal	$ 2	$160,000
Phosphoric acid	AB22A06	30,000	Gal	1	30,000
Paint, miscellaneous	AB22A13	54	Drums	500	27,000
Aerosol cans	CN74A01	24	Drums	500	12,000
Trichloroethylene	AB22A02	40	Drums	259	10,360
Battery acid	AB21CV02	12	Drums	89	1,068
Plating sludge	AB31H01	8	Drums	63	504

verted into a liquid fertilizer for use at the facility. It also was recommended that waste battery acid be evaluated as a pH adjustment chemical in the waste treatment plant. In comparison to other plating waste treatment plants, the amount of sludge produced here was negligible because the platers rinsed parts over the plating tank.

Perform Feasibility Study of Alternatives

Pollution prevention techniques identified in the previous step need to be evaluated for technical and economic feasibility. The technical evaluation should incorporate such criteria as whether the proposed change will produce an acceptable level of product quality and worker safety, amount of effort required to install and operate new equipment or processes, and potential environmental problems created by the change. This evaluation also should take into account the effects on meeting applicable product specifications and the commitment of management and workers to implement the change.

The economic evaluation identifies the major cost items associated with each proposed alternative and estimates the potential cost savings resulting from a reduction in waste treatment and disposal. Where reasonable, economic feasibility can be measured in terms of payback period; that is, the time period in years required to realize a savings equal to the initial capital investment.

A second way of ranking projects is by capital investment required per annual unit of waste reduced. Implementing the highest ranking projects would minimize the quantity of waste generated for a fixed investment.

Identify and Support a Champion

Ideally, a champion should be the individual who originated the idea and who has extensive experience with the process. This person should be given the responsibility and the authority to make the change work.

Fund Projects

Combining production benefits with pollution prevention savings can often justify the costs of a project that cannot be justified on either basis alone. Providing funding for (and help in justifying) pollution prevention projects separately from the normal budget for capital improvements can expedite implementation of waste reduction projects.

Example. The PMB paint stripping facility at Hill AFB was funded by a Productivity Enhancement Capital Investment fund that allowed construction in less than 2 years, rather than the usual 5-year budget cycle for military construction projects.

Fund a Demonstration Project

A demonstration project can help to determine if the change is practical for an entire production unit. The demonstration should include testing to determine the actual reduction of wastes, effects on production costs, ease of operation and maintenance, and effects on product quality.

Reward Successes; Learn from Failures; Publicize Both

Recognition is a powerful motivator. Rewarding successes is a means of affirming an innovator's decision to do something different; rewards encourage others to put in extra effort to reduce waste so that they also can be recognized and rewarded. At the same time, failures should be advertised (anonymously) so that others can avoid the same mistakes.

Example. Westinghouse Hanford Company created a *Pollution Prevention Accomplishments Book* that publicizes the accomplishments of individual employees in successfully implementing pollution prevention projects. The book consists of color photos and writeups crediting the employees with the waste reductions and cost savings. Copies of the book are provided to the individual employees as well as their supervisors.

Provide for Technology Transfer and Training

A company should encourage the transfer of successful pollution prevention methods within its organization. Hands-on workshops or informal training programs at successful production units are particularly effective.

CHAPTER 4

Total Quality Environmental Management: Managing Corporate Change

*Kathi Futornick**

"The world we have created today as a result of our thinking thus far has problems which cannot be solved by thinking the way we thought when we created them"

Albert Einstein

INTRODUCTION

There is no waste. There are only materials for which uses have yet to be determined. Is this concept fantasy or is it the inevitable outcome of industry seeking to improve its environmental performance and also obtain a competitive advantage?

Although all natural and industrial processes produce waste materials, waste results in pollution when it exceeds the assimilative capacity of the environment. A commonsense approach to the problem of pollution is to eliminate the source of pollution and to find beneficial uses for waste materials that cannot be eliminated.

We are already seeing the influence of environmental issues on international standards, on trade agreements, and in global accords on environmental topics. **How does industry get ahead of the curve?** Companies seeking to minimize materials and wastes and lower their energy costs will improve their environmental performance. Some markets are already demanding clean products and technologies. Companies pioneering the "no waste" concept can realize significant cost savings in their own production and will be at the cutting edge of new marketable technologies.

Management consultant Tom Peters has described the profitable company of the year 2000 as "lean, green, and clean," with a "massive competitive advantage for those who get there the firstest with the mostest." For more than 20 years, U.S. companies have been subjected to strict environmental regulations, and some have argued that environmental protection reduces economic competitiveness. According

---

*Ms. Futornick is Manager, Environmental Affairs for the Port of Portland, Oregon, and may be reached at (503) 731-7236.

to strategist Michael E. Porter, ". . . this is a false dichotomy. It stems from a narrow view of the sources of prosperity, and a static view of competition. Strict environmental regulations do not inevitably hinder competitive advantage against foreign rivals; indeed, they often enhance it. Tough standards trigger innovation and upgrading."

Traditionally, U.S. companies have implemented environmental programs as a reaction to laws and regulations. Environmental programs have typically been placed at the periphery of the company organization. As regulations become more complex and the costs of compliance consume a significant part of corporate budgets, leading companies have changed their approach from reacting to regulatory change to anticipating and even influencing regulatory change.

Governments around the world are imposing broad new environmental requirements that affect not only companies but also products and manufacturing processes. Companies worldwide are realizing that reducing waste and conserving resources inevitably improves their competitiveness and profitability. With global awareness of environmental issues increasing, companies are feeling growing pressures from not only government regulators but also from customers, investors, and grass-roots environmentalists to reduce the environmental impacts of their products and processes. Environmental pressures are affecting business strategies as well as corporate core values. Environmental programs are no longer relegated to the periphery of industry but are being elevated to the highest corporate levels.

This chapter describes how a systems approach to integrating environmental stewardship with total quality management (TQM) principles provides a framework for creating environmentally responsible products while remaining competitive in a global marketplace. Leading the shift in environmental thinking is the concept of pollution prevention. The chapter describes the relationship of pollution prevention and TQM and how they both contribute to a total quality environmental management (TQEM) program. It also presents strategic approaches and operational practices that contribute to "holistic" environmental management: assessing environmental performance, life-cycle analysis and product design, industrial ecology, and sustainable development.

ELEMENTS OF CULTURAL CHANGE

The role of environmental management is one of the most rapidly changing areas of management. Its focus has shifted from reacting to crises and solving problems to complying with regulations. The next stage of environmental management goes beyond compliance to pollution prevention. For leading companies, the role of environmental management has evolved further to managing on behalf of the environment. The objective of these changes is the "greening" of companies. Leading companies recognize that industry will be expected to be at the forefront of this evolution/revolution during the shift from:

- public apathy to public concern
- local interest to global interest
- end-of-pipe treatment to pollution prevention
- top management insulation to involvement
- regulatory compliance to continual improvement

- environmental costs to competitive advantage
- adversarial and isolationist relations with regulators to cooperation and participation[1]

As environmental regulations multiply, the broadening responsibilities of environmental management place new demands on an organization's resources. This has caused a major organizational shift, elevating environmental management to the strategic business planning level of the organization and broadening its influence across the firm.

Environment management has emerged as an element in the competitive marketplace. Markets are dynamic: they are constantly in flux and affect costs and revenues in ways that traditional business planning cannot always foresee or control. Environmental responsiveness may be the most important strategic move for some industries over the next two decades. "Green" consumerism is changing market patterns. Product-oriented policies strike at the core of a business.

The uncertainty of the environmental requirements and influence of diverse groups underscores the importance of implementing a systems approach to environmental management and its newly found sphere of influence that includes product design; marketing; and the processes used to assemble, transport, service, and recycle products.

TOTAL QUALITY ENVIRONMENTAL MANAGEMENT: WHAT IS IT?

Before discussing TQEM, it is important to understand TQM. TQM was introduced in Japan following World War II, and during the late 1970s was applied to environmental management by Japanese industry.[2] Although TQM has been practiced in the U.S. for years, its status was enhanced when the Malcolm Baldrige National Quality Award program was launched.

Industry has embraced TQM's customer- and quality-focused principles in order to remain competitive. The TQM process requires a commitment to continuous improvement and to constant reevaluation of performance and redefinition of purpose. The only constant is change. Companies that have successfully implemented TQM programs have had increased revenue and profits, reduced customer complaints, and a motivated work force.

The basic elements of a TQM program include:

- commitment to customer satisfaction
- commitment and leadership of management
- measurement of quality and continuous improvement
- employee empowerment and teamwork

Implementation of a TQM program is a shift from traditional top-down management to participation by all employees in improvement efforts. TQM serves as a method for integrating quality in all business activities, using a fact-based approach to make improvements.

The Malcolm Baldrige Award program consists of seven interrelated areas organizational excellence: leadership, human resources utilization, information a analysis, strategic quality planning, quality assurance, quality results, and customer satisfaction. The strength of the Baldrige criteria lies in their flexibility and comprehensiveness.

The Malcolm Baldrige Quality Award program is viewed as the "holy grail" of TQM, with emphasis on customer satisfaction and continuous improvement. The detailed criteria of the Baldrige categories further provide an approach to quantifying an organization's strengths, weaknesses, and opportunities for improvement. Quantification of the benefits associated with introducing "quality" into an organization has been difficult but necessary for corporate support.

Integration of TQM and the Environment

Managers are beginning to realize that their companies will remain competitive only if they can demonstrate environmental responsibility. Customers make demands that go beyond environmental compliance—desiring products and manufacturing processes that protect and improve the environment. Integration of environmental management principles with TQM is known as TQEM.

TQEM integrates environmental issues into corporate thinking. TQEM can take an environmental management program, which traditionally has been viewed as a cost center, and by perceiving its value, ultimately turn it into a profit center.

Quality and customer satisfaction are at the heart of TQM. Corporate environmental staff typically have been seen as internal regulators, soaking up profits and presenting costly obstacles to production. By applying TQM principles, environmental management can be viewed as a service supplier to its customers. Its customers are internal (e.g. production, accounting, human resources, and marketing) and external (consumers, regulatory agencies, environmental groups, and the local community).

Leadership and Environmental Management

TQEM cannot be implemented without the commitment of top management. Achieving fundamental change requires an examination of the roles and responsibilities of various parts of the company in achieving customer satisfaction. Only the commitment of senior management to TQEM can alter these responsibilities and provide the vision to transform an organization into a "lean, green, and clean" company.

The TQEM process embraces individual initiative and responsibility and also redistributes ownership to create partnerships throughout the organization. In TQEM, the concept of leadership moves from center stage, and the spotlight shifts to the people doing the work, the focus being on individual and team ownership and responsibility. Peter Block describes top management as "defining the mission, playing field, and vision while serving as a caretaker guiding cultural values rather than acting to control and define purpose for others."[3]

One of the tangible manifestations of TQEM is the company's environmental policy statement. Environmental policy statements provide guidance for business decisions and operating practices. An effective environmental policy statement elevates environmental issues from the compliance arena to the strategic arena.

Top management is also responsible for communicating policy and providing the framework for all functional areas to benefit. Managers that have successfully integrated environmental management into their business planning process have provided equal support for both production and environmental decisions.

Information and Analysis

Information is needed for measurement systems to implement a corporate environmental policy. Information will, at a minimum, include identification of customer needs and expectations; risks associated with products, services, and operations; toxic raw materials used; regulatory drivers; efficiencies of the production processes; quantities of pollutants released; compliance issues; and approaches taken by competitors. Critical measures need to be incorporated into information systems to notify staff early about potential problems so that they can monitor and control the effects. An effective information management system provides not only retrospective analyses but also forecasts problems before they become crises.

Strategic Quality Planning

Strategic planning translates goals into specific plans. In pursuit of environmental excellence, leading companies must develop new products and processes. These processes and products require significant capital investment and are included in the corporate business plan. Incorporating environmental objectives early in the planning process can lead to the development of "green" products. These efforts are becoming more and more significant in light of global environmental standards.

Human Resources

TQEM is a team effort. Production staff, operators, maintenance staff, trainers, and accountants are all pulled together for a common purpose—to identify environmental issues and prevent problems leading to waste—and ultimately, to improve the company's profitability. The success of this effort depends on empowerment and reward programs that encourage individuals to take responsibility for everything that affects their process or product. As an example, the individual responsible for the production of batteries would take responsibility for the battery itself, for the waste streams the manufacturing process generates, for compliance-related problems, and for the product design.

Quality Assurance

Quality assurance is a system of checks and balances that compares what should have happened with what was actually done. A corporate culture of continuous improvement requires a change in traditional quality assurance programs. Quality assurance should focus on processes and process documentation as well as on products. Quality assurance is key to anticipating and preventing problems instead of reacting to problems.

Environmental Results

Environmental results are the bottom line for environmental performance. The concept of environmental results has expanded from measurements of compliance and waste reduction into areas such as product design and customer satisfaction.

Compliance and waste reduction measures lend themselves to quantitative evaluation more readily than do product design and customer satisfaction. Nevertheless, measuring performance in all areas is important in the continuous improvement process.

A significant obstacle to effective environmental management is the absence of accounting information that accurately reflects the true costs, lowered costs, and revenue generation **(yes, revenues)** from marketing environmental products to "green" consumers. With this type of accounting data, companies can minimize adverse environmental impacts and improve product quality and processes. The cost savings and revenues associated with environmental programs are more likely to sustain a TQEM program over the long term than a program that focuses solely on compliance.

Enviro-Feedback

The merging of environmental programs and TQM presents a new challenge for top management: the integration of nonmanufacturing operations into business planning. Environmental programs have been perceived as taking away from a company's profitability. The challenge for companies is to turn environmental management from a cost center into a profit center, and integrate it into the core of their business practice.

There are many obstacles to integrating environmental management into a corporate TQM program. Determining both the cost of environmental improvement and the value it adds to products is key to implementing a successful TQEM program, but is difficult to accomplish.

Before environmental management can be integrated successfully into a business, top management needs to establish corporate objectives and standards that can be measured. Remember, "what gets measured gets done."

Approaches to and tools for measurement vary, but typically include an assessment matrix typically scored through compliance audits; an evaluation of the management systems; and documentation of the assigned scores. The questions to be answered are: Has the program really improved bottom-line and environmental performance? and, Has the program affected customer satisfaction and changed his or her purchasing behavior?

The Assessment Matrix

An assessment matrix is commonly used to measure performance. Xerox, AT&T, DuPont, Dow, and other leading industries apply an assessment matrix to measure environmental performance. Besides being a self-assessment tool, a matrix is a roadmap for implementing TQEM.

In general, establishing an assessment matrix program includes four steps: develop the program measures; establish a matrix scoring system weighted toward measurement of management systems performance; provide supporting documentation; calculate scores; and use the results. Selecting appropriate measures is critical because systems tend to be optimized on the basis of these key measures. The Baldrige evaluation criteria can serve as the headings for the matrix: Leadership; Information & Analysis; Strategic Planning; Human Resource Development; QA of Environmental Performance; Environmental Results; Customer/Stakeholder Satisfaction.

Generally, the elements within the matrix reflect corporate environmental policy. The trend is to go beyond compliance to committing the company to minimal or zero pollution through waste reduction, applying the same principles or standards globally, engaging in partnerships with governments and citizens, and adopting a life-cycle approach to product development.

Benchmarking

Benchmarking is "the practice of comparing programs or processes with the intent of establishing reference points for continuous improvement."[4] Benchmarking, as described by Marcia Williams, "allows a company to step outside its own culture to see new solutions that might be available."[4] Benchmarking is used to assess a company's environmental performance both against internal success factors and against the environmental programs of other companies or organizations.

External benchmarking usually needs input from external sources. Information on performance can be obtained from suppliers and customers through direct contact or through surveys. Competitor benchmarking can be done through a literature search, federal and state agency data systems, a company's annual reports, or studies from consulting firms.

Results from benchmarking are then compared to targets established in the assessment matrix, and together with other factors are used to develop goals, recommendations, and strategies. In the context of TQEM, a cross-functional team with top management support and stewardship is typically involved in the overall assessment program and development of implementation strategies.

In 1991, AT&T and Intel joined forces to identify the "best-in-class" corporate pollution prevention programs and develop benchmarks for targeting their own company's efforts at continuous improvements in pollution prevention. The conclusions from their study identified several consistent trends: measurement systems give meaning to corporate environmental policy statements; measurement programs must be implemented and provide useful feedback; the right measures will draw from mainstream business data systems; and measures should be as lean and clean as possible.[5]

For the assessment program to be successful, the assessment must occur at two levels: both at the corporate level, and at the facility level. The objective at the facility level is to identify the day-to-day activities that can affect corporate goals and can be rolled up to the corporate level.

ENVIRONMENTAL PROGRAMS AND COMPETITION

European and Japanese companies are preparing to supply the global market with alternative technologies and green products and processes. Leading U.S. companies are also seizing environmentally related business opportunities such as more efficient technologies and marketing of "green" products.

In recent years, pollution prevention has become the dominant theme in environmental legislation and regulation. In a 1990 survey conducted by Abt Associates of Fortune 200 nonservice firms, senior environmental executives perceived their top four management challenges to be "pollution prevention, product development, regulation, and marketing."[6]

While environment as competitive advantage is more than just pollution prevention, this is a significant and necessary start. In the context of TQM principles, waste discharges are viewed as defects that should be reduced or eliminated. Applying the principles of pollution prevention by using substitute materials or changing processes can prevent waste generation. When wastes are generated, they are recycled back into the process that produced them or reclaimed and reused in other products.

Pollution prevention has been the catalyst for industry to begin managing environmental programs as a competitive issue, and in recent years has become the focus of most corporate activity. With increasing frequency, top managers have ranked pollution prevention as their company's most important environmental activity, providing the greatest opportunity to improve environmental performance.

Successful pollution prevention requires that environmental professionals work as a team with professionals from operations, engineering, quality, information management, marketing, accounting, and human resources. This teaming of environmental activities across functional areas of industry increases profitability by reducing costs and liability, and generates environmentally based opportunities.

Market forces expand TQM's focus on the customer. Customer satisfaction is at the core of quality programs—the Baldrige award criteria assign 300 out of 1,000 points to achieving and measuring customer satisfaction. An output has value only when a customer sees value in it. The measurement for environmental quality broadens the customer base to include legislators, regulators, and activist groups.

THE THREE PHASES OF THE ENVIRONMENTAL (R)EVOLUTION

The environmental (r)evolution can be divided into three phases (see Figure 4.1). The first phase, up to the 1960s, was largely based on a voluntary effort to protect the environment from degradation. When voluntary efforts failed to adequately protect the environment, mandated controls were placed on industry. Although voluntary efforts have continued, industry has seen an exponential growth in regulations (see Figure 4.2). During the second phase, from the 1960s to the 1980s, the nearly exponential increase in environmental laws and regulations increased the expense of compliance and resulted in companies addressing contamination problems but not in preventing the problems from occurring. Measures for controlling

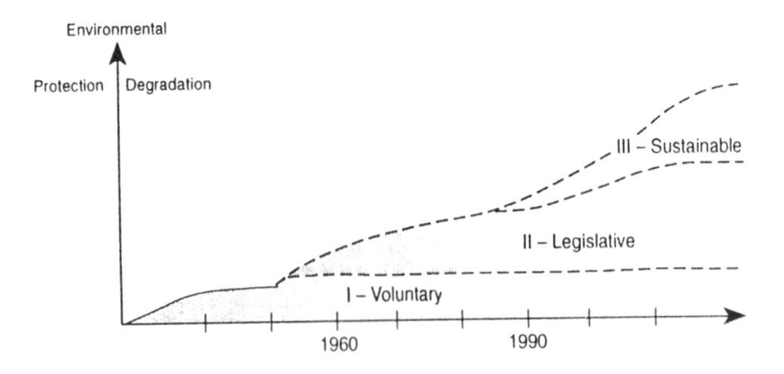

Figure 4.1. Environmental evolution's three phases. (Adapted from Strombold, J. ABB Brown Boveri, Limited. Presentation, Proactive Environmental Strategies for Industry, Massachusetts Institute of Technology, Cambridge, MA, November 17, 1993.)

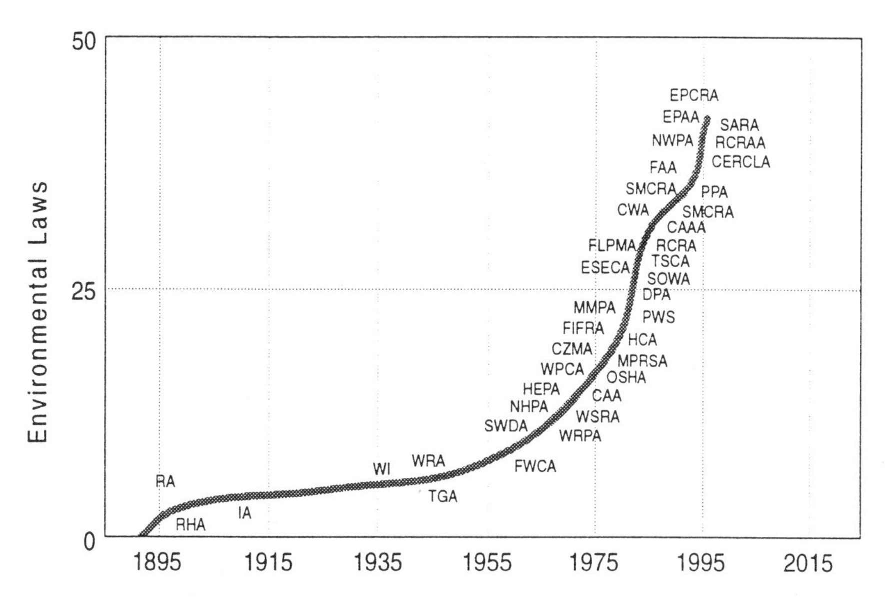

Figure 4.2. Federal regulations. (Source: Adapted from Ferrone, R. Digital Equipment Corporation. Presentation, Proactive Environmental Strategies for Industry. Massachusetts Institute of Technology, Cambridge, MA, November 17, 1993.)

pollution focused on end-of-the-pipe command-and-control methods, and environmental management became firmly established as a cost center. It has been estimated that the U.S. EPA has promulgated more than 11,000 pages of regulations and imposed more than $1.4 trillion (1990 dollars) in compliance costs on industry since the Agency's founding in 1970. These costs are increasing. In 1990, the Agency estimated that complying with federal pollution-control regulations costs American industry $115 billion a year—2.1% of the gross national product. During the 1990s, total costs for compliance are estimated at $1.6 trillion, excluding the effects of the Clean Air Act Amendments of 1990, which could add another $25 to $40 billion annually.[7]

With the costs of environmental management on the rise, leading industrial firms during the early 1980s began to consider an alternative strategy—reducing their costs by reducing the quantity of waste generated. Only since the mid-1980s have pollution prevention programs gained a significant role in environmental management. At this time, leading companies began to consider environmental costs as a justification for improving production processes. Major companies such as the 3M Corporation, which introduced the 3P program (Pollution Prevention Pays), and Chevron, with its Save Money and Reduce Toxics Program, sent a strong message that companies can improve the efficiency of their operations and procedures to reduce waste and cut costs.

While pollution prevention programs were beginning, the Emergency Planning and Community Right to Know Act was passed in the mid 1980s. EPCRA included a new provision, the Toxics Release Inventory (TRI), which required industry to report all waste released to the environment. The TRI informed the public of the volume of pollutants entering the environment, and provided them a platform from which to influence future regulations and industry performance. TRI was the beginning of instantaneous accountability and provided the context for the phrase: "Industry lives in a fishbowl and sits on a gas stove."

In 1990, Congress passed the Pollution Prevention Act, and in May 1993, U.S. EPA Administrator Carol Browner set forth pollution prevention as the central policy of her administration. Pollution prevention set the stage for environmental management to move into a strategic planning role. From the lessons and benefits learned from focusing on pollution prevention, many companies instituted chemical inventory management programs and changed the design of products and manufacturing procedures.

Sustainable development, the third phase, now dominates the international debate on the environment and development. In 1987, the World Commission on Environment and Development appointed by the United Nations General Assembly defined sustainable development as progress that "meets the needs of the present without compromising the ability of future generations to meet their own needs." The Brundtland definition states that sustainable development is the "process of change in which the exploitation of resources, the direction of investments, the orientation of technological development, and institutional arrangements all are in harmony for the enhancement of both current and future potential to meet human needs and aspirations."

One of the greatest challenges facing industry and consumers in the next decade is to stimulate market forces to protect and improve the quality of the environment. Recognizing the link between environment and economic viability, in 1991 the International Chamber of Commerce (ICC) created a Business Charter for Sustainable Development. The charter, which by 1992 had been endorsed by more than 600 companies, has 16 key principles (see Figure 4.3).

In 1990, Stephan Schmidheiny, a fourth-generation Swiss industrialist with worldwide business interests, urged 48 other business leaders to develop a blueprint for "green" industry for distribution at the 1992 "Earth Summit" in Rio. After more than 50 meetings, the Business Council for Sustainable Development (BCSD) submitted its report, *Changing Course*. Among the proposals included in the report, council members advocated that companies gradually adjust their prices to reflect the environmental costs of making, using, recycling, and disposing of a product. Only with such pricing will corporations be able to internalize the environmental costs of doing business. That change, in turn, should prompt companies to use natural resources more efficiently and develop innovative methods of reducing waste and pollution.

Sustainable development will require more than pollution prevention and tinkering with environmental regulations. Consumers are the real day-to-day environmen-

Figure 4.3. ICC: 16 Principles for Environmental Management

- Corporate priority
- Integrated management
- Process of improvement
- Employee education
- Prior assessment
- Products and services
- Customer service and advice
- Facilities and operations
- Research and development
- Precautionary approach
- Contractors and suppliers
- Emergency preparedness
- Transfer of technology
- Common effort
- Openness to concerns
- Compliance and reporting

tal decision makers. Sustainable development requires their support through their preferential purchase of "green" products.

GLOBAL COMPETITION

The European Community (EC) in 1957, through the Treaty of Rome (Article 130r), specified that an objective of the EC is to "preserve, protect, and improve the quality of the environment, and to contribute towards protecting human health and to ensure a prudent and rational utilization of natural resources." Until recently, the EC has had limited effectiveness in getting member nations to adhere to environmental policies, even though it has passed more than 200 legislative measures since 1972. EC members often have disagreed about environmental commitments, arguing that requiring specific practices can place some member countries at an economic disadvantage in relation to more prosperous nations. Nevertheless, in recent years individual European countries and the EC have established environmental visions that emphasize pollution prevention and recycling, and deemphasize management of wastes after they are generated. Their environmental agendas have included monitoring manufacturing processes using life-cycle analysis; labeling products; implementing environmental management systems and performing audits; and protecting employee health and safety.

New Environmental Standards for Competing Abroad

As a result of an initiative by the BCSD, the International Organization for Standardization established a Strategic Advisory Group on the Environment (SAGE) that is proposing to extend international standards beyond product quality to ensure ecological efficiency. International standards have been prepared in several areas, including environmental auditing, labeling, methods for assessing environmental performance, and life-cycle analysis.

Auditing

The U.S. EPA and the EC are developing voluntary plans to institutionalize environmental management and auditing. Their goal is to urge companies to go beyond minimal compliance with environmental laws and regulations.

The EC's voluntary environmental management and audit plan (CEMAS) establishes a standard approach to environmental management. Companies are required to set up environmental management systems, audit those systems, and make a public statement on performance. The target date for implementation is early 1995.

Larger U.S. companies have environmental audit programs, but because of the profusion of environmental regulation in the U.S., the focus is still on compliance. European companies are expected to take a broader approach in evaluating overall environmental performance.

CEMAS sites will be able to display their certification at the factory gates, and peer pressure is expected to ensure widespread adoption of the program. However, unlike the Ecolabel, described later, the audit logo may not be displayed on a product.

Although, on the whole, companies abroad are embracing these initiatives, the lack of uniformity of some of these initiatives and the burdensome documentation they require are of concern. Whereas the intent of the initiatives is to supply consumers with information about the products they are purchasing and about the environmental record of a particular company, some of the aspects of the initiatives may restrict market opportunities, especially to U.S. industries hoping to compete abroad.

Labeling

The administration of the EC recently finalized a uniform "ecolabeling" scheme for products sold throughout the EC. The scheme is modeled on Germany's Blue Angel program, which is based on a life-cycle approach. Canada's Environmental Choice program is similar, based on an analysis of the environmental impact of production, use, and disposal of a product. Businesses should take note that the EC Ecolabel and Canadian program address manufacturing processes, and that third-party certification is required. In the future, U.S. companies wishing to sell products in countries with ecolabel regulations will have to meet the regulating country's certification requirements.

Official ecolabeling programs, such as Germany's Blue Angel, Canada's Environmental Choice, and Japan's Eco-mark, assist consumers and stimulate companies to design better products. By the end of 1991, nine member companies of the Organization for Economic Cooperation and Development had introduced labeling schemes, and plans for 13 more labeling schemes were being developed. Requirements for ecolabels can seriously affect the marketability of a business's products.

Compared with Europe, Canada, and Japan, ecolabeling is not widely practiced in the United States. United States initiatives include Scientific Certification Systems (formerly Green Cross) and Green Seal Labels. These environmental labels supplement claims for recycled content, recyclable, etc. However, these labels are not standardized. A 1993 U.S. EPA regulation requires manufacturers to provide consumers with information to distinguish between products containing ozone-friendly and ozone-depleting chemicals (ODCs), and products containing ODCs must carry a warning label. Other labels such as "recyclable" or "biodegradable" and "environment friendly" are ambiguous and do not conform to an established standard. In May 1991, the U.S. Federal Trade Commission published the *Green Report II* in the *Federal Register*. The *Green Report II* proposed guidelines listing examples of when labels such as "recyclable," "recycled," "source reduction," and "reusable" would and would not be considered deceptive.

Without clear standards, legal action can arise from questionable environmental claims. U.S. industry is faced with a dilemma: face potential legal challenges to environmental claims or respond to customer preference for environmentally friendly products. According to a 1989 *New York Times*-CBS poll, 80% of Americans believe that progress in protecting the environment must continue regardless of cost. With such survey results showing that consumers are interested in purchasing environmentally friendly products, the trend for the future will be a form of green label standards.

Standards for Environmental Performance

Companies registering under the EC's audit scheme or licensed under the Ecolabel program must maintain their products and industrial activities in conformance with their established standards of practice. Documentation will need to emphasize, at a minimum, manufacturing procedures, strict adherence to specifications, and legal requirements. Compliance with the new initiatives requires that environmental management systems be established that are heavily oriented toward TQM.

The International Standards Organization (ISO) published the ISO 9000 standard in 1987 as a set of guidelines for developing a documented quality system. A quality system is the organizational structure, responsibilities, procedures, processes, and resources for implementing quality management.

The ISO did not specify compliance dates for ISO 9000 certification; however, many companies are complying now. Pressure to comply will build as individual buyers begin to purchase only from ISO 9000 registered vendors. The ISO 10011 series also presents standards for conducting quality audits. In the few years since their first publication in 1987, the international standards have been adopted as national standards in more than 60 countries, including all of the developed nations. They have become the accepted basis of quality systems requirements for product conformity assessment in the global marketplace. The standards primarily affect manufacturing companies. More than ever, companies are interested in reducing costs, eliminating supplier quality audits by providing a common language for quality, and reducing liability.

The EC realized that in order to facilitate free trade, environmental programs must be standardized. SAGE was established by the ISO and the International Electrotechnical Committee (IEC) in 1991 to make recommendations about the environment. SAGE recommended establishing an international standard for environmental management.

A technical committee (TC207) was set up under ISO and is currently involved in generating ISO 9000E as the model for Environmental Management Systems. This is expected to strongly resemble the British Standard (BS 7750). Already, France and Germany have a standard similar to BS 7750.

BS 7750 was prepared under the direction of the Environmental and Pollution Standards Policy Committee of the British Standards Institution (BSI). BS 7750 provides for an independent professional assessment for all aspects of a company's environmental performance. It is a voluntary management tool and in many ways parallels the ISO 9000 and the EC audit scheme.

The environmental management system certification programs are supported by companies operating under strict environmental standards, export constraints, and customer requirements. The benefits that industries may see directly include potential reductions in energy use, waste disposal, and raw material costs through use of recycled materials. Companies certified under the Environmental Management Standard will be able to show that they are "environmentally friendly" and assure those interested in the environmental effects of the companies' operations.

In the United States, the American National Standards Institute (ANSI) has been active in SAGE and has developed a standard for the U.S., ANSI/ASQC E4, "Quality Systems Requirements for Environmental Programs." The standard details the quality elements that should be considered when planning, implementing, and assessing environmental programs. Many of the ISO 9000 elements are intertwined in the standard, which also resembles BS 7750.

In 1993, the U.S. EPA introduced the Environmental Leadership program, which targeted leading companies implementing prevention-oriented environmental management. The program, although heralded as a major step forward in providing incentives for companies to establish environmental management systems, has also been criticized for its inflexible approach. Many U.S. companies struggling to comply with command-and-control regulations have little capital left for embracing noteworthy programs such as the Environmental Leadership Program.

The senior business group in Japan, the Keidanren, adopted an Environmental Charter in 1991 that sets standards for environmental behavior. Malaysia has established a corporate environmental policy that calls on companies "to give benefit to society; this entails . . . that any adverse effects on the environment are reduced to a practicable minimum." The Confederation of Indian Industry has also promoted an "Environment Code for Industry" among its members.

Chemical industry associations in several countries have agreed to a "responsible care" program to promote continuous improvement in environmental health and safety. Begun in Canada and adopted by the United States, Australia, and many European countries, the program encourages associations to draft codes of conduct in many areas of operations, and it recommends that large companies help smaller ones with environmental and safety improvements.

International standards are a trend for the future. Companies whose operations are under strict environmental requirements will most likely be actively seeking certification. Businesses outside the EC must be aware of the mounting pressures and effects of environmental standards on their products, production facilities, and operations. Although the standards are now voluntary, the standards may become mandatory, and exporters would need to conform. Businesses in the U.S. will need to become familiar with EC standards for environmental management systems, labeling, and auditing.

Joseph Cascio, program director of Product Safety and Chemical Management for IBM, predicts that "four years from now [in 1997], American plants will have to be certified to environmental standards in order to sell their products in Europe."[8] Already DuPont's European customers are asking for or requiring an environmental management system certification. In addition to market pressure, an environmental management system will likely produce the same types of results that the ISO 9000 program has.

Life-Cycle Analysis

As companies become better at reducing waste, attention is shifting from pollution created during manufacturing processes to environmental impacts caused by the product during its entire life cycle. The dividing line between process and product is artificial; many products are inputs of other processes. Furthermore, the industrial ecosystem of the manufacturing process is only part of a more comprehensive ecosystem, which contains all flows of materials and wastes produced through raw material acquisition, material and product manufacture, use (including energy consumption), and disposal of goods and services. Life-cycle analysis accounts for all the stages of product processing, use, and disposal (see Figure 4.4).

The first life-cycle analyses (LCAs) were performed in the late 1960s and early 1970s and tended to focus on the comparative energy consumption of different materials, particularly packaging. Today, LCAs go beyond energy balances and are used to evaluate resource requirements and environmental impacts. An LCA has three parts:

- an inventory of energy, resource use, and emissions during each step of the product's life

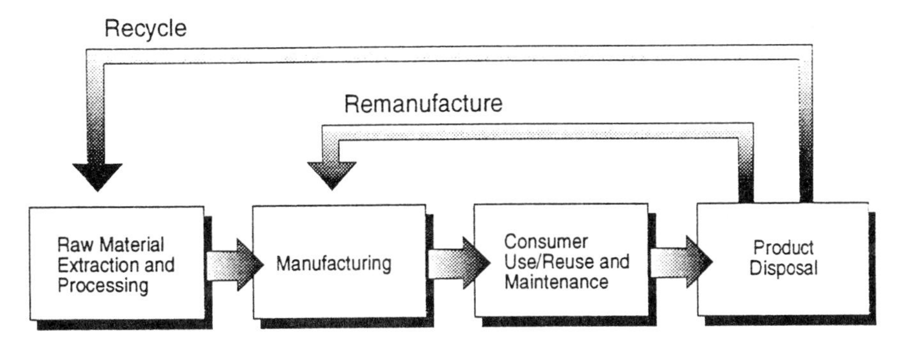

Figure 4.4. Product life cycle. (Source: Adapted from Oakley, B. "Total Quality Product Design—How to Integrate Environmental Criteria into the Product Realization Process," *Total Quality Environmental Management* 2(3):309–321, Spring 1993.)

- an assessment of the impact of these components
- an action plan for improving the product's environmental performance

The tools used in performing LCAs reflect an increase in not only environmental concerns but also in concerns about a product's extended life-cycle. Corporate environmental responsibility for its products no longer ends at the factory gate—it extends from cradle to grave. Ultimately, this will lead to manufacturing products that minimize environmental impacts and maximize environmental efficiency, and will require new relationships between producers, consumers, and governments.

LCA has several other names, such as "design for environment," "life-cycle inventory," "product life-cycle investment," and "product stewardship." Whichever term is used, the relative success in reducing pollution and demand from consumers for "green" products is turning business's attention toward improving the environmental performance of products. Functional groups within a company typically involved in life-cycle approaches include marketing; product design and development; research; health, safety, and environmental impacts; and manufacturing. Companies using LCA can stimulate product innovations through alternative design and influence research priorities. Some use LCA for marketing purposes.[9]

The shift from mass production to mass customization has greatly increased the number and variety of products, posing new challenges. According to an OECD report, " . . . if present trends continue, 50% of the products that will be used in 15 years' time do not exist yet . . ."

Product development is the source of sustainable profit. Product design accounts for about 5% of a product's cost but affects between 70% and 80% of the product's life-cycle cost (materials, manufacturing, distribution, and servicing).[10] Product quality begins with product design. Environmental considerations integrated into the design cycle can result in high-quality products, with waste treated as an environmental defect and eliminated from a product.

Under the pressure of tightening regulations, increasingly "green" consumer expectations, and new management attitudes toward extended corporate responsibility, companies are recognizing that environmental management now requires minimizing risks and impacts throughout a product's life cycle. Focusing on the total life cycle will ensure a net reduction in a product's environmental impacts. This in turn is leading to the industrial ideal of an economic system based on reconsumption – that is, the ability to use and reuse goods in whole or in part over several generations.

A number of companies have used LCA early in the manufacturing process, during the planning stage. Companies such as AT&T, General Electric, IBM, DEC, Procter & Gamble, Whirlpool, and Xerox are finding that designing "greenness" into their products offers unique opportunities for improving their performance in environmental management as well as in the marketplace.

The Industrial World as a Natural System

The Ecofactory

The concept of the ecofactory began in Japan. The ecofactory is defined as a manufacturing infrastructure that maximizes productivity, performance, and economy, and minimizes ecological impacts such as greenhouse-gas emissions, hazardous wastes, and air emissions. Japanese researchers at the Industrial Science and Technology Agency submitted a model for the ecofactory that integrated design of production systems technology – including design at product and process levels that addressed environmental issues – with disassembling, reuse, and materials recycling technology.

The ecofactory depends not only on energy-efficient manufacturing but also on product design, with a goal of minimizing ecological impacts for the life of the product. The success of the ecofactory is measured in terms of productivity, economy, performance, market needs, and, most importantly, ecological impacts.

Industrial Ecology

Industrial ecology is a "systematic organizing framework for the many facets of environmental management. It views the industrial world as a natural system, embedded in local ecosystems and the global biosphere. It provides a fundamental understanding of the value of modeling the industrial system on ecosystems to achieve sustainable environmental performance."[11] Industrial ecology uses biosphere constraints to understand industrial system constraints.

The present design of the industrial ecosystem is flawed; rather than acting according to the circular principles of natural ecosystems, the flow of goods and services is essentially linear. Many products are manufactured, purchased, used, and discarded with little regard to environmental efficiency or impact. Recycling materials after products have reached the end of their useful life can bring considerable savings, as it avoids the stages of extraction and processing, which use energy and produce waste. For example, about 45% less energy is used to produce steel from scrap material than from iron ore, and air and water pollution are cut by more than 75%.

Industrial ecology provides a broad organizational framework for TQEM and waste reduction programs. One of the basic principles of the ecosystem is that there is no waste. Industrial ecosystems are a logical extension of life-cycle thinking. They involve "closing loops" by optimizing use of raw materials and maximizing use of recycled materials and energy, minimizing waste generation, and using wastes as raw material for other processes.

In Denmark, Novo-Nordisk, the Asnaes coal-fired power station, a Statoil refinery, and a Gyproc plasterboard factory are involved in a complex material flow.

Novo buys "waste" steam from the power station, while the power stations buys "waste" cooling water from the refinery. Known collectively as industrial symbiosis, this system has evolved over time to become commercially viable and meet environmental quality standards. Asnaes will soon buy the flare gas from the oil refinery, while Gyproc is evaluating the possibility of using the gypsum produced by the desulfurization scrubber in the power station.

An integration of three companies is under evaluation in the United States involving a steel company, a high-tech manufacturing facility, and a cement manufacturer. The cement manufacturer is considering using waste from the high-tech manufacturer as fuel in its cement kilns. The steel manufacturer may soon be supplying by-products as ingredients for the production of cement. The cement kiln is capable of recycling various kinds of waste-derived fuel, including hazardous wastes, waste tires, and sludge.[12]

Cooperative initiatives among industries and between government and industries are being launched. Several utilities in Europe and in the United States are planting forests to offset emissions or are working to encourage sustainable forestry practices. The International Network for Environmental Management in Germany has launched another type of offset program that is based on the assumption that industry has "plucked the low-hanging fruit" in their own plants, and the most cost-effective environmental policy would be to invest in pollution prevention programs in developing countries.

From Cradle-to-Grave to Cradle-to-Cradle

The key to creating industrial ecosystems is to think of wastes as products. This suggests that companies not only search for ways to reuse waste, but also that regulations change their orientation from waste management to offering incentives to companies for reusing waste. Industry is moving toward "remanufacturing"—that is, recycling the waste materials in their products and thus reducing the use of raw materials and energy to convert those raw materials. Not only is "remanufacturing" technically feasible, it is also profitable. Successful and competitive companies are at the forefront of what is known as "eco-efficiency." Industry and government regulators must work together to make the cradle-to-cradle concept more accessible.

Some examples of remanufacturing are:

- U.S. Gypsum is making 100% recycled content sheetrock from synthetic gypsum generated as a by-product of utility flue gas desulfurization.
- Wisconsin Public Service and Rhinelander Paper Co. are creating a materials and energy loop whereby the power company will burn paper-mill waste, which, together with low-sulfur coals, will power two counties and the paper mill itself. Waste sent to the landfill will be cut by 25%.
- The Boeing Longacres Plant in Renton, Washington, will use wastewater from a neighboring wastewater treatment plant for cooling, thus reducing energy and water costs.
- Oregon Steel Mills, Inc. (OSM), a steel manufacturer in Portland, Oregon, generated approximately 8,000 tons of electric arc furnace (EAF) dust, a listed hazardous waste. After several years of research, OSM developed a process known as Glassification, which uses EAF dust as an ingredient in the

manufacture of glass-based products such as roofing granules and abrasive blasting media.

- Inland Steel has a commercial operation in which pickle liquor (an acidic solution of iron and heavy metals) is used to produce high-tech magnetic materials. A second, smaller operation recovers fine graphite dust (called kish) from effluent gases from steel plants to produce high-tech graphite lubricants.

These are just a few ways that industrial systems can mirror natural systems by implementing no-waste concepts. By looking at wastes as potential raw materials, companies can profit in ways not envisioned just a few years ago.

INDUSTRY: FROM VILLAIN TO HERO

Since the 1970s, waste reduction has been gaining ground in policy and corporate circles as the most cost-effective way of improving environmental efficiency. To Stephan Schmidheiny, a Swiss industrialist and chairman of the BCSD, the changing course of business is logical. "The common denominator between environment and development is efficiency—efficiency in using resources and creating wealth."[13] More and more companies are realizing that the waste they produce is a sign of inefficiency, and that waste represents raw materials not sold in final products.

The economies of industrial countries have grown while the resources and energy used to produce them have declined. Chemical companies in industrial nations have doubled output since 1970 while more than halving energy consumption per unit of production.[13] In West Germany, the chemical industry managed to cut emissions of heavy metals by 60% to 90% between 1970 and 1987, while boosting output by 50%.[14] These improvements have been matched on individual company levels as growing numbers of companies improve their environmental efficiency—that is, the ratio of resource inputs and waste outputs to final product. At Nippon Steel Corporation, producing a metric ton of steel in 1987 emitted 75% less sulfur dioxide and 90% less dust than in 1970. Since 1960, Dow Chemical has cut production of hazardous wastes from 1 kg per kg of salable product to 1 kg per 1,000 kgs. At Ciba-Geigy, finished products represented only 30% of all outputs in 1979, the rest being waste. By 1988, the company's efficiency had increased 62%; a goal of 75% efficiency has been set for the end of this decade.[15]

- The 3M Company has pioneered corporate pollution prevention since 1975 with its Pollution Prevention Pays (3P) program. With total sales of over $13 billion in 1990, 3M has undertaken more than 4,000 3P projects in the past 15 years, cutting air pollutants by 120,000 tons, wastewater by 1 billion gallons, and solid waste by 410,000 tons. In the process, the company saved more than $600 million. 3M has further called for slashing all releases to the environment by 90% and generating 50% less solid waste by 1995.

- In 1975, S.C. Johnson Wax president Samuel Johnson voluntarily removed chlorofluorocarbon propellants from the company's aerosol products 3 years before the first legislated ban and 12 years before the Montreal Protocol. In 1990, he initiated a drive that reduced wastewater from the company's U.S. manufacturing plants by 55% and solid waste by 30%. His most recent directive should reduce air and water pollution as well as solid waste generation at manufacturing plants by another 50%.[16]

- The German steel industry has cut dust emissions from 9.3 kg/ton crude steel in 1960 to below 1 kg/ton in 1992.
- Chemical manufacturing has increased production 200% over the past 25 years while reducing air emissions by 60% and water emissions by 90%.
- The Swedish pulp and paper industry has reduced per unit BOD emissions and sulfur emissions by 90% each, almost eliminated dioxins used in bleaching, and increased recycled fiber to a one-third share of resource inputs.

Waste reduction is best seen as a process of continuous improvement. U.S. companies, such as Monsanto and General Dynamics, have established goals of "zero pollution" for certain substances, mirroring the "zero defects" pledge in the total quality management field. Monsanto's chief executive officer Richard Mahoney has stated that although a goal of zero emissions is technically impossible to achieve, it is "the only goal that will keep us stretching for ever greater improvement."[13]

THE CHALLENGE TO BUSINESSES

Lester Thurow stated in his recent book *Head to Head: The Coming Economic Battle Among Japan, Europe, and America*, that the faster an industry "grows today, the easier it is to grow faster tomorrow. The slower one grows today, the more certain it is that one will grow more slowly tomorrow." Industries that seize the moment and implement pollution prevention programs, integrate environment into TQM and strategic business planning, and view the environment as part of a holistic system leading toward sustainability, will be in a better position for economic growth.

The requirement for clean, equitable economic growth remains the biggest challenge for environmental management. Business and industry must devise strategies to maximize added value while minimizing resource and energy use.

The second greatest challenge to business is that processes and capital investment must perform well on an environmental scorecard to meet increasingly stringent controls on pollution worldwide. Despite billions of dollars being spent worldwide on environmental programs, levels of industrial waste continue to increase, outpacing economic growth. In France, 1% growth in the economy generates 2% more waste. The U.S. EPA estimates that in the United States, the generation of hazardous waste is increasing at an annual rate of 7.5%. For the next few years, pollution prevention will be an integral part of a major change in how industry does business and in how government oversees and regulates industries.

A third challenge to business is the need to respond to the effect of worldwide environmental standards and trade agreements on products, manufacturing facilities, and operations. The current momentum for using environmental initiative to seize a competitive advantage favors the EC countries, Japan, and multinational companies, which are moving ahead vigorously to meet consumer requirements for "green" products.

The fourth challenge to business is the role of management. Management must constantly rethink its approach and be willing to relearn the fundamentals of every aspect of business. The ability to tolerate uncertainty, design new strategies, and use tools for managing processes is required to manage change for environmental quality. The future, though it cannot be predicted, can be shaped. Managers will need to be better prepared to cope with unexpected change and promote flexible organizational responses.

REFERENCES

1. DeWitt, T., Vice President, Honda Automobiles/U.S. Presentation at Proactive Environmental Strategies for Industry, Massachusetts Institute of Technology, November 17, 1993.
2. Wever, G. and G. Vorhauer. "Kodak's Framework and Assessment Tool for Implementing TQEM." *Total Quality Environmental Management*, August 1993.
3. Block, P. Stewardship: Choosing Service Over Self Interest. Berrett-Koehler Publishers, Inc., 1993.
4. Williams, M.E. "Why—And How to—Benchmark For Environmental Excellence," *Total Quality Environmental Management*, Winter 1992/93.
5. Fitzgerald, C. "Selecting Measures for Corporate Environmental Quality: Examples From TQEM Companies," *Total Quality Environmental Management*, Summer 1992.
6. Lent, T., and R.P. Wells. "Corporate Environmental Management Study Shows Shift From Compliance to Strategy," *Total Quality Environmental Management*, Summer 1992.
7. Thomson, R.P., C.H. Le Grand, and J. Dresser. "The Importance of Environmental Auditing in the 1990's," *Industrial Wastewater*, August/September 1993.
8. Kirchenstein, J.J. and R.A. Jump. "The European Ecolabels and Audits Scheme: New Environmental Standards for Competing Abroad," *Total Quality Environmental Management*, Autumn 1993.
9. Sullivan, M., and J.R. Ehrenfeld. Reducing Life-Cycle Environmental Impacts: An Industry Survey of Emerging Tools and Programs.
10. Oakley, B. "Total Quality Product Design—How to Integrate Environmental Criteria Into The Product Realization Process," *Total Quality Environmental Management,* Spring 1993.
11. Tibbs, H. "Industrial Ecology, An Environmental Agenda for Industry," Arthur D. Little Inc., Technology and Product Development Directorate and the ADL Center for Environmental Assurance, 1991. (Contact Tibbs at 510-547-6822 for copies.)
12. Szeleky, J. "Proactive Environmental Strategies for the Basic Metals Industry," Conference on Proactive Environmental Strategies for Industry, Massachusetts Institute of Technology, November 1993.
13. Schmidheiny, S. *Changing Course. A Global Business Perspective on Development.* (Cambridge, MA: The MIT Press, 1992).
14. Robins, N. Managing the Environment: The Greening of European Business. London. Business International, 1990.
15. Hirschorn, J.S., and K.U. Oldenberg. *Prosperity Without Pollution* (New York: Van Nostrand Reinhold, 1991).
16. Siwolop, S. "Pollute less, profit more," *International Wildlife* 23(s):48–51 September/October 1993.

CHAPTER 5

Performing a Pollution Prevention Assessment

*Robert Harries**

MAKING IT HAPPEN

Pollution prevention is not only a prudent step for industries to take, but also is required by environmental regulations in many countries. Examples include the Resource Conservation and Recovery Act (RCRA) in the United States and the system of Integrated Pollution Control in the United Kingdom.[1,2] But how does a company go about assessing the status of its operations and developing a pollution prevention program? The best way to start is to conduct a pollution prevention assessment. Until recently, there has been little guidance on conducting a pollution prevention assessment; however, in the past few years a number of government departments, institutions, and individuals have published documents on recommended approaches.[3-7] The author was involved in developing one of the first of these practical guides, which have many principles in common with other guidance documents for conducting a pollution prevention assessment.[3]

One of the first and most crucial aspects of conducting a pollution prevention assessment is to obtain the commitment and support of senior management at the facility to undertake what amounts to a comprehensive and ongoing program to reduce waste at the source, recycle wastes, and adopt efficient methods for waste treatment and disposal. The assessment can represent a major undertaking, affecting many functional groups within a company. Therefore, the support of senior management and good preparation and organization are prerequisites for ensuring the overall success of the program.

The assessment should be performed by a small team led by a manager with good organizational and communication abilities and the knowledge, authority, credibility, and skill to obtain the required information. The team should include personnel with direct responsibility and knowledge of the processes, waste streams, and areas of the plant under consideration, and should solicit and incorporate input from plant operators. Production and operating personnel know how processes are actually run (not necessarily how they can and should be run) and can often provide valuable and pertinent input. Openness and trust are required in these circumstances

*Dr. Harries is a senior chemist in Engineering Science's Lincoln, England, office and may be reached at 44/522/575-857.

to build team spirit; therefore, it is essential that no recriminations result if poor operating practices are identified.

It can be helpful to include at least one team member with pollution prevention experience from outside the plant personnel, either a consultant or corporate employee. This team member can act as a catalyst by seeing things with fresh eyes and providing unconstrained input.

SETTING GOALS AND MILESTONES

Once company management buys into the pollution prevention concept, it is imperative to plan and organize the program. First define the goals for the assessment. Reducing all wastes, maximizing the use of all raw materials, and increasing treatment and disposal efficiency all at once would be impossible goals. Setting specific and achievable goals is important so that programs can be monitored and their success quantified. Typical goals are listed below.

- Reduce the quantity of RCRA wastes that are shipped off the site to reduce disposal costs and long-term liability.
- Achieve compliance with discharge or emissions regulations.
- Reduce water usage in a process because of a community-wide shortage.
- Find a replacement solvent for chlorofluorocarbons (CFCs) because of the planned phaseout of their production.
- Reduce solid waste generation because of decreasing landfill space.

Example: Setting specific goals and defining the level of detail required to meet the objectives of a project is illustrated by a waste assessment carried out at a printed circuit board manufacturing facility in Ontario, Canada.[7] The immediate problem related to the copper levels being discharged to the municipal sewer system; therefore, the waste assessment focused specifically on copper in wastewater. The assessment achieved the following objectives:

- provided a sound understanding of all the sources of waste copper at the manufacturing plant
- identified and quantified the major sources of waste copper
- permitted evaluation of processing efficiencies using assembled information on unit processes, products, raw materials, water usage, and waste generation
- identified opportunities for waste reduction
- eliminated some wastes along with their associated disposal problems
- identified problem wastes requiring special attention
- enabled the development of an efficient, integrated system for waste segregation and wastewater treatment and recovery

This thorough and systematic approach of conducting a pollution prevention assessment enabled the company to develop an effective waste management system capable of compliance with existing and proposed discharge regulations, which resulted in improved public relations.

POLLUTION PREVENTION ASSESSMENT

This section describes an approach for conducting a pollution prevention assessment. Because the approach is generic, in order to apply to as broad a spectrum of industries as possible, certain parts of the procedure may not apply to a particular plant. Therefore, it is important to modify and develop the approach to meet the individual needs of each facility.

The assessment can be divided into four major phases as follows:

1. data collection and audit preparation
2. audit of waste
3. identification and ranking of pollution prevention alternatives
4. feasibility analysis

Data Collection and Audit Preparation

After the team leader is in place, he or she should issue a policy statement to disseminate information and present the company's commitment to the program, thereby increasing awareness and enthusiasm for the program in the plant. Then, in preparation for the audit, the other members of the audit team should be chosen. The audit team members should be selected on the basis of their experience in the following areas:

- a detailed knowledge of the industrial operations
- a working knowledge of applicable pollution prevention and other environmental regulations
- an understanding of potential pollution prevention alternatives
- an ability to make technical and economic appraisals of pollution prevention alternatives

A pollution prevention team could include, for example, the team leader (engineering manager), site process engineer, production supervisor, and external environmental consultant. Specialists and staff from different areas should be consulted or drafted as necessary. Table 5.1 shows the areas of expertise that should be considered for input or inclusion in the assessment team, depending on the objectives of the assessment.

The second important pre-audit step is for the team to access, assemble, and review site information to gain a good understanding of the processing operations and build a picture of waste generation. Information should be obtained on the processes, raw materials, production schedules, waste streams, operating costs, environmental reports, permits, and company policies. Table 5.2 shows facility infor-

Table 5.1. Expertise Needed for Pollution Prevention Assessment

Expertise	Information/Function
Environmental	Regulations Pollution control Waste disposal costs Hazards and risks
Design and process engineering	Plant and processes Impacts of changes
Production and maintenance	Plant descriptions Process descriptions Feedback on proposed changes Operating details
Legal	Environmental liability
Accounting, finance, and purchasing	Costs Inventory controls
Health and safety	Hazards and risks
Research and development	Suggestions for modifications Generation of options Technical options Modifications
Operators, supervisors, and transport department	Operational suggestions Assessment of operational, procedural, or equipment changes
External consultants	Insight/"catalyst"

mation that can be collected in preparation for the waste audit and assessment. It is unlikely that the plant data will be complete, and information gaps will need to be identified and filled during the audit stage.

Audit of Waste

The audit phase of the pollution prevention assessment involves going into the plant to confirm information, collect additional data, and develop a detailed and practical understanding of how and where the waste streams are generated. This is accomplished in four major steps:

1. conduct site investigation
2. construct process flow diagrams
3. define process inputs and outputs
4. construct a material balance

For the site investigation, unit operations in the plant are examined one at a time, and process flow diagrams developed as necessary. Although it will be obvious that a reactor or a plating bath is a unit operation, it is equally important to examine and

Table 5.2. Information Needed for Pollution Prevention Assessment

1. **Design Information**
   Process flow diagrams
   Process descriptions
   Material and energy balances
   Operating manuals
   Equipment specifications and data sheets
   Piping and instrumentation diagrams
   Plot and elevation plans
   Equipment layouts and work flow diagrams

2. **Environmental Information**
   Waste manifests and disposal records
   Waste characterization
   Wastewater discharge records
   Air emission records
   Emission limits and discharge consents
   Environmental reports

3. **Raw Material/Production Information**
   Material safety data sheets
   Product and raw-material inventory records
   Operator data logs
   Production schedules

4. **Other Information**
   Treatment and disposal costs
   Water and sewer charges
   Company environmental policy statements

describe ancillary operations such as storage, materials handling, pumping, and treatment, because these processes can all potentially contribute waste.

After all the production and ancillary operations are designated and interconnections established, the next step is to account for inputs and outputs of materials. Inputs can include raw materials, chemicals, reused materials, process water, steam, and power; outputs can include primary products, co-products, waste to be reused, waste requiring disposal, and emissions. Inputs and outputs must be quantified. If sufficient information is not collected in the pre-audit, then in-plant measurements and characterization will be required.

Inputs are generally well recorded (e.g., raw materials, water use, and power), as are products. However, in-plant characterization of wastewater, solid waste, and air emissions may be required to determine the fate of chemicals, metals, and other materials and to facilitate the construction of a material balance.

A wastewater sampling program should be planned carefully and cover the full range of plant operating conditions (e.g., startup, changeover, and shutdown). Wastewater flows can be determined easily if water meters are installed; otherwise, pump curves and times and manual measurements can be used. Overall, waste quantities are probably recorded in waste manifests, but additional measurements may be required in the plant to establish the amounts and characteristics of wastes generated in specific areas. Emissions inventories and emissions factors also are used to establish gaseous outputs.

Once the material inputs and outputs around a process have been established, it is useful to construct a material balance to confirm the data and highlight information gaps and inaccuracies. Developing a material balance for some complex process units might be difficult, but trying to account for all inputs and outputs generally encourages a sound understanding of the operations and ensures the use of accurate material flow data.

Example: A printed circuit board facility was having problems with elevated levels of copper in its wastewater discharge.[7] Figure 5.1 shows the general schematic process flow diagram that was developed for the circuit board plant. By studying the processing operations and reviewing background information, it was possible to identify the areas in the plant where waste copper was generated. The next step was to quantify the inputs to and outputs from the copper processing areas. This information was obtained from chemical data, production data, flow measurements, and a wastewater sampling program. Once all the process input and output information was collected for the four major copper-generating areas, the information was recorded on process flow diagrams such as the one shown in Figure 5.2, which was developed for the sensitizing area.

Material balances were subsequently developed for the major copper processing areas. Figure 5.3 shows the material balance for the sensitizing line. The results clearly illustrated the sources of copper, and the overall copper balance validated the data. An additional material balance was constructed on the basis of the discharges from the individual copper processing areas and the collecting drain that fed the wastewater treatment system. The material balance for copper discharges to the process drain is shown in Figure 5.4. The material balance again provided good confidence in the data and clearly showed that the microetch line, which was not considered by the company to be a significant source of copper, was the major source of copper loading to the process drains.

The pollution prevention audit data enabled pollution prevention alternatives to be identified and ranked.

Identification and Ranking of Pollution Prevention Alternatives

From the information accumulated during the pollution prevention audit and the observations made by the assessment team while examining the plant in detail, a number of pollution prevention alternatives can be identified. These can range from obvious and easily implemented measures to longer-range and more-sophisticated alternatives. Furthermore, problem waste streams also can be identified. These can include wastes that are particularly expensive to dispose of and wastes that can affect treatment plant efficiency (e.g., chelated metals or biocides).

The assessment team is responsible for evaluating and screening the pollution prevention alternatives identified from the audit and ranking them in relation to the assessment objectives and other selected criteria such as cost and ease of implementation. The pollution prevention options should then be allocated into priority groupings to guide the effort for the subsequent feasibility analysis.

Source reduction is a good operating practice because it avoids or minimizes the generation of waste. Therefore, options for source reduction generally have a high priority. Recycling allows waste materials to be put to beneficial use but frequently involves significant effort or cost and therefore generally warrants a medium-level

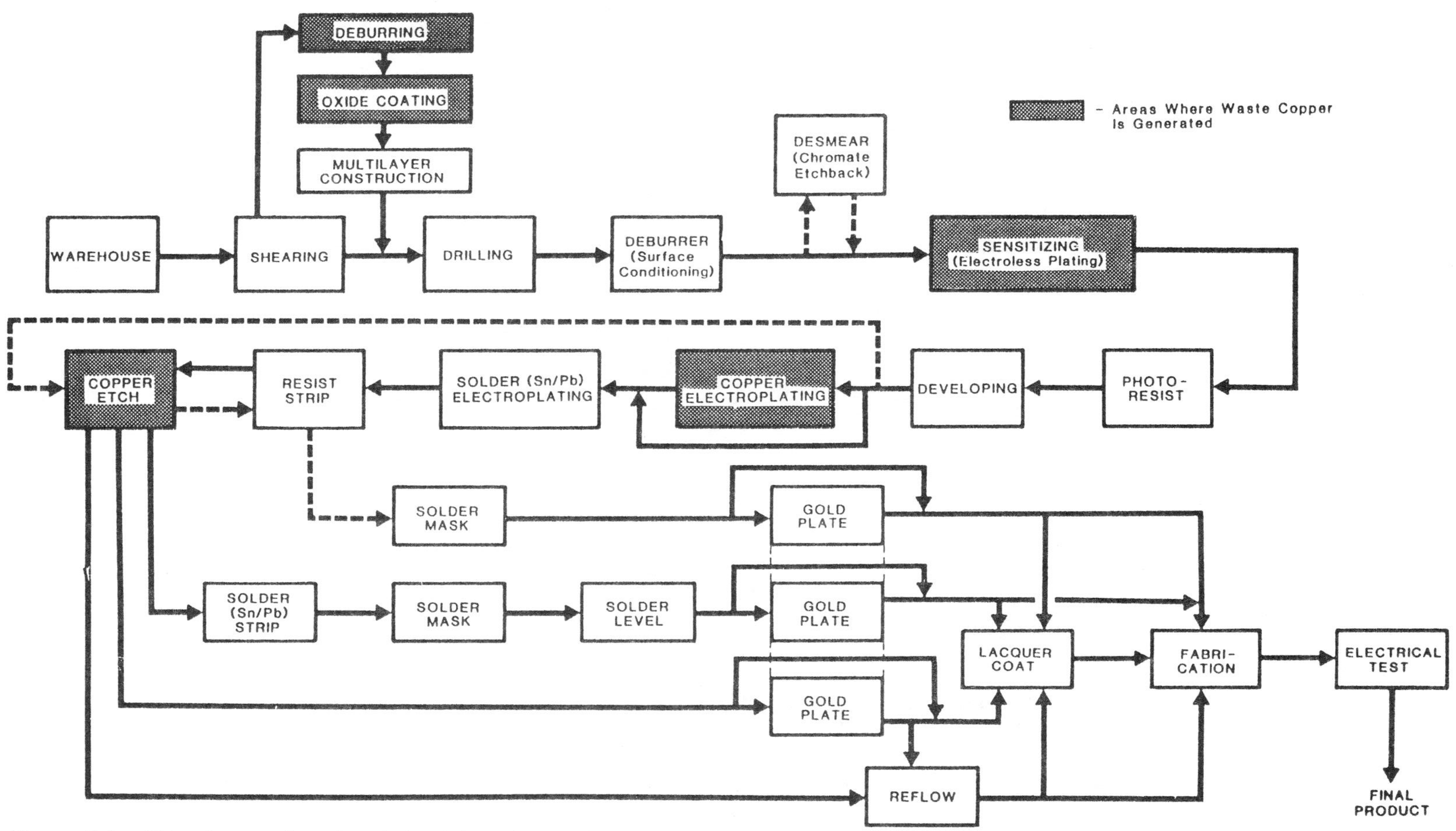

Figure 5.1. Flow diagram for printed circuit board operations.

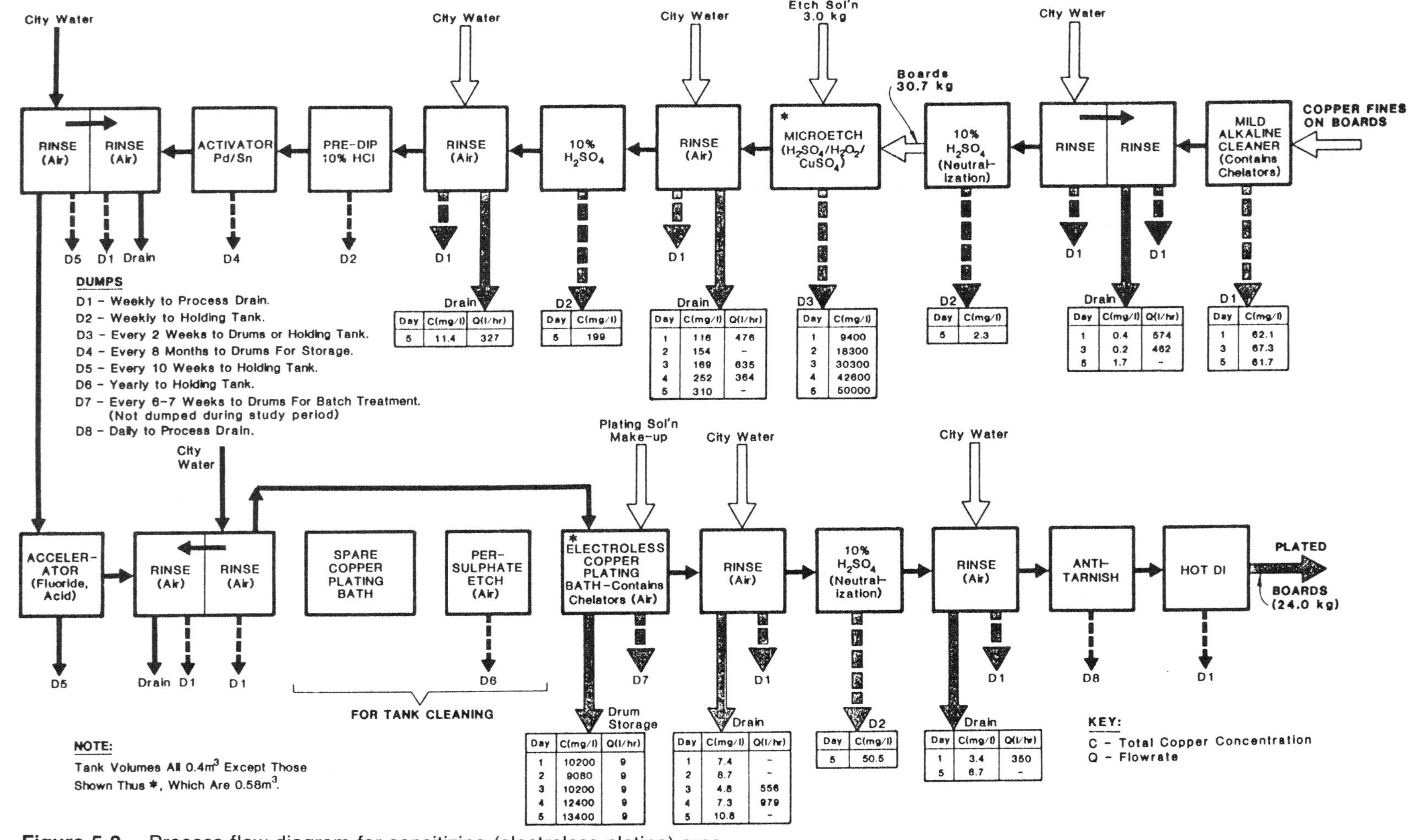

Figure 5.2. Process flow diagram for sensitizing (electroless plating) area.

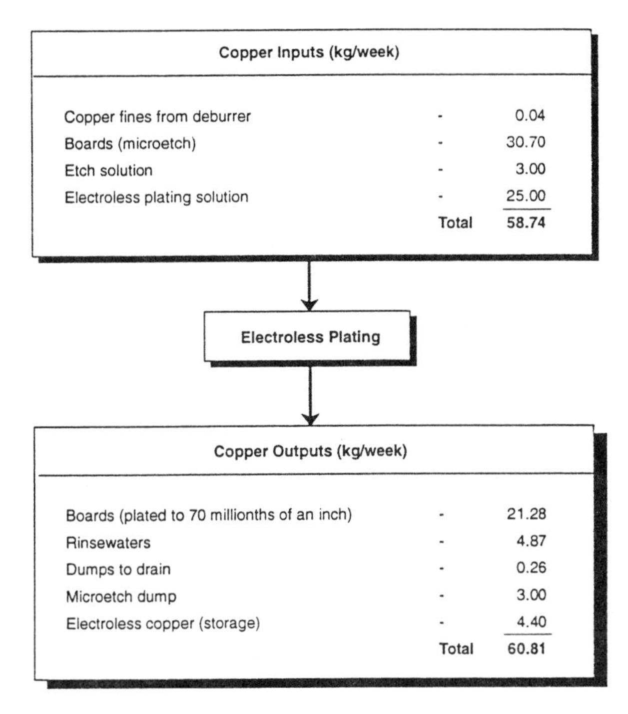

Figure 5.3. Material balance for electroless plating area.

priority. Treatment options are generally the lowest on the priority scale and therefore would be considered only if source reduction and recycling would not achieve the required objectives.

Feasibility Analysis

The number of pollution prevention alternatives taken to the feasibility analysis stage will depend on the time and resources available for the program. Alternatives should be evaluated in terms of technical and economic feasibility. Procedural or housekeeping changes can often be implemented directly, at minimal cost, after appropriate review and training. Options that involve process or equipment changes are likely to be more expensive and can affect production rate and quality. These options therefore would require a more detailed analysis and may need field testing to demonstrate their effectiveness and impacts on production.

Technical criteria that need to be considered at this stage are:

- worker health and safety
- product quality
- space requirements
- compatibility with existing production processes
- labor requirements
- environmental effects

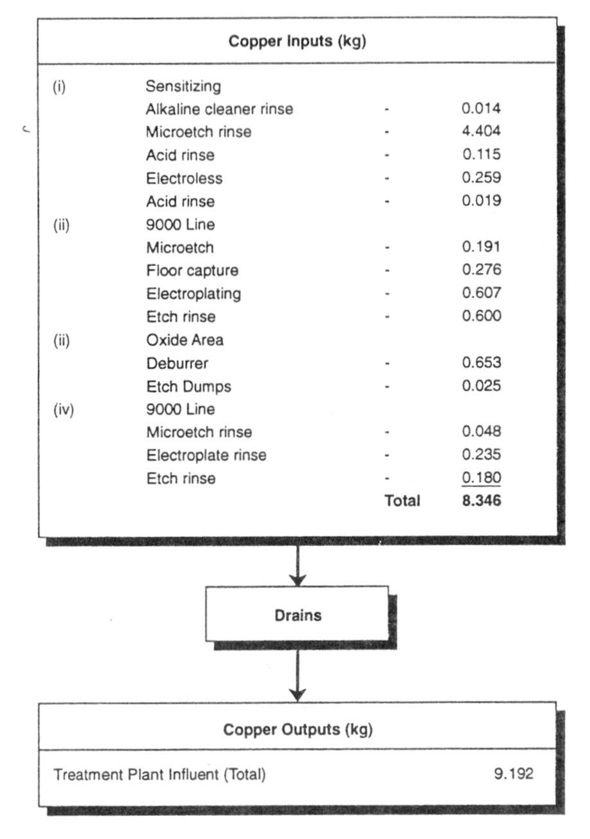

Figure 5.4. Material balance for process drains.

The economic evaluation of pollution prevention alternatives should be carried out using the company's preferred methods, but generally will involve a cost/benefit analysis. Payback periods are commonly used to assess the economic feasibility of pollution prevention projects.

Example: The prioritization of alternatives and feasibility analysis steps can again be illustrated by the example of the printed circuit board facility. Table 5.3 shows some of the source-reduction measures that were identified at the facility. The pollution prevention audit identified the microetch rinsing operation as the major single source of copper waste, and an improved drip time and rinsing operation resulted in significant reductions in copper loadings on the wastewater treatment system.

In addition to the pollution prevention measures, targeting problem waste streams and developing a waste treatment strategy to ensure long-term compliance for the plant were priorities. The wastewater management strategy involved both recovery and treatment. The sampling performed during the waste audit indicated that the copper discharged to the sanitary sewer was primarily dissolved (75% to 95% of total copper concentration) and regularly exceeded the sewer discharge limit of 2 mg/L. Previous experience with the treatability of printed circuit board wastewaters has established that electroless copper wastewaters are particularly difficult to treat

Table 5.3. Examples of Obvious Measures for Pollution Prevention

Processing Area	Existing Practice	Waste Reduction Measures
Electroless plating	No drip time allowed after microetch tank leading to high concentrations of copper in the microetch rinsewater.	Allow one minute drip time over microetch tank.
Deburring operations	Backwashing of electroless plating deburrer sand filter with deburrer return water leading to entrainment of copper fines in the filter bed. This led to copper fines being introduced into the electroless plating line.	Backwash filter with clean water.
Electroplating line	Mechanical seals on copper electroplating recirculation pumps leak directly onto the floor. Waste copper is subsequently discharged to the floor drain.	Implement regular maintenance program to prevent leaks and install drip trays.

because of the presence of chelating agents in electroless copper-plating solutions. Treatability tests confirmed that the chelating agents in the electroless rinse and other rinses (e.g., resist strip) affected copper hydroxide precipitation and therefore these rinses needed to be segregated and treated separately. Knowledge gained from the pollution prevention assessment allowed the following features to be designed into the wastewater treatment system.

- segregation of all the chelate-containing wastewaters from the existing metal hydroxide precipitation system
- segregation and separate treatment or recovery of all the chelate-containing rinsewaters and concentrated bath dumps using copper-selective chelating ion exchange and electrolytic recovery
- replacement of the existing resist products with aqueous resist products that could be stripped using a straight caustic solution instead of the chelate-containing proprietary strippers
- collection of all general bath dumps (nonchelate-containing) in a holding tank for metering back to the conventional treatment system at a controlled rate (to prevent surges in copper loadings)
- use of existing pH adjustment, polymer addition, clarification, and sand filtration systems for efficient metal hydroxide precipitation and subsequent discharge of high-quality effluent

The cost of upgrading the treatment and recovery system was $350,000. It was estimated that approximately $25,000 per year would be saved on waste disposal costs, and $5,000 worth of copper would be recovered per year from the electrolytic recovery unit. More importantly in this example, however, the plant avoided legal action for discharge violations, and considerably less time had to be spent by management on environmental concerns.

ASSESSMENT REPORT

At the end of a pollution prevention assessment, documenting the information and conclusions of the study in a formal assessment report is extremely important. This report should include:

- a summary of the pertinent background information
- results of the pollution prevention audit
- results of the pollution prevention assessment
- technical and economic analyses of the pollution prevention alternatives
- recommendations and priorities for implementing pollution prevention measures

MONITORING, FEEDBACK, AND REEVALUATION

The pollution prevention effort does not end with issuing the assessment report. Monitoring the success of the measure taken, giving feedback to the assessment team and employees, and generally maintaining awareness and enthusiasm for the program are also necessary.

Finally, the pollution prevention program should be dynamic. It needs to be reviewed and reevaluated in response to changes in production, new regulations, and advances in technology. The successful team leader will continue to champion pollution prevention in meeting these changing needs.

REFERENCES

1. Resource Conservation and Recovery Act, Public Law 94-580 (1976), as amended by the Hazardous and Solid Waste Amendments, Public Law 98-616 (November 1984).
2. United Kingdom Environmental Protection Act, London HMSO (1990).
3. *Industrial Waste Audit and Reduction Manual*, Ontario Waste Management Corporation, Ontario, Canada (1987).
4. *Waste Minimization Opportunity Assessment Handbook*, U.S. Environmental Protection Agency, EPA 625/7-88-003 (July 1988).
5. *New York State Waste Reduction Guidance Manual*, N.Y. State Department of Environmental Conservation (March 1989).
6. *Waste Minimization Guide*, United Kingdom Institution of Chemical Engineers (1992).
7. Harries, R. C., K. C. Bradley, and D. Gardiner. "Using a Waste Audit Approach to Determine Waste Management Alternatives at a Printed Circuit Board Manufacturing Plant," in *Proceedings of 43rd Purdue Industrial Waste Conference* (May 1988).

CHAPTER 6

Recycling Programs for Solid Waste

*Ellen Bogardus**

INTRODUCTION

Many state and local governments have set aggressive recycling goals to divert waste from solid waste landfills. To achieve these goals, cities and counties have had to make substantial changes in the way they manage their solid waste. What does this mean for commercial and industrial enterprises?

Some cities and counties are asking or, in some cases, requiring businesses to submit source reduction and recycling plans as part of land-use permit applications for new developments, business license applications, or business license renewals. Other cities and counties are providing financial incentives through variable fees for waste collection—the more waste disposed of, the higher the unit cost; the more material diverted, the lower the unit cost. The high cost of solid waste disposal and the perception by business that customers prefer firms with environmental consciences have helped to reinforce the importance of developing source reduction and recycling plans. The State of Rhode Island, for example, requires businesses and institutions to recover materials that are on the state's mandatory recycling list and to file a recycling plan with the state. The City of Los Angeles, on the other hand, has formed a partnership with commercial and industrial waste generators and has set voluntary recycling goals.

A major barrier for businesses and industry alike in developing source reduction and recycling plans is their inability to quantify their solid waste management practices. As disposal costs continue to rise, as interest in recycling as a cost-avoidance strategy increases, and as model programs are publicized, waste generation data will become available, and this barrier will be eliminated. This chapter provides guidance on developing a source reduction and recycling plan, suggests actions that waste generators can consider for further pollution prevention and recycling, and presents case studies of several successful source reduction and recycling programs.

---

*Ms. Bogardus is the director of HDR's Solid Waste Division for Southern California, located in Irvine, California, and can be reached at (714) 756-6800.

DEVELOPING A SOLID WASTE SOURCE REDUCTION AND RECYCLING PLAN

The Pollution Prevention Assessment

An assessment is the crucial first step in planning and implementing a successful source reduction and recycling program. By studying the existing solid waste stream, identifying ways to reduce waste, and identifying target materials for recycling, a workable program can be developed that reduces waste disposal costs *and* contributes to environmental protection.

The purpose of a pollution prevention assessment is to study the waste stream to find answers to six simple questions about materials discarded from a facility:

- How much waste is generated?
- What types of material are generated?
- Where, how, and why is waste generated?
- How are discarded materials collected and disposed of?
- How much does collection and disposal cost?
- How much waste in the facility can be reduced or recycled?

The answers to these questions will provide a better understanding of existing waste generation practices and serve as a starting point for a source reduction and recycling program. With information about the amount and types of materials generated, identification of specific recyclable materials in the waste stream can begin. Knowing how and where material is generated can lead to a plan for recovering materials for recycling. Recyclables can be found in many facility locations, from the cafeteria, to a workshop, to a loading dock. Each location may require a different program for recovering the recyclable material.

Estimating Material Quantities

The first step is a "walk-through" of the facility to identify what types of materials are generated and where they are generated. The next step is to estimate the quantities of material generated. The facility's disposal records or those of the waste hauler can sometimes provide useful information. However, if the disposal contract is by number and/or size of containers, not by weight, other methods must be used to estimate the amount of discarded materials.

Estimating Quantities Without Knowing Materials Content. One method is illustrated in Table 6.1. The method uses a simple calculation based on the number and size of containers, frequency of collection, and an estimate of how full each container is at the time of collection. If several different container types are used (for example, open roll-off, compactor, 3 cubic yard), the volume must be calculated for each container type and then a total volume must be calculated.

Another method of estimating waste quantities is by multiplying the total number of employees by daily material generation rates per employee. Typical rates are given in Table 6.2.

Other rates that can be used in estimating waste quantities are listed in Table 6.3.

Table 6.1. Amount of Discarded Material

No. of Containers (A)	Size of Containers (B)	Frequency of Collection (C)	Percent Filled (D)	Total (E) A x B x C x D = E
1	40 cubic yds	4 pick-ups/ month	80 percent	1 x 40 x 4 x 0.8 = 128 cu. yds/month

Table 6.2. Typical Solid Waste Generation Rates for Commercial Facilities

Facility Type	Daily Rate
Restaurant	12.5 lb/employee
Office	1.32 lb/employee
Retail Store	2.68 lb/employee
School	.49 lb/person (not including cafeteria)
Light Industry	2.1 lb/employee

Source: Figures adapted from Washington Department of Ecology, 1988.

Table 6.3. Typical Solid Waste Generation

Facility Type	Daily Rate
Apartments	2.5 lb per person or 4 lb per bedroom or 8 lb per unit
Cafeteria	1 lb per meal served
Department Store	75 lb corrugated paper per $1,000 of sales; 15 lb other waste per $1,000 of sales
Discount Store	60 lb corrugated paper per $1,000 of sales; 10 lb other waste per $1,000 of sales
Fast-Food Restaurant	200 lb per $1,000 of sales
Hospital	16 lb per occupied bed
Hotel (first class)	3.2 lb per room and 2 lb per meal served
Hotel (mid-range)	1.7 lb per room and 1.2 lb per meal served
Manufacturing	
100–399 employees	3 lb per person
400–3,000 employees	7 lb per person
Motel	2 lb per person
Nursing/Retirement Home	5 lb per person
Office	1 lb per 100 square feet
Restaurant	1.5 lb per meal served
School	0.5 lb per person without cafeteria; 1 lb per person with cafeteria
Shopping Mall	2.5 lb per 100 square feet
Supermarket	100 lb corrugated paper per $1,000 of sales; 65 lb other waste per $1,000 of sales
Warehouse	1 lb per 100 square feet

Source: National Solid Waste Management Association, 1991.

It should be noted that the rates given in Tables 6.2 and 6.3 are only approximate; waste quantities can vary significantly from one facility to another. For example, fast-food restaurants often generate considerably more waste per employee than traditional restaurants.

Estimating Quantities of Specific Materials. Several methods are available for estimating quantities of specific materials. If a collection container is filled with just a few different materials, you can perform the calculation illustrated in Table 6.1, estimating the percent filled with each type of material instead of a total for all materials. Volume estimates can be converted to gross weights using the conversion factors presented in Tables 6.4 and 6.5.

Table 6.4 provides conversion factors to calculate gross weights from gross volumes. Your chosen methods for processing and handling materials, as well as how tightly you pack loads of recyclables, can affect the weight. Also, moisture or humidity can affect weights for paper, leaves, and yard waste. The amount of material in each container can be affected by the container material or design.

The density figures provided in Table 6.5 are useful for estimating the weight and volume of materials generated at a facility. Variations from these figures are common, as the density depends on a number of factors (e.g., the density is higher for dented than for undented plastic soda bottles). Since the density of materials can be extremely variable, the figures in the table should be viewed as a range within which the actual density of the materials would probably fall.

Evaluating Source Reduction Programs

With information in hand about the types of materials generated in a facility, and where, about how, and why they are generated, you can begin identifying different

Table 6.4. Volume-to-Weight Conversion

Materials	Unit Quantity	Weight (lbs)
Newsprint, loose	1 cubic yard	360–800
Newsprint, compacted	1 cubic yard	720–1,000
Newsprint	12-inch stack	35
Corrugated cardboard, loose	1 cubic yard	300
Corrugated cardboard, baled	1 cubic yard	1,000–1,200
Glass, whole bottles	1 cubic yard	600–1,000
Glass, semicrushed	1 cubic yard	1,000–1,800
Glass, crushed (mechanically)	1 cubic yard	800–2,700
Glass, whole bottles	one full grocery bag	16
Glass, uncrushed to manually broken	55-gallon drum	125–500
Aluminum cans, whole	1 cubic yard	50–74
Aluminum cans, flattened	1 cubic yard	250
Aluminum cans	one full grocery bag	1.5
Aluminum cans	one 55-gallon plastic bag	13–20
Ferrous cans, whole	1 cubic yard	150
Ferrous cans, flattened	1 cubic yard	850

Source: The National Recycling Coalition Measurement Standards and Reporting Guidelines, October 31, 1989.

Table 6.5. Density of Recyclable Materials

Material	Uncompacted (lb/yd³)	Compacted (lb/yd³)
Aluminum cans	50–70	430
Metals cans, ferrous	145	405–485
PET soda bottles	30–40	515
HDPE milk jugs	25	270
Odd plastic	50	700
Glass	515–600	1,000–2,000
Newsprint	360–505	720–1,010
Corrugated containers	40–45	405–575
White ledger paper		
Flat	380	755–925
Crumpled	110–205	325
Laser printout	430	865
Computer printout	655	1,310
Mixed ledger/office paper		
Flat	380	755
Crumpled	110–205	610
Waste paper	70–90	215–270
Residential waste	220	
Mixed food and beverage containers	175	

Source: Modified from Rhode Island Ocean State Cleanup and Recycling (OSCAR), 1988.

ways to reduce material generation at the source. The fundamental goal of source reduction is to enact changes in consumption, use, and waste generation patterns associated with products. This means that less waste is created in the first place by changing the behavior of suppliers who provide a company with goods and services, educating employees regarding the use of materials, and offering customers opportunities for source reduction. Common approaches to source reduction include:

- Establish purchasing practices that favor products that are reusable, durable, and have a reduced volume and/or weight. Examples include:
  - Require suppliers to ship orders in reusable or returnable packaging such as wooden pallets and polystyrene peanuts; these reusable items may then be used for the facility's packaging and shipping operation.
  - Install hot-air dryers or cloth towel rolls instead of using paper towels.
- Educate employees to reuse and/or use less materials. Examples include:
  - Use central bulletin boards for memos, reports, and announcements instead of making one copy for each employee.
  - Use double-sided copying.
- Donate or sell usable materials. Examples include:
  - Donate leftover and surplus food to a food bank.
  - Identify and donate old inventory and/or capital equipment to a charity, school, or theater group or offer it for sale to employees.

- Use material exchanges. Please refer to the chapter on waste and material exchanges for information.
- Offer opportunities for source reduction to customers. Examples include:
  - Train staff to minimize the packaging of products sold to customers.
  - Offer products in bulk at a lower cost.
- Reduce the amount of yard waste generated (if applicable). Examples include:
  - Recycle grass clippings by leaving them on the lawn and allowing them to decompose naturally.
  - Convert clippings, brush, and pruned branches to yard mulch.

Identifying Recycling Programs: Source Separation versus Mixed-Waste Processing

With information in hand about the types of materials generated in your facility, where and how they are generated, and local market conditions for recyclables, you can begin identifying different methods for recovering recyclables.

There are two basic approaches to recovering or separating recyclables from a waste stream. The first is a front-end approach in which recyclables are source-separated; this means recyclables are separated at or close to the source or point of generation (for example, an employee's desk, receiving area, lunchroom, or stock room). For this approach to work, it is important to know **where** recyclables are generated within your facility.

The second approach, back-end or mixed-waste recycling, relies on separating recyclables from a mixed-waste stream and is typically done at a centralized processing facility. Such a facility would route a mixed-waste stream through a series of manual and/or mechanical separation processes to remove recyclables. Remaining organic materials (yard waste, food waste, etc.) would be composted, and other remaining wastes would be disposed of in a landfill.

Both approaches have strong supporters. The following advantages and disadvantages should be considered before deciding which approach is best for you.

- Source separation provides cleaner, less contaminated recyclables and thus a more marketable material.
- Mixed-waste processing tends to recover greater amounts of recyclables, especially if composting is part of the process.
- A modified back-end approach in which various types of waste are kept as separate as possible (for example, containing damp wastes in a plastic bag) increases the market value of materials, but decreases overall recovery of recyclables.
- Source separation offers the potential for receiving revenue from selling recyclables and educating employees about the benefits of hands-on participation in recycling.
- Mixed-waste processing is easier to implement because few or no changes are needed in current material-handling procedures.

Blending the two methods is often the best approach. Source separation can be used for higher-value or more easily separated recyclables at the front end, and lower-value or bulkier recyclables can be recovered at the back end. The optimal approach will be determined largely by the types and quantities of recyclables, local market conditions, and program goals.

Source-Separation Approach. The most important aspect of the source-separation approach is determining the degree to which targeted recyclables will be separated from one another. Choices range from collecting recyclables separately to collecting them fully commingled (mixed together). Commingled recyclables will be separated at a materials recovery facility.

Several other important considerations in source-separation programs are presented below. They include options for collection, immediate storage, final storage, and handling of solid waste recyclables.

Options for Collection and Intermediate Storage. After deciding on the level of source separation, you can then focus on options for collection and intermediate storage. (Intermediate storage consists of containers that may or may not be located near recyclable generation points. The containers are used for consolidating recyclables before they are picked up by a refuse hauler or recycler or delivered to a recycling facility by your staff.) When considering options for collecting recyclables, ask the following questions:

- Where are recyclables generated?
- Can you collect recyclables, separate from your trash, at or close to the point of generation (for example, at an employee desk, receiving area, lunchroom, or stock room)?
- If recyclables cannot be collected near the point of generation, can they be collected at a location central to several generation points?
- If lack of space, fire safety, or other constraints preclude the solutions listed above, can recyclables be collected at a limited number of points even though they may be isolated from generation points?

Answers to these questions should be apparent after the walk-through of your facility. In addition, custodial staff and other employees are an invaluable source of information. Given the constraints and opportunities unique to your facility, it is important to remember that the key consideration in selecting collection locations is to get them as close to recyclable generation points as possible. Convenient locations optimize collection efficiency, thereby maximizing overall recovery of materials.

Now that you have decided where to collect recyclables, you can decide how to collect and store the recyclables. Obviously, your choice of the level of separation and where to collect recyclables will determine the types and sizes of collection containers to use. A wide variety of containers is available for collection and intermediate storage. Storage alternatives range from using or modifying existing containers (for example, cardboard boxes, paper or plastic bags, canvas bags, plastic or metal bins or boxes, file folder boxes, or trash receptacles) to purchasing specialized containers from a recycling vendor. The selection should be consistent with facility operations and policies.

Options for Final Storage. Final storage containers hold the recyclables until they are either delivered to a recycling facility or picked up by a refuse hauler or recycler. Intermediate storage containers may also serve as final storage containers. Alternatively, a refuse hauler or recycler may provide drums, barrels, or 2- or 3-cubic yard containers for storage.

In surveying local market conditions, you will learn whether or not haulers or recyclers provide storage containers. If a hauler or recycler will not provide a storage container because you generate insufficient quantities of recyclables, you should look into sharing a storage container with neighboring businesses. In some cases, haulers and recyclers may also provide in-house collection containers. As with the location, types, and sizes of collection containers, key considerations for final storage options include the types and quantities of targeted recyclables, and your facility operations and policies.

Options for Handling Recyclables. Decisions about the degree of separation and the location and type of container(s) for collection and storage will influence how the recyclables will be handled. Answers to the following questions will help determine the best option for handling recyclables:

- Who will be responsible for source separation?
- Who will be responsible for handling recyclables (i.e., emptying recyclables from collection containers and/or intermediate storage containers and transferring them to final storage containers) within your facility?
- If applicable, who will be responsible for transporting recyclables to a recycling facility?

In response to the first question, in almost every case the best option is for each employee to be responsible for separating recyclables that he or she generates. Typically, it is not cost-effective or practical for custodial or housekeeping staff to separate recyclables at each generation point or at a central disposal location such as a loading dock.

In determining whether employees or custodial staff should handle and/or transfer recyclables, you need to consider employee convenience, potential costs of employee or custodial labor, consistency with facility operations and policies, and other factors. Although not common, some haulers and/or recyclers will provide in-house collection and intermediate storage containers and also service these containers. Another option is to pursue a joint employee/custodial approach in which employees empty their recyclables into central collection or intermediate storage containers and then custodial staff empty these containers and transfer recyclables to final storage containers.

In selecting someone from your facility for transporting recyclables to a recycling facility, it is important, once again, to consider employee convenience, potential costs of employee or custodial labor, and consistency with facility operations and policies.

Mixed-Waste Processing. If you choose to implement a mixed-waste processing recycling program, no further modifications are required inside your facility. For a viable program, you will need to identify a centralized processing facility and a hauler who will transport recyclables to the processing facility.

Identifying Markets for Recyclables

Once the types and quantities of recyclables in your waste stream have been identified, the next step is identifying markets for the recyclables. Contact local recycling companies and ask the following questions:

- What materials do they accept?
- Will they rent or provide any equipment or supplies for collecting and handling recyclables? What are their procedures for collecting recyclables?
- What minimum quantities of recyclables are required for pick-up?
- Are there any processing (for example, baling, shredding, or compacting) requirements for recyclables to be accepted? What are the "do's" and "don'ts" of handling (for example, delivering mixed bottles and cans or mixed-color glass containers)?
- What are the allowable levels of contamination for the recyclables? For example, can a few sheets of colored paper be mixed in with white paper? Can staples or paper clips be included with white paper?
- Can long-term contracts be negotiated?
- What are the current prices for specific recyclables? Are the markets for such recyclables expected to be stable or become stable during the next year or two?

Answers to the above questions will help you choose the recycling company that best fits your needs. Typically, you will choose the company that maximizes total recyclable value for the materials you collect.

Evaluating Source Reduction and Recycling Programs

Once you have identified recyclables within your facility and viable markets for these materials, you can then evaluate different ways to recover materials for recycling. The criteria presented below can be used to evaluate potential recycling programs. The criteria are grouped under four categories: cost-effectiveness, technical feasibility, administrative feasibility, and logistical feasibility. These criteria also can be used to evaluate methods of reducing waste generation within your facility.

Cost-Effectiveness. Three criteria related to cost-effectiveness are as follows:

- potential for diverting waste
- total costs
- potential for reducing disposal cost

Diverting Waste. The following factors influence the potential for diverting waste:

- the amount of material in your waste stream targeted for source reduction and/or recycling

- the expected level of participation by employees in reducing or recycling targeted materials
- the expected rate for material recovery (that is, how much you can potentially reduce or recycle)
- the expected level of contamination of material or percentage of rejected material

In evaluating the criteria, programs having high diversion rates should receive a high rating; programs having low diversion rates should receive a low rating.

Total Costs. The evaluation of total costs includes the following:

- capital costs for recycling bins and other equipment (for example, a baler)
- operation, maintenance, and administrative costs
- educational and promotional costs
- expected revenues from recyclables

In evaluating the criteria, programs that are relatively inexpensive to implement and operate should receive a high rating; programs that are expensive to implement and operate should receive a low rating.

Reducing Disposal Costs. The potential for reducing disposal costs depends on the extent to which a recycling program diverts materials that would otherwise incur significant disposal costs. Thus, programs diverting materials that contribute a large part of the total volume or weight of your facility's total waste stream should receive a high rating; programs diverting materials contributing a small volume or weight to the waste stream should receive a low rating.

Technical Feasibility. In evaluating technical feasibility, three criteria are used:

- requirements for new or expanded equipment
- technical reliability
- marketability of collected materials

Equipment Requirements. The need for new or expanded equipment such as ecycling bins, storage containers, balers, or shredders will affect the implementability of any program. If new and expensive equipment is needed, the program will be nore difficult to implement than if no new equipment or only relatively inexpensive quipment is required. Programs requiring minimal or no new equipment should eceive a high rating; programs requiring extensive new equipment should receive a ow rating.

Technical Reliability. Programs using technology (that is, diversion methods nd/or equipment) that has been successfully implemented in a number of comunities or businesses for many years, with little incidence of failure, have low risks nd should receive a high rating. Programs using technology that is still experimenal or has been unsuccessful have high risks and should receive a low rating.

Marketability. Markets are the key to the success and dependability of a recycling program. The marketability of the materials you target for collection directly affects the feasibility of different programs. Materials that can be marketed easily and dependably should receive a high rating; materials that are difficult to market should receive a low rating.

Administrative Feasibility. Administrative feasibility refers to the "people" side of program evaluation. In evaluating administrative feasibility, two criteria are used:

- compatibility with your company's plans, policies, and procedures
- availability of personnel

Compatibility with Company Policies. Programs that are compatible with your company plans, policies, and procedures are more easily implemented and should receive a high rating. Programs that require procedural changes that are not compatible with your company policies or plans will be difficult to implement and should receive a low rating.

Availability of Personnel. The feasibility of a program will depend on the availability of personnel and level of personnel commitment needed for implementation. Programs for which personnel are readily available, or for which personnel needs are minimal, are more easily implemented and should receive a high rating. Programs requiring personnel that are not readily available or requiring extensive commitments of personnel are more difficult to implement and should receive a low rating.

Logistical Feasibility. Logistical feasibility refers to an assessment of the ease of implementing a program. In evaluating logistical feasibility, three criteria are used:

- compatibility of the program with facility operation
- implementability (short- and medium-term)
- acceptability by facility employees

Compatibility with Operation. Programs that are compatible with existing facility operations are more easily implemented and should receive a high rating. Programs requiring changes in operations or requiring persons with specialized skills should receive a low rating.

Implementability. California's recycling law (Assembly Bill 939) requires that cities and counties meet a 50% diversion goal by the year 2000 and make adequate annual progress toward this goal. Companies can help cities and counties with source reduction and recycling goals by implementing programs that can be started up as quickly as possible. Programs that can be implemented quickly should receive a high rating; programs that would be slow to implement should receive a low rating.

Acceptability by Employees. Programs that are readily accepted by employees, because of convenience or other factors, are more easily implemented and should receive a high rating. Programs that are likely to be opposed by employees will be more difficult to implement and should receive a low rating.

IMPLEMENTING A SOURCE REDUCTION AND RECYCLING PROGRAM

Once the source reduction and recycling program has been identified, evaluated, and selected, implementation can begin. Typical steps in implementing a program include:

- Choose a coordinator (or coordinators) to manage the programs.
- Select a vendor for collecting recyclables, or identify a recyclable processing facility to which you will deliver recyclables.
- If necessary, buy equipment needed for the programs.
- Set up a recycling collection and storage system.
- Develop an employee education program to promote the benefits of the source reduction and recycling programs, describe how the programs work, and emphasize the importance of employee participation.
- Coordinate and hold ongoing educational events and distribute informational materials to reinforce source reduction and recycling habits and educate new employees.
- Establish a monitoring system for source reduction and recycling rates, material revenues, and program costs.
- Perform ongoing monitoring and evaluation, and change the programs as needed.

CASE STUDIES

Case Study 6.1: *Reader's Digest* Recycles

The world corporate headquarters of *Reader's Digest* in Pleasantville, New York, began recycling in 1979. Two waste materials were targeted, high-grade paper, which produced revenue, and low-grade paper, which cost $50,000 per year to recycle. Recycling was performed by a nearby paper packer that provided and serviced collection bins.

In 1987, waste disposal costs for paper began to increase. The cost of waste disposal was $23,000 per month, or $276,000 per year, plus $50,000 for recycling, for a total annual cost of $326,000. *Reader's Digest's* eight-member Facilities Management Team examined ways to reduce costs. They decided to work on reducing their waste disposal bills by increasing their recycling effort.

To supplement their existing recycling program, they:

- switched to a new recycling contractor who did not charge for recycling the low-grade paper
- renegotiated their hauling contract so that a reduction in volume would translate into savings in cost
- set up a pilot program in a department that generated a large quantity of

paper, and allowed the program to spread to other departments through employee initiatives

- allowed each department to sign up by choice and to implement programs that work best for them
- installed two additional balers (the company owns two and the private recycler has installed a third)
- gave the custodial staff at *Reader's Digest* responsibility for collecting and preparing recyclables for carting.

The changes that *Reader's Digest* made in its recycling program have saved money in several ways:

- By renegotiating their hauling contract, *Reader's Digest* realized savings of approximately $50,000 per year ($4,170 per month) for trash removal, a 15.3% cost savings.
- By selecting a new recycler that does not charge to recycle their low-grade paper, the company is saving another $50,000 per year.
- *Reader's Digest's* contract with the recycler stipulates that revenues from the high-grade paper and corrugated cardboard will first go to the recycler to offset the expenses of hauling and container rental. Subsequent revenues are split 50/50 between the recycler and *Reader's Digest*.
- *Reader's Digest* can renegotiate the contract if paper prices increase, but the hauler cannot renegotiate the contract if prices drop.
- By installing two new balers, staff members no longer had to compact cardboard manually. The labor savings paid for the new balers immediately.

As one Facility Management staff person at *Reader's Digest* said, "Every penny saved is added back into our retirement and profit-sharing programs. Everybody realizes it and everybody benefits." A similar contract was adopted by Westinghouse Hanford (Figure 6.1).

Reader's Digest management is committed to recycling and to other programs that benefit the environment. For example, concern about plastic waste prompted the cafeteria to switch to paper plates and cups for take-out service despite a $30,000 cost to the company.

Case Study 6.2: Hughes Aircraft Corporation

In 1991, in response to the California Integrated Waste Management Act of 1989 (AB 939) and consistent with its corporate environmental policies, Hughes Aircraft Company initiated a comprehensive program for reducing its solid waste. Headquartered in the Westchester section of Los Angeles, Hughes hired a consultant to conduct a solid waste reduction study at Hughes's 900-person corporate facility. The focus of the study was to develop programs to reduce waste and recover recyclables. The 9-month study now serves as the Hughes organization's master plan for waste reduction and recycling activities through the year 2000. The study comprised four integrated tasks.

Figure 6.1. Office paper collected by Basin Recycling, Inc. for Westinghouse Hanford Company with the only payment being the profit from use of the paper collected. (Photo courtesy of Westinghouse Hanford Company/BCSR, Richland, WA).

- A **waste stream audit** was performed to determine the quantity and types of materials generated, recycled, and disposed of. Material types were determined by hand sorting and weighing of randomly selected bagged waste. The waste was sorted into 22 different categories.
- A **market survey** was conducted of current and future markets for recycled materials.
- **Source reduction, recycling, and composting programs** were identified and evaluated.
- A **final report** was prepared addressing program implementation, including goals and objectives, employee education and outreach requirements, need for new equipment or waste-handling procedures, monitoring, assignment of responsibility, costs, revenues, and schedule.

A unique aspect of the final report included the development of a computerized database management and reporting system to track financial benefits and waste diversion resulting from implementation.

In addition, a manual was prepared to assist all Hughes' facilities in California in developing source reduction and recycling programs. Hughes Aircraft has facilities in five counties and 15 cities in California, and employs more than 45,000. Programs recommended for implementation could yield a 60% reduction in waste. The following elements are included in the programs:

- A source reduction policy guides all of its activities, including a requirement that Hughes' vendors adopt a source reduction policy.
- A procurement program favors the purchase of reusable and durable products, materials, and equipment.
- An employee education and outreach program highlights ways to use products and materials more efficiently, including greater use of scrap paper, double-sided copying, shared magazine subscriptions, greater use of electronic mail, and reuse of envelopes and other supplies.
- Source separation is practiced for recyclable papers—such as white ledger paper, computer printout, mixed paper (commingled newspaper, magazine, and colored ledger) and corrugated containers—through either a desktop or central collection system.

These programs are projected to reduce Hughes' disposal costs by 50% per year and yield annual material revenues of more than $10,000. Combined savings and material revenues will more than offset program costs within 1 year of startup.

Close cooperation of the Corporate Environmental Affair staff at Hughes Aircraft was essential to the successful completion of the study. Other assistance was provided through interviews with company staff from departments as diverse as shipping and receiving, purchasing, and legal. Additional information was gathered during facility walk-throughs, which included observation of custodial procedures and all other waste-handling operations.

Hughes plans on continuing its efforts to improve waste diversion efforts in a cost-effective manner and maintain its corporate commitment to resource conservation. "Our corporate commitment to resource conservation aims to go beyond meeting minimum compliance with environmental laws to real environmental protection," says Greg Taylor, a corporate environmental specialist at Hughes.

GENERAL REFERENCES

Bogardus, E.R. "A Look to the Future in Solid Waste Management." *Proceedings of Sixteenth International Madison Waste Conference*, Madison, WI, 1993.

INFORM, Inc., and Recourse Systems, Inc., *Business Recycling Manual*, 1990.

"CH2M HILL Designs Recycling Program for Hughes Aircraft," *California Consulting Engineer*, Winter 1991.

Steinberger, M.F. "Government Recycling Programs: Experiences with the Commercial Sector," *Resource Recycling Magazine*, April 1990.

CH2M HILL. "Volume 4, Specific Strategies for Targeted Generators; Targeted Materials, City Departments." *City of Los Angeles Source Reduction and Recycling Element*. Prepared for the City of Los Angeles by the CH2M HILL Project Team, April 1993.

CHAPTER 7

Waste and Material Exchanges and Recyclers

*Alison Gemmell**

Eggshells are used for tile pigment . . . Fish waste is used for soil cleanup . . . Oil sludge is used for asphalt blending . . . And a Freddy Krueger movie set is recycled for use as a haunted house . . . What do all these activities have in common? Reuse and recycle! As these examples show, materials that are wastes to some companies are valuable commodities to others. Who connected these waste generators and waste users? Waste and material exchanges and recyclers.

WHY USE AN EXCHANGE OR RECYCLER?

Successful waste exchanges** and recyclers can help industries save money, comply with environmental regulations, and recover valuable resources. Benefits include:

- avoiding costly disposal fees at hazardous and nonhazardous waste landfills
- recovering valuable resources from waste products
- reducing operating and maintenance costs through sale of surplus materials and wastes
- saving energy by using wastes as fuel
- satisfying requirements for hazardous waste reduction
- achieving goals for solid waste diversion***

---

*Ms. Gemmell works for the Environmental Programs Department of Del Monte Foods in Walnut Creek, California, and may be reached at (510) 944-7274.

**The term "waste exchange" is used here to mean both material and waste exchanges.

***"Solid waste," as used here, refers to nonhazardous waste only.

Waste exchanges and recyclers help provide these benefits to industries by finding buyers or users for traditional and unusual waste streams and by recycling wastes for immediate reuse or for use by others.

INTRODUCTION

Waste exchanges and recyclers offer different services to waste generators and waste users. In general, waste exchanges transfer information between industries and do not take possession of wastes. They facilitate the exchange of materials or wastes from one party, which has no use for them, to another party that views them as valuable commodities. Waste exchanges provide listings of both "wastes available" and "wastes wanted," and a means of matching waste generators with waste users. Most waste exchanges issue quarterly catalogs with updated listings; however, some issue catalogs monthly, bimonthly, or annually. Some waste exchanges offer on-line computer databases. Some visit businesses and actively seek generators and users of waste materials. Most waste exchanges advertise both hazardous and nonhazardous materials and wastes. However, some operate separate exchanges, one for solid wastes and one for hazardous wastes.

Recyclers take possession of wastes and then analyze, process, package, transport, and ultimately sell recycled products made from these wastes. A wide variety of recyclers provide opportunities for recycling a full range of waste types, including both hazardous and nonhazardous materials.

WASTE EXCHANGES

History

Waste exchanges have been operating in North America since the early 1970s, and even longer in Europe. In the late 1970s and early 1980s, the primary driving force for waste exchange growth was the high cost of collection and disposal of hazardous waste. Hazardous waste permitting, recordkeeping, and disposal fees caused industries to look for waste management options. Currently, solid waste diversion laws, regulatory emphasis on small-quantity generators, and implementation of hazardous waste land bans are prompting further growth in the number of waste exchanges and increasing the complexity of waste exchange activities.

More than 40 waste and material exchanges are currently active in North America. A partial list is presented in Table 7.1. Waste categories typically advertised by exchanges are summarized in Table 7.2.

Users

Waste exchanges are used by every component of the marketplace: industries, government agencies, nonprofit organizations, and for-profit materials brokers. Use of a waste exchange may be beneficial if one or more of the following conditions apply

- waste is too expensive to manage via onsite treatment or disposal
- waste contains constituents that can be reused or recycled

Table 7.1. North American Waste Exchanges[a]

Name of Exchange	Address and Phone Number	Contact
Exchanges in the United States		
Alabama Waste Materials Exchange[b]	411 E. Irvine Avenue Florence, AL 35630 205/764–6830	Linda Quinn
Arizona Waste Exchange	4725 E. Sunrise Drive, Suite 215 Tucson, AZ 85718 602/299–7716	Barrie Herr
Arkansas Industrial Development Commission[c]	#1 Capitol Mall Little Rock, AR 72201 501/682–1370	Ed Davis
B.A.R.T.E.R. Materials Exchange	Minnesota Public Interest Group (MPIRG) 2512 Delaware Street, S.E. Minneapolis, MN 55414 612/627–6811	Jamie Anderson Steve Moroukian
California Materials Exchange (CALMAX)[b]	Calif. Integrated Waste Management Board 8800 Cal Center Drive Sacramento, CA 95826 916/255–2369	Joyce Mason
California Waste Exchange[d]	Dept. of Toxic Substances Control P.O. Box 806 Sacramento, CA 95812–0806 916/322–4742	Claudia Moore
Hawaii Materials Exchange	P.O. Box 1048 Paia, HI 96779 808/579–9109	Jeff Stark
Hudson Valley Materials Exchange and Buy Recycled! Consortium	P.O. Box 550 New Paltz, NY 12561 914/246–6181	
Indiana Waste Exchange	c/o Recycler's Trade Network, Inc. P.O. Box 454 Carmel, IN 46032 317/574–6505	Jim Britt
Industrial Materials Exchange (IMEX)	506 2nd Avenue, Suite 201 Seattle, WA 98104 206/296–4899	Bill Lawrence
Industrial Materials Exchange Service	P.O. Box 19276, #34 Springfield, IL 62794–9276 217/782–0450	Diane Shockey

Table 7.1. North American Waste Exchanges[a] (Continued)

Name of Exchange	Address and Phone Number	Contact
Iowa Waste Reduction Center	By-Product and Waste Search Service (BAWSS) University of Northern Iowa Cedar Falls, IA 50614–0185 319/273–2079	Susan Salterberg
Intercontinental Waste Exchange[e]	5200 Town Center Circle, Suite 303 Boca Raton, FL 33486 800/541–0400	Vincent Militano
Kentucky Department of Environmental Protection, Division of Waste Management[c]	18 Riley Road Frankfort, KY 40601 502/564–6716	Charles Peters
Minnesota Technical Assistance Program (MnTAP)[e]	1313 5th Street, Suite 307 Minneapolis, MN 55414 612/627–4646	Helen Addy
Missouri Environmental Improvement Authority[c]	325 Jefferson Street Jefferson City, MO 65101 314/751–4919	Thomas Welch
MISSTAP	Post Office Drawer CN Mississippi State, MS 39762 601/325–8454	Caroline Hill
Montana Industrial Waste Exchange	Montana Chamber of Commerce P.O. Box 1730 Helena, MT 59624 406/442–2405	Manager
New Hampshire Waste Exchange	122 North Main Street Concord, NH 03301 603/224–5388	Emily Hess
New Jersey Industrial Waste Information Exchange	50 West State Street, Suite 1110 Trenton, NJ 08608 609/989–7888	William Payne
New Mexico Material Exchange	Four Corners Recycling P.O. Box 904 Farmington, NM 87499	Dwight Long
Northeast Industrial Waste Exchange, Inc.	620 Erie Boulevard West Suite 211 Syracuse, NY 13204–2442 315/422–6572	Carrie Mauhs-Pugh

Table 7.1. North American Waste Exchanges[a] (Continued)

Name of Exchange	Address and Phone Number	Contact
Oklahoma Waste Exchange Program	P.O. Box 53551 Oklahoma City, OK 73152 405/745–7100	Fenton Rude
Olmsted County Materials Exchange	Olmsted County Public Works 2122 Campus Drive, S.E. Rochester, MN 55904 507/285–8231	Jack Stansfield
Pacific Materials Exchange	1522 North Washington, Suite 202 Spokane, WA 99205 509/325–0551	Bob Smee
RENEW	TNRCC P.O. Box 13087 Austin, TX 78711–3087 512/463–7773	Hope Castillo
Rhode Island Department of Environmental Management	Box 1943 Brown University Providence, RI 02912 401/863–2715	Marya Carr
Rocky Mountain Materials Exchange	418 South Vine Street Denver, CO 80209 303/692–3009	John Wright
SEMREX	171 West 3rd Street Winona, MN 55987	Anne Morse
Southeast Waste Exchange	Urban Institute, UNCC Station Charlotte, NC 28223 704/547–2307	Maxie May
Southern Waste Information Exchange (SWIX)	P.O. Box 960 Tallahassee, FL 32302 904/644–5516, 800/441-SWIX	Eugene Jones
Transcontinental Materials Exchange	1419 CEBA Baton Rouge, LA 70803 504/388–4594	Rita Czek
Wastelink, Division of Tencon, Inc.	140 Wooster Pike Milford, OH 45150 513/248–0012	Mary E. Malotke
Wisconsin Department of Natural Resources, Bureau of Solid and Hazardous Waste Management[c]	101 South Webster Street P.O. Box 7921 Madison, WI 53707–7921 608/267–9523	Sam Essak
Exchanges in Canada		
Alberta Waste Materials Exchange	6815 8th Street, N.E. Digital Boulevard, 3rd Floor Calgary, Alberta T2E 7H7 403/297–7505	Cindy Jensen

Table 7.1. North American Waste Exchanges[a] (Continued)

Name of Exchange	Address and Phone Number	Contact
Bourse Quebecoise des Matieres Secondaires, Quebec Waste Exchange	14 Place Du Commerce, Bureau 350 Le-Des-Squers, Quebec H3E 1T5 514/762–3333	Francois Lafortune
British Columbia Materials Exchange	1525 West 8th Avenue Suite 102 Vancouver, B.C. V6J 1T5 604/731–7222	Jill Gillett
Canadian Waste Materials Exchange	ORTECH 2395 Speakman Drive Mississauga, Ontario L5K 1B3 905/822–4111 x265	Robert Laughlin
Durham Region Waste Exchange	Region of Durham, Works Department P.O. Box 605 105 Conaumers Drive Whitby, Ontario L1N 8A3 905/668–7721	Elaine Collis
Essex-Windsor Waste Exchange	Essex-Windsor Waste Management Committee 360 Fairview Avenue West Essex, Ontario N8M 1Y6 519/776–6441	Steve Stephenson
Manitoba Waste Exchange	c/o Recycling Council of Manitoba, Inc. 1812–330 Portage Avenue Winnipeg, Manitoba R3C 0C4 204/942–7781	Todd Lohvinenko
Ontario Waste Exchange	ORTECH 2395 Speakman Drive Mississauga, Ontario L5K 1B3 905/822–4111 x512	Mary Jane Hanley
Waterloo Waste Exchange	Region of Waterloo 925 Erb Street West Waterloo, Ontario N2J 3Z4 519/883–5137	Mike Birett

[a]This list is constantly changing. This information was obtained in January 1994.
[b]Deals in solid wastes only.
[c]Uses and distributes Industrial Materials Exchange Service catalog.
[d]Deals in hazardous waste only.
[e]For-profit exchange.

Table 7.2. General Categories of Wastes Listed in Waste Exchange Catalogs[a]

Wastes Available/Wanted[b]	Recent Categories[c]
Acids	Laboratory Chemicals
Alkalis	Glass
Other Inorganic Chemicals	Construction Materials
Solvents	Container and Pallet
Other Organic Chemicals	Durable and Electronic[d]
Oils and Waxes	Paint and Coating
Plastics and Rubber	
Textiles and Leather	
Wood and Paper	
Metals and Metal Sludges	
Miscellaneous	

[a]These categories vary somewhat between waste exchanges; however, these are the basic categories used.
[b]These categories have been used for many years. The ratio of "wastes available" to "wastes wanted" varies; however, an approximate ratio is 3:1 (available/wanted).
[c]Solid waste diversion laws have resulted in increased activity in these categories. In the past, items in these categories were listed under "miscellaneous" or "wood" or "plastic."
[d]Durables include items such as used furniture and old appliances.

- no in-house expertise is available for onsite waste treatment
- waste is routinely generated, and volume and characteristics are predictable
- waste is generated one time only

Waste exchanges also may be used when regulatory agencies mandate source reduction and a facility does not have the resources or finances to implement process modifications, onsite recycling, or input changes. Advertising in a waste exchange is sometimes considered a waste reduction activity, and may be part of a compliance program for fulfilling mandatory source-reduction requirements. Increasingly, government agencies are requiring facilities that violate such requirements to advertise wastes in waste exchanges as part of the compliance process.

Funding

Typically, waste exchanges are nonprofit and are supported to some degree by state or regional government funding. Some exchanges charge a nominal fee for their catalogs; some offer their publications free of charge; still others sell advertising space or solicit private donations. Some exchanges charge a fee for accessing their on-line database services. In general, all of the existing waste exchanges have difficulty raising the support dollars they require. For government agencies, an investment in waste exchange is a cost-effective way of achieving waste diversion goals. Costs for waste exchanges are typically in the range of $1 to $2/ton of waste diverted. The cost of such programs is much lower than that of other waste diversion programs in the public sector.

Computerization

Many exchanges are starting to offer on-line computer database listings in addition to hard-copy catalogs. Computer programs are written to match wastes that are wanted with wastes that are available, and an on-line system can tally and analyze records of inquiries more easily than a manual system. One such database, the National Materials Exchange Network, was a cooperative project of more than 40 industrial waste exchanges, materials brokers, and government waste-reduction programs. Users dialed in by modem for 24-hour access to more than 6,000 catalog listings. Advantages of computer access included faster access to information (no waiting for catalogs to be published and mailed) and increased visibility in large geographical areas. As of December 1993, the National Materials Exchange Network was seeking funding to continue its services.*

Regional computer databases also have been developed by three waste exchanges, the Florida-based Southern Waste Information Exchange, the North Carolina-based Southeast Waste Exchange, and the Northeast Industrial Waste Exchange. These exchanges are also listed in Table 7.1.

Procedures

In the past, wastes were listed in catalogs, using confidential codes that forced parties to deal through the exchange. Increasingly, exchanges are listing names, addresses, and telephone numbers of waste buyers and sellers to facilitate timely transfers. Although this practice makes it harder for exchanges to determine their success rate, many users still contact waste exchange staff and work closely with them to find uses for unusual wastes. Some exchanges send staff to business sites to facilitate material transfers. Other exchanges routinely contact all listers for information on successful transfers. Users are asked to contact waste exchanges when a successful transfer has occurred.

An exchange that limits its activities to publishing a catalog and that relies on readers to make their own matches is a "passive" waste exchange. "Active" exchanges work to make the match between generators and users. Typically, this is done by telephoning or faxing contacts to potential users once a waste is referred. The Ontario Waste Exchange is an active exchange, and does not publish a catalog. For wastes not transferred, the Ontario Waste Exchange refers listings to the passive Canadian Waste Exchange, whose catalog and bulletin are published and distributed throughout Canada.

Listings include information such as primary waste constituents, amount of waste available or wanted, frequency of specific waste availability, and packaging details (e.g., drum, tank truck, bags). Listings also often include general concentrations of the recoverable constituents. Most listings mention that samples are available to interested parties.

Figure 7.1 shows examples of waste listings from the Industrial Materials Exchange Catalog. Figure 7.2 shows the waste listing application form used by the California Waste Exchange.

---

*Information on the National Materials Exchange Network is available from Bob Smee, Pacific Materials Exchange (see Table 7.1).

W A N T E D

▼ —DENOTES A NEW LISTING

W1100154 - FISH PROCESSING WASTE

Wanted in White Salmon, WA:
Large quantities (tons) of fish processing waste is wanted monthly for use in manufacturing liquid fish fertilizer. Some oyster processing waste may also be of use.
CONTACT: Terry Horn, H & H Ecosystems (509) 427-7353

W1100813 - FLUORESCENT LAMP AND BALLAST

Wanted in Anywhere, USA:
Fluorescent lamps and ballasts. We recycle all materials except PCB's, which are sent to incineration. Any amounts accepted.
CONTACT: Korey Johnson,
Resource Recovery (701) 234-9102

W1100494 - LASER PRINTER CARTRIDGES

Wanted from Puget Sound area:
Most laser printer cartridges; including Canon, Hewlett-Packard, Apple. Each cartridge collected will be turned into funding for our programs. Will pick up.
CONTACT: David Bobanick, Northwest Harvest(206) 625-0755

W1100194 - MAGNESITE

Wanted in Pittsburgh, PA:
By-products or residues; used or scrapped refractory brick containing magnesium oxide. Any amount/frequency/location/packaging.
CONTACT: Joe Quigley, B.P.I., Inc. (412) 371-8554

W1100544 - MISC BUILDING MATERIALS

Wanted in Roseburg, WA:
Lumber, paint, or any usable surplus wood/hardware to be used for retraining labor displaced workers in Southwestern Washington. Wanted within 100 miles of Olympia.
CONTACT: C.B. Smith,
Grays Valley New Start Program (206) 465-2625

W1100624 - MISCELLANEOUS MATERIAL

Wanted in Seattle, WA:
Will accept materials to use in arts and crafts with children. Stickers, paper, plastics, fabrics, woods, metals, and wires. Must be non-toxic.
CONTACT: CC Leonard,
The Creation Station (206) 775-7959

W1100121 - MISCELLANEOUS THEATRE SUPPLIES

Wanted in Seattle, WA:
Lumber, fabric, lighting or electrical supplies, tubing or sheets of plastic or rubber, paint, latex paint and paper all are needed for live theatre groups. Need is sporadic. Will pick up within 60 miles of Seattle.
CONTACT: Jonathan Harris (206) 323-5826

▼ W1100893 - OFFICE FURNITURE AND MACHINES

Wanted in Seattle, WA:
We need various amounts of wooden chairs, folding 6' and 8' step ladders, wood tables, copier machines, answering machines, lighting fixtures (except fluorescent). Use in theatre company.
CONTACT: Tony Soper, Book-It (206) 767-8005

W1100646 - PCB BALLASTS

Wanted Anywhere in USA:
Pre-1979 ballasts removed from fluorescent light fixtures contain PCB's. Fulcircle incinerates the PCB material and recycles uncontaminated metals.
CONTACT: Mitchell Dong,
Fulcircle Ballast Recyclers (617) 876-2229

W1100449 - ROOFING MATERIALS

Wanted in Everett, WA:
Non-asbestos bearing roofing materials. Will arrange collection.
CONTACT: D. Campos, Cycle II (206) 745-4210

W1100020 - SILICON CARBIDE

Wanted in Pittsburgh, PA:
Scrap, used silicon carbide refractory or silicon carbide by-products. Any amount/frequency/location/packaging.
CONTACT: Joe Quigley, By-Product Industries (412) 371-8554

W1100273 - SOD, DIRT, HORSE & COW MANURE

Wanted in Kent, WA:
Iddings, Inc will accept sod, dirt, horse manure and cow manure. Material must be source separated. Fee covers processing but is substantially lower than landfills. Acceptable range is South King County.
CONTACT: Laure Iddings, Iddings, Inc. (206) 630-0600

LABORATORY CHEMICALS

W1200046 - CHEMICALS, SURPLUS LABORATORY

Wanted in Oakland, CA:
A variety of laboratory and industrial chemicals accepted. Must be unused and uncontaminated.
CONTACT: Doug Young, Chemicals for
Research & Industry, Inc. (510) 893-8257

W1200264 - CHEMICALS, SURPLUS LAB

Wanted in Eugene, OR:
Willing to take small quantities of a variety of organic laboratory chemicals. Chemicals must be unopened. Acceptable range is western United States and Canada.
CONTACT: Larry Guadagno,
Molecular Probes Inc. (503) 344-3007

▼ W1200903 - LABORATORY CHEMICALS

Wanted in Pacific NW,:
Triwaste will accept and recycle laboratory chemicals. Any amounts considered.
CONTACT: Paul Beauchemin, Triwaste Treatment (604) 984-8767
Portland (503) 682-3234

Figure 7.1. Examples of catalog listings.

CALIFORNIA WASTE EXCHANGE
LISTING APPLICATION

CWE USE ONLY

Disposition: ______ Status ______ Date ______

Code ______

Received: ______ M M D D Y Y

Form D ______

Please read the information on the reverse side before completing form.

1. Company Name:	2. SIC Code:
3. Mailing Address:	
4. City, State, Zip:	Phone ( )
6. Company Contact:	7. Title
8. Signature	9. Date:

Information in shaded boxes will NOT be listed in *Listing of Hazardous Wastes Available.* NOTE: Area Code WILL BE released.

MATERIAL DESCRIPTION
PLEASE LIST ONLY ONE WASTE STREAM PER APPLICATION

1. Material: ☐ Available ☐ Directory
2. Type of Material: (Generic name and/or main constituent)
3. Description: (Describe the material as you want it listed; use specific terms and percentages) (For Directory Listings include Authorization Status and EPA Identification Number)
4. Contaminants and Percentages:
5. Physical State: ☐ Liquid ☐ Slurry ☐ Sludge ☐ Solid ☐ Other ______
6. Present Quantity: No. ______ ☐ Tons ☐ Gallons ☐ Pounds ☐ ______
7. Frequency: ☐ One-time only ☐ Daily ☐ Monthly ☐ Yearly ☐ Other ______
8. Packaging: ☐ Drums ☐ Tank Truck ☐ Bales ☐ Roll-off ☐ Other ______
9. Location Where Material is Available/Directory Facility: County ______ State ______ Zip ______
10. Process That Generated Material or Process Used for Recycling:
11. Available to Interested Party: ☐ Sample ☐ Analysis ☐ Independent Analysis
12. Additional Information Attached: (Please include company name on each page) ☐ Yes ______ (No. of Pages) ☐ No

Figure 7.2. Example of a catalog listing application.

Other Services

Other services provided by waste exchanges include regulatory interpretations; notices of new regulations and regulatory deadlines; notices of local workshops and conferences; directories of local recyclers, recycled products dealers and distributors, and environmental consultants; notices of grants for waste treatment, reuse, or recycling studies; and information about cooperative purchasing of recycled products. These services are provided via a newsletter section of the exchange catalog or as separate publications sponsored by the exchange.

Some exchanges provide general consulting services for recycling and waste reduction or can put a company in contact with low-cost or free consulting for small businesses through a network of college students or retired environmental professionals.

Waste exchange personnel and customers meet at an annual 3-day conference held at various locations. Conference proceedings are made available and exhibitors are sometimes present. (See General References.)

Success Rate

It is often difficult for waste exchanges to determine success rates because of the confidential nature of transfers (particularly sales prices). However, it is estimated that a 10% to 30% successful exchange rate is achieved nationwide. The most frequently exchanged wastes include spent plating solutions, precious metal solutions, mercury, antifreeze, solvents, used oils, plastics, and paper. Table 7.3 lists examples of successful exchanges.

As of September 1993, CALMAX, the California Materials Exchange, which is restricted to nonhazardous materials, received an average of more than 330 phone calls per month, with an ongoing subscribers' mailing list numbering 7,000. In less than 2 years since its creation, it has diverted more than 152,000 tons of materials from landfills at an estimated savings of $833,000 to participants. Its listings have grown from 336 in the first catalog to 727 in the September/October 1993 issue. Exchanges in other parts of the country are showing similar patterns of growth and success.

RECYCLERS

Recyclers are independent for-profit businesses that take possession of a waste or used material and process it for resale. In the case of hazardous materials recyclers, the types of wastes usually handled are single-waste streams from predictable, routinely generated sources. Hazardous materials recyclers generally deal in waste oils, solvents, precious metal solutions, highly concentrated metal-waste sludge, and spent etching baths. Specialty items such as vehicle batteries, antifreeze, and fluorescent bulbs also are recycled by specialty recyclers. Solid waste recyclers handle a wide variety of materials such as fiberboard, aluminum, glass, paper, some plastics, and ferrous metals. Technologies for recycling additional solid waste and hazardous waste are developing rapidly, in part as a result of mandatory state source reduction and diversion laws and of state-sponsored incentive programs.

Recyclers advertise through normal marketing channels (including waste exchange catalogs) and charge a fee for their services. Market conditions vary with regard to the demand for specific waste materials. In a favorable market, sale of a waste material to a recycler can result in a profit for the waste seller. More typically, the seller's main benefits are lower costs for waste disposal and less paperwork.

Table 7.3. Examples of Waste Uses

Waste	Use
Refinery spent caustic	Wood pulping process
Electronic circuit manufacturing plating baths	Plating applications or for metals recovery
Solvent from electronics industry	Paint manufacturing
Oil sludges	Asphalt manufacturing
Blast furnace slag	Aggregate
Dimethyl acrylamide	Manufacturing of vinyl fiber
Waste with fuel value and metallurgical coke fines from electrode mfg.	Cement kiln fuel
Hydrochloric acid in steel pickle liquor	Flocculating sediment and removing phosphates in wastewater
Methanol used to clean electronics parts	Auto windshield cleaning fluid
Waste oil	Rerefined as lube oil; reprocessed as fuel oil
Solvent	Reused
Pickle liquor	Iron salts recovery
Metal sludges	Precious metals and other heavy metals recovery (depending on current market conditions)
Spent foundry sands	Reuse or use as aggregate in portland cement
Activated carbon	Reactivate and reuse
Grease and fats	Render and use in soaps
Lead acid batteries	Recover lead in secondary smelters
Scrap bronze, brass, plumbing materials	Recover tin
Styrene tar (10% S) and sulfuric acid wastes	Recover sulfuric acid
Used drums	Recondition and reuse
Fluorescent lightbulbs	Recover mercury and glass, and recycle the aluminum in the ends of the tubes
Tires	Fuel or recondition/reuse
Animal carcasses	Fertilizers, bone meal, and soap
Hog hair	Nondigestible protein supplement in livestock feed
Fiber roll cores	Animal bedding
Excess laboratory chemicals	Donate to college or high school laboratories

Users

Recyclers frequently work closely with waste generators, often providing advice on proper storage, packaging, labeling, and recordkeeping. In some cases, a chemical manufacturer may sell a virgin material with a purchase price that includes the assumption that spent material will be taken back by the seller and recycled (for example, solvents or oils).

Recyclers may sell their recycled product to the same industry that generated the waste (for example, antifreeze recyclers in the auto repair industry or spent etching bath reclaimers in the printed wiring board industry), or they may sell their recycled product to an industry that can use a recycled product instead of virgin material. For example, semiconductor waste solvents may be recycled and used as cleaning solvents.

Procedures

Recyclers advertise in waste-related, environmental engineering, science, and regulatory newsletters or magazines; in trade association publications; and in journals of professional societies. Most waste exchanges assist recyclers by publishing separate directories of recyclers in their geographical regions. Technical assistance branches of regulatory agencies also provide general listings of recycling companies. Recyclers can be found in local telephone book yellow pages and at environmental trade shows and conferences.

Wastes that are typically handled through recyclers are listed in Table 7.4. Examples of recycling trade associations are listed in Table 7.5.

Selecting a Recycler

Because waste management is highly regulated, with potential liability for the waste generator in the case of mismanagement, it is wise to carefully select a recycler. Following these suggestions will help in evaluating recyclers:

Table 7.4. General Categories of Wastes Handled by Recyclers

Acids
Antifreeze
Catalysts
Coolants (Metal Working Coolants and Machine Oils)
Dry Cleaning Wastes
Metals and Metallic Salts (including batteries, sludges, filter cake, ash)
Metals (Lead) (including batteries)
Metals (Precious) (including jeweler and photoprocessing wastes)
Metals (Mercury) (including fluorescent lamps)
Oils
Post-consumer Plastics
Solvents
Tires
Transformers

Table 7.5. Examples of Recycling Trade Associations

Solvent Recyclers

National Association of Solvent Recyclers (NASR)
202/463–6956

Metal Scrap Recyclers

Institute of Scrap Recycling Industries (ISRI)
202/466–4050

Association of Container Recycling (ACR)
202/543–9449

Steel Can Recycling Institute (SCRI)
800/876–7274

Solid Waste, General

American Plastics Council
Society of Plastics Industry (SPI)
800/243–5790

National Recycling Coalition (NRC)
202/625–6406

Textiles

Council for Textile Recycling (CTR)
301/656–1077

Miscellaneous

National Association of Chemical Recyclers (NACR)
202/986–8150

Canadian Association of Recycling Industries
416/510–1244

International Cartridge Recycling Association (ICRA)
202/857–1154

- Request proof of adequate permit or approval status (most important if the waste is hazardous).
- Contact applicable regulatory agencies with authority over the recycling process (e.g., wastewater, air, and hazardous waste agencies) and ask about recent violations or any problems the recycler may have had.
- Tour and audit the recycling facility to make sure the facility meets industry standards for cleanliness, waste tracking, and environmental compliance.
- Interview facility personnel to make sure they understand current environmental requirements and regulatory implications for waste generators and recyclers.
- Ask for a reference list of satisfied customers.
- Review the recycler's qualifications by repeating the above activities at least annually.

A list of waste recyclers is shown in Table 7.6.

Table 7.6. Examples of Waste Recyclers[a]

Waste	Recyclers[b]
Solvents	Safety-Kleen Corporation 8795 Folsom Boulevard, Suite 108 Folsom, CA 95826 916/387–6335 800/769–5845
	Oil and Solvent Process Company 1704 W. First Street Azusa, CA 91702 818/334–5117
Oils	Evergreen Oil, Inc. 6880 Smith Avenue Newark, CA 94560 510/795–4400
	USPCI Solvent Services 1040 Commercial Street, Suite 109 San Jose, CA 95112 408/451–5000
Metal Solutions/Sludges	Encycle/Texas 5500 Up River Road Corpus Christi, TX 78407 512/289–0035
	Cypress Miami Mining Corp. P.O. Box 4444 Claypool, AZ 85532 602/473–7135
Precious Metals	Drew Resource Corp. 1717 4th Street Berkeley, CA 94710 510/527–7100
	Eastern Smelting and Refining 3739 Bubier Street Lynn, MA 01901 800/343–0914
Acids/Caustic	Eticam 2095 Newlands Drive Fernley, NV 89408 800/648–9931
	Hevmet Recovery Ltd. P.O. Box 278; Port Colburne Ontario, Canada L3K 5W1 416/834–0034
Vehicle Batteries	Kinsbursky Brothers, Inc. 1314 N. Lemon Street Anaheim, CA 92801 714/738–8516

Table 7.6. Examples of Waste Recyclers[a] (Continued)

Waste	Recyclers[b]
Vehicle Batteries (Continued)	RSR Quemetco, Inc. 720 South 7th Avenue City of Industry, CA 91746 800/527–9452 818/330–2294
Antifreeze	Romic Chemical Corp. Antifreeze Environmental Service 2081 Bay Road East Palo Alto, CA 94303 415/325–2666
	DeMenno/Kerdoon 2000 N. Alameda Compton, CA 90222 310/537–7100

[a]Companies listed should be evaluated for compliance with current environmental laws and regulations. Inclusion here does not constitute endorsement by the author or publisher.
[b]Contact your local waste exchange for a more detailed listing of recyclers in your geographical area.

MATERIALS BROKERS

One type of materials exchanger, also an active user of the National Materials Exchange Network (computer database), is the materials broker. Most materials brokers have been in business for many years and, although they do not typically refer to themselves as "materials exchanges" or "recyclers," they are instrumental in diverting surplus inventories, off-specification materials, and out-of-date chemicals from disposal in regulated landfills.

Brokers are independent, for-profit businesses that may or may not take possession of surplus chemicals, containers, or equipment. In every case, a broker buys the material from the original seller and resells the material for a profit. Brokers often make arrangements for transportation of a material directly from a seller to an end user, but also take possession of and store the material in warehouses until the material is resold. Unlike recyclers, brokers do not physically alter materials.

For brokers, most deals involve transferring unused, sealed-in-original-container materials (for example, surplus inventory, out-of-date inventory, and off-specification inventory). Brokers deal in chemicals or materials that are never labeled "waste."

Examples of materials transferred through brokers are listed in Table 7.7. Two brokers are listed here as examples. Contact your local waste exchange for a more detailed listing of brokers in your geographical area. Brokers are generally listed in the Surplus Materials Sections of the Industrial Resources Directories.

Table 7.7. General Examples of Materials Handled by Brokers[a]

Borax	Lubricants
Boric acid	Metallic oxides
Carbon	Oils
Catalysts	Paints
Dyes	Photosensitive film/paper
Electroplating and circuit board chemicals	Pigments
Epoxy	Plastics
Fiberglass	Powder charcoal
Fixer, developer, stabilizer	Reagents
Flake caustic	Scrap electronic components
Fuel additives	Soda ash
Graphite	Solvents
Inks	Typesetter chemicals
Ion-exchange resins	Unused bottles
Kevlar	Urethane materials
Lab chemicals	Waxes

[a]Brokers generally require that materials be unused and sealed in original containers.

Park Trading Co.
P.O. Box 9521
Providence, RI 02940
401/467-3100

Canadian Chemical Exchange
P.O. Box 1135
Ste-Adele, Quebec JOR1L0
514/229-6511
800/561-6511
Philippe LaRoche

ACKNOWLEDGMENTS

Special thanks are extended to the waste and materials exchange experts who contributed information for use in this chapter:

Dr. Robert Laughlin, Canadian Waste Materials Exchange
Joyce Mason, CALMAX
Bill Lawrence, Industrial Materials Exchange (IMEX)
Bob Smee, Pacific Materials Exchange, National Materials Exchange Network
Eugene Jones, Southeast Waste Information Exchange (SWIX)
Claudia Moore, California Waste Exchange
Steve Moroukian, B.A.R.T.E.R. Materials Exchange
Susan Salterberg, Iowa Waste Reduction Center
Rita Czek, Transcontinental Materials Exchange
Diane Shockey, Industrial Materials Exchange Service
Jill Gillett, British Columbia Waste Exchange
Maxie May, Southeast Waste Exchange

GENERAL REFERENCES

Center for Environmental Studies, Arizona State University. *The Third National Conference on Waste Exchange*. Proceedings of a conference held in Tempe, Arizona, on March 4–6, 1986, sponsored by the Western Waste Exchange.

Government Institutes, Inc., and the Northeast Industrial Waste Exchange, Inc. *1992 North American Conference on Industrial Recycling and Waste Exchange*. Proceedings of a conference held in Syracuse, New York, on September 9–10, 1992.

Jones, E.B., et al., "An Assessment of the Effectiveness of North American Waste Exchanges," *Waste Minimization & Recycling Report* (December 1989): 7–9.

Jones, E.B., et al., "Incentives Influencing Participation in a Waste Exchange Program," *1992 North American Conference on Industrial Recycling and Waste Exchange* (September 1992), pp. VIII-1 to VIII-11.

1991 North American Waste Exchange Conference on Industrial Recycling: Waves of Change. Proceedings of a conference held in Spokane, Washington, on May 5–8, 1991, sponsored by the Pacific Materials Exchange, Eastern Washington University, and the North American Waste Exchanges.

Ontario Research Foundation. *1988 North American Conference on Waste Exchange: Waste Minimization Through Technology and the Marketing of Wastes and Surpluses*. Proceedings of a conference held in Toronto, Ontario, Canada, May 15–18, 1988, sponsored by the Canadian Waste Materials Exchange, the Ontario Waste Exchange, and the Ontario Research Foundation.

Powell, J., and M. Lynch, "Successfully Simple: The Waste Exchange," *Resource Recycling* (April 1990): 82–85 et seq.

"A Review of I/C/I Material ('Waste') Exchanges in North America." Prepared for the Recycling Council of British Columbia by Robert Smith and Associates. Vancouver, B.C., May 1992.

Southeast Waste Exchange. *The Fourth National Conference on Waste Exchange*. Proceedings of a conference held in Charleston, South Carolina, on March 3–5, 1987, sponsored by the Southeast Waste Exchange, South Carolina Department of Health and Environmental Control, and U.S. EPA Region IV.

Texas Water Commission. *1989 North American Conference on Waste Exchange. Waste Minimization: The Industrial Challenge for the '90s*. Proceedings of a conference held in San Antonio, Texas, on September 17–20, 1989, sponsored by RENEW.

CHAPTER 8

Treatment to Reduce Disposal

*Phil Benson and Thomas Higgins**

INTRODUCTION

Wastes are generated by virtually every industrial enterprise. These wastes can be solids or liquids. For example, the pharmaceutical industry generates waste solvents when purifying ethical drugs and discarding manufactured products or raw materials that do not meet specifications. Paper mills generate waste from wood pulping, trimmings from the paper-making machines, and from roll ends. Dry cleaners generate waste oil, dirty filters used to recycle solvents, and spent solvents. The list and variety of waste products is virtually endless.

There are three main methods of waste disposal: land disposal, which has been and continues to be predominant; waterborne disposal with eventual drainage out to sea; and dispersal to the atmosphere. Treatment before disposal is desirable for reducing the toxicity, mobility, or volume of the waste.

Techniques are available for recycling parts of these wastes and developing new, marketable products. To be commercially viable, these products must meet minimum standards. For instance, to be saleable, recycled lubricating oil must meet standards for lubricating quality and temperature stability set by the Society of Automotive Engineers. Recycled solvents must meet commercial criteria for industrial solvents, such as cleaning effectiveness and corrosion inhibition. Recycled paper must be suitable for cardboard manufacture, writing paper, or cellulosic insulations. If these commercial constraints are not met, the recycled product itself will be discarded as waste.

In a waste reduction project, the first goal is to eliminate the waste as early in the production process as possible. It is, however, rarely feasible to optimize a production process to the extent that no waste is produced. "Zero discharge" is a laudable goal, but it is seldom practical. However, even when the volume of waste from a production process is minimized, waste reduction opportunities do not end. The waste can still be processed to recover useful materials or treated to reduce its volume, toxicity, or mobility, thus reducing its impact on human health and the environment. This chapter discusses and presents examples of a few of the treatment

*Mr. Benson is a chemical engineer and Dr. Higgins is a vice president in CH2M HILL's Reston, Virginia, office, and both may be reached at (703) 471-1441.

technologies for reducing the environmental and health effects of materials that are disposed of as waste.

POSSIBLE USES OF TREATMENT

Treatment can be used to reduce the volume of waste requiring disposal, to render a hazardous waste nonhazardous, or to recover a useful product or some other resource. The following examples illustrate potential uses of treatment for waste reduction.

Volume Reduction

Volume reduction can be used to reduce treatment costs or to reduce the handling and disposal costs for residues remaining after treatment. Volume reduction can be accomplished by using a variety of methods, including:

- reuse of treated wastewater or other wastes
- treatment modifications to reduce the generation of solid residues
- segregated treatment to reduce hazardous waste mixtures

In addition, incineration can be used to reduce waste volume or to render a hazardous waste nonhazardous.

Wastewater Reuse

The most extreme example of wastewater reduction is a zero-discharge wastewater management system. In such a system, wastewater is treated to sufficient quality that it can be reused in place of raw water within the facility that generated the wastewater. Typically, wastewater that cannot be reused in the original process is evaporated or used for some other beneficial use, such as spray irrigation.

Zero discharge has been practiced for some time in locations where water is scarce. Zero discharge may involve the use of technologies for removing suspended solids, such as clarification and filtration, and technologies for removing dissolved solids, such as reverse osmosis and evaporation. The actual technologies required depend on the quality of the wastewater as well as the water-quality requirements of the process that reuses the treated water. Complete demineralization is relatively expensive; however, in some cases, wastewater discharges can be reduced significantly using less-expensive selective removal technologies.

Case Study 8.1: Zero Discharge at an Integrated Manufacturing Facility

Background and Objectives. A corporation located a new manufacturing facility in a rural area. This facility was to be highly integrated, with many product components manufactured onsite. Minimizing the impact of the facility on the local environment was an important consideration; in addition, local water supplies and wastewater disposal capacity were limited.

Because of these factors, the company decided to investigate the feasibility of in-plant recycling of wastewater. The ultimate goal was zero discharge of wastewater from the site.

Water Management Computer Model. To assist in evaluating water and wastewater management alternatives, a computer model was developed of the manufacturing facility's water use and wastewater production. The water management model included water usage and quality requirements for more than 50 manufacturing processes in 9 separate manufacturing areas, taking into account the chemical constituents and pollutants added to the water. In addition to manufacturing processes, 20 water treatment processes were included in the model.

Approaches to Zero Discharge. Approaches to achieving zero discharge include at-source treatment and recycle and end-of-plant treatment and recycle, or some combination of these in an integrated water management system. At-source treatment and recycle was appealing because it would treat waste streams in small, possibly modular, systems, and would be designed for only those parameters not meeting water reuse requirements. The primary disadvantage was that the large size of this facility would necessitate multiple treatment and recycle systems. An end-of-pipe treatment system would have to be extensive and contain many unit processes in order to accommodate the various flows at the facility. An integrated approach combined the best features of at-source and end-of-pipe treatment.

Recommended Water Management System. The water quality model was used to analyze numerous configurations of an end-of-plant wastewater treatment facility. An end-of-pipe treatment system was recommended to receive all process effluents, treat the wastewater, and recycle the treated water back to the manufacturing units. This was supplemented with some at-source treatment to reduce the number of unit processes required in the end-of-pipe facility. A flow diagram of the recommended water management system is shown in Figure 8.1. An estimate of the capital costs for this approach is presented in Table 8.1.

Case Study 8.2: Wastewater Reuse at a Coal-Fired Power Plant

Background and Objectives. An electric utility located in the eastern United States operates a large coal-fired power station. Recently, state regulators required the station to meet new stringent discharge limits for metals. Meeting the limits would require very expensive treatment of the entire discharge from the station.

Wastewater Management Planning. The client contracted CH2M HILL to develop a wastewater management plan for the station. Figure 8.2 presents a flow diagram of the power station. CH2M HILL conducted in-plant sampling and flow monitoring to identify flow and pollutant sources. This information was used to develop a mass-balance diagram of the power station. Significant sources of flow include cooling-tower blowdown, ash-hopper overflow, coal pile runoff, and miscellaneous waste streams. Significant concentrations of metals are present in enough waste streams so that end-of-pipe treatment will be required.

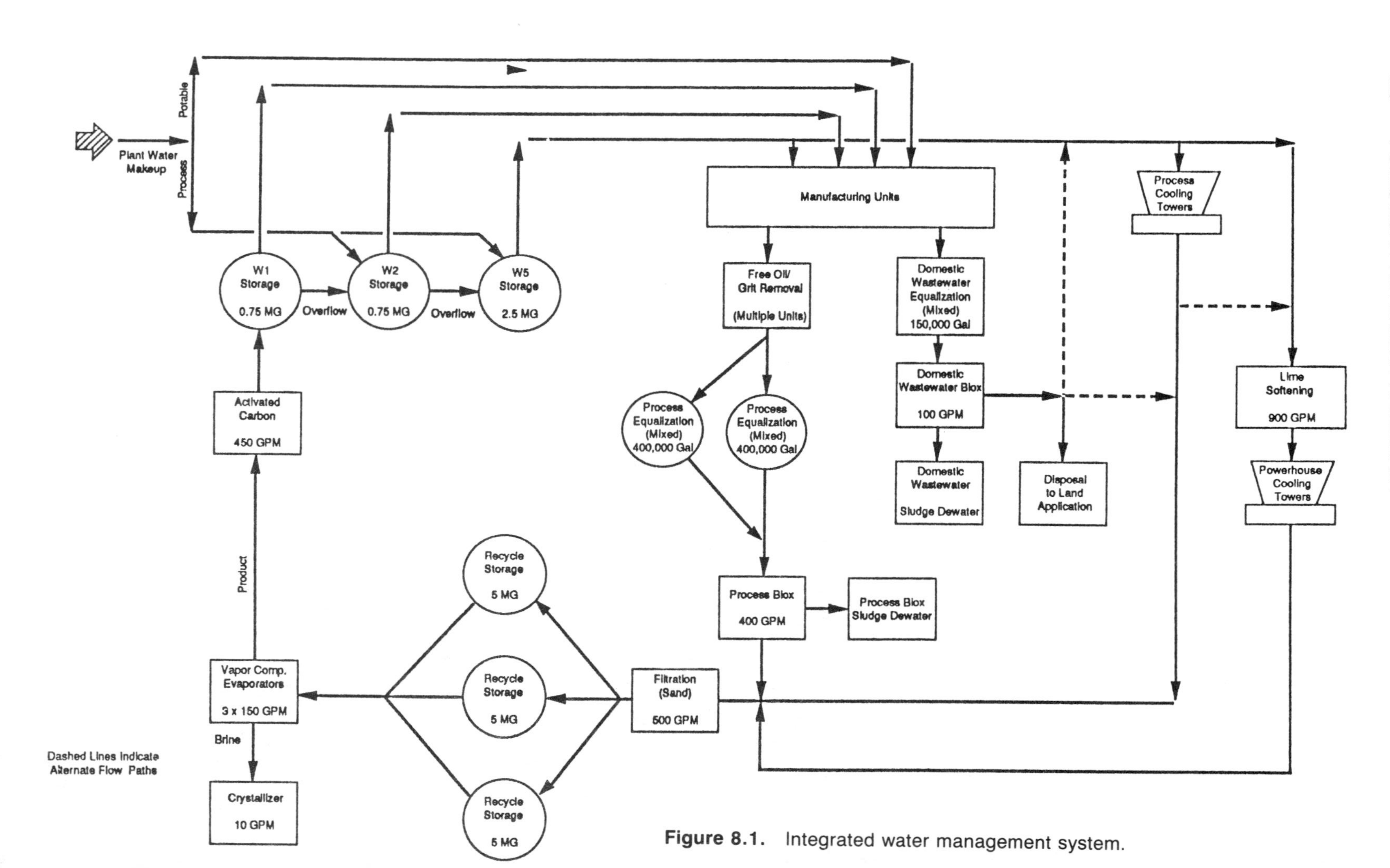

Figure 8.1. Integrated water management system.

Table 8.1. Capital Cost Estimate for Integrated Water Management System

Equipment Used	Size (gal)	Design Flow (gal/min)	No. Used	Estimated Installed Cost ($)
Chemical feed systems			3	144,000
Ultrafiltration			2	780,000
Oil/water separators				333,000
Metals removal—lime prec.		250		589,000
Process equalization (mixed)	400,000		2	697,000
Process biological treatment		400		2,102,000
Domestic biological treatment	150,000	100		677,000
Lime-soda softening		900		430,000
Recycle storage	5,000,000		3	3,358,000
Filtration		500		824,000
Vapor compression evaporation with crystallizer		150	3	9,904,000
Activated carbon		450		1,421,000
Plant water storage			3	1,193,000
Building				379,000
TOTAL ESTIMATED EQUIPMENT COSTS				**22,831,000**
Piping (20% of modified equipment)				5,707,750
Electrical (10% of modified equipment)				2,536,778
Instrumentation and control (7% of modified equipment)				1,718,462
Site work (10% of modified equipment)				2,536,778
Mobilization and insurance (7% of all above)				2,473,154
Subtotal				37,803,922
Scope allowance (30% of subtotal)				11,341,177
TOTAL ESTIMATED CONSTRUCTION COSTS				**49,145,099**
Engineering (15% of construction costs)				7,371,765
Subtotal				56,516,864
Contingency (10% of construction costs)				5,651,686
TOTAL CONCEPTUAL LEVEL CAPITAL COSTS				**62,168,550**

Recommended Treatment Process. The power station currently generates an average of 7 million gallons per day (mgd) of wastewater. The required size of an end-of-pipe treatment facility for this volume of effluent is approximately 9 mgd. Many of the waste streams are contaminated with suspended solids. Settling of solids from these waste streams can produce an effluent having a water quality rivaling that of the raw river water used at the facility. These waste streams represent more than 50% of the wastewater currently discharged from the facility. The settled wastewater can be reused as cooling-tower makeup water, ash-hoppers makeup water, and for other power plant processes. The client and CH2M HILL reached a consensus that average wastewater discharge flows could be reduced to approximately 3 mgd. A flow diagram of the selected wastewater management system is shown as Figure 8.3.

Table 8.2 presents order-of-magnitude capital costs for the end-of-pipe treatment facility with and without wastewater reuse. As shown, reuse of wastewater can save

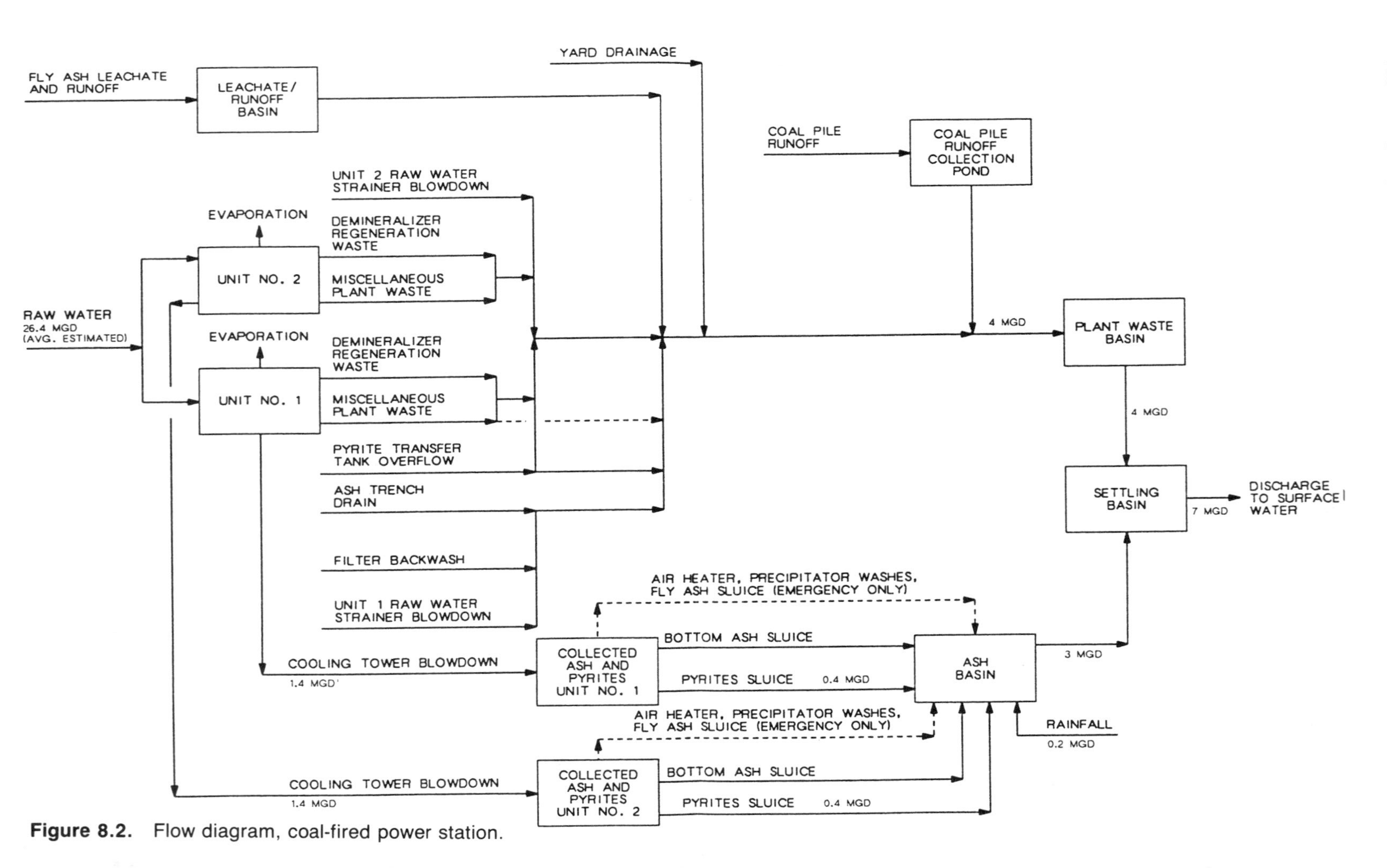

Figure 8.2. Flow diagram, coal-fired power station.

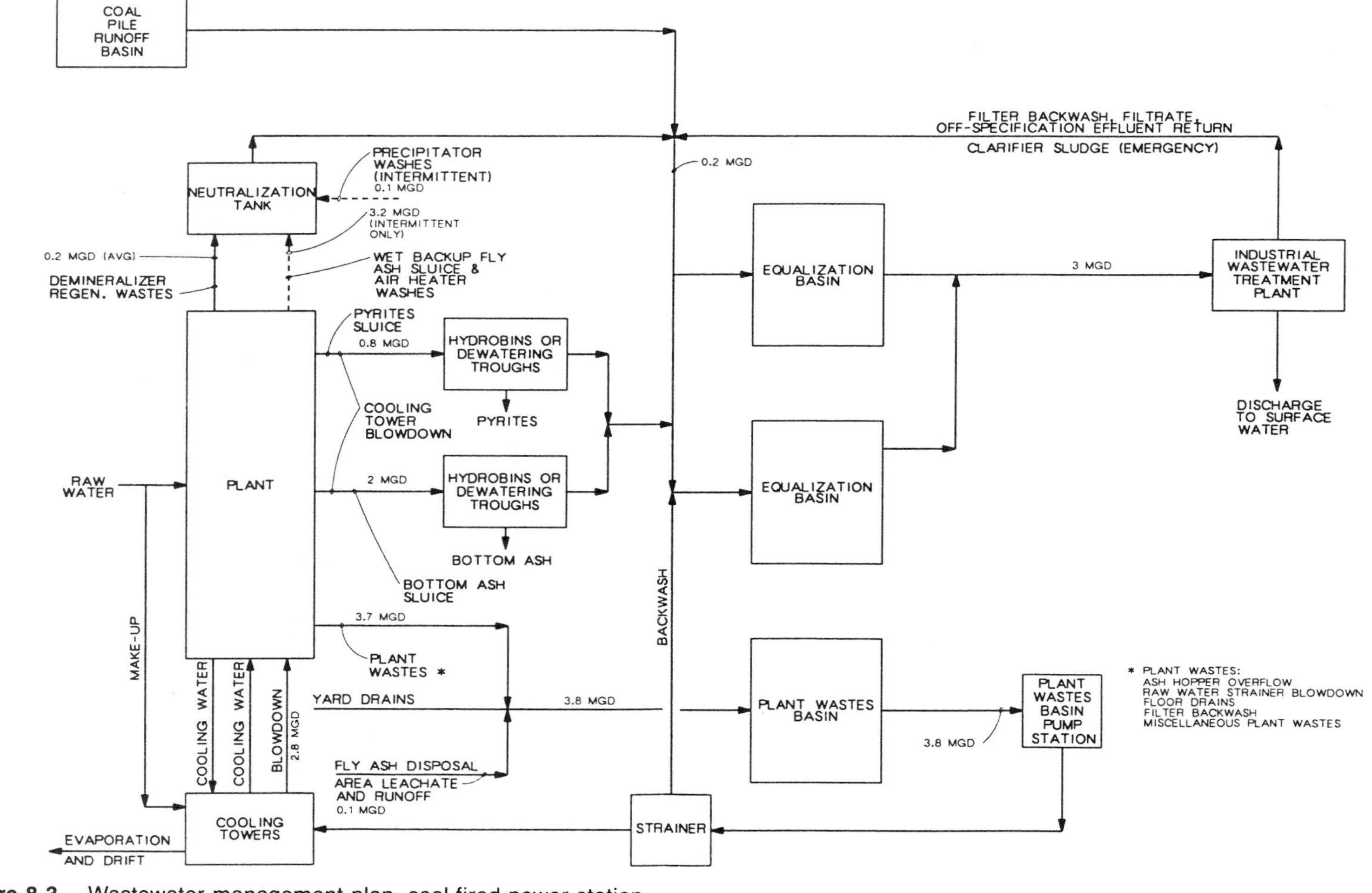

Figure 8.3. Wastewater management plan, coal-fired power station.

Table 8.2. Comparison of Capital Costs for Treatment at a Coal-Fired Power Station

Treatment Process	With Wastewater Flow Reduction	Without Wastewater Flow Reduction
Settling basins	1,150,750	1,430,630
Pump station	136,730	143,630
Flocculation tanks	82,590	106,680
Clarifiers	1,144,250	1,669,000
Filter package	900,490	1,016,230
Sludge handling	792,670	1,077,520
Building	268,530	399,140
Chemical feed systems	70,250	80,700
Subtotal	**4,546,260**	**5,923,530**
General conditions	248,000	323,100
Finishes	165,300	215,400
Misc. undefined mech.	826,600	1,077,000
Electrical/I&C	1,239,900	1,615,500
Sitework	413,300	538,500
Yard piping	826,600	1,077,000
Subtotal	**8,265,900**	**10,770,100**
Contingency	2,479,800	3,231,000
Subtotal	10,745,700	14,001,100
Engineering	3,761,000	7,900,000
Mobilization/insurance/bonding	537,300	700,100
Total Capital Cost ($)	**15,044,000**	**19,601,200**

the client $4.6 million in treatment costs. The estimated capital cost of the piping modifications required to reuse wastewater is approximately $1 million.

Reducing Generation of Solid Residues

Modifying waste treatment practices can reduce the amount of solid residues requiring disposal. This can be achieved by modifying treatment chemistry to produce less sludge, or by removing water from the sludge to produce a drier cake.

Modification of Treatment Chemistry to Reduce Sludge

In conventional treatment of a mixed-metal waste containing hexavalent chromium, the pH of the waste stream is reduced to below 3 with a mineral acid (usually sulfuric). A reducing agent (sulfur dioxide or sodium metabisulfite) is then added to convert hexavalent chromium to the reduced trivalent state. After the reaction is complete, lime is added to raise the pH of the combined wastewater to approximately 9.5 in order to precipitate heavy metals as hydroxides. A disadvantage of this treatment scheme, however, is that it produces large quantities of sludge, in excess of the metals targeted for removal. The sludge is produced because calcium carbonate and calcium sulfate are precipitated out of the wastewater, particularly when treating a hard alkaline wastewater.

In research sponsored by the Air Force, it was found that chrome reduction could be accomplished at neutral to slightly alkaline pH using a combination of ferrous sulfate and sodium sulfide as reducing agents and sodium hydroxide for precipitating heavy metals.[1] The study found that the resulting iron hydroxide precipitate was effective in removing other heavy metals, such as cadmium and nickel, at a neutral to slightly alkaline pH. The advantages of these modifications include eliminating an acidic chrome reduction step, eliminating the addition of acid and reducing the need for lime, and precipitating metals without also producing calcium sulfate or carbonate solids.

This process has been adopted at the Tinker AFB industrial wastewater treatment plant to remove chrome, copper, nickel, cadmium, and other metals from a 1-mgd combined industrial waste stream. Operating at a pH of 7.5 to 8.5, the treatment system achieves the same chrome reductions as the old process operating at a pH of 2.5 to 3, while reducing the sludge volumes by two-thirds. The Air Force estimated that this process modification would save $1,000 per day in chemical usage and sludge disposal costs.[2]

In a similar treatment process, ferrous iron is generated electrochemically using sacrificial iron electrodes. Equipment has been installed at a Navy ordnance plant in Pomona, California, to remove copper and traces of chromium and nickel.[3] The pH of the reactor is held to a range between 6 and 9 using caustic soda (sodium hydroxide). The precipitated metals are settled out in a clarifier and dewatered in a filter press. The process was reported to produce 75% less sludge when compared with acidic chrome reduction and lime precipitation.

Sludge Thickening

Combined industrial wastewater treatment facilities typically use hydroxide precipitation for removing toxic metals. The metal hydroxide solids are usually removed by clarification. Metal hydroxide sludges withdrawn from a clarifier typically have solids concentrations ranging from 1% to 2%. A pound of copper precipitated with hydroxide produces 1.54 lb of solid copper hydroxide. In a 1% sludge, this pound of copper produces 154 lb of sludge. Disposal of such a large volume would be expensive, even if permitted (liquid waste disposal in hazardous waste landfills is banned). Therefore, most waste treatment plants dewater their sludge before disposal.

Use of a thickener can be useful before the sludge is dewatered. Adding a thickener after or as part of the clarifier provides additional sludge storage volume in addition to increasing the sludge solids to as much as 5 to 6%. (In our example, this equates to reducing the 154 lb of copper hydroxide sludge to 30 lb.) Increasing solids by thickening assists subsequent sludge dewatering in two ways. First it reduces the time required for dewatering, and second, it usually results in a drier sludge cake after dewatering.

Sludge Dewatering

Sludge dewatering has traditionally been accomplished in open sand drying beds, especially in areas of the country with warm climates, low rainfall, and cheap land. However, because sludge from industrial wastewater treatment plants is frequently hazardous, sludge drying beds have all of the design (and, potentially, the regulatory) requirements of a hazardous waste landfill, with collection and treatment of leachate required. Therefore, mechanical dewatering is most frequently used because of regulatory and cost advantages.

Three types of mechanical dewatering devices are typically used for treating sludge from industrial wastewater treatment plants: vacuum filters, belt presses, and plate-and-frame filter presses.

Vacuum filters are used infrequently because they produce the least-dry cake of the three mechanical methods. Vacuum filters frequently employ a precoating process, using diatomaceous earth, to improve dewatering. This is a disadvantage because some of the precoat is scraped off during the process, adding to the volume of solids to be disposed of. In addition, vacuum filters generally require higher energy use than the other mechanical dewatering processes. In some applications (for example, in dewatering aluminum hydroxide sludge), vacuum filters are preferred, because they are gentler to gelatinous sludges and can produce a dry cake without blinding the filter cloth, as occurs with the other filters.

Capital costs for the smallest belt press are higher than for the smallest plate-and-frame press. However, belt presses can be economical for large waste treatment plants. Operating costs are lower because belt presses operate continuously, as opposed to plate-and-frame presses, which operate in a batch mode. Belt presses generally produce a cake having a range of 20% to 30% solids content.

Plate-and-frame filter presses are generally used for facilities that produce a small volume of waste, because they are simple to operate and are available in a broad range of sizes. Sludge from the thickener or clarifier is pumped, typically using an air diaphragm pump, to the chambers of a filter press. The solids are retained by the synthetic cloth filter media, and the liquid flows through the media and is returned to the wastewater treatment plant for retreatment. After the pressure required to pump sludge to the press reaches a maximum value, the hydraulic press that holds the plates together is released, and a filter cake is discharged. Plate-and-frame filter presses generally produce the driest sludge cakes (30% to 40% solids), which is an advantage when disposal is based on weight or volume.

Filter press operation generally requires little operator attention except at the beginning and end of a cycle. Automatic plate shifters greatly reduce the manual labor required to remove the filter cake. Adequately dewatered sludge will literally fall out of the press when it opens.

The major maintenance cost of a mechanical sludge filter is replacing filter cloths, which can be frequent when handling abrasive wastes having an extreme pH. However, metal hydroxide sludges generally have a moderately alkaline pH and are nonabrasive.

Sludge Drying

Typically, mechanical dewatering can increase the solids content of a sludge from 30% to 40%. Using the 1 lb of copper as an example, an unthickened 1% sludge weighing 154 lb would be reduced to 5 lb of filter cake (30% solids). This filter cake still contains 3.5 lb of water. In addition, even though this sludge appears to be dry on the surface, free water typically will escape during shipment. Landfill operators may then either reject the sludge (because of a ban on landfill disposal of free liquids) or require that cement kiln dust be added to react with or soak up the excess moisture.

Further drying of the filter cake to 80% to 90% solids content is feasible. The process reduces the weight of our hypothetical sludge from 5 to 2 lb (80% solids), eliminating the need for adding kiln dust to prevent the generation of free water during shipment.

SEGREGATED TREATMENT TO REDUCE HAZARDOUS MIXTURES

In the past, operating a centralized treatment plant for industrial waste has been more cost-effective than providing individual treatment systems for each production unit in a large manufacturing facility. However, hazardous waste regulations have classified the residues from the treatment of wastes from certain industrial operations (i.e., electroplating or chromate-conversion coating of aluminum) as listed hazardous wastes, regardless of their actual composition or toxicity. This is complicated by the RCRA mixture rule, which states that mixing any quantity of a hazardous waste with a nonhazardous waste renders the entire mixture hazardous.

The combination of these two regulations makes it imperative that a manufacturer investigate whether combined treatment of all waste streams makes sense. It is often economical to provide treatment for hazardous waste separately from all other wastes, especially when the hazardous waste is only a small portion of total waste flow in the facility or produces only a small portion of the total sludge when treated.

Following is an example in which an aerospace manufacturing facility plans to provide separate treatment (and recycle) of the wastewater from chromate conversion coatings operations. Segregating the wastes will result in the industrial wastewater treatment plant sludges becoming classified as nonhazardous. Reducing the amount of waste to be disposed of justifies the construction and operation of two separate treatment plants.

Case Study 8.3: Separate Treatment of Waste From Chromate Conversion Coating

Background and Objectives. CH2M HILL performed a waste reduction study for a major aerospace manufacturing facility in the western United States. The study identified rinsewater from chromate conversion coating (Iriditing) of aluminum as the main target for waste reduction efforts because it is responsible for the sludge from the entire industrial wastewater treatment plant being classified as hazardous. This rinsewater (approximately 15,000 gal/day) is mixed with other industrial wastes (212,000 gal/day) for treatment at a central industrial wastewater treatment plant.

The U.S. EPA lists "wastewater treatment sludges from the chemical conversion coating of aluminum" as hazardous under the classification F019, specifically because they typically contain hexavalent chromium (RCRA, Appendix VII). In addition to being used in the chromate conversion coating process, hexavalent chromium compounds are used at the facility to deoxidize surfaces (remove oxide surface coating) when preparing parts for chromate conversion coating. Table 8.3 shows the amounts of chromium discharged to the waste treatment plant from the three locations that employ chromate conversion coating at the facility. These data were estimated from composition of the process baths involved and from approximations by company personnel of the production loads on these processes and typical dragout rates.

In 1987, 170 tons of dewatered sludge was produced at the industrial wastewater treatment facility. The waste did not contain toxic concentrations of heavy metals, but because it was a listed waste, it was disposed of as hazardous at a cost of

Table 8.3. Chromium Discharges from Conversion Coating Facilities

Location	Discharge (lb/day)
Chem Mill Facility	0.90
Remote Location A	0.05
Remote Location B	1.01
Total	0.96

$210,000. Thus, less than 1 lb of chromium discharged to the treatment plant resulted in more than 1,000 lb of sludge per day being classified as hazardous. The cost of disposal was a significant incentive for eliminating chromium-containing wastewater from the industrial treatment plant.

Eliminating the F019 classification of the industrial wastewater treatment sludge could be achieved only by attaining zero discharge of rinse waters and baths from the chromate conversion coating and the deoxidation processes. To achieve this objective, a closed-loop rinsewater system would have to be installed, with all residues being hauled off the site for disposal as hazardous waste.

Discussion of Alternatives. Four unit processes were considered necessary for a closed-loop system:

1. rinsewater recycle
2. chemical recovery
3. volume reduction
4. bath purification

Rinsewater recycle is necessary because rinsewater is usually maintained at a low concentration of contamination to effect good rinsing. The existing operation produced approximately 15,000 gal of rinsewater waste per day (125,000 lb) in the process of removing the 1 lb of chromium per day. Alternatives considered for rinsewater cleanup were:

- ion exchange (IX)
- reverse osmosis (RO)
- ion transfer membranes
- electrodialytic processes

Recovering chromium from the rinsewater would eliminate the major toxic constituent from this waste stream and simplify disposal of any blowdown stream. Alternatives considered for process chemical recovery were:

- IX
- electrodialytic processes

Volume reduction was considered necessary for reducing the size of the rinsewater recycle system and the quantity of blowdown required to prevent buildup of salts in the system. Alternatives for volume reduction were:

- innovative rinsing
- evaporation

Bath purification is necessary if process chemicals are to be recovered from the rinsewater and returned to the bath, inasmuch as such systems will return contaminants and useful chemicals. Otherwise, impurities are concentrated in the bath until the bath is no longer functional, necessitating disposal of a large quantity of hazardous material. It makes little sense to recover 1 lb of chromium per day if this recovery necessitates the disposal of a 20,000-gal tank of Iridite solution containing more than 800 lb of chromium. Also, because the concentration of impurities in the process solutions is much higher than that in the rinsewater, bath purification is easier than rinsewater cleanup. Furthermore, cations are removed in the bath and are not carried over to the rinsewater; therefore, the volume of cation regenerant is reduced significantly. Alternatives for purification of chrome-containing process solutions were:

- electrodialytic processes
- porous pot
- IX

The alternative technologies were evaluated for their potential for waste reduction, effects on production, and relative cost. Existing users of the technologies were contacted to discuss their operating experiences. Sites were visited to collect information on the most-promising technologies.

Recommended System. IX was recommended for rinsewater cleanup. IX is an established technology and can also segregate chromic and nitric acid from cations and concentrate them for reuse. Deox solutions contain high concentrations of nitric acid. Reduced rinse flows are accomplished by providing some counterflow rinsing, so that four individual recycle streams are used for the seven tanks. A schematic of the rinsewater recycle system is shown as Figure 8.4. An electrodialytic bath maintenance unit was recommended for the principal deox bath (Figure 8.5). Finally, evaporation was recommended for concentrating IX regenerant solutions, either to enhance recovery or to reduce cost of disposal.

A process flow diagram for the rinsewater treatment system is shown in Figure 8.6. Contaminated rinsewater is passed through a cartridge filter to remove particles that could plug the IX resin. A cartridge filter was recommended because suspended solids loading is low in these acidic rinsewaters. Also, a cartridge filter does not require backwashing, so less wastewater is generated. Cation exchange is used for removing metallic impurities such as trivalent chromium, aluminum, sodium, and zinc. Then an anionic exchange resin is used to remove hexavalent chromium and nitrate ions. Finally, the rinsewater is treated with activated carbon to remove traces of organics that could foul IX resins if allowed to build up in the closed-loop rinsewater system. The carbon is placed after, rather than in front of, the IX resin, because carbon removes hexavalent chromium, reducing the potential for chromium recovery.

The cation exchange resin is regenerated with sulfuric acid, removing the metals and returning the resin to its acid form. The volume of this acidic regenerant solution is reduced by evaporation and disposed of off the site.

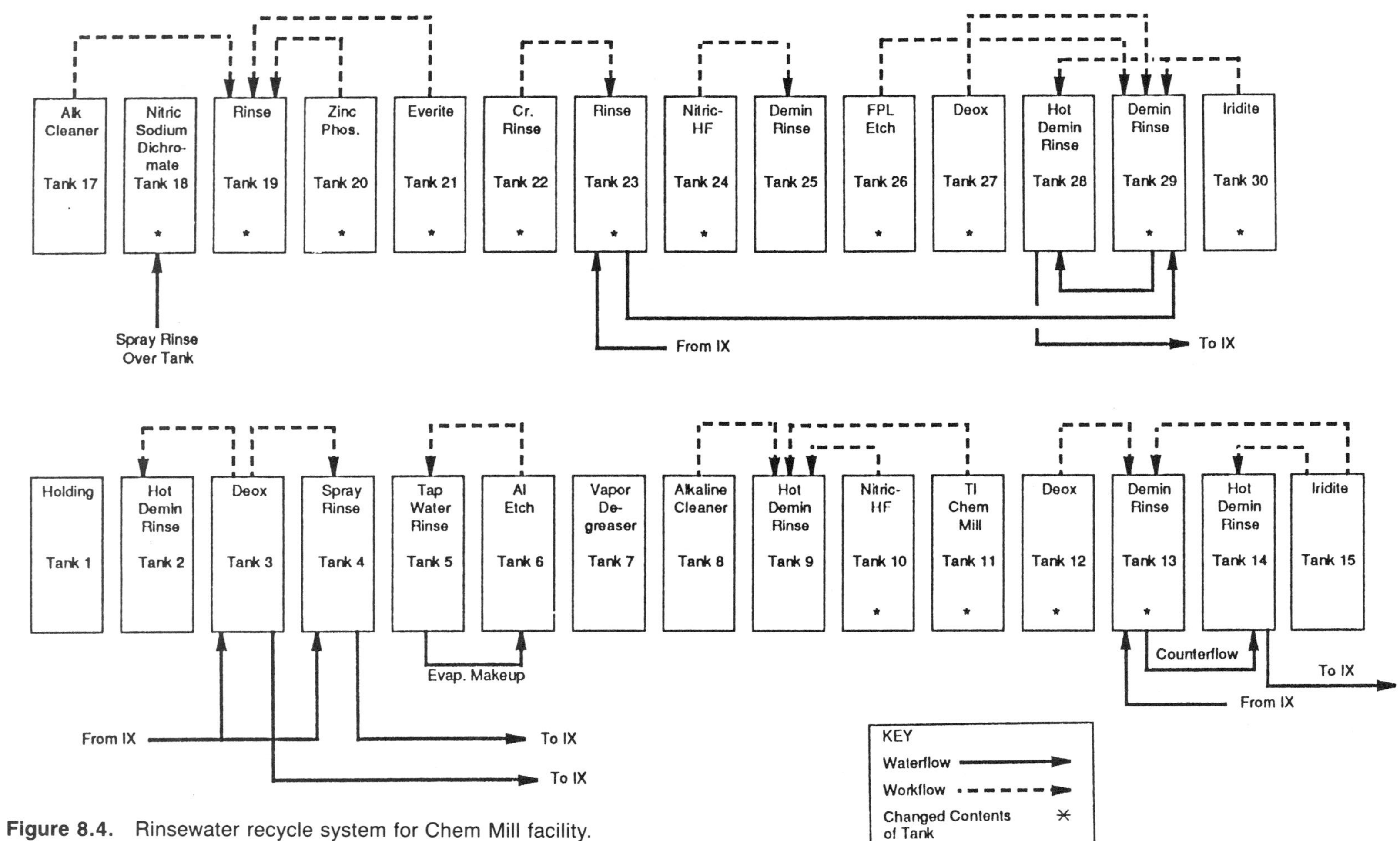

Figure 8.4. Rinsewater recycle system for Chem Mill facility.

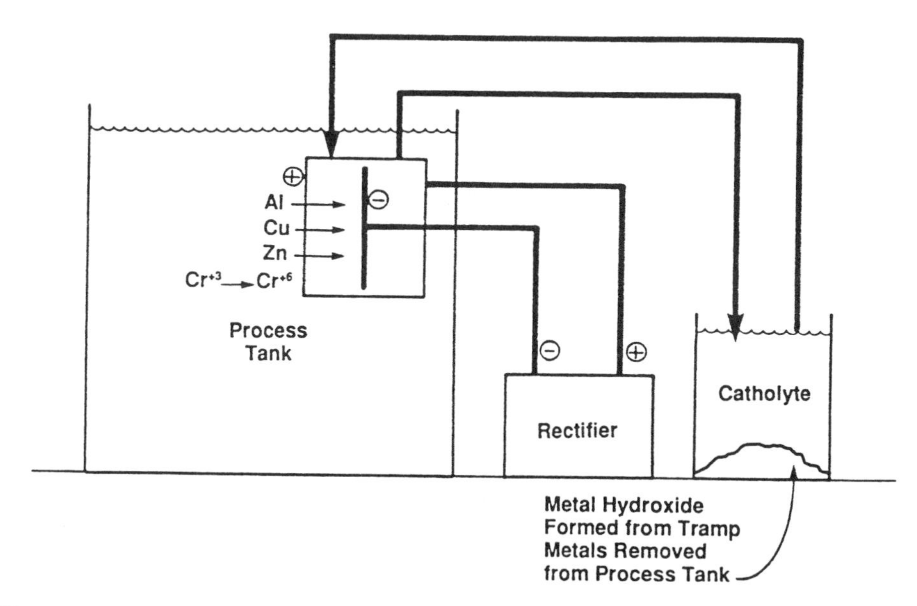

Figure 8.5. Iridite and deoxidizer bath maintenance system.

The anion exchange resin is regenerated with caustic soda (sodium hydroxide). The resulting regenerant solution is a caustic mixture of sodium chromate, sodium nitrate, and sodium hydroxide. This regenerant is then concentrated by evaporation. (A discussion of potential recovery of nitric acid and chromic acid is included later in this chapter.)

Because of the low production of contaminants at Remote Sites A and B, it is not cost-effective to provide these locations with completely independent demineralizer units. It is more cost-effective to provide these locations with portable IX units, which are returned periodically to the Chem Mill facility for regeneration. A typical portable IX unit is shown as Figure 8.7.

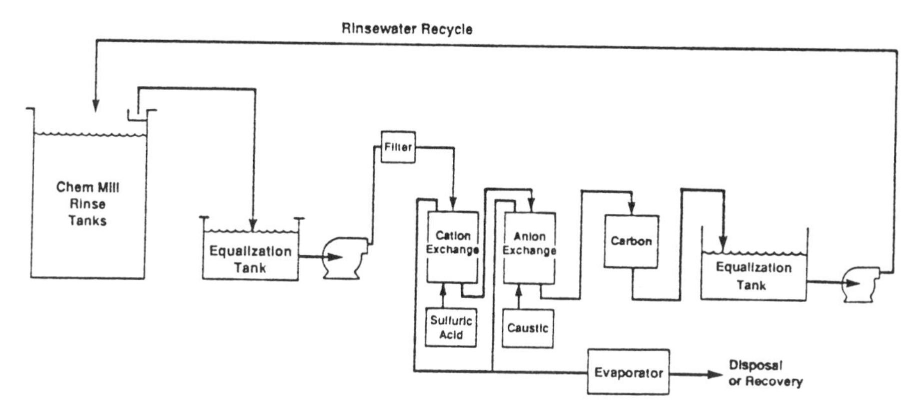

Figure 8.6. Rinsewater recycle system for Chem Mill facility.

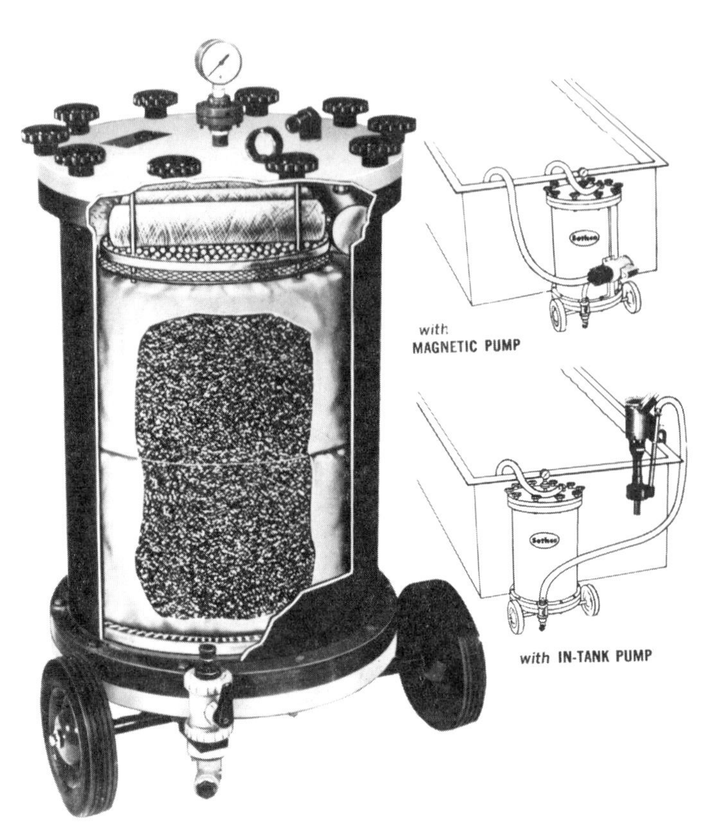

Figure 8.7. Remote ion exchange unit for Chem Mill facility (Photo courtesy of Sethco Manufacturing Corporation, Freeport, New Jersey).

Estimates of Segregated Treatment. Projected capital and operating costs for the IX systems (a central system with regeneration facilities and two remote systems) are provided in Tables 8.4 and 8.5. The cost of installing this system is approximately $294,000 and would result in a savings of $148,300 per year, with a resulting payback period of less than 2 years for this investment.

Installing a bath maintenance system on Chem Mill Tank 3 (deox tank) would cost approximately $16,000. This would decrease cation loads to the demineralizer system by approximately 90% from this source, reducing overall cation loading to the demineralizer system by approximately 80%. The annual savings from installing a bath maintenance system for this tank are provided in Table 8.6. The table shows an annual savings of $4,400, for a projected payback period of approximately 3.6 years.

An additional production benefit not included in this analysis is improved consistency and increased operating life of the process solution, which reduces the need for disposal. The nonquantified production benefits (and low capital cost) were sufficient to convince management to adopt this improvement, despite the relatively long period projected for payback.

Table 8.4. Capital Cost Estimate for Ion Exchange System

Item	Cost (1988 $)
Central IX system	40,000
Cartridge filters with feed pumps	6,600
Discharge pumps (2)	3,000
Equalization tanks (2–1,000 gal FRP)	9,200
Spray rinse storage tank (1,000 gal FRP)	4,500
Carbon adsorption (15 ft^3)	4,500
Remote IX canisters (4 sets)	2,000
Remote pumps (2)	2,000
Hydrostat surge tank (1,500 gal FRP)	6,000
Remote cartridge filters (2)	1,000
Remote carbon adsorption tanks (2)	3,000
Anion exchange resin (18 ft^3 @ $100/$ft^3$)	1,400
Cation exchange resin (18 ft^3 @ $80/$ft^3$)	1,400
Evaporators (2–15 gal/hr capacity)	9,500
Immersion heaters (2)	1,000
Subtotal equipment cost	95,100
Installation labor (640 hr @ $39.70/hr)	120,500
Allowance for nonquantified items (mechanical, 10%; I&C, 5%; electrical, 8%; and miscellaneous, 10% = 33% of constructed cost before contingency)	59,400
Estimated constructed cost	180,000
Contingency (25% of estimated constructed cost)	45,000
Engineering	50,000
ESTIMATED TOTAL INSTALLATION COST	275,000

Table 8.5. Estimated Operating Costs and Payback for Ion Exchange System

Item	Cost (1988 $)
Anion regenerant chemicals (8,000 lb NaOH)	2,400
Cation regenerant chemicals (5,500 lb H_2SO)	400
Power for evaporation ($.10/gal)	10,000
Other power (10 hp)	3,800
Disposal cost (10,000 gal @ $1.00/gal)	10,000
Carbon use (1 change/year)	1,100
Filter cartridge replacement and disposal	2,000
Labor (1,000 hr @ $11.00/hr)	11,000
Annual operating cost	40,700
Reduction in sludge disposal cost	189,000
Annual savings	148,300
System installed cost	275,000
Payback period (years)	1.9

Table 8.6. Estimated Costs Savings Due to Bath Maintenance System

Item	Cost (1988 $)
Regenerant chemicals (4,400 lb H_2SO)	300
Power for bath maintenance (1 kW)	(600)
Evaporation power (12,000 gal)	1,200
Disposal cost (20% reduction)	2,000
Bath chemicals savings	1,500
Total cost savings per year	4,400
Installed cost of system	16,000
Payback period (years)	3.6

RENDERING A HAZARDOUS WASTE NONHAZARDOUS

Disposal costs are considerably higher for disposal in a hazardous waste disposal facility versus disposal in an industrial solid waste disposal facility. Thus, there is a strong economic incentive to treat hazardous waste to render it nonhazardous.

A waste can be rendered nonhazardous by changing the characteristics of the waste that make it hazardous (i.e., corrosivity due to high or low pH can be eliminated by neutralization). An effective method is to recover and recycle the constituent that would render the waste hazardous (see recovery of a useful product later in this chapter). Tying up a hazardous constituent in a nonleachable solid matrix (stabilization or solidification) can permit the mixture to pass the appropriate leaching test and thus be classified as nonhazardous. Another treatment method is destruction of the hazardous constituent in a waste mixture (usually associated with destruction of an organic contaminant in a chemical or thermal reaction).

Case Study 8.4: Mixing Foundry Wastes to Produce a Nonhazardous Waste

Background and Objectives. A casting company operates three foundry facilities. Two of the foundries (Plants A and C) are located on adjacent sites in the city, and the third (Plant B) is located in a nearby town. Each of these plants uses corrosive materials that require special disposal when spent.

Plants A and B use a strong (12 normal) potassium hydroxide (KOH) solution for removing shell material from castings. As the KOH is used, solids build up in the form of a sandy KOH sludge. Spent KOH solutions are pumped out in bulk and sold as a neutralizing agent or disposed of as a caustic liquid waste at a chemical waste disposal facility. Sludge remaining after the liquids are removed is placed in drums for disposal at the chemical waste disposal facility.

Plant C uses a solution of hydrofluoric and nitric acids to chemically mill (Chem Mill) cast titanium parts. Spent acid is neutralized with purchased sodium hydroxide (NaOH), producing a sludge. Neutralized liquids are decanted to a sanitary sewer, and the sludge is disposed of in an industrial landfill. Plant C also uses an NaOH solution to remove shell material from castings. The shell material contains a radioactive constituent that causes the NaOH sludge to be classified as a low-level radioactive waste, thus requiring disposal at a radioactive waste disposal facility. The NaOH liquid is not considered a radioactive waste and is sold to local industry as a neutralizing agent.

The casting company recognized the desirability of using its spent alkaline material to neutralize spent Chem Mill acid waste instead of purchasing virgin sodium hydroxide and then paying to dispose of its spent alkaline materials. However, tests showed that mixing the wastes resulted in precipitation of solids. The facilities had no equipment to deal with the solids; therefore, they had not implemented such a program.

The company retained CH2M HILL to develop a conceptual design of a facility capable of neutralizing its spent acids with its own spent caustic materials and handling the resulting solids. A key objective of this study was to determine if the facility would require a RCRA permit as a treatment, storage, and disposal facility, and whether the resulting neutralized solids could be disposed of in an industrial landfill, instead of a RCRA facility.

Laboratory Testing. Samples of the spent chemicals (and mixtures) were characterized (Table 8.7). Since all of the materials fell outside the pH range of 2 to 12.5, they were considered RCRA hazardous wastes. In addition, three of the samples exceeded the extraction procedure (EP) toxicity limits for certain metals.

Mixing KOH wastes with the Chem Mill acids produced a solid with a pH range of 7 to 11, which passed the EP toxicity test for heavy metals and therefore was suitable for disposal in a conventional landfill.

Regulatory Issues. According to RCRA, materials used as a substitute for a pure commercial product are not considered a solid waste and are therefore exempt from RCRA regulation. The preamble to the regulation that allows this exemption discusses preventing shams; in particular, the use of materials that are not nearly as effective as the commercial product or are so highly contaminated with toxic materials that their use is primarily for commingling and dilution. In this company's application, the KOH wastes are as effective as the commercial products they are replacing and already are being sold as a commercial product; therefore, they should be eligible for exemption.

Process Description. Table 8.8 shows the quantities of chemicals produced at the three facilities and relates the caustic materials to the quantities of acidic waste they can neutralize. As can be seen, an excess of caustic liquids will be available for sale.

Figure 8.8 is a schematic for the neutralization/solidification facility. The system is designed to be batch operated because of the small quantities produced. Batch neutralization systems are also easier to control than continuous systems. NaOH sludge at Plant C will be solidified and will continue to be shipped for disposal as low-level radioactive waste.

Figure 8.9 is a layout of the neutralization facility. Figure 8.10 is a schematic of the neutralization and solids-handling equipment. Table 8.9 lists the major systems components and the basis for their sizing. Table 8.10 shows the production rates for waste products from the facility.

Spent liquid waste is transported to the facility in tanker trucks. Chemical sludge is collected in dumpable containers transported on flatbed trucks. Liquids are pumped from the tanker trucks into storage tanks, and sludge is stored in shipping containers. Radioactive sludge is segregated from other sludges.

KOH liquid is pumped from the storage tank to the mixer and mixed with Chem Mill acid waste until it is neutralized. The resulting thick slurry is collected in a hopper and dewatered using a plate-and-frame filter press. The filtrate is disposed

Table 8.7. Characteristics of Foundry Spent Chemicals

	Plant A KOH		Plant B KOH		Plant C			Combined KOH[c]	Combined KOH[d]	
Characteristic	Sludge	Liquid[a]	Sludge	Liquid	NaOH[b] Sludge	NaOH Liquid	Chem Mill Solution	Sludge and Chem Mill Solution	Liquid and Chem Mill Solution	RCRA Limits[e]
Total Solids (%)	86	–	85	–	65	–	–	–	23	–
Wet Density (g/mL)	2.19	–	2.16	1.51	2.77	–	1.11	1.53	1.19	–
pH	~14	–	~14	~14	~14	~14	~ 1	~11	~ 7	2–12.5
Normality (eq/L)	6.5	–	7.7	11.8	4.2	8.9	3.4	–	–	–
EP Metal Concentration[f] (mg/L)										
Arsenic (As)	<0.005	–	0.310	1.40	–	0.38	4.60	0.28	0.37	5
Barium (Ba)	<1.50	–	0.11	6.77	–	<9.4	2.0	<1.0	<1.3	100
Cadmium (Cd)	0.0008	–	0.0005	0.01	–	1.69[g]	0.150	0.022	0.029	1
Chromium (Cr)	<0.005	–	<0.005	6.77[g]	–	2.81	21.0[g]	<0.10	<0.13	5
Lead (Pb)	<0.125	–	<0.125	6.18[g]	–	<0.09	0.32	<0.10	<0.13	5
Mercury (Hg)	<0.00010	–	0.00028	0.0022	–	0.0015	0.0006	0.0004	0.0005	0.2
Selenium (Se)	0.450	–	0.540	15.1[g]	–	<0.09	<0.1	0.24	0.32	1
Silver (Ag)	0.029	–	0.100	0.10	–	0.094	<0.02	<0.02	<0.03	5

[a]Plant A KOH liquid was not analyzed. It was assumed to be similar to Plant B KOH liquid.
[b]Plant C NaOH sludge was not analyzed for solids or metals because of its radioactive character. Solids determination would have concentrated the radioactive material, and the EP metals determination would not have contributed any significant information.
[c]The Plant A KOH sludge and Plant B KOH sludge were combined at a 5:1 ratio and then neutralized with Chem Mill solution to pH 11.
[d]The metals concentrations are 1.32 times those for the KOH sludge and Chem Mill solution combined. The only difference in these two materials is the sand in the sludge, which was assumed not to contribute any metals.
[e]Limits exist under RCRA for determining characteristics of corrosivity and extraction procedure (EP) toxicity.
[f]Metal concentrations shown for Plant A KOH sludge and Plant B KOH sludge are for the filtrate after neutralization.
[g]Metal concentration is higher than EP toxicity standard.

Table 8.8. Quantities and Balance of Foundry Spent Chemicals

Quantity	Plant A KOH		Plant B KOH		Plant C		
	Sludge	Liquid[a]	Sludge	Liquid	NaOH[b] Sludge	NaOH Liquid	Chem Mill Solution
Batch size (gal)	14	6,000	53	1,670	220	4,800	9,000
Frequency (batches/yr)	780	10.3	780	26	4	9	29.3
Annual amount (gal)	10,800	62,000	41,300	43,400	900	43,200	264,000
Chem Mill neutralization capacity (gal)	10,800	186,000	41,300	130,000	900	130,000	

Acid/Base Balance
Excess caustic liquid: 78,000 gal/yr (available for sale)

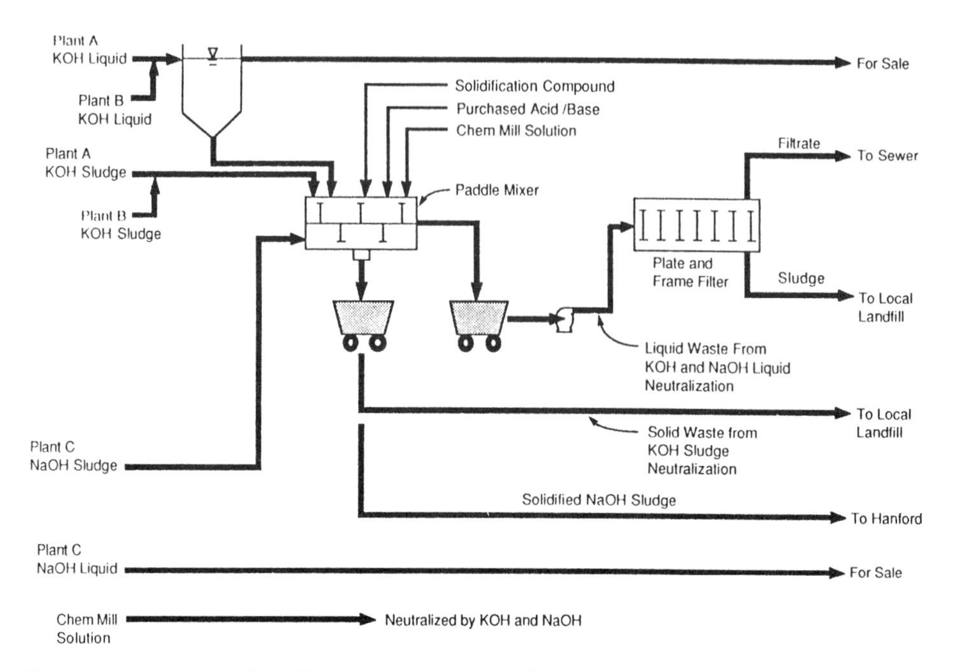

Figure 8.8. Schematic of foundry waste neutralization facility.

of in a sanitary sewer, and the filter cake is dumped into a bin for disposal in a local landfill.

KOH sludge is brought to the mixer with a forklift equipped with rotating grippers for dumping the containers. Chem Mill acid waste is added to the sludge until the mixture is neutralized. When the mixture does not form a thick solid, a solidifi-

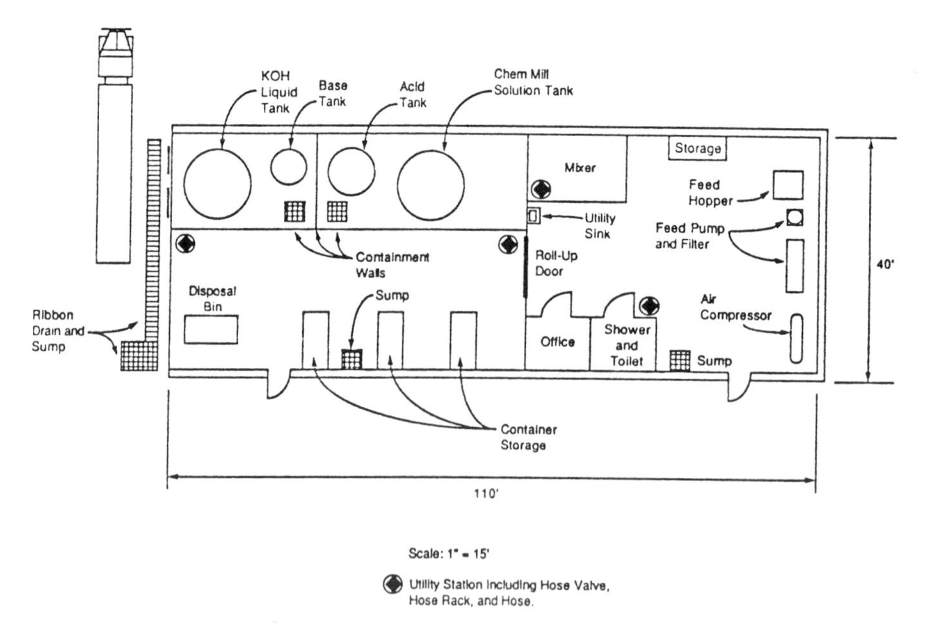

Figure 8.9. Layout of foundry waste neutralization facility.

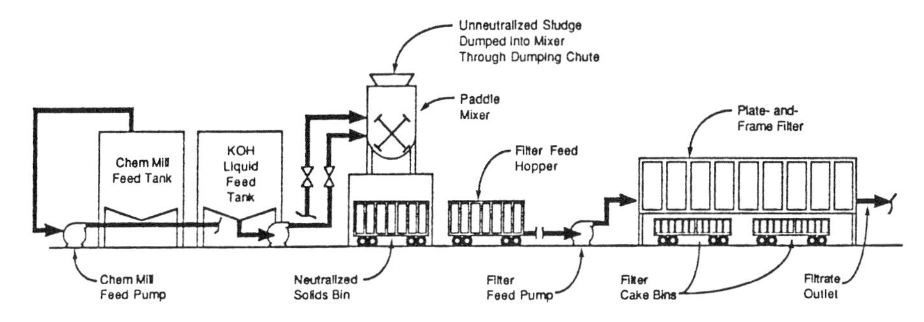

Figure 8.10. Flow schematic for foundry waste neutralization and solids-handling equipment.

Table 8.9. Basis of Design for Foundry Waste Neutralization System

Treatment System Components	Size	Sizing Basis
KOH liquid tank volume	10,000 gal	Tank sized to hold KOH liquid produced in 4-week (maximum batch period) plus 700 gal extra volume
Chem Mill solution tank volume	12,000 gal	Tank sized to hold Chem Mill solution produced in 2-week batch period plus 3,000 gal extra volume
Supplementary acid tank volume	5,000 gal	Tank sized to hold enough 12N supplementary acid to neutralize 2-week production of KOH waste liquid and sludge
Supplementary base tank volume	6,000 gal	Tank sized to hold one batch of NaOH liquid plus 1,200 gal extra volume
Mixer volume	50 ft^3	Mixer sized to hold two 55-gal containers of Plant B sludge, one 55-gal container of Plant A sludge, and the Plant C Chem Mill solution required to neutralize the sludges, with approximately 6 ft^3 of extra volume
Neutralized solid bin volume	50 ft^3	Bin sized to hold one mixer volume
Filter feed hopper	50 ft^3	Hopper sized to hold one mixer volume
Plate and frame filter		
filtering volume	4 ft^3	Filter sized to process the daily production of neutralized KOH liquid in 6 hours
filter area	90 ft^2	
Building area	4,400 ft^2	
Forklift capacity	5,000 lb	

Table 8.10. Production Rates for Waste Products of the Foundry Waste Neutralization System

Waste Products	Rate	Production Rate Basis
Filtrate	950 gal/day	
Maximum fluoride in filtrate	145 lb/day	The amount of fluoride shown assumes that only the Chem Mill solution is used to neutralize the liquid KOH to pH 7. The resulting fluoride concentration in the plant effluent will be approximately 25 mg/L at peak flow. Filtrate concentration will be approximately 18,400 mg/L fluoride.
Filter Cake		
Solids content	60%	
Volume	700 yd^3/yr	
Weight	900 tons/yr	
Neutralized KOH sludge		
Volume	500 yd^3/yr	
Weight	630 tons/yr	
Solidified NaOH Sludge		
Volume	16 barrels/yr	Amounts shown assume the sludges are mixed with water, then solidified with bentonite clay.
Weight	9 tons/yr	

cation agent, such as bentonite, is added to produce the desired consistency. The neutralized sludge is then collected in the hopper and disposed of with the filter cake.

NaOH liquid that is not sold as a commercial product is stored in the supplementary base tank and used as an alternative neutralization agent replacing KOH. NaOH sludge is processed separately from all of the other materials because it is radioactive. After processing NaOH sludge, the equipment is cleaned to prevent contamination of processed materials.

Capital and Operating Costs. Table 8.11 shows a budget-level estimate of approximately $700,000 (in 1986 dollars) for the construction costs, including engineering, of the neutralization facility. Table 8.12 shows annual operating costs for the system and compares those costs with the annual costs of treating and disposing of spent chemicals. As shown in the table, the estimated annual operating cost of the neutralization system was $179,000, for an annual savings of $137,000, which would result in a 5-year payback of the initial investment.

Incineration to Reduce Volume, Toxicity, and Mobility

The only method that simultaneously reduces volume, toxicity, and mobility is burning in incinerators, boilers, or power plants or supplementing fuel used in cement kilns or light aggregate kilns. When a waste is burned (yielding ash or exhaust gas), its chemical makeup is altered and its toxicity is generally reduced.

Table 8.11. Budget-Level Cost Estimates for Foundry Waste Neutralization Facility

Basic Equipment	Size	Cost ($)
Paddle mixer[a]	50 ft^3	55,000
Plate and frame filter[b]	4.1 ft^3	15,000
Tanks[c]		
Chem Mill solution	12,000 gal	24,000
KOH liquid	10,000 gal	9,000
Supplementary acid	5,000 gal	10,000
Supplementary base	6,000 gal	5,000
Forklift attachment[d]		6,000
Process pumps[e]		8,000
Sump pumps[f]		35,000
Miscellaneous equipment[g]		36,000
Equipment installation[h]		33,000
TOTAL BASIC EQUIPMENT		236,000
Site work and building		3,000
Site preparation		11,000
Yard piping[i]		3,000
Roads and walks		3,000
Site improvements		21,000
Concrete		21,000
Metal building (4,400 ft^2)[j]		73,000
Miscellaneous building appurtenances[k]		25,000
Miscellaneous metals		5,000
TOTAL SITEWORK AND BUILDING		144,000
Process piping		18,000
Electrical (approx. 10%)		40,000
Instrumentation (approx. 10%)		40,000
Mobilization/demobilization (approx. 5%)		20,000
SUBTOTAL		498,000
Contingency (20%)		100,000
TOTAL CONSTRUCTION COST		598,000
Estimated Engineering Services Cost		
Geotechnical investigation		7,500
Engineering design		73,600
Services during construction		20,000
TOTAL ENGINEERING SERVICES COST		$101,100
TOTAL FACILITY CAPITAL COST		$699,100

[a]The paddle mixer is constructed of stainless steel without a cooling jacket. It will be equipped with a 30-hp motor.
[b]The plate-and-frame filter press is manual-type hydraulically operated and equipped with an air manifold, cake carts, and feed pump.
[c]The Chem Mill solution and acid tanks will be constructed of FRP, and the KOH and base tanks will be constructed of steel.
[d]A forklift equipped with rotating forks will be purchased to move hoppers, bins, and barrels of material.
[e]Air diaphragm pumps are used for corrosion resistance and equipment uniformity. An air compressor will be required for operation of these pumps.
[f]Sump pumps will be installed in the sumps with special pH control loops and level switches.
[g]Miscellaneous equipment includes the hoppers, bins, air compressor, and other minor items.
[h]Equipment installation is estimated at 15% of the equipment cost.
[i]Water and sanitary sewer piping will be brought to the structure from a street approximately 250 ft from the building.
[j]The metal building will enclose only the processing area. The storage area will be provided.
[k]HVAC equipment will include an exhaust fan for the mixer, but not a fume scrubbing system. Only the office and restroom areas will be heated or air conditioned. The building will be ventilated, and freeze protection will be included as required.

Table 8.12. Cost Savings of Proposed Foundry Waste Neutralization Facility

Item	Annual Cost ($) Current System	Annual Cost[a] ($) Proposed System
Labor[b]	–	45,000
Maintenance[c]	–	25,000
Forklift rental	–	15,000
Materials[d]	–	19,000
Hauling	58,000	23,000[e]
Disposal charge	258,000	36,000[e]
Utilities	–	16,000
TOTAL	316,000	179,000

[a]Costs based on material quantities provided by client in December 1986.
[b]Assumes 1½ operators to run the system at $30,000 per year per operator.
[c]Assumes 10 percent of original equipment cost per year.
[d]Includes 40,000 gallons of supplementary acid and 40,000 pounds of solidifying compound per year.
[e]Assumes disposal of filter cake and neutralized KOH sludge at industrial landfill and NaOH sludge at a low-level radioactive waste landfill.

Volume is reduced by evaporation of water, and simple organics in the wastes are oxidized to water and carbon dioxide. Mobility is reduced because only inorganic metals and silicates remain in the ash, which is a solid with minimum volume. The other product of burning is exhaust gas, which contains water vapor, carbon dioxide, and the excess air required to support the combustion process. Exhaust gas still containing entrained particulates, acid gases, and any organics that did not burn will require extensive further treatment to remove these particulates and acid gases and to complete the combustion of the organics. The following description focuses on incineration as a waste management technique. The principles, however, also apply to boilers, industrial furnaces, or light aggregate kilns.

Feed Storage and Metering. The feed to an incinerator must be collected and controlled to remain within the allowable limits of the combustion device. For instance, too much water in the feed stream will cool the incinerator, and the waste will not burn completely. Chlorinated organic compounds fed to an incinerator must also be controlled, because chlorine will form hydrochloric acid (HCl). HCl can cause corrosion problems, but it can readily be removed from the gas stream either by water absorption or by reaction with an alkaline scrubbing solution. Other important feed characteristics to be considered are:

- ash content
- heating value or heat of combustion
- sulfur content
- phosphorus content
- nitrogen content
- physical state (solid, liquid, or gas)
- melting point

- boiling point
- particle size distribution of solids
- packaging (drums, pails, bulk)
- metals content

An incinerator can be designed to handle wastes with a wide range of physical properties. Once designed and constructed, however, the feeds to an incinerator must be kept within a particular operating range to ensure adequate incinerator performance.

After the requisite feed control is established, the feed materials should be collected and stored, at least temporarily. Liquids can be collected and stored in tanks. These tanks can also provide surge capacity to allow for some mixing and blending to meet the feed-control criteria. Liquid wastes are then pumped to the incinerator at a controlled rate to be atomized and burned. Flow-control valves are typically used to control the feed rate.

Solids can be collected in drums, in a pit, or on a tipping floor. Solids must be inspected either on the receiving floor or in the delivery vehicle to ensure that they can be safely and effectively burned in the incinerator. Drums of solids can be loaded onto conveyors and periodically charged into the incinerator. Bulk solids in pits can be picked up with cranes and loaded into hoppers to be charged at a controlled rate. Solids on tipping floors can be picked up with a front end loader and loaded into hoppers to be charged to the incinerator. A metering conveyor with a variable-speed drive can be used to control the amount of material fed to the incinerator.

Gases can be fed to the incinerator directly from the transport vessel or through a pipeline. Gas flow rate can be controlled by a flow-control value.

Combustion Control. The various feeds to the incinerator generally react according to the following formula:

$$\text{Combustibles} + \text{Air} + \text{Water} = \text{Ash} + \text{Flue Gases} + \text{Water Vapor} + \text{Excess Air}$$

and

$$\text{Heat Input} + \text{Heat Generated by Combustion} = \text{Heat Out}$$

The operator of the incinerator can control a few variables around the combustor. These typically are:

- feed rate of waste
- excess air
- temperature inside the incinerator

These variables are interactive. Increasing the feed rate of waste at a given total air-flow rate will result in a higher temperature. Increasing the excess air at a constant waste-feed rate will reduce temperature. The effects of these variables can be compensated for by adding an auxiliary fuel. By supplementing the fuel, the feed rate and air rate are set and the temperature increased above that which can be achieved by combustion of the waste alone. The auxiliary fuel is then added to create additional heat and raise the temperature of the incinerator to the desired set point. As feed rate of waste, heating value of the waste, and air rates vary about the nominal condition, the flow of auxiliary fuel is varied to maintain the constant temperature required to destroy toxic organic compounds.

Once the balance of feed rates, air rates, and temperature is achieved, an incinerator operates at steady state, burning away the combustibles in the wastes, reducing the volume of the ash to a minimum, and destroying the organic compounds contained in the waste.

Treatment of Flue Gases. The gases created in the incinerator will require treatment to remove particulates, acid gases, and trace organic compounds that may remain from the incineration process. Trace organic compounds are further oxidized either by providing sufficient reaction time at an adequate temperature in the primary combustor or by passing the flue gas exiting the primary combustion chamber through an afterburner to complete the combustion process.

After combustion, the gas stream is quenched to reduce its temperature, and particles are removed. Several technologies are used in incinerator installations, including:

- venturi scrubbers
- inertial-impact scrubbers
- baghouse filters
- electrostatic precipitators
- plate-scrubbing towers

Acid gases can be removed either by absorption in water to generate an aqueous acid stream or by reaction with an alkaline scrubbing solution to generate a salt stream. Acid gas can be absorbed using:

- venturi scrubbers
- packed-tower or tray scrubbers
- multistage absorption towers

All of these technologies have been applied to incinerators.

Scrubbing systems can be either wet or dry. Wet systems generate a wastewater stream, which must be treated to remove suspended solids and perhaps adjust the pH before being discharged. Dry scrubbers generate a dry solid waste, which can be disposed of with the ash from the incinerator, although the latter systems require closer control and more operator attention.

RECOVERY OF A USEFUL PRODUCT

Many treatment technologies concentrate waste materials into smaller volumes or convert wastes into less mobile forms (such as metal hydroxide precipitation from wastewater). Useful products can be recovered from waste streams. In some cases, that product is a raw material that can be used as feed stock. In other cases, the recovered product is energy, such as when wastes are used to fuel a power-generation facility.

Case Study 8.5: Recovery of Copper at a Printed-Circuit-Board Facility

Background and Objectives. An aerospace manufacturer operates a circuit-board fabrication facility in New England. CH2M HILL was contracted to design a waste treatment plant for this facility. The principal objective was to minimize the volume of sludge generated by the treatment plant.

Effluent from the treatment plant discharges to a publicly owned treatment works. The effluent is required to comply with federal and local pretreatment discharge limits. Wastes treated in the facility consist mainly of continuous overflow rinses from scrubbing, cleaning, electroless plating of copper, electroplating of copper, photoresist processes, and etching. These wastes contain a mixture of regulated metals (principally copper and lead) and complexing agents from electroless copper plating and etching processes.

The current maximum flow is approximately 90 gal/min. The treatment facility is designed for 50 gal/min with a maximum hydraulic capacity of 90 gal/min. The presence of complexing agents renders conventional hydroxide precipitation treatment ineffective. One effective technique for treating this type of waste has been the addition of large doses of ferrous iron, which displaces copper and lead from their complexes, allowing their precipitation as metal hydroxides. This process, however, produces voluminous sludge because of the precipitation of large quantities of ferrous hydroxide.

Process Description. To minimize hazardous waste, CH2M HILL recommended treating the combined rinsewater waste by chelating IX, and recovering elemental copper from the spent regenerant solution using electrowinning. A process flow diagram for the treatment system is shown as Figure 8.11.

Rinsewaters are pumped directly to a pH adjustment tank. Concentrated acidic and caustic wastes are pumped to holding tanks to be metered into the mixing tank to prevent slug loading of acids, alkalis, or metals to the treatment system. In the pH adjustment tank, the waste is adjusted to a pH between 4 to 5 using sulfuric acid or caustic soda (sodium hydroxide), because this pH range is optimal for selective IX removal for copper.

The waste is then pumped through duplex cartridge filters. It is estimated that weekly replacement of filters will be required. The effluent from the filter passes through two activated carbon vessels in series. Activated carbon adsorption removes organic compounds that could foul the IX resins, reducing the resins' capacity and resulting in more frequent replacement. Two carbon adsorption columns are provided in series, thus allowing complete use of the carbon in the first vessel, with the second providing polishing. Following complete breakthrough of organic compounds from the first carbon column, that carbon should be replaced and the roles of the two vessels reversed. It is estimated that annual replacement of carbon will be required. The carbon vessels will be backwashable, with the backwash water to be returned to the initial pH adjustment tank.

The wastewater then flows through dual IX columns, operated in series, using a chelating cationic IX resin. Following exhaustion, the lead IX column resin is regenerated with sulfuric acid, rinsed with city water, and returned to service as the lag, or

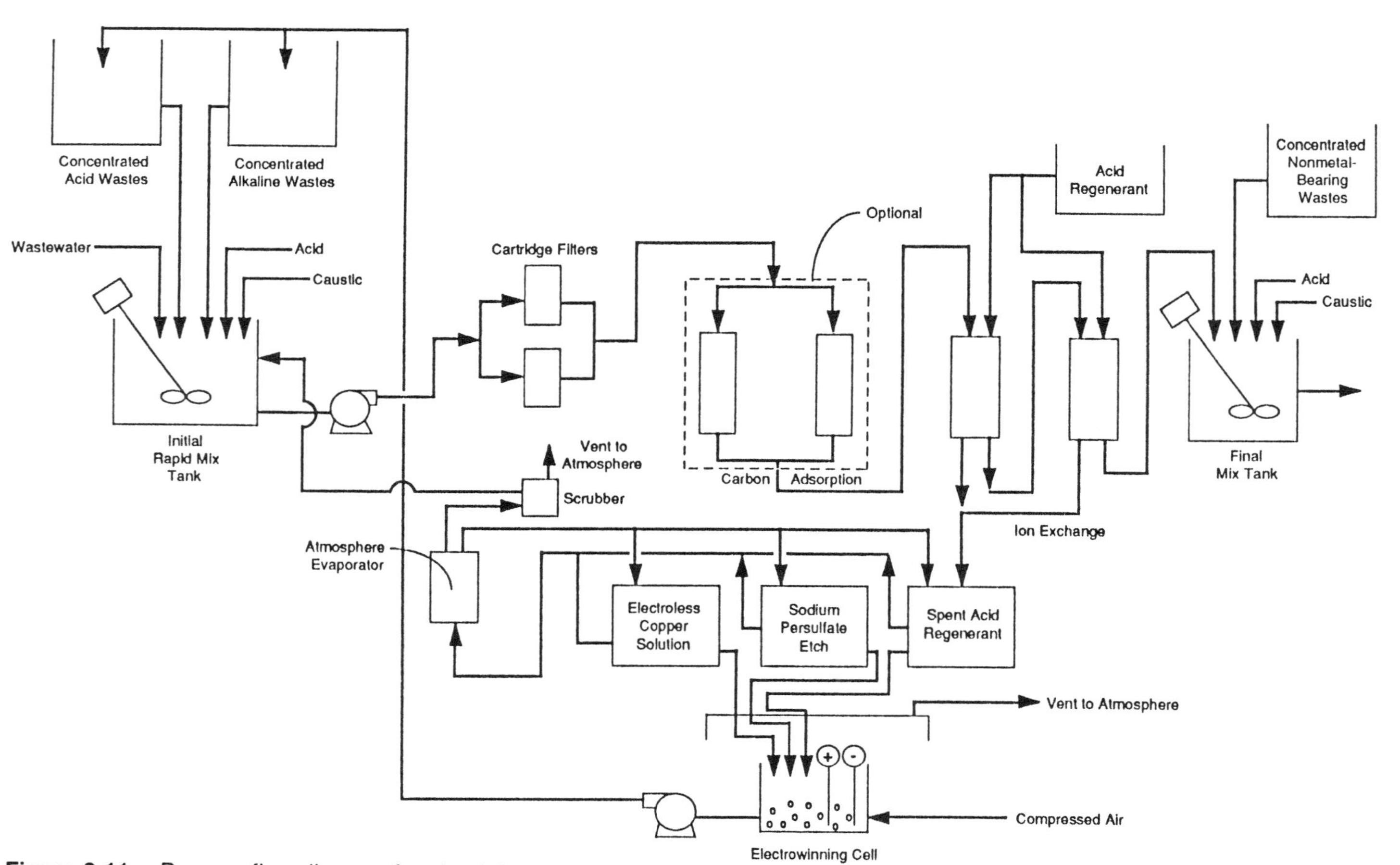

Figure 8.11. Process flow diagram for circuit-board waste treatment.

polishing, unit. The rinsewaters are returned to the rapid-mix tank for treatment. The waste regenerant is piped to a holding tank for copper recovery by electrowinning.

The effluent from the IX units discharges to the final rapid-mix tank for pH adjustment with sulfuric acid or caustic soda before discharge to the city sewer.

The spent IX regenerant, electroless copper plating bath growth, and sodium persulfate etching solutions are stored in separate holding tanks. These solutions are pumped to a low-surface-area (LSA) parallel-slate, electrowinning unit for recovery of copper by electroplating onto flat stainless steel plates. The electrowinning unit is batch operated. After a batch is treated, the effluent is pumped to the acid or alkaline holding tank and bled into the influent pH adjustment tank for additional treatment. This effluent is expected to contain 1 or 2 grams of copper per liter.

A high-surface-area, high-mass-transfer (HMT) electrowinning system is being considered for use instead of an LSA system. The advantage of an HMT system is its reported ability to reduce effluent copper concentrations to a few milligrams per liter. This equipment would reduce the load of copper on the IX system resulting in less frequent regeneration. Disadvantages of the HMT are increased power consumption and lower rates of metal recovery (resulting in longer electrowinning times).

Another alternative to HMT electrowinning is an atmospheric evaporator, which would increase the concentration of metal in the IX regenerant prior to LSA electrowinning. Benefits include volume reduction, improved electrowinning efficiency from increased metal concentration, resulting in reduced frequency of IX regeneration, and potential reuse of IX regeneration acid following dilution to the original acid concentration. This reduction in volume benefits operation of the system in that fewer batches are required to be electrowinned. The required electrowinning cathode area and the time for electrowinning are unchanged because they are dependent on the mass of metal to be removed rather than hydraulic volume throughput. Disadvantages of evaporation include the requirement for heating an acidic solution, safety problems associated with handling a heated concentrated acid, and additional power and maintenance requirements.

Cost Estimates. Two systems were considered for installation at this facility: LSA electrowinning with an atmospheric evaporator and HMT electrowinning. Order-of-magnitude cost estimates for installing the two systems are provided in Table 8.13. Operating costs are listed in Table 8.14.

Thermal Treatment for Resource or Energy Recovery. Thermal processes can be used for recovering minerals from waste streams and for recovering energy in the form of steam or electricity.

Chlorinated organic compounds, when burned, will generate HCl gas and a small amount of chlorine gas. The HCl can be absorbed in a multistage absorption tower to manufacture hydrochloric acid at strengths varying from 6 to 24% HCl. These acid streams have been used for pickling acid in steel manufacture.

Waste containing more than about 20% sulfur can be burned. The sulfur dioxide can then be catalytically oxidized to sulfur trioxide and absorbed to manufacture sulfuric acid. An alternative process can be used to reduce the sulfur dioxide to make elemental sulfur.

The previous examples are of well-established industrial processes that have been modified to recover commercial products from wastes. The use of these methods has

Table 8.13. Order-of-Magnitude Capital Cost Estimates for Circuit-Board Treatment Facility

Category	Items	LSA Elect. Alternative	HMT Elect. Alternative
EQUIPMENT			
Tanks		$ 50,200	$ 50,200
Mixers		13,200	13,200
Pumps		47,300	47,300
Duplex cartridge filters		10,000	10,000
Activated carbon system		40,000	40,000
Ion exchange system		109,000	109,000
Electrowinning system		58,300	88,300
Atmospheric evaporator		4,000	0
Evaporator heaters		3,000	0
Evaporator scrubber		4,000	0
EQUIPMENT TOTAL		$ 339,000	$ 358,000
EQUIPMENT INSTALLATION		$ 168,000	$ 156,000
BUILDING MODIFICATIONS		131,000	131,000
PIPING		81,000	81,000
SUBTOTAL		$ 719,000	$ 726,000
ALLOWANCES FOR NONQUANTIFIED ITEMS			
Mobilization, bonding, insurance	5%	$ 36,000	$ 36,000
Painting	2%	14,000	15,000
Electrical, I&C	19%	137,000	138,000
SUBTOTAL		$ 906,000	$ 915,000
CONTINGENCY	25%	$ 227,000	$ 229,000
SUBTOTAL		$1,133,000	$1,144,000
AREA ADJUSTMENT	15%	170,000	172,000
TOTAL ORDER-OF-MAGNITUDE ESTIMATE		$1,303,000	$1,316,000

Table 8.14. Operating Cost Estimates for Circuit-Board Treatment Facility

Items	LSA Elect. Alternative	HMT Elect. Alternative
Labor (2,080 hr/yr @ $50/hr)	$104,000	$104,000
Chemicals	8,000	6,500
Electricity	8,000	9,000
Filters (52 sets/yr @ $38)	2,000	2,000
Activated carbon	1,600	1,600
Ion exchange resin	4,000	4,000
Maintenance	20,000	20,000
TOTAL ORDER-OF-MAGNITUDE ESTIMATE	$147,600	$147,100

been limited, thus far, to situations where the waste streams are generated in a chemical production plant, and the furnace has been added to the production process for waste treatment or air pollution control.

Recently, work has been done on the feasibility of treating ash produced by solids incinerators to recover iron from the ash. These methods have not yet proved fruitful because of the low price of iron. However, reclaiming noble metals from the manufacture of printed circuit boards and electronic parts is being actively considered as an alternative to disposal. As with all recycling/reclamation projects, the attractiveness depends on the value of the reclaimed product and the cost of reclamation.

Recovering energy in the form of steam and/or electricity from a thermal treatment process has been much more widespread. The exhaust gases from the furnace are cooled in a boiler, and the heat is used to boil water under pressure. The typical limit for steam generation in these applications is less than 400 pounds per square inch gauge (psig), and 150- to 200-psig steam with nominal superheat is common practice. The steam can be exported to nearby consumers or used to drive turbine-generators to generate electricity. In addition, the steam can be used to chill water through absorption refrigeration units to provide cooling for nearby consumers.

REFERENCES

1. Higgins, T. E., and S. Termaath, "Treatment of Plating Wastewaters by Ferrous Reduction, Sulfide Precipitation, Coagulation and Upflow Filtration," in *Proceedings of the 36th Purdue Industrial Waste Conference* (Ann Arbor, MI: Ann Arbor Science Publishers, Inc., 1982), pp. 462–471.
2. "A New Waste-Treatment Method is Saving the US Air Force $1,000/day," *Chem. Eng.* (April 28, 1986), pp. 10–11.
3. Roberts, R. M., J. L. Koff, and L. A. Karr, "Hazardous Waste Minimization Initiation Decision Report," Technical Memorandum 71-86-03, Naval Civil Engineering Laboratory, Port Hueneme, CA (January 1988).

CHAPTER 9

Economic Evaluation in Pollution Prevention Programs

*Mike Matichich and Dikran Kashkashian**

INTRODUCTION

During the past 10 years, many commercial and industrial companies have established pollution prevention programs with the goals of reducing costs and liabilities and improving compliance with state and federal regulations. The processes for evaluating projects under these programs vary significantly from company to company and from industry to industry. In some cases, deliberate and extensive studies have been conducted of waste generation processes and waste reduction options; these studies often consider the effects of pollution prevention programs on waste generation and operations in the company. In other cases, "paper" programs have been developed to simply comply with regulatory requirements.

The purpose of this chapter is to provide the economic tools needed by companies to make decisions about funding pollution prevention projects. In addition to aiding companies in considering options, a better understanding of the dynamics of industrial decisionmaking in this area may be of use to legislators and regulators interested in developing guidelines, regulations, or incentive programs to accomplish pollution prevention objectives.

COMPANY CHARACTERISTICS

The free-market approach to achieving pollution prevention that is recommended throughout this book will result in companies considering different types of pollution prevention projects on the basis of such factors as:

- the size of the company
- the economic condition of the company (e.g., profitability)
- the geographic reach of the company (e.g., does it extend across state or other jurisdictional boundaries)

*Mr. Matichich and Dr. Kashkashian are economists in CH2M HILL's office in Reston, Virginia, and can be reached at (703) 471-1441.

- the economic condition of the company's industry
  - —Is the market for the industry's products increasing or decreasing?
  - —Is the industry facing significant technological change?
  - —Are there many new entrants into the industry (e.g., the microcomputer industry)?
  - —Is the industry experiencing or expecting to experience substantial price competition (e.g., the airline industry)?
- the degree of regulation of the company by local, state, and federal agencies
- the company's internal organizational structure

THE IMPORTANCE OF COMPANY GOALS

Variations in the factors listed above will lead to differing goals and objectives for pollution prevention programs; these goals can have a significant effect on the projects that are considered and how they are evaluated by the company. For some companies, the primary objective is to minimize the amount of money spent on waste management programs in the near term; others are looking for long-term cost reductions. For yet other companies, the goal is to reduce the total amount of waste being generated. Other goals include achieving environmental objectives in order to be good corporate citizens, improving health and safety for their employees, and reducing corporate liabilities.

The company's goals and priorities will affect the rating of certain types of pollution prevention projects (e.g., housekeeping improvements over process modifications, recycling rather than treatment). They also will influence the level and timing of corporate funding for implementing these projects.

CORPORATE DECISION MODELS

The various corporate goals identified above are not mutually exclusive. However, the primary goal or goals are likely to control the approach taken to evaluate and implement projects. A review of corporate practices has identified five primary sets of goals and related decision models. The remainder of this chapter will describe the context and resulting implications of the five models, which are based on the following goals:

- Minimize near-term costs.
- Minimize long-term costs.
- Reduce total waste.
- Limit total expenditures to a fixed amount.
- Achieve other corporate objectives (e.g., public image, reduced liability).

Tools needed to evaluate projects on the basis of these goals will also be discussed.

Table 9.1 provides a summary of the typical characteristics, sample industries, and analysis and decision frameworks for each of these decision models. The table

Table 9.1. Summary of Decision Models

Decision Model	Typical Characteristics of Firms Likely to Employ This Model	Sample Firms, Industries	Typical Analysis and Decision Frameworks
1. Minimize near-term cost	Highly competitive industry Poor recent firm financial performance Poor near-term profitability prospects Emerging technologies and industries	Computer, electronics	Payback analysis Internal rate of return (short time-frame) Present value of costs/ benefits (short time-frame)
2. Minimize long-term costs	Moderately competitive industry Established firms Industry leaders Favorable recent financial performance Favorable financial prospects Established technologies and industries	Established commodities, such as electric utilities and oil refineries	Payback analysis Internal rate of return (longer time-frame) Present value of costs/ benefits (longer time-frame) Artificial intelligence modeling
3. Reduce total waste	Highly visible waste Waste disposal a significant part of operating expenses Locating in highly regulated jurisdictions	Chemical industry	Analyses of the effectiveness of various technologies Optimization models related to waste reduction Net present value of projected waste streams over time
4. Limit total expenditures to a fixed amount	Near-term cost conscious (see characteristics for Model 1) OR Interested in making substantial investment to achieve other objectives (e.g., environmental leadership)	Government-owned, such as TVA	Benefit cost analysis Cost-effectiveness analysis
5. Achieve other corporate objectives	Variable (see text discussion)	Chemical industry	Analysis will depend on goals established (see text discussion)

and figures presented in the next section illustrate how alternatives for a sample project might be ranked for these differing models. Each of these models can range in complexity, from simplified to highly sophisticated.

Model 1: Minimize Near-Term Costs

One corporate decision model that is currently used in many highly competitive industries focuses on reducing near-term costs and maximizing near-term income. A current example is the computer industry, where highly competitive pricing practices and other factors have resulted in minimum profit margins for many companies.

Tools and Analytic Framework

For companies where cost minimization is the primary goal (or one of a few primary goals), the tools employed in analyzing waste reduction options would consist of standard economic analysis tools, such as the present value of costs and benefits, internal rate of return, and payback analysis.

Net present value and internal rate of return are both techniques in which cash flows are discounted to reflect the time value of money. For example, in net present value analysis, with a 10% discount rate, a dollar spent 10 years from today would count as 39 cents in the analysis. Payback analysis is the calculation of the time required, usually measured in years, to recoup the company's initial investment in a project. It is calculated simply by dividing the investment by the annual net savings accrued as a result of the project. No consideration is made for interest rates or the time value of money.

The costs included in the analysis would depend on such factors as the industry and size of company. Capital costs frequently include such items as site development, process equipment, construction, and indirect capital costs. Operation and maintenance (O&M) costs would include such items as changes in utilities costs, changes in O&M labor, changes in O&M supplies, and changes in overhead costs. Tables 9.2 and 9.3 provide listings of the factors that can be considered in estimating the capital and O&M costs of a pollution prevention project. Tables 9.4 and 9.5 are "fill-in-the-blank" worksheets that are useful in tabulating these costs for a particular project. Table 9.6 is a worksheet for calculating net present value and internal rate of return for a project. This table includes consideration of items such as depreciation and income taxes. These items may be ignored, as in our case study below, for preliminary or more simplified analyses.

Key considerations for companies with a primary goal of reducing near-term costs are the accuracy of cost estimates being used in the analysis and the uncertainties about future costs. Estimating the performance reliability (effectiveness) of proposed pollution prevention programs and estimating the effect of these programs on the company's industrial processes are also important.

In performing these economic analyses, the company must make explicit assumptions about the time value of money (i.e., the discount rate to use in the analyses of net present value and internal rate of return). In addition, assumptions must be made about the costs and benefits of competing projects and the operating margin available for project expenditures, in light of such factors as the anticipated price competition with other companies in the industry.

Table 9.2. Capital Costs of a Typical Large Pollution Prevention Project

Direct Capital Costs

Site development

- Demolition and alteration work
- Site clearing and grading
- Walkways, roads, and fencing

Process equipment

- All equipment listed on flow sheets
- Spare parts
- Taxes, freight, insurance, and duties

Materials

- Piping and ductwork
- Insulation
- Coatings and painting
- Electrical
- Instrumentation and controls
- Buildings and structures

Connections to existing utilities and services (water, HVAC, power, steam, refrigeration, fuels, plant air and inert gas, lighting, and fire control)

New utility and service facilities (same items as above)

Other non-process equipment

Construction/installation

- Construction/installation labor salaries and burden
- Supervision, accounting, timekeeping, purchasing, safety, and expediting
- Temporary facilities
- Construction tools and equipment
- Taxes and insurance
- Building permits, field tests, and licenses

Indirect Capital Costs

- In-house engineering, procurement, and other home office costs
- Outside engineering, design, and consulting services
- Permitting costs
- Contractors' fees
- Startup costs
- Training costs
- Contingency
- Interest accrued during construction

Total Fixed Capital Costs

Working capital

- Raw materials inventory
- Finished product inventory
- Materials and supplies

Total Working Capital

Total Capital Investment

Source: Adapted from Perry, *Chemical Engineer's Handbook* (1985) and Peters and Timmerhaus, *Plant Design and Economics for Chemical Engineers* (1980).

Table 9.3. O&M Costs and Savings of a Typical Large Pollution Prevention Project

Reduced waste management costs
This includes reductions in costs for:
- Offsite treatment, storage, and disposal fees
- State fees and taxes on hazardous waste generators
- Transportation costs
- Onsite treatment, storage, and handling costs
- Permitting, reporting, and recordkeeping costs

Input material cost savings
An option that reduces waste usually decreases the demand for input materials.

Insurance and liability savings
A waste reduction option may be significant enough to reduce a company's insurance payments. It also may lower a company's potential liability associated with remedial cleanup of TSDFs and workplace safety. (The magnitude of liability savings is difficult to determine.)

Changes in costs associated with quality
A waste reduction option may have a positive or negative effect on product quality. This could result in higher (or lower) costs for rework, scrap, or quality control functions.

Changes in utilities costs
Utilities costs may increase or decrease. This includes steam, electricity, process and cooling water, plant air, refrigeration, or inert gas.

Changes in operating and maintenance labor, burden, and benefits.
An option may either increase or decrease labor requirements. This may be reflected in changes in overtime hours or in changes in the number of employees. When direct labor costs change, then the burden and benefit costs also will change. In large projects, supervision costs also will change.

Changes in operating and maintenance supplies
An option may increase or decrease the use of O&M supplies.

Changes in overhead costs
Large projects may affect a facility's overhead costs.

Changes in revenues from increased (or decreased) production
An option may result in an increase in the productivity of a unit. This will result in a change in revenues. (Note that operating costs also may change accordingly.)

Increased revenues from by-products
An option may produce a by-product that can be sold to a recycler or sold to another company as a raw material. This will increase the company's revenue.

Source: U.S. Environmental Protection Agency, *Waste Minimization Opportunity Assessment Manual* (EPA 1625/7–88/003) July 1988.

Table 9.4. Capital Cost Worksheet for Pollution Prevention Project

Categories/Items	Item Cost ($)	Category Cost ($)
Delivered Process Equipment Cost (itemize)		
Initial spare parts		
Installation cost		
Total Equipment Installed Cost		
Construction Costs		
Contractor-supplied equipment (itemize)		
Piping & mechanical		
Electrical		
Instrumentation & controls		
Structural		
Demolition		
Site clearing & grading		
Walkways, paving, drainage & fences		
Mobilization		
Contractor profit		
Total Construction Cost		
Subtotal This Page		

Decision Process

Considering the data needs for the analyses defined above, as a minimum, the companies following this model should include members of the production staff, cost estimators, and the company's director of financial operations in the decision-making process. Along with increasing the quality of information during this process, including production staff increases the likelihood for successful implementation by increasing the level of staff "buy-in" to the program. Depending on the level of visibility and significance of the costs involved to the company as a whole, the study team also might include higher levels of company management, up to the chief executive officer.

Table 9.4. Capital Cost Worksheet for Pollution Prevention Project (Continued)

Categories/Items	Item Cost ($)	Category Cost ($)
Engineering and Procurement Costs		
Planning		
Design		
Permitting		
Procurement		
Construction inspection & startup		
Total Engineering & Procurement Costs		
Other Startup Costs		
Vendor		
Contractor		
Training		
Initial charge of chemicals & other expendables (itemize)		
Total Startup Costs		
Permitting Costs		
Consultants fees		
In-house staff costs		
Total Permitting Costs		
Subtotal This Page		
Subtotal Previous Page		
Contingency		
Total Capital Cost		

Alternatives might be identified in brainstorming sessions, in which the potential effectiveness and cost implications of various options are discussed by the parties identified above.

Model 2: Minimize Total Costs Over Time

Another corporate decision model for pollution prevention would be applicable to companies seeking to reduce long-range costs and to maximize long-range income. Examples include electric utilities and oil companies seeking to position themselves for long-term global competition.

Table 9.5. O&M Cost-Savings Worksheet for Pollution Prevention Project

Categories/Items	Unit Cost ($/Unit)	Decrease (or Increase) in Use (Units/Year)	Decrease (or Increase) in Cost ($/Year)
Utilities			
Electricity			
Steam			
Cooling water			
Process water			
Fuel			
Total Utility Savings (Losses)			
Waste Disposal Costs			
Transportation			
State Fees and Taxes			
Onsite treatment & handling			
Waste disposal			
Total Waste Disposal Savings (Losses)			
Raw Materials Use (list items)			
Total Raw Materials Savings			
Subtotal This Page			

Tools and Analytic Framework

The capital and operating cost components identified for Model 1 also would be relevant for this model, except that some elements related to a longer-term perspective would be included in the evaluation. For example, costs of complying with enacted or anticipated changes in regulations might be important. Long-term liability issues also may be considered.

The tools identified for Model 1 (net present value, internal rate of return, payback analysis) also would assist decisionmakers here. Depending on the size of the company and the significance of the pollution prevention strategies to the company's bottom line, it may be worthwhile to analyze the implications of broader-scale factors such as:

- broad demographic shifts and implications for the company's long-term profitability

Table 9.5. O&M Cost-Savings Worksheet for Pollution Prevention Project (Continued)

Categories/Items	Unit Cost ($/Unit)	Decrease (or Increase) in Use (Units/Year)	Decrease (or Increase) in Cost ($/Year)
Labor Savings (or Losses) (List Items)			
Labor Savings (Losses)			
Revenue From Byproduct Sales			
Estimated Revenues From Byproduct Sales			
Maintenance Savings (or Losses) (Itemize)			
Maintenance Savings (or Losses)			
Other Operating Savings (or Losses) (itemize)			
Other Operating Savings (or Losses)			
Subtotal This Page			
Subtotal Previous Page			
Contingency			
Total O&M Cost Savings			

- effects of emerging or anticipated technological shifts on the company's competitiveness and profit margin
- implications of long-term inflation scenarios, escalation possibilities for waste-reduction-related costs, and other costs and revenues
- implications of differing program options with regard to long-term productivity, specific resource requirements, and related factors

Decision Process

The decision process for this model could be structured in a manner similar to that described for Model 1, above. It would include all of the parties identified for the near-term cost minimization model; in addition, the company's strategic planning

Table 9.6. Profitability Worksheet for Pollution Prevention Project Cash Flow for Net Present Value, Internal Rate of Return

Line	Constr. Year	Operating Year[a]							
	0	1	2	3	4	5	6	7	8
A. Total Capital Cost									
B. Salvage Value[b]									
C. Operating Savings									
D. –Interest on Loans									
E. –Depreciation									
F. Taxable Income									
G. –Income Tax[c]									
H. Aftertax Profit[c]									
I. + Depreciation									
J. –Repayment of Loan Principal									
K. –Capital Cost (line A)									
L. + Salvage Value (line B)									
M. Cash Flow									
N. Present Value of Cash Flow[d]									
O. Net Present Value (NPV)[e]									
Present Worth[f] (5% discount)	1.0000	0.9524	0.9070	0.8638	0.8227	0.7835	0.7462	0.7107	0.6768
(10% discount)	1.0000	0.9091	0.8264	0.7513	0.6830	0.6209	0.5645	0.5132	0.4665
(15% discount)	1.0000	0.8696	0.7561	0.6575	0.5718	0.4972	0.4323	0.3759	0.3269
(20% discount)	1.0000	0.8333	0.6944	0.5787	0.4823	0.4019	0.3349	0.2791	0.2326
(25% discount)	1.0000	0.8000	0.6400	0.5120	0.4096	0.3277	0.2621	0.2097	0.1678

[a] Adjust table as necessary if the anticipated project life is less than or more than 8 years.
[b] Salvage value includes scrap value of equipment plus sale of working capital minus demolition costs.
[c] The worksheet is used for calculating an aftertax cash flow. For pretax cash flow, use an income tax rate of 0%.
[d] The present value of the cash flow is equal to the cash flow multiplied by the present worth factor.
[e] The net present value is the sum of the present value of the cash flow for that year and all of the preceding years.
[f] The formula for the present worth factor is $1/(1 + r)^n$ where n is years and r is the discount rate.
[g] The internal rate of return (IRR) is the discount rate (r) that results in a net present value of zero over the life of the project.

Source: U.S. Environmental Protection Agency, Waste Minimization Opportunity Assesment Manual, (EPA/625/7–88/03), July 1988.

and long-range planning group should be included, as well as the legal department (to identify liability and related considerations that should be evaluated).

Model 3: Reduce Total Waste

Because of regulatory requirements, legal mandates, or internal company objectives, the goal of some companies may be to reduce the total amount of waste being generated by a specified percentage or to an identified level. Although costs associated with various waste reduction strategies would be developed and reviewed, this exercise would focus on identifying alternatives that achieve a specified reduction in waste.

Tools and Analytic Framework

This pollution prevention model could be designed for short-term or long-term waste reduction. Depending on the time frame, this model could be considered similar to either Model 1 or Model 2, as described above.

The tools and framework for this model would assist decisionmakers by identifying strategies for reducing wastes to a specified level. Such factors as the effectiveness of various technologies in accomplishing the objectives, and potential roadblocks to successful implementation of the identified strategies, would be identified and evaluated. Optimization models could be used for evaluating the effectiveness of various waste reduction methods.

Another tool that could be employed would be a comparison of streams of waste resulting from implementing various program options. For example, projected waste streams from implementing various programs can be discounted to a present value for comparative purposes, using the discounting methodologies presented in the cost analyses described under Models 1 and 2, above.

Decision Process

Companies following this model should include members of the production staff, cost estimators, and the director of operations of the various processes for identifying alternatives, gathering data, and making decisions. A process similar to that identified for the cost-reduction models could be employed, with increased emphasis on the waste-reduction implications of the alternatives.

Model 4: Limit Total Expenditures to a Fixed Amount

The goal of this corporate decision model would be to limit expenditures on pollution prevention to a fixed amount (either overall or expressed on an annual basis). This model could be used by companies wishing to restrict their expenditures, for immediate profit-related reasons, to a limited commitment. In this case, the model would closely parallel the cost minimization models (Models 1 and 2) described above. This model also could be used by companies wishing to make a specific financial commitment to accomplish objectives, such as public relations or good corporate citizenship objectives. In the latter case, this model would more closely parallel Model 5 (Other Corporate Objectives) described below.

Tools and Analytic Framework

The tools that would be most useful for this decision model would include benefit-cost analysis and cost-effectiveness analysis, which can be tailored to measure the performance of various techniques for achieving a goal, in this case, pollution prevention.

For this model, the decision makers are assumed to have identified a funding level that they are willing to commit to the pollution prevention program; therefore, the primary purpose of the analysis is to determine how much pollution can be prevented at that level of expenditure. The anticipated performance (effectiveness) of the various pollution prevention techniques would be key variables for companies using this decision model.

Depending on the time frame for the analysis, discounting the measurements of performance, cost, or both may be worthwhile.

Decision Process

The participants in the process would be expected to parallel those identified for the cost minimization models (Models 1 and 2), depending on the time frame involved. If this technique is being used by a company wishing to accomplish public relations objectives, the public affairs or public relations department also may be involved in the decision process.

Model 5: Achieve Other Corporate Objectives

A key goal driving the development of a pollution prevention program may relate to matters other than program cost or specific amount of pollution prevented. Examples of these goals could be establishing the company as an environmental leader to enhance its public image, accomplishing substantial modernization of the company's operating processes and plants, or achieving a change in the company's location or industrial mix.

Tools and Analytic Framework

Elements of the decision tools described for each of the other four decision models may be employed, depending on the nature of the goals identified. In addition, it may be appropriate for this model to employ a goals achievement matrix in which the various identified goals (e.g., public image, cost reduction) are weighted for significance and their performance ranked for anticipated effectiveness.

Decision Process

The decision process could involve any of the corporate actors identified in Models 1 through 4, above. In addition, depending on the goals identified, the process could involve others, such as the company's public affairs or public relations department, members of the public or public groups (whose input may be sought through a public involvement program, focus groups, or other means), regulatory agency representatives, industry organizations, or panels.

Example. A simplified example is provided to illustrate how the various decision models might lead to selection of different pollution prevention projects. The case study is about an aircraft components remanufacturing company that currently pays significant disposal costs for solvents used to clean components before refurbishing. The solvents currently disposed of by the company are considered to be hazardous because they are predominantly trichloroethylene (TCE).

During a brainstorming session, the company's manager and operations staff identify three alternatives to reduce waste from the production process:

- continue to dispose of solvent off the site
- recycle solvent off the site
- purchase a still to process the solvents for onsite recycling

For each of the alternatives, cost estimates and estimates of the potential effectiveness in achieving waste reduction goals would be developed, including estimates for such items as:

- major capital facilities that would be required
- rolling stock or other equipment purchases
- capital replacement costs
- annual labor costs
- annual energy costs
- annual savings (labor, fuel, solvent costs) resulting from implementing each alternative

For the example, we have made the simplifying assumption that capital and annual net operational costs for the three alternatives are as shown in Table 9.7.

Table 9.7. Cost Assumptions

	Alternative 1 Offsite Disposal	Alternative 2 Offsite Recycling	Alternative 3 Onsite Recycling (Purchase Still)
Initial Capital Cost			
Still	$0	$0	$52,000
Other Equipment	$0	$0	$38,000
Installation	$0	$0	$10,000
Total Initial Cost	$0	$0	$100,000
First Year O&M Cost			
Labor	$0	$0	$14,000
Power	$0	$0	$20,000
Virgin Solvent	$0	$0	$10,000
Incineration	$76,800	$0	$0
Recycling	$0	$28,800	$0
Residuals Disp.	$0	$0	$8,000
Miscellaneous	$96,000	$96,000	$20,000
O&M Total	$172,800	$124,800	$72,000

For Alternative 3, which involves purchasing a still, it is assumed that additional capital cost for replacement will be required in years 6 and 11. Annual O&M costs are assumed to increase by 5% each year. The resulting projected costs per year for each of the alternatives is illustrated in Table 9.8 for a 15-year time period. Net present values are shown for expenditures through each of the 15 years of the study period. The equation used to calculate present value for each year is:

$$P = \frac{S}{(1 + i)^n}$$

where P = the present worth of the year's expenditures
s = the future payment for the single year
i = the interest rate expressed as a ratio (e.g., .05 = 5%)
n = the number of years from the beginning of the evaluation period, t^0

Figure 9.1 shows the resulting annual expenditure streams that would result if all costs would be paid through direct annual expenses. The company's policy regarding capitalization should be taken into consideration when conducting economic or financial analyses of waste reduction projects.

If the hypothetical aerospace company were following decision Model 4 (limit expenditures to a specified amount per year) and wanted to keep expenditures below $150,000 per year during the first 5 years of the program, that would rule out Alternative 3 (purchase a still) because of the combined capital and O&M cost during year 1.

Figure 9.2 indicates the net present value, using a relatively high discount rate, of the total costs through each of the 15 years of the analysis period. As indicated in the figure (and corresponding data shown in Table 9.8), purchasing a still is significantly more cost-effective than offsite recycling if the decision is made on a long-term basis

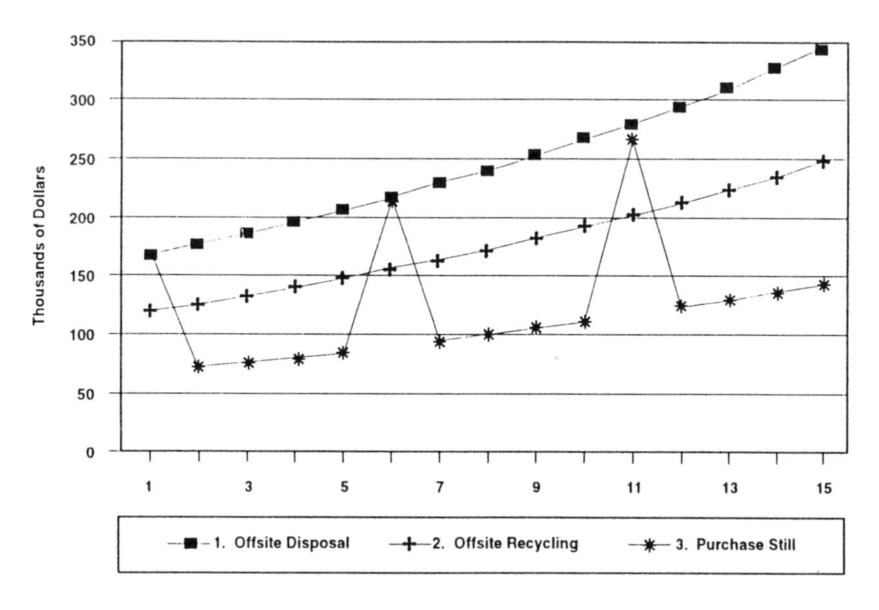

Figure 9.1. Comparison of total annual cost.

Table 9.8. Results of Sample Economic Analysis

Year	1	2	3	4	5	6	7	8	9	10	11	12	13	14	15
Alternative 1: Offsite Disposal															
Capital Cost	0	0	0	0	0	0	0	0	0	0	0	0	0	0	0
Annual O&M Cost															
Incineration	77	81	85	89	93	98	103	108	113	119	125	131	138	145	152
Virgin Solvent	96	101	106	111	117	123	129	135	142	149	156	164	172	181	190
O&M Total	173	182	191	200	210	221	232	243	255	268	281	295	310	326	342
Total Annual Cost	173	182	191	200	210	221	232	243	255	268	281	295	310	326	342
Present Value															
Low Discount	169	344	524	709	899	1,096	1,298	1,505	1,718	1,938	2,164	2,397	2,636	2,883	3,138
Mid Discount	165	330	495	659	824	989	1,154	1,318	1,482	1,647	1,811	1,975	2,140	2,305	2,469
High Discount	154	299	435	562	682	794	898	997	1,089	1,175	1,256	1,331	1,402	1,469	1,532
Alternative 2: Offsite Recycling															
Capital Cost	0	0	0	0	0	0	0	0	0	0	0	0	0	0	
Annual O&M Cost															
Recycling	29	30	32	33	35	37	39	41	43	45	47	49	52	54	57
Virgin Solvent	96	101	106	111	117	123	129	135	142	149	156	164	172	181	190
O&M Total	125	131	138	144	152	159	167	176	184	194	203	213	224	235	247
Total Annual Cost	125	131	138	144	152	159	167	176	184	194	203	213	224	235	247
Present Value															
Low Discount	122	248	378	511	649	790	936	1,086	1,240	1,399	1,562	1,731	1,904	2,082	2,266
Mid Discount	119	238	357	475	594	713	832	951	1,070	1,189	1,307	1,426	1,545	1,664	1,783
High Discount	111	216	314	406	492	572	648	719	785	848	906	961	1,012	1,061	1,106

Alternative 3: Onsite Recycling (Purchase Still)															
Capital Cost															
Still	52	0	0	0	0	65	0	0	0	0	78	0	0	0	0
Other Equipment	38	0	0	0	0	48	0	0	0	0	57	0	0	0	0
Installation	10	0	0	0	0	13	0	0	0	0	15	0	0	0	0
Capital Total	100	0	0	0	0	125	0	0	0	0	150	0	0	0	0
Annual O&M Cost															
Labor	14	15	15	16	17	18	19	20	21	22	23	24	25	26	28
Power	20	21	22	23	24	26	27	28	30	31	33	34	36	38	40
Virgin Solvent	10	11	11	12	12	13	13	14	15	16	16	17	18	19	20
Residuals Disp.	8	8	9	9	10	10	11	11	12	12	13	14	14	15	16
Miscellaneous	20	21	22	23	24	26	27	28	30	31	33	34	36	38	40
O&M Total	72	76	79	83	88	92	96	101	106	112	117	123	129	136	143
Total Annual Cost	172	76	79	83	88	217	96	101	106	112	267	123	129	136	143
Present Value															
Low Discount	169	241	316	393	472	665	749	835	924	1,016	1,231	1,328	1,428	1,531	1,637
Mid Discount	164	232	301	370	438	600	669	737	806	874	1,031	1,099	1,168	1,236	1,305
High Discount	154	214	270	323	373	483	526	567	606	642	719	750	780	808	834

Note: All figures in thousands of dollars

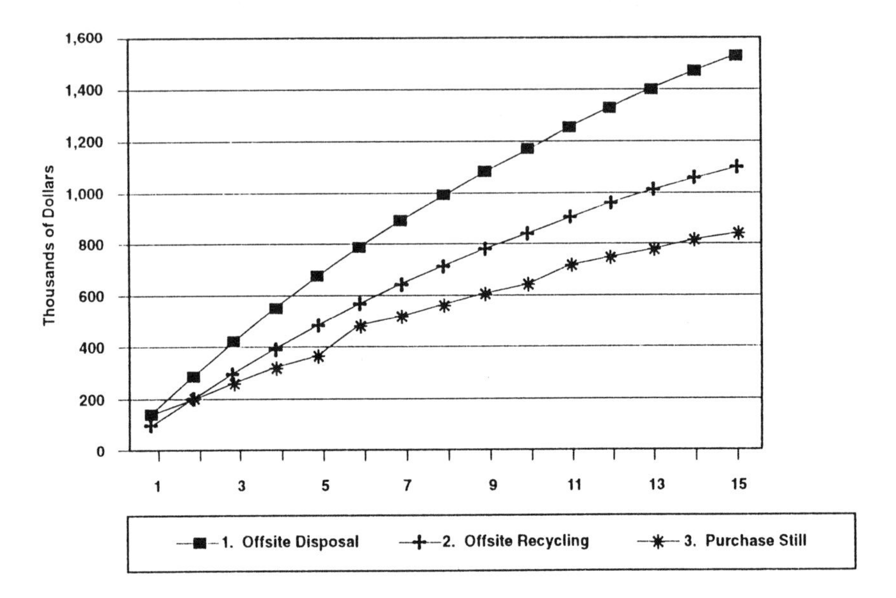

Figure 9.2. Present values with high discount rate.

(e.g., 15 years), using the high discount rate. If the decision is made on the basis of short-term financial outlays (e.g., the first 5 years of the analysis period), then the cost of the two alternatives is comparable.

Figure 9.3 indicates that the relative ranking of alternatives in this case is not significantly different over time if the hypothetical company used a lower discount rate in conducting the analysis (i.e., if the company placed a lower overall value on the costs that are projected to occur during the early years of the study period).

A company that is short-term oriented would need a high rate of return because of

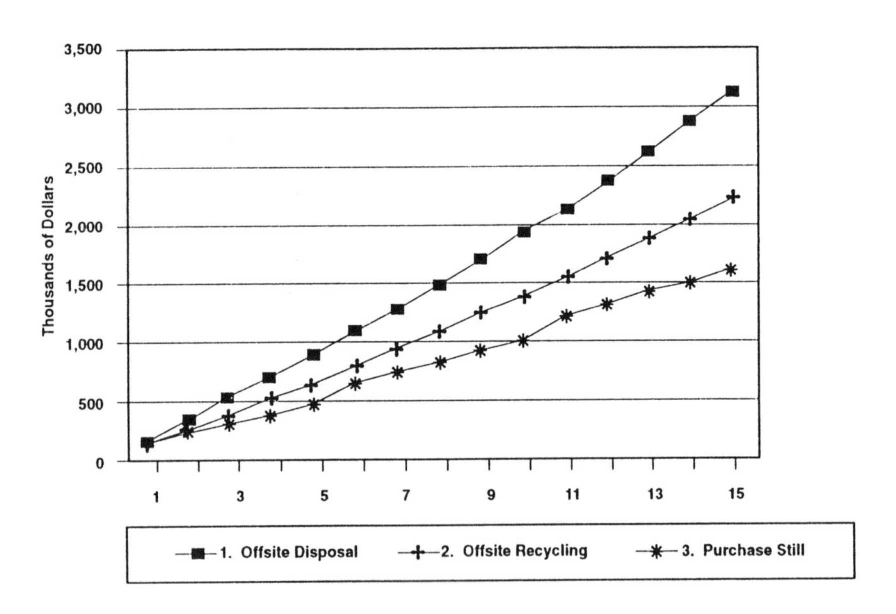

Figure 9.3. Present values with low discount rate.

market uncertainty and the resulting high cost of money. A company that is long-term oriented may require a lower rate of return. This is one of several reasons for considering the effect of differing rates of return in the evaluation.

In conducting a detailed analysis such as that indicated in the worksheets (Tables 9.4, 9.5, and 9.6), such factors as a salvage value at the end of the study period may be factored into the present-worth calculations.

GENERAL REFERENCES

Alexander, H.D., and J.V. Interrante. "Waste Minimization: Case Study," in *Proceedings of the 21st Mid-Atlantic Industrial Waste Conference,* 1989.

Allerton, H.G. "Hazardous Waste Minimization—Source Reduction Alternatives in the Aerospace Industry," *Journal of Environmental Health*, July/August 1990.

Apel, M.L., J.S. Bridges, M.F. Szabo, and S.H. Ambekar. "Three Case Studies of Waste Minimization Through Use of Metal-Recovery Processes," presented at the International Symposium on Metals Speciation (2nd), Separation and Recovery, Rome, Italy, May 14–19, 1989, prepared in cooperation with Illinois Institute of Technology.

Benefiting from Toxic Substance and Hazardous Waste Reduction, Oregon Department of Environmental Quality, Portland, OR, October 1990.

Bryant, J. S. "Strategies for Source Abatement of Industrial Hazardous Wastes," in *Proceedings [of the] 11th Canadian Waste Management Conference*, Ottawa, Canada, 1989.

Byers, R.L. "Regulatory Barriers to Pollution Prevention: A Position Paper of the Implementation Council of American Institute for Pollution Prevention," *Journal of the Air and Waste Management Association*, April 1991.

Carroll, D. W. "Chemical Industry Waste Reduction and Management," in *Proceedings of the Air and Waste Management Association 82nd Annual Meeting and Exhibition*, Anaheim, CA, June 25–30, 1989.

Chang, L.Y., and B.J. McCoy. "Alternative Waste Minimization Analyses for the Printed Circuit Board Industry: Examples for Small and Large Manufacturers," *Environmental Progress*, May 1991, pp. 110–121.

Ciccolo, V.M. "Seattle Metro's Industrial Pretreatment Program: A Case Study for Regulation of the Boeing Commercial Airplane Company," *Proceedings of the 42nd Industrial Waste Conference*, West Lafayette, IN, May 12–14, 1987.

Clearwater, S.W., and J.M. Scanlon, "Legal Incentives for Minimizing Waste," *Environmental Progress*, August 1991, pp. 169–174.

Dharmavaram, S., J.B. Mount, and B.A. Donahue. "Automated Economic Analysis Model for Hazardous Waste Minimization," *Journal of the Air and Waste Management Association*, July 1990.

Duke, L.D. "Hazardous Waste Minimization. A Site Specific Risk Assessment Approach for Evaluating Industrial Waste Management Measures," in *Proceedings of the 1990 Specialty Conference*, Arlington, VA, July 8–11, 1990, pp. 791–798.

Duke, L.D. "Industrial Hazardous Waste Minimization Incentives and Impediments: An Industry Case Study of Metal Platers and Metal Finishers," paper presented at the Annual Meeting of the Air and Waste Management Association, Vancouver, British Columbia, June 16–21, 1991.

"Environmentally Proactive Firms Set the Pace for the Chemical Industry," *Environmental Business Journal*, July 1991.

Fitzpatrick, M. "Waste Minimization in the Petroleum Industry," presented at the Prevention and Treatment of Groundwater and Soil Contamination in Petroleum Exploration and Production Conference, Calgary, Canada, May 9–11, 1989.

Freeman, H.M. *Hazardous Waste Minimization*. McGraw-Hill, January 1990.

Goldbaum, E., D. Rotman, and L. Tantillo. "Hazardous Waste: Faced with Dwindling

Choices, Companies Must Seek New Ways to Manage It," *ChemicalWeek*, August 23, 1989, p. 18.

Huisingh, D., L. Martin, H. Hilger, and N. Seldman. "Proven Profits from Pollution Prevention: Case Studies in Resource Conservation and Waste Reduction," Institute for Local Self-Reliance, 1986.

Industrial Waste Audit and Reduction Manual. (Ontario, Canada: Ontario Waste Management Corporation, September 1987) with the assistance of CANVIRO consultants.

Industrial Waste Reduction Program, report sponsored by the Office of Industrial Technologies, U.S. Department of Energy, DOE/CE-0344, October 24, 1991.

Jendrucko, R.J. "Development of a Comprehensive, Multi-Step Procedure for Identifying, Quantifying and Minimizing Waste in Industrial Facilities," presented at World Energy Engineering Congress (WEEC) and World Environmental Engineering Congress, Atlanta, GA, October 23–25, 1991.

Karam, J.G., C. St. Cin, and J. Tilly. "Economic Evaluation of Waste Minimization Options," *Environmental Progress*, August 1988.

Lorton, G.A., C.H. Fromm, and H. Freeman. *The EPA Manual for Waste Minimization Opportunity Assessments*, Hazardous Waste Engineering Research Laboratory, U.S. Environmental Protection Agency, EPA-600/S2-88-025, August 1988.

Merton Allen Associates, "Succeeding at Waste Minimization," *Industrial Health and Hazards Update*, November 1991.

Nichols, A.B. "Industry Initiates Source Prevention," *Water Pollution Control Federation*, January 1988, pp. 36–44.

"North Carolina Promotes Waste Minimization Program," *Hazardous Waste News,* April 11, 1988.

Overcash, M.R. *Techniques for Industrial Pollution Prevention: A Compendium for Hazardous and Nonhazardous Waste Minimization* (Chelsea, MI: Lewis Publishers, Inc., 1986).

Peters, R.W., C.A. Wentz, S.Y. Chiu, L.E. Martino, and L.J. Habegger. "Application of a Hazardous Waste Management Model for Selected Industries," presented at the Mid-Atlantic Industrial Waste 20th Conference, Washington, DC, June 19–21, 1988.

Saraswat, N., and P. Khanna. "Waste Minimization in Industry—Issues and Prospects," *Industry and Environment*, January-March 1989, pp. 45–47.

"Total Cost Assessment: Accelerating Industrial Pollution Prevention Through Innovative Project Financial Analysis," Office of Pollution Prevention and Toxics, U.S. Environmental Protection Agency, EPA-741/R-92-002, May 1992.

Trattner, R., and C. Wagner. "Waste Minimization/Source Reduction: Current Practices in U.S. Industry," *Advances in Environmental Technology and Management*, 1990, p. 13.

"Waste Minimization—Issues and Options," Vol. 1, Office of Solid Waste, U.S. Department of Commerce, PB87-114351, October 1986.

Waste Minimization Opportunity Assessment Manual. Hazardous Waste Engineering Research Laboratory, U.S. Environmental Protection Agency, EPA-625/7-88-003, July 1988.

"Waste Minimization Workshop," Office of Research and Development, U.S. Environmental Protection Agency, CERI-89-38, March 1989.

White, A.L., and M. Becker. "Total Cost Assessment: Revisiting the Economics of Pollution Prevention Investments," presented at the Conference on Pollution Prevention in the Chemical Process Industries, April 6–7, 1992, sponsored by the Chemical Engineering Industry Forum.

CHAPTER 10

Machining and Other Metalworking Operations

*David Drake and Thomas Higgins**

PROCESS DESCRIPTION AND SOURCES OF WASTES

Machining and Metalworking Operations

Machining and metalworking operations are used to change the shape of metal by either removing some metal from the surface of the piece as it is worked or by bending, pressing, or stretching the metal into the desired shape.

In machining, a cutting tool or abrasive materials remove metal from a metal workpiece to produce a desired shape and dimension. Thus, machining is but one aspect of metalworking, which includes sawing, milling, grinding, drilling, boring, reaming, turning, stamping, forging, shaping, and heat treating.

Most metalworking operations involve high-pressure, metal-on-metal moving contacts between tools and workpieces. These contacts produce friction and generate heat. Uncontrolled, heat and friction can cause excessive wear on tools and can produce undesirable metallurgical transformations in the workpieces. To reduce these effects, metalworking fluids are circulated over working surfaces, reducing friction, cooling the tool and the workpiece, and removing metal chips from the work face. A variety of cutting oils and coolants have been developed to perform these functions. These metalworking fluids are typically stored in open reservoirs, located either in the individual machine or in a central reservoir serving several similar machines.

Metalworking Fluid Types

Metalworking fluids exhibit a variety of characteristics, some of which may be contradictory and none of which should be considered in isolation. A metalworking fluid is chosen as the optimum compromise among the abilities to:

*Mr. Drake is a chemical engineer for Industrial Design Corporation in Albuquerque, New Mexico, and may be reached at (505) 821-7115. Dr. Higgins is a vice president in CH2M HILL's Reston, Virginia, office, and can be reached at (703) 471-1441.

- control friction to reduce force and power requirements and provide more homogenous deformation in the metal as it is shaped
- separate surfaces between metal and tool to prevent metal-to-metal contact
- control metal pickup on tool surface that can become self-accelerating if not controlled
- control wear of the tool surface that could reduce quality or efficiency
- protect old and new surfaces by wetting and spreading to cover the new surface that is exposed during the machining process as well as the old surface that is undergoing the machining process
- adapt to a variety of working conditions, such as pressure, temperature, and relative sliding velocities that may change gradually during the course of a metalworking operation
- provide compatibility with tool, workpiece, and surrounding atmosphere without degradation of performance
- generate a protective film rapidly on the newly exposed metal surface during the metalworking process
- provide a durable lubricant film that withstands continued and repeated encounters with the tool
- control the surface finish of the workpiece so that it is suitable for subsequent operations
- provide thermal insulation for hot workpieces to protect the tool from excess heat and to protect the workpiece material from oxidation or gas pickup
- provide rapid heat removal from metal worked at high rates to prevent lubricant breakdown, thus avoiding excessive and sudden tool wear
- control stability in order to maintain the required fluid properties over time and with use
- control reactivity in order to minimize chemical interaction between the fluid and either the tool or the worked material
- leave only harmless residues on the part after machining to prevent chemical or metallurgical changes in the product
- be easy to apply and remove
- be easily recycled or recovered through an appropriate reconditioning process
- be safe to handle, avoiding exposure of workers to skin irritation, toxicity, carcinogenic effects, odor, or flammability
- be relatively inexpensive because the materials that are made into parts are typically inexpensive and the major cost of the part is that associated with processing
- provide an integrated approach that regards the metalworking fluid not in

isolation but as part of a process from supply through usage to waste management

Modern metalworking fluids are of five basic types: mineral oils; natural oils, fats, and derivatives; synthetic lubricants; compounded fluids; and aqueous fluids.[1]

Mineral oils are fluids derived from crude oil by distillation. They have been available since the 19th century and represent a significant proportion of all metalworking fluids. Typically, mineral oils contain 10 to 70 carbon atoms per molecule. The physical properties of mineral oils depend on such factors as the length of the molecular chain of carbon atoms, the structural complexity of the carbon atoms in the molecule, and the degree to which the crude oil is refined.

Straight mineral oils are the least-used metalworking fluids because they pose health and safety problems, such as fire hazards, slippery floors, and potential respiratory problems for workers breathing oily mists. Other considerations include higher cost, poor appearance of work area, and difficulties with treatment and disposal of used oils.[2]

Natural oils, fats, and derivatives are water-insoluble substances of animal and vegetable origin. The distinction between oils and fats is that oils are liquid at room temperature whereas fats are solid. Natural oils and fats are believed to have been the first metalworking lubricants, and they have retained some significant usefulness over the years.[1] The most widely used metalworking fluids in this category are derivatives created by special processing, such as sulfonation, chlorination, hydrolysis, or reaction with a metal hydroxide, to produce a desired property.

Synthetic lubricants are chemical compounds manufactured to have certain desired properties as a metalworking fluid. Synthetic lubricants include synthetic mineral oils and derivatives, synthetic esters, silicon compounds, and polyphenyl ethers. Some of these compounds are the synthetic equivalent of what is found in nature, and some are not found in nature at all. Most synthetic fluids contain no oil at all but rather water and a wide range of water-soluble chemicals that provide the desired lubricating properties and rely on the water to cool and clean the work face.

Typically, **compounded fluids** are used in a broad range of metalworking applications, because no single material has all of the properties desired for a particular application. The compounded fluids are generally formulated from a base of one of the fluids described above, with additives included to enhance some property lacking in the base material. Base materials include compounded oils; compounded fats, waxes, and soaps; and greases. The compounded fluids typically have additives that make them more stable and give them a longer shelf-life than the straight fluids.

Aqueous fluids, or soluble oils, are the most commonly used metalworking fluid today because they are of comparable quality and applicability as other metalworking fluids and are less costly. They are prepared by mixing a water-soluble oil concentrate with water to produce a working fluid with a water content of 90% to 98%.

The use of water as the primary ingredient in soluble oils accounts for their lower cost when compared to straight oils. A disadvantage of using soluble oils, however, is that they require more maintenance than other fluids. For example, the concentration of soluble oil must be measured and adjusted by adding concentrate daily to maintain the proper oil-to-water ratio.

In addition to these contact operations, oils are also used in metalworking operations for noncontact purposes, such as to hydraulically transfer energy and to lubricate gear boxes and moving parts in metalworking machines. These hydraulic fluids or lubricating oils and greases are contained within enclosed reservoirs in the individual machines. Because these fluids do not come in contact with the workpiece, they are less prone to contamination than contact fluids such as cutting oils or coolants, although leaking hydraulic or lubricating oils can contaminate metalworking fluids.

Straight oils and greases are typically used as hydraulic fluids and for machine lubrication. Soluble oils could also be used, but equipment manufacturers usually specify that certain grades or types of petroleum oils be used to ensure optimal efficiency and minimize wear, and their warranties may depend on following these recommendations. As a result, machine owners are wary of using hydraulic fluids or lubricating oils other than those specified for hydraulic, gear box, or similar applications. In addition, machine owners may be reluctant to change from something that works.

Fluid Degradation

Fluid degradation is the single greatest contributor to the waste disposal burden associated with machining and metalworking operations. When metalworking fluids no longer meet performance requirements, they are removed from their reservoirs and disposed of. The degradation of the fluids is caused by accumulation of metal particles or shavings (swarf), grease, tramp oil, and dirt in the coolants or contact oils. In addition, heat can cause loss of water and depletion or breakdown of additives, and bacteria can grow in the fluid and cause it to degrade. Leaks in hydraulic seals can result in contamination of the metalworking fluids by incompatible hydraulic or lubricating oils.

Hydraulic and lubricating oils also require replacement or restoration, although less frequently than metalworking fluids. Although manufacturers' recommendations need to be considered, frequency of fluids replacement also should be based on operating conditions. Typical waste oil generation rates for representative machine tools are provided in Table 10.1.

ENVIRONMENTAL REGULATIONS

Environmental regulations do not specifically address the wastes generated by machining and other metalworking operations. Such wastes are generally regulated as either wastewater with high oil content or waste oils. Discharge of oily wastewaters to a POTW or to surface waters of the United States is greatly restricted. Waste oils are not regulated as hazardous wastes. The use of waste oil as a fuel is regulated, as is the recycling of used oil.

Wastewater with High Oil Content

Wastewaters discharged to a POTW typically must contain less than 100 ppm of total oil. This limit is exceeded by most waste fluids from machining and other metalworking operations. The National Pollution Discharge Elimination System (NPDES) provides even tighter oil limits on wastewater or stormwater discharges to surface water.

Table 10.1. Representative Waste-Oil Generation

Example Operation	Typical Machine	Oil Use and Purpose	Typical Waste-Oil Generation
Forging	5,000-ton hydraulic press	Soluble oil—hydraulics	1,000 gal/yr
		Soluble oil—coolant	500 gal/yr
		Lubricating oil—moving parts	150 gal/yr
Grinding	Vertical grinder	Hydraulic oil—hydraulics	70 gal/yr
		Gear oil—gear box	70 gal/yr
Metal cutting	Orbital saw (abrasive blade)	Soluble oil—coolant liquid; sludge	7,500 gal/yr
			10 yd^3 (dry)
		Hydraulic oil—hydraulics, gear box	
		Oil—metal parts lubricating	100 gal/yr
	Horizontal lathe	Soluble oil—coolant	750 gal/yr
		Straight oil—coolant, gearbox	20 gal/yr
		Hydraulic oil—hydraulics	75 gal/yr
Die sinking/milling	Milling machine, vertical lathe	Soluble oil—coolant	2,500 gal/yr
		Hydraulic oil—hydraulics	50 gal/yr
		Penetrating oil—cutting operation	200 gal/yr
		Gear oil—gear box	15 gal/yr
Stamping	Stamping	Soluble oil—coolant	500 gal/yr
		Hydraulic oil—hydraulics	25 gal/yr

Hazardous Waste Exclusion

Since 1985, the U.S. EPA has considered whether to regulate used oil as hazardous waste.[2] RCRA and CERCLA sections that define hazardous waste specifically exclude "petroleum, including crude oil or any fraction thereof" from the hazardous material definition. Hazardous substances that are added during the petroleum refining process are also protected under the petroleum exclusion sections of the hazardous waste legislation. However, petroleum waste material that is contaminated with hazardous substances not normally found in crude oil or refined petroleum products must be handled and disposed of according to hazardous waste regulations. If metalworking fluids have been protected from contamination by hazardous materials, under state and federal laws, they can often be classified as waste oil rather than as hazardous waste.

Waste Oil as Fuel

On June 17, 1991, the U.S. EPA amended the RCRA regulations to define used oil as "any oil that has been refined from crude oil, used, and, as a result of such use, is contaminated by physical or chemical impurities" and suitable to be burned for energy recovery in any boiler or industrial furnace unless such used oil has been mixed with hazardous waste and is regulated as a hazardous waste fuel. The regulations that cover the burning of used oil as a fuel place specific requirements on the generator for performing analysis, providing the proper notification, and preparing invoices; therefore, any waste-oil generator considering burning waste oil as a disposal option is advised to fully identify, evaluate, and comply with the latest regulations.

Waste-Oil Recyclers

Congress declared in the Used Oil Recycling Act of 1980 that used oil is a valuable energy and materials resource; technology exists to refine, process, reclaim, and recycle used oil; and used oil poses a threat to the public and the environment unless reused or disposed of properly. Consistent with those findings, Congress authorized the U.S. EPA to investigate how used-oil recycling should be regulated to preserve this energy resource while ensuring that the used-oil recycling is conducted properly. Generators of used oil are advised to check in their areas for reputable used-oil recyclers and consider that option as a route to manage their waste-oil material.

METHODS FOR REDUCING METALWORKING WASTES

The generation of waste metalworking fluids can be reduced through operational changes and by investing in equipment to recondition and recycle waste fluids. Since equipment to recondition waste oils can be expensive, methods that do not require capital expenditure should be explored first. A program to reduce the disposal of waste fluids should, therefore, proceed in four steps:

1. optimize material usage
2. maintain fluid quality
3. maintain regulatory exemption
4. treat to meet discharge requirements

Optimize Material Usage

Prepare a current inventory of the metalworking, hydraulic, and lubricating fluids being used. Confirm that the materials and procedures being used conform with plant engineering policy or manufacturer's specifications. One goal of this evaluation should be to reduce the number of types of fluids used. Consolidating fluids can significantly reduce the volume of waste produced and simplify recycling or disposing of the remaining fluids.

Each fluid application should be reviewed to determine specific fluid requirements. The following factors should be considered.[2]

- type of operation
- central reservoir system or sumps on individual machines
- tooling and setup requirements or physical constraints

- type of metals being worked
- quality of water used for fluid preparation
- production part requirements (e.g., tolerances, finish, rust protection)
- machine maintenance (e.g., lubrication, seals, paint, cleanliness of work area)

Minimize Variety in Use

Prepare an inventory of existing metalworking fluids and waste production and determine if the number and types of different fluids being used can be reduced. Consolidating fluids can significantly reduce the volume of waste being produced and the methods needed for waste disposal. Further reduction can result if a new fluid is adopted that produces less waste than fluids currently in use.

The quantity, type, and intended performance of each metalworking fluid, along with reasons and locations for its use, should be catalogued. It is not unusual to find dozens of different oils and other metalworking fluids being used within a single plant or on one factory floor. These fluids are selected by plant personnel on the basis of preference, cost, quality, performance, or longevity. Plant personnel making these choices may include purchasing agents, process engineers, shift supervisors, maintenance staff, and production managers.

A complete inventory of oil type and practice is recommended for each machine. If not already developed, a computerized database can be compiled that would include fluid used, manufacturer, quantity used, frequency of purchase, and cost. This inventory provides for objective analysis of current operations, comparisons of costs, and ranking of areas of concern. In preparing this inventory, interview staff members who are involved in decisions on fluid use and who are most familiar with their routine effectiveness. Staff knowledge and training should influence later decisions involving process changes.

Laboratory or screening tests and vendor-supplied information are useful in selecting a fluid, but the best information is obtained by testing prospective fluids on individual production machines. Production management, maintenance (machine overhaul and lubrication), purchasing, laboratory, environmental, and health and safety personnel should assist in the fluid selection process, but the ultimate responsibility should rest with one person or group.

Other measures can be taken to reduce the complexity of coolant management. The primary goal should be to reduce the number of coolants used; one is ideal. Fluid management is simplified if a central reservoir is used for all machines in a particular department or section of the plant floor, instead of using individual reservoirs for each machine.[3]

Case Study 10.1: Purchase Problems Negate Coolant Reconditioning Program

A machine shop at the Pensacola, Florida, Naval Air Station Rework Facility operates approximately 75 drilling, grinding, milling, and lathing machines to recondition Navy A-4 jet aircraft and H-3 and H-53 helicopters. The machine shop uses an emulsified coolant, diluted to 4% concentration with deionized water. A centrifuge was purchased to remove tramp oil from the coolant, which was then being changed approximately every 2 to 4 weeks. With the new centrifuge, new coolant purchase was reduced considerably, and coolants could be used for four to six times the former life. Machine tool life also was extended.

Unfortunately, the centrifuge could be used only sporadically because the unit is effective for treating water-soluble coolants only, and the new coolants, supplied by low-bid purchase, were frequently oil soluble. Even the same types of coolant, when supplied by different manufacturers, were found to be incompatible, so that the inventory of coolants in the sumps had to be disposed of when the new material was introduced into the system. Modifying purchasing requirements to require a compatible coolant would have eliminated this wasteful practice.

Consider Eventual Disposal

An inventory of waste fluids generated from each machine should be prepared. An important part of the inventory should be a laboratory characterization of the waste fluids. Work history or suspected composition of a waste fluid is no substitute for an accurate analysis. This information also is needed for proper regulatory classification and disposal of waste fluids. In addition, it provides a basis for further waste reduction projects.

It has been recommended that the following performance and waste treatment factors be considered in selecting or reevaluating metalworking fluids.[2] These factors also apply to noncontact hydraulic and lubricating fluids.

- Performance—Ensure that the fluid meets production and maintenance objectives by using laboratory screening tests.
- Health and safety—Avoid risks to operators by evaluating material safety data sheets and other information supplied by the manufacturer.
- Waste treatment—Require screening tests, field tests, or supplier assistance to define treatment requirements, if any.
- Quality of fluid—Ask for evidence that the supplier can provide a consistent product quality.
- Technical service support—Request the suppliers' assistance in providing internal control and troubleshooting of degradation problems with fluids.
- Cost—Consider the overall or life-cycle cost of a fluid, not just its purchase price. Include labor for mixing, transporting, and controlling the fluids; machine downtime for adjusting and disposing of fluid; and waste treatment and disposal. In addition, consider effects on operator safety and health.
- Delivery—Have supplier provide "just-in-time" delivery to minimize deterioration of fluids during storage.

Install Full-Service Programs

Full-service programs are users' attempts to maintain a competitive edge in their market by minimizing the number of suppliers with whom they must deal, and ensuring compatibility among the materials they do use. In concept, each plant is supplied by a single vendor, who is responsible for providing the material and managing the fluid.

Example. A good example of this approach is the fluid management program implemented at General Motors.[4] General Motors uses its considerable power as a large user to select suppliers who meet its criteria for technological expertise, price, and the ability to coordinate all chemicals in the plant. In return, the supplier is compensated on a parts-produced basis, which fairly rewards the supplier for the service it provides. Through this system, General Motors has forced suppliers to form alliances to provide all the chemicals required in each factory and then to participate in a supplier-user partnership. In 1989, five General Motors plants had successfully implemented the fluid management partnership program. Because of the high capital risk associated with the system, only the larger publicly held fluid suppliers have been able to win the contracts.

An example of effective integrated service is when one supplier provides both the fluid concentrate and waste treatment chemicals. The fluid concentrate includes an additive to help form a chemical emulsion when the fluid concentrate is mixed with water. The waste treatment chemical from the same chemical supplier should have been designed and selected specifically to overcome the effectiveness of the chemical emulsifier. The lubricant emulsion is then separated into fluid concentrate suitable for reclamation and aqueous waste suitable for disposal through a POTW.

Maintain Fluid Quality

In the process of inventorying and reviewing usage of metalworking fluids, operational or maintenance deficiencies can be identified. These deficiencies can often be remedied with minimal labor or capital expense. The goal is to reduce contamination of fluids and extend their useful lives to reduce waste generation.

Additives

Machine operators can extend the useful life of soluble coolants by using high-quality makeup water. Water with greater than 100 ppm of dissolved solids can contribute to degradation because chlorides and sulfates may react with the solids, enhance bacterial growth (sulfates are biochemically reduced to hydrogen sulfide, causing a rotten egg odor), create foaming problems, increase concentrate use, and affect product quality.[2]

Additional steps can be taken to maintain coolant quality. An individual responsible for maintaining metalworking fluids should perform the following tasks daily:

- Add makeup water to replace soluble coolant losses (5% to 15% per day) resulting from evaporation.
- Check the fluid pH for signs of bacteria buildup.
- Remove swarf and dirt from the work area.

The maintenance worker also should routinely:

- Enforce good housekeeping practices to prevent or minimize contamination of the fluids by oil, dirt, or metal.
- See that machines are overhauled periodically, paying particular attention to

maintaining hydraulic, lubricating, and gear oil seals and keeping reservoirs tightly covered.

- Discourage operators from disposing of rags, paper, or other debris in coolant reservoirs or sumps.
- Control the addition of concentrate and makeup water (machine operators tend to overconcentrate or overdilute).
- Prevent the use of bactericides or other chemicals that can cause health and safety problems or can result in spent fluids that are unnecessarily toxic.

Maintenance

Proper machine maintenance is particularly critical to keeping waste fluid volumes low. A leaking hydraulic fluid reservoir, for example, can generate 3 to 10 times the waste volumes considered to be normal (Table 10.1). It is not uncommon for 5% to 10% of the hydraulic mechanisms in a high-production machine shop to leak, where production is emphasized over maintenance.

Daily checks of soluble coolant using a refractometer or by titration are warranted. These tests should replace the operators' test of "good feel" to the hands. Avoiding physical contact is particularly important, because many modern coolants contain bactericides that are irritating to the skin. The pH can be measured with litmus paper or a pH meter (fairly inexpensive, compact, battery-operated pH meters allow testing in seconds).

The most fundamental method for minimizing waste from metalworking operations is fluid conservation. The following extra costs will be incurred when fluid usage is excessive:[5]

- costs for replacing material that is lost through leakage
- costs associated with containing and collecting leakage, including sumps, pumps, booms, piping, and labor associated with servicing the hardware
- waste disposal costs associated with the leakage that is captured with oil spill absorbent material and is therefore not reclaimable
- environmental costs associated with less than 100% success at capturing spilled fluids
- extra costs associated with handling fluid to replace that which is lost
- extra costs to dispose of empty containers associated with replacing lost fluids
- temptation to use less expensive fluids to save money since purchase cost of replacing lost fluids is high
- extra labor associated with maintaining an adequate inventory of fluids in the leaky machines
- shorter equipment life due to poor maintenance because leaking equipment typically takes precedence over equipment scheduled for preventive maintenance

- shorter equipment life because empty reservoirs are not providing lubricating fluids to gears and bearings

- substitution of inferior lubricant because it is the only one available when an emergency fill is required

- housekeeping associated with machinery losing fluids outside the containment reservoir

Maintain Regulatory Exemption

Waste oils are exempt from state and federal hazardous waste regulations, provided they have not been contaminated by being mixed with regulated wastes. Waste oils are still controlled under other regulations, such as the NPDES permit system that regulates discharges to the nation's surface waters. Maintaining the regulatory exemption will result in cost savings because the cost of treating and safely disposing of nonhazardous oily wastes and the expense of paperwork preparation are lower than if the wastes are classified as hazardous.

Proven methods of maintaining the regulatory exemption for waste oils include:

- segregate waste metalworking fluids from hazardous constituents such as chlorinated solvents or polychlorinated biphenyl (PCB)

- clearly label waste storage containers

- train operators to avoid mixing hazardous wastes with exempt wastes

Select Disposal Method

One route for identifying opportunities to recycle machining and metalworking wastes is through a waste exchange clearinghouse publication serving the area of the country where the facility is located. Individuals desiring to dispose of wastes as well as individuals desiring to receive wastes would be listed in this publication. Generators of wastes are advised to check the reputation of those requesting the wastes to ensure that the waste is handled properly.

State waste disposal regulators are another source of information on waste disposal options in a particular area. They can identify those who are licensed to reclaim or burn waste hydrocarbon material.

Treat to Meet Discharge Requirements

Metalworking fluids are generally treated to separate water and contaminants from the fluid concentrate. The water that is separated is typically suitable for discharge to a sewage system although some separation processes produce water suitable for reuse as makeup water. The contaminants separated from metalworking fluids typically must be disposed of. The concentrate separated from metalworking fluids may be reconditioned or used as waste-oil fuel.

Reconditioning Fluids for Reuse

Reconditioning waste metalworking fluid consists of removing impurities and then restoring the fluid to its original condition by adding concentrate and individual constituents such as surfactants, bactericides, emulsifiers, conditioners, antioxidants, or other chemicals that make the fluid effective in metalworking operations. Frequently, a supplier can recondition waste fluid more cost-effectively than a single shop. The supplier can arrange to pick up waste fluid when delivering new concentrates.

Many metalworking fluids can be reused for months or even years before they require reconditioning. For these fluids, contaminants such as dirt, metals, or bacteria can be removed, and concentrate can be added to restore the fluid to near-original condition. Soluble oils usually require the addition of water. Straight oils, however, generally require reconditioning much sooner than soluble oil or synthetic fluids because of more rapid depletion of essential constituents.

Contaminants such as tramp oils, dirt, metals, and bacteria can be removed through a variety of reconditioning processes. Treatment processes available for individual machine sumps or central reservoirs are presented in Table 10.2. Some of these processes are waste treatment methods, not recycling processes, inasmuch as the waste is concentrated to facilitate disposal rather than reuse.

Selecting the most applicable fluid-reconditioning system depends on several site-specific factors. Chief among these are the following:

- favorable economics, considering capital, operation and maintenance costs, and savings from reduced raw material purchase and waste disposal
- equipment effectiveness at removing contaminants such as tramp oil, dirt, and bacteria
- availability of an automatic makeup system for reconstituting the fluid
- operational simplicity and maintenance requirements
- requirement for floor space
- quality and durability of construction and warranty protection
- availability of spare parts and services[2]

Table 10.2. Recycling Systems for Metalworking Fluids[a]

	Removes		
Recycling Process	**Oil**	**Dirt & Metals**	**Bacteria**
Settling/drag-out		X	
Cartridge filtration		X	
Basket strainers		X	
Gravity settling	X	X	
Cyclone separator		X	
Centrifuges	X	X	
Magnetic separator		X	
Pasteurization/distillation			X
Ultrafiltration		X	X

[a]Adapted from Reference 1.

Reconditioning equipment can be supplied in capacities to match the requirements of individual machine sumps, although it is normally more economical to service a central reservoir system.

Cleanup of contaminated metalworking fluids can involve physical separation processes (settling, flotation, straining, filtration, centrifugation, and ultrafiltration), as well as thermal processes (pasteurization and distillation). Each of these processes is described briefly. A more-detailed description of these processes is presented in other literature (see References 6 through 15).

Physical Separation

The physical removal processes remove fine solids and produce a clean fluid. The removal processes rely on capital expenditures to provide for equipment and on an operating budget to pay for power to overcome head loss, which is accomplished by pumping. For example, simple settling tanks or oil/water separators can separate swarf and unemulsified oils from coolant; basket strainers can remove coarse solids from a recirculating fluid; magnetic and cyclone-type separators can remove ferrous and nonferrous metal fines, respectively; and cartridge filters can remove finer solids and metal fines.

Filtration

Filtration is effective for removing fines from metalworking lubricants and extending the service life of such fluids. Various filtration media are used, depending on the size of particles to be removed from the fluid.[1]

- Wire filters can remove particles 100 microns and larger. A sand bed may be built up on the wire filter to prevent rapid clogging of the filter media.
- Woven or nonwoven filter media of controlled pore size can be used to filter particles 10 microns and larger. Larger particles accumulate on the filter material to form a filter cake that is effective for removing the smaller material without a high pressure drop. When the pressure drop becomes high, then the filter media must be cleaned.
- Disposable filter media are used to capture particles smaller than 10 microns (Figure 10.1). The application of such technology is limited to a combination of filter media and lubricant fluids that will not react adversely with each other. Cellulose fiber filters will swell excessively in these applications, and weak emulsions will be disrupted and quickly clog fine filter media.

Centrifuges employ induced high gravitational forces to separate solid/liquid or liquid/liquid mixtures that would take too long to separate under normal gravitational conditions. To operate successfully, centrifuges usually require that the two materials to be separated have differences in specific gravity of at least 2% to 3%.

Case Study 10.2: Reconditioning of Heat-Treating Quench Oil Using Settling and Cartridge Filtration

In a metal manufacturing facility in upstate New York, waste oil was generated in a high-volume heat-treating operation in which heated metal parts were quenched

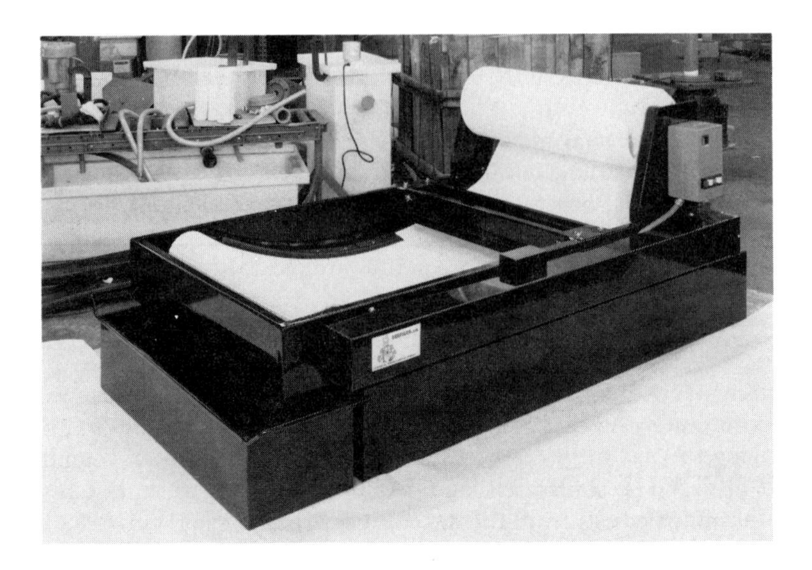

Figure 10.1. Automatic gravity filtration system using a disposable filter fabric (Photo courtesy of Serfilco, Ltd., Northbrook, Illinois).

(rapidly cooled) from temperatures as high as 1,750°F. After cooling, the parts were spray-washed with water. The combined oil/water mixture was treated in an oil/water separator, which removed waste oil and sludge for disposal. This waste-oil mixture contained 4% water and 1,400 ppm of suspended solids, consisting of sediment and metal fines.

Pilot testing by Hilliard, an oil reclamation system manufacturer in Elmira, New York, demonstrated that this waste oil mixture could be treated to sufficient purity that it could be reused in the quench tank. On the basis of this testing, a complete treatment system was installed. Treatment processes begin with batch settling in an oil/water separator for 24 hr. Following settling, water and solids are removed from the bottom of the tank, and the oil is recirculated at a rate of 210 gal/hr through a 3-micron Hilco polypropylene fiber cartridge filter and returned to the dirty oil tank. After filtration is completed, the oil is pumped at 25 gal/hr through an oil reclaimer. In the reclaimer, the oil is heated under a vacuum to reduce the water content and kill bacteria, and is then filtered through a 0.5-micron Hilco cartridge filter and pumped into a clean oil tank for dispensing to the quench tank.

Settling effectively reduced the water content of the oil to less than 0.3% and reduced suspended solids to less than 120 ppm. Filtration and heat treatment reduced the concentrations of these contaminants to 75 and 5 ppm, respectively. The 3,000-gal/yr waste-oil recycling system paid for itself in less than 18 months, on the basis of a virgin oil cost of $3.82/gal and an O&M cost for the recycling system of $0.29/gal.

Case Study 10.3: Reconditioning of Honing Oil at a Midwestern Auto Engine Plant Using Filtration, Settling, and Centrifuge

An automobile manufacturer generates waste metalworking fluids in the process of machining engine bores. The final step in machining involves use of a "honing" fluid, in this case, kerosene, to achieve a clean, mirror-like finish on the engine bore surface. During subsequent engine cleaning operations, the honing oil becomes mixed with water, producing a mixture that is approximately 30% water, 70% kerosene, and less than 2% solids. To be acceptable for reuse, honing fluid can contain no more than 0.5% water and no solids larger than 5 microns.

Kerosene is difficult to recycle. When new, it is lighter than and insoluble in water, although both have similar viscosities. Having a low flash point (100 to 200°F), it is combustible and therefore presents a fire hazard.

Enhanced gravity settling and low-speed (less than 2,000 G) centrifuges were used unsuccessfully to reclaim this waste honing fluid. These systems failed because of inability to meet specifications (by gravity settling) or a buildup of solids (by centrifuge).

After consulting with Sanborn, an equipment vendor in Wrentham, Massachusetts, a high-speed centrifuge was installed. Pretreatment consists of magnetic separation and filtration through a paper roll filter, gravity oil/water separation, and disposal of the water phase. The oil phase is then heated to 150°F and fed at a rate of 3 gal/min to a centrifuge to remove remaining fines and water from the honing fluid. Cleaned fluid is pumped through a heat exchanger to recover energy, then into a clean tank for reuse.

This system has been operating successfully, with 500 to 700 gal of honing oil recovered each week. Payback for the system, including reduced waste disposal costs, was less than 8 months (on the basis of a purchase and handling cost for fresh kerosene of $1.60/gal).

Evaporative Separation

Pasteurization and distillation involve heating contaminated fluids to between 140° and 250°F. Higher temperatures can cause degradation of the fluid. Pasteurization is used to kill bacteria. Distillation (Figure 10.2) also kills bacteria, but in addition, it removes water from the fluid or oil. Although distillation is commonly employed for oil/water mixtures, pasteurization is generally used for straight oil wastes.

Case Study 10.4: Distillation of Hydraulic Oils at an Auto Engine Assembly Plant in the Midwest

An automotive manufacturer operates more than 3,000 pieces of hydraulically operated power equipment and machinery at a single large-engine plant. Hydraulic fluids had to be replaced periodically because of contamination by water and dirt. Careful plant maintenance practices were employed to prevent mixing the hydraulic fluids with coolants or other fluids. Replacing hydraulic fluids at the plant cost more than $60,000 per year. Faced with this cost, the manufacturer decided to install an onsite recycling system to recover waste hydraulic fluids.

An oil-recycling system manufactured by Aquanetics, Farmingdale, New York, was installed to handle all of the plant's hydraulic fluids at a central facility. Dirty hydraulic fluids are delivered to one of three 2,000-gal, steam-heated (150°F) tanks. In these tanks, water and grit are removed by gravity settling. Heated oil is then pumped through a 50-micron cartridge filter to another tank. From this tank, the oil is pumped through a 100-mesh strainer and a 3-micron cartridge filter to a vacuum

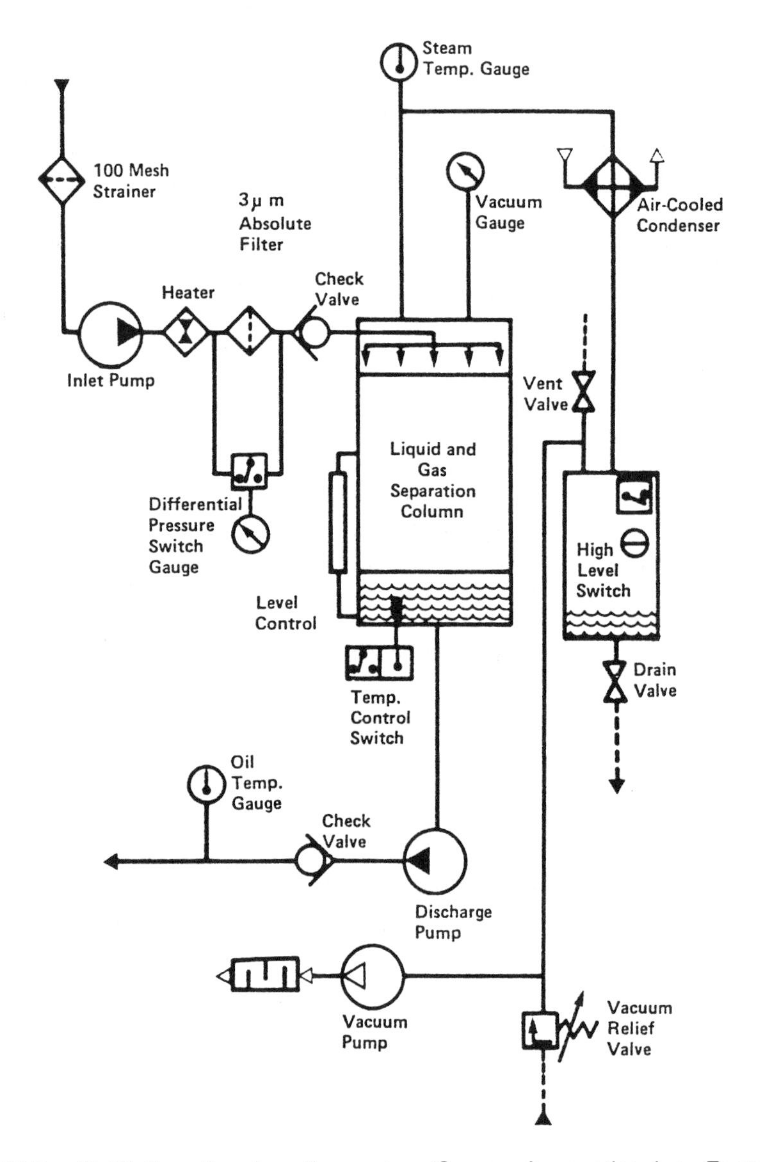

Figure 10.2. Distillation oil reclamation system (Source: Aquanetics, Inc., Farmingdale, New York).

distillation unit operated at a temperature of 200° to 250°F. Water is discharged to sewers, and the cleaned oil is pumped to a 2,500-gal holding tank for cooling. The cooled oil is tested for quality and then recycled.

The resulting cleaned oil has met or exceeded specifications for fresh oil. In 5 years of operation, the recycling system has recovered an average of 300 gal/day of hydraulic oil, significantly reducing fresh oil purchase and waste-oil disposal. The system paid for itself in less than a year.

Chemical Treatment

Chemical treatment is used for more metalworking fluids than all other waste treatment processes combined.[16] Chemical separation processes use either inorganic materials or polymers to neutralize surface charges and disrupt oily emulsions. Frequently, acid is added to neutralize the weak acid molecules of oil, breaking the attraction between the polar water molecules and the charged oil molecules. When a polymer is used, the basic process is for the polymer to overcome the charge on the surface of the emulsified oil and bridge several emulsified oil droplets to form a larger droplet that will then float because of the action of buoyant forces.

The equipment associated with the chemical treatment system includes the following:

- storage, addition, and control system for each of the chemicals used
- rapid-chemical-mix systems to disperse the chemicals into the waste solution
- slow-speed flocculation system to grow the dispersed oil droplets into larger oil particles that can float
- some type of enhanced gravity settler such as a dissolved-air flotation device to enhance the gravity separation.

Membrane Separation

Ultrafiltration (UF) (Figure 10.3) employs membranes that have pores small enough to remove bacteria, microscopic solids, and oils. UF provides excellent separation (90% to 98% efficiency) of contaminants from a metalworking fluid. UF is frequently used as an end-of-pipe treatment process for oily wastes. Wastes containing from 0.1% to 10% oil, with temperatures up to 150°F and having a wide pH range, are suited to UF. A concentrate of 40% to 60% oil and a permeate of 50 ppm oil can be achieved. Pretreatment by filtration and pasteurization is normally recommended to keep the membranes from becoming plugged or fouled with dirt and bacteria.

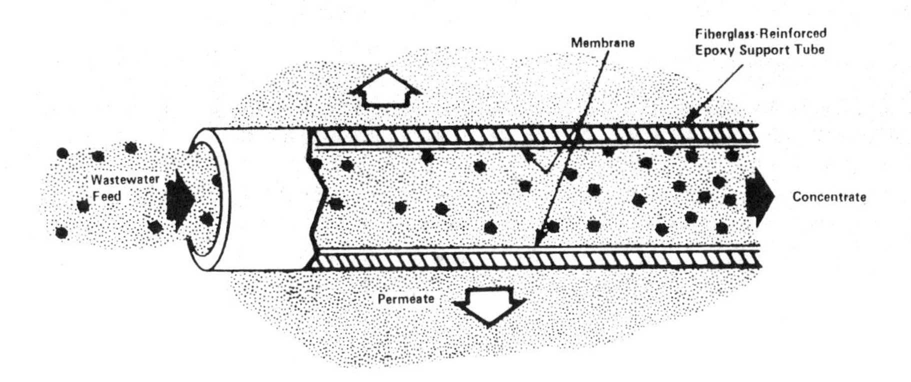

Figure 10.3. Tubular membrane ultrafiltration (Source: Membrane Systems, Inc., Wilmington, Massachusetts).

Case Study 10.5: Ultrafiltration for Waste Oil at an Auto Body Stamping Plant in Ontario, Canada

An automobile manufacturer operates a high production metal-stamping operation in Ontario to produce automobile frames, axle housings, and body extensions. A forming oil is used in the stamping process. Following stamping, the parts are cleaned with a hot (150°F) dilute (1.5%) alkaline aqueous cleaner containing emulsifying agents. After becoming contaminated with forming oil (11,000 ppm) and metal particles, this cleaner was previously disposed of without treatment in a municipal sewer.

Changing regulations prohibited continued discharge of this untreated waste to the municipal sewer. Because of the high cost of offsite disposal for this waste (more than $300,000/yr in 1983 Canadian dollars), the feasibility of treatment to meet sewer discharge requirements was evaluated.

Laboratory-scale treatability testing proved that conventional chemical and physical treatment methods would not achieve discharge standards. UF proved more successful and was adopted for full-scale implementation.

A treatment system was installed, consisting of a heat exchanger for cooling, a bag filter for removing large dirt and metal particles, and a UF system for separating the forming oil. A second batch UF system was used to further concentrate the forming oil to 55%. The permeate from the UF system, consisting of water, alkaline cleaner, and emulsifier with less than 200 ppm of oil, was returned to the wash tank for reuse. The volume of waste discharged to the sewer was significantly reduced and was limited to waste from backflushing the UF membranes and cleaning the tank.

No major difficulties were encountered in startup and operation of the system. The major components of the system are:

- two Romicon UF units, one with 20 cartridges and 15 square meters of total filter area and the second with 10 cartridges and 7.5 square meters of filter area
- three Gould 1,800-liter-per-minute feed pumps
- one Tranter 1.9-square-meter heat exchanger
- four Gould 110-liter-per-minute transfer pumps
- two 8.5-m^3 process tanks
- two 1.5-m^3 tanks to store permeate for UF membrane cleaning
- automatic level controls and associated equipment to control the process

The system operates continuously, except for batch concentration of the forming oil. Economics of the system are shown in Table 10.3.

On the basis of installed costs, an annual O&M cost of $30,000 for the new facility, and recovered costs, the payback period was less than one year.

Recycling systems for metalworking fluids can use either a single separation process or several processes operated in series. Simple settling or magnetic separation may be sufficient for removing metal turnings from a coolant, but removal of dirt or fines can require filtration. Separation of tramp oil or bacteria may require distillation or UF. Typically, a combination of separation processes is used. A complete

Table 10.3. Cost Saved by Installing an Ultrafiltration Oil Treatment System (Canadian Dollars, 1983)

Previous Annual Cost:	
Alkaline cleaner	$100,000
Waste-oil hauling	293,000
Total	$393,000
Study, System Design, and Installation:	
Treatability studies	$ 24,000
Detailed engineering	36,000
Romicon UF system	126,000
Support equipment and installation	96,000
Total	$282,000
Annual Cost Savings:	
Alkaline cleaner (50%)	$ 50,000
Waste-oil hauling (90%)	260,000
Recovered forming compound (20%)	70,000
Total	$380,000

coolant recycle system, employing a settling tank, strainer, cyclone, pasteurizer, and centrifuge, is shown as Figure 10.4.

Typical capital costs, O&M costs, and payback periods for individual components of a fluid-recycling system are listed in Table 10.4. The annual volumes shown represent the upper and lower extremes for waste oil generation from small- to medium-size metalworking operations. From these ranges, it is possible to compare costs of the different recycling equipment.

A rule of thumb in waste-oil recycling is that operations generating more than 25,000 gallons per year (gal/yr) can best afford the time and staff for proper operation and maintenance of recycling equipment. Most manufacturers do not offer systems sized for less than 10,000 gal/yr, although often one unit can accommodate both ranges shown in Table 10.4. As can be seen, the simpler techniques are

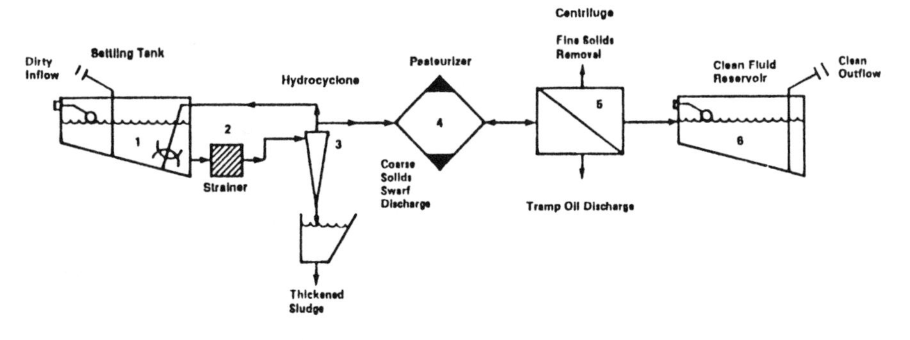

Figure 10.4. A coolant reclamation system using pasteurization and centrifugation (Source: Sanborn Inc., Wrentham, Massachusetts).

Table 10.4. Typical Fluid Recycling Systems and Costs

	25,000 gal/yr			100,000 gal/yr		
Fluid Recycling System	**Capital Cost, $**	**O&M Cost ¢/Gal Processed**	**Payback Period, Months**	**Capital Cost, $**	**O&M Cost ¢/Gal Processed**	**Payback Period, Months**
Cartridge filters	300–1,000	3–8	< 6	600–2,000	2–6	< 6
Solids separators	2,000 5,000	4–10	< 6	5,000–8,000	3–7	< 6
Centrifuge	10,000–20,000	5–15	<24	20,000–30,000	3–12	<18
Distillation	10,000–20,000	5–15	<24	25,000–35,000	3–12	<18
Ultrafiltration	10,000–20,000	15–30	<24–36	20,000–30,000	7–15	<18

Figure 10.5. A mobile unit used to clean metalworking fluids at a customer's plant (Photo courtesy of Safety-Kleen, Elgin, Illinois).

less capital intensive and have shorter payback periods, but the more sophisticated equipment merits consideration.

Safety-Kleen, a recycling and waste disposal company, is currently testing a mobile service unit for those machine shops that produce such low volumes of used coolants that it is uneconomical to purchase and operate recycling equipment (Figure 10.5). Waste coolant is placed into the portable unit and recycled onsite, and the user is charged per gallon of recovered oil.

In evaluating which system would be most appropriate, the machine tool operator or owner will first need to characterize the oil and its intended application. Payback periods of 1 to 2 years or less are typical and usually warrant introduction of recycling systems into plants where further waste reduction is desired. Perhaps the most important benefit of waste reduction for the plant, and ultimately for the environment, is that less waste oil requires treatment or disposal off the site. When waste oil is disposed of off the site, the owner retains responsibility under federal law(s) and is liable for environmental damage caused by improper or inadequate actions by a licensed treatment, storage, or disposal facility.

Case Study 10.6: Machining-Coolant Management Program at Pratt & Whitney in Connecticut

Pratt & Whitney, a manufacturer of jet engines in Connecticut, developed a machining-coolant management program to reduce end-of-pipe treatment requirements for disposal of the spent fluids.[17] The production facility fulfills the requirements of 23 different specifications for milling, turning, and grinding by operating machine tools having a total of approximately 4,000 individual sumps for various coolants. The coolants were generally water-extendable machining fluids composed of approximately 95% water, 5% oil, and trace additives. Coolant degradation, requiring disposal of the entire sump contents, was caused by accumulation of a tramp oil layer, growth of bacteria, contaminant fouling, and decrease in performance. Coolant disposal requirements were 24 million lb per year of straight oil or synthetic concentrates. These materials were chemically disrupted to crack the emulsions, with the water treated through a wastewater treatment plant and the separated layer incinerated.

Performance testing helped to reduce the number of different coolants used in the facility, which helped narrow the scope of the coolant management program. A

Table 10.5. Portable Sump Pilot Test

Cost Component	Old System: Disposal	Pilot Test: Reconditioning
Disposal of sump contents	3 times	0
Gallons for disposal	3,600	100 (from leaks)
Treatment costs	$310	$10
Coolant purchase (gal)	210	70
Coolant costs ($3 per gal)	$630	$210
Labor costs	approximately equal	
Total operating costs	$940	$280
Capital costs	–	$8,000
Payback period	–	24 months

recycling treatment system for the contents of a portable sump was pilot tested for 2 months on an 8-machine system sharing a common 1,200-gal sump. The results of the pilot test are shown in Table 10.5.

The part of the pilot test that could not be quantified in terms of cost was the reduction in risks associated with managing the coolant sump material versus letting the sump contents go out of specification and then disposing of the entire sump contents. Pratt & Whitney favors the coolant management approach as having fewer associated environmental risks and is extending the coolant management pilot program to other machining-coolant sump systems.

REFERENCES

1. Schey, J.A. *Tribology in Metalworking, Friction, Lubrication and Wear*. (Metals Park, OH: American Society for Metals, 1983).
2. Dick, R.M. "Metalworking Fluid Management," Cincinnati Milacron, Cincinnati, OH.
3. Minetola, A.J. "Bulk Lubricant Storage and Distribution—Some of the Problems," *Lubrication Engineering* 37(5):268–271 (May 1981).
4. Childers, J.C. "Metalworking Fluids—A Geographical Industry Analysis," *Lubrication Engineering* 45(9):542–546 (September 1989).
5. Lantz, T.L. "Lubricant Conservation," *Lubrication Engineering* 44(5):408–411 (May 1988).
6. Perry, R.H., and C.H. Chilton. *Chemical Engineer's Handbook*, 5th ed. (New York: McGraw-Hill Book Company).
7. Dick, R.M. "Ultrafiltration for Oily Wastewater Treatment," in *Proceedings of the 36th Annual American Society of Lubricating Engineers Meeting*, May 11–14, 1981. Available from the Society of Tribologists and Lubrication Engineers, Chicago, IL.
8. Hachadoorian, R.H. "Recycling Industrial Oil," *Engineering Digest*, December 1984.
9. Springborn, R.K., Ed. "Cutting and Grinding Fluids: Selection and Application," American Society of Tool and Manufacturing Engineers, Dearborn, MI, 1967.
10. Napier, S. "Waste Treatability of Aqueous-Based Synthetic Metal-Working Fluids," *Lubrication Engineering* 41(6):361–365 (June 1985).
11. Zelnio, L.L. "Reclamation of Metalworking Fluids from Individual Machine Tools," *Lubrication Engineering*, January 1987.
12. Brinkman, D.W. "Technologies for Re-Refining Used Lubricating Oil," *Lubrication Engineering*, May 1987.
13. Swain, J.W., Jr. "Conservation, Recycling, and Disposal of Industrial Lubricating Fluids," *Lubrication Engineering,* September 1983.
14. Webb, H.R. "Establishing Oil Recovery and Reclamation Programs," *Lubrication Engineering* 39(10):626–630 (October 1983).
15. Davies, R., R. Curtis, and R. Laughton. "Recycling Metal Stamping Plant Wastes," *Water and Pollution Control*, September/October 1985.
16. Burke, J.M. "Waste Treatment of Metalworking Fluids, A Comparison of Three Common Methods," *Lubrication Engineering* 47(4):238–246 (April 1991).
17. Zavodjancik, J.P. "Pratt & Whitney Coolant Management Program," in *Proceedings of the Seventh Annual Aerospace Hazardous Materials Management Conference*, October 27–29, 1992.

GENERAL REFERENCE

Code of Federal Regulations (*CFR*). Section 40, Chapters 261 to 270: RCRA Regulations.

CHAPTER 11

Solvents Used for Cleaning, Refrigeration, Firefighting, and Other Uses

*James Thom and Thomas Higgins**

SOLVENT USES AND WASTES

Chlorinated and nonchlorinated hydrocarbon solvents have been used for many years in cleaning, removal of coatings, painting, refrigeration, fire suppression, and foam blowing applications. The use of many of these solvents, however, is being severely restricted because of a desire by regulators and industry to minimize worker exposure and reduce the harmful effects associated with the release of solvents into the environment. A worldwide effort is being undertaken to develop alternative solvents or application methods that do not have negative impacts and can maintain or improve current capabilities.

CLEANING AND DEGREASING

In solvent cleaning and degreasing, an organic solvent is used to remove unwanted grease, oil, and other organic films from surfaces of manufactured parts. The pollutants generated include (1) liquid waste solvent and degreasing compounds containing unwanted film material; (2) air emissions containing volatile solvents; (3) solvent-contaminated wastewater from vapor-degreaser-water separators or subsequent rinsing operations; and (4) solid waste from distillation systems, consisting of oil, grease, soil particles, and other film material removed from manufactured parts.

Solvent cleaning and degreasing is used (1) before applying protective coatings, such as paint, enamel, lacquer, or electroplate; (2) before inspection; (3) before assembly of parts; (4) before and after metal working operations, such as welding, heat treatment, or machining; and (5) before packaging.[1]

Three distinct operations fit the category of solvent cleaning and degreasing: cold

*Mr. Thom is an industrial engineer in CH2M HILL's Corvallis, Oregon, office and may be reached at (503) 752-4271. Dr. Higgins is a vice president in CH2M HILL's Reston, Virginia, office and may be reached at (703) 471-1441.

cleaning, vapor degreasing, and precision cleaning. These operations and waste sources, as well as associated equipment and controls, are described below.

Cold Cleaning

Cold cleaning is the simplest, least costly, and most common type of solvent cleaning. The solvent is usually applied at or slightly above ambient temperature. It is applied either by brush or by dipping the items to be cleaned in a solvent dip tank. The most common solvent used in cold cleaning, a highly flammable mineral spirit, is known by proprietary commercial names—Stoddard Solvent or Varsol—or as petroleum distillate (PD 680). Generally the least expensive solvent used for cleaning, PD 680 costs between $1.50 and $2.50 per gallon.[2]

Vapor Degreasing

Vapor degreasing uses nonflammable chlorinated hydrocarbons in a vapor phase to clean metallic and other suitable surfaces. Compared with cold cleaning, vapor degreasing cleans faster because of the higher temperature. Some of the solvents commonly used in vapor degreasing and applications are:

- 1,1,1-trichloroethane (methyl chloroform)
- perchloroethylene (PCE)
- trichloroethylene (TCE)
- Freon TF (chlorofluorocarbon [CFC])
- methylene chloride

The simplest form of vapor degreasing employs only solvent vapor for degreasing operations and is called a straight vapor-cycle degreaser. This type of degreaser uses a tank that is one-tenth full of liquid solvent (Figure 11.1) to provide solvent vapor for degreasing. Typically, steam coils heat the solvent to its boiling point, producing solvent-saturated vapor in the upper portion of the tank. The item to be cleaned is either inserted manually or automatically into the vapor region, where hot solvent vapor immediately condenses onto the surface of the item. The condensed solvent dissolves grease and removes dirt, then drips back into the liquid bath, taking with it the dirt and grease. The cleaning continues until the temperature of the part being cleaned reaches the temperature of the vapor, and condensation ceases. Cleaning time depends on the type of solvent used, the operating temperature, the size and weight of the part, and the specific heat of the material used in the part. Solvent vapor that does not condense on the parts condenses on cooling coils in the upper part of the tank and returns to the liquid sump, thereby preventing solvent from escaping into the atmosphere.[3]

Straight vapor-cycle degreasing is not very effective on small parts or parts made of material with low specific heat, because the temperature of the part being degreased will quickly equilibrate with the vapor temperature without sufficient condensation of solvent to clean the surface. These parts can be effectively cleaned using a vapor-spray-cycle degreaser, which is equipped with a spray nozzle that directs cooler solvent from the condensate trough to mechanically wash oil and

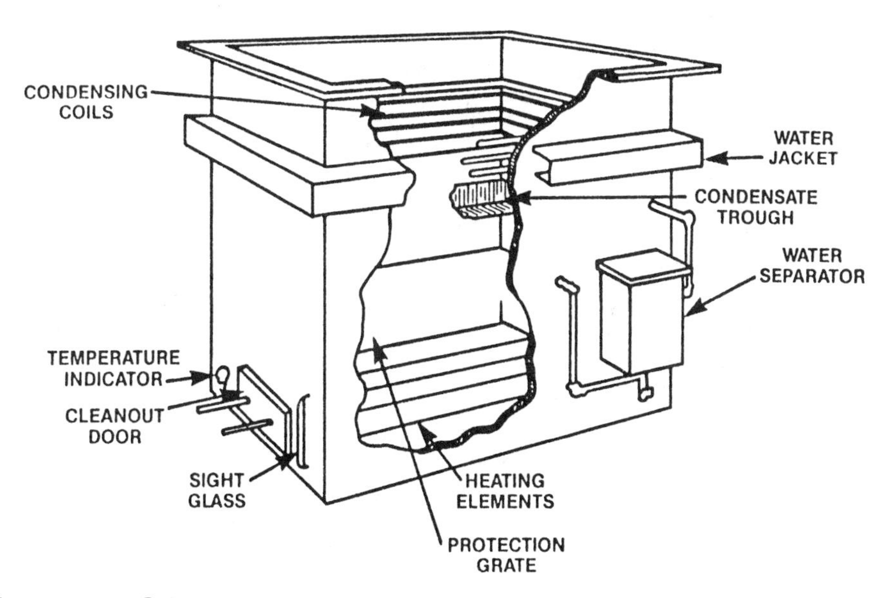

Figure 11.1. Solvent vapor degreaser.

grease from the surface of the part and partially cool the part, permitting final cleaning by condensing vapor.[3]

Several other types of degreasers are also used throughout the industry. Vapor-spray-cycle degreasers have a warm solvent compartment adjacent to a vapor area. Liquid-liquid-vapor-cycle degreasers have separate compartments for warm solvent, boiling solvent, and vapor. These types of degreasers typically are employed for small or heavily soiled parts.[3] Ultrasonic degreasers have a transducer mounted at the base of the tank that produces sound in the 20 to 40 kHz range.[3] The sound waves induce vibrations that cause the rapid formation and collapse of gas bubbles in the solvent medium, a phenomenon known as cavitation. Cavitation bubbles implode during compression, resulting in the production of small shock waves that radiate from the point of implosion. These implosions disturb the soil adhering to the parts and provide a cleaning action. Once used only for critical cleaning, these degreasers are increasingly being chosen for routine cleaning.

To reduce solvent losses to the atmosphere (air emissions) and reduce worker exposure, most vapor degreasers incorporate many of the following features:

- additional freeboard (the vertical distance from the top of the highest normal vapor level to the top of the degreasing tank)
- sliding or roll-top covers
- refrigerated freeboard or secondary chillers
- integral recovery stills

TCE used to be the most popular vapor degreasing solvent; however, worker health and safety concerns have reduced its use over the last decade. The compound 1,1,1-trichloroethane (TCA) was a favorite substitute for TCE in cleaning applications, but with the global environmental concerns over ozone depletion, this compound and Freon (CFC) cleaners are being rapidly withdrawn from use. Operations continuing to use these solvents are enhancing their air emissions controls. Some facilities are switching to hydrochlorofluorocarbons (HCFCs) and hydrofluorocarbons (HFCs).

Precision Cleaning

Precision cleaning typically consists of precleaning and cleanliness verification. Precleaning may include one or more of the following: alkaline cleaning, solvent vapor degreasing, and liquid solvent cleaning. Cleanliness verification is accomplished by flushing the article being cleaned with a solvent in a cleanroom or enclosed system to minimize recontamination. The final portion of flushed solvent is captured and analyzed for contamination. The degree of contamination is determined by filtering the solvent sample and microscopically counting and classifying the particulates on the filter material. In conjunction with filtration, the solvent sample is tested for nonvolatile residue (NVR), which typically is oil. Following cleaning, the part is double-bagged in polyethylene or Aclar bags and closed with a tamperproof seal.

The term "fine cleaning" is sometimes applied to a precision cleaning process that does not include the final cleanliness verification step.

Precision cleaning of instruments and electronic components requires solvents of high purity, high solvency, and rapid evaporation rates. Freon TF (CFC 113) and isopropyl alcohol are the two most common solvents used for the final flush in precision or fine cleaning. The choice of solvent is dictated by ignitability concerns, material compatibility, and final use of the product.

REFRIGERANTS

CFCs, HCFCs, HFCs, and blends of these compounds are used as heat-transfer fluids in the refrigeration cycle of air conditioners, freezers, and refrigerators. Some of the commonly employed refrigerants include CFC-12, HCFC-22, HCFC-123, HFC-134a, and CFC-500 and 502 (cryogenic applications). Because the refrigerant remains in a sealed system throughout the refrigeration cycle, solvents are not emitted until the unit is disposed of, although small quantities may leak from the system or be emitted during servicing. Losses during servicing have been significantly reduced as a result of the Clean Air Act Amendments, which require capture and recycling of refrigerants during these operations (Figure 11.2).

FIRE SUPPRESSANTS

Halons have long been employed as fire-suppressant agents because of their extinguishing ability, clean evaporation, electrical nonconductivity, and low toxicity. Halon 1301 is commonly used as a total flooding agent, while Halon 1211 is used as a streaming agent in fire-fighting applications. Unfortunately, halons also have one of the highest relative ozone-depleting potentials (ODP) of any solvent.

Figure 11.2. Refrigerant Management Centers store refrigerant R-12, R-22, or R-134 while vehicle is being serviced, and pumps refrigerant back into vehicle when repairs are complete. (Photo courtesy of Westinghouse Hanford Company/BCSR, Richland, WA).

Propellants and Blowing Agents

CFC-11 is commonly employed as a blowing agent in the manufacture of polyurethane foam. Polyurethane is used as insulating foam in most household refrigerators and freezers and in building insulation and roofing. CFCs captured in the foam during the manufacturing process are bound in the foam and are reported to leak out slowly (less than 2% per year) during their aging process. CFCs used in the blowing of foam that are not bound to the foam are typically not captured but are released to the environment.

PROBLEMS WITH CURRENT PRACTICES

In the past, hydrocarbons were used as cleaning agents; ammonia was used as a refrigerant; and carbon dioxide or water was widely used for firefighting. However, in many cases, these compounds were replaced by CFCs, chlorinated solvents, and halons, which are nonflammable, have high solvency, have higher thermodynamic efficiency, and are compatible with most substrate materials. Unfortunately, CFCs, chlorinated solvents, and halons also exhibit high ODP and pose a threat to human health.

Several of the solvents are carcinogenic and are regulated as hazardous substances. In addition, the costs associated with using these solvents have risen dramatically over the last decade because of increasingly strict management requirements. These requirements include engineering controls to minimize releases of these compounds, regulatory reporting requirements, worker training requirements, secondarily contained storage facilities, and disposal restrictions.

While there is much concern over the depletion of ozone in the upper atmosphere (stratosphere), equal concern exists over the formation of ozone in the lower atmosphere. Many of the solvents employed as substitutes to ozone depleters are photochemically reactive and contribute to both smog formation and global warming.

REGULATION OF OZONE DEPLETING CHEMICALS (ODCs)

To halt depletion of the earth's protective stratospheric ozone layer, an international treaty titled the "Montreal Protocol on Substances that Deplete the Ozone Layer" was ratified by the United States and 25 other countries in 1987. This agreement calls for production limits on CFCs and halons and for a scheduled phasing out of the production of these substances.

In June 1990, the parties to the Montreal protocol met in London and decided to completely phase out production of the CFCs and halons already subject to the protocol control requirements by January 1, 2000. Methyl chloroform (TCA), carbon tetrachloride, and HCFCs, which were not originally regulated by the protocol, were added. Under the recent amendments, the current manufacturing levels of CFCs are frozen at 1986 levels and are to be reduced to 50% of the 1986 levels in 1995, further reduced to 15% of 1986 levels in 1997, and completely eliminated in the year 2000.

At the June 1990 meeting, the parties also passed a resolution regarding the use of HCFCs as transitional or interim substitutes for CFCs. HCFCs generally have a much lower ODP than CFCs. A phasing out of the production of HCFCs is scheduled for 2020, if feasible, and not later than 2040.

The production phaseout schedules for ODCs per the Montreal protocol and Title VI of the Clean Air Act Amendments are provided in Table 11.1.

Table 11.1. Production Phaseout Schedule for Ozone-Depleting Substances (Percent of Base Year)

	CFCs and Halons		Methyl Chloroform (TCA)	
Base Year	Montreal Protocol (% of 1986 Production)	Title VI of the CAA (% of 1986 Production)	Montreal Protocol (% of 1989 Production)	Title VI of the CAA (% of 1989 Production)
1991	100	85	100	100
1992	100	80	100	100
1993	80	75	100	90
1994	80	65	100	85
1995	50	50	70	70
1996	50	40	70	50
1997	15	15	70	50
1998	15	15	70	50
1999	15	15	70	50
2000	0	0	30	20
2001	—	—	30	20
2002	—	—	30	0
2005	—	—	0	0

Note: 1986 is the base year for all Class I substances except TCA, for which the base year is 1989. All effective dates are January 1 of the base year.

Title VI of the 1990 Clean Air Act Amendments enables the United States to fulfill its obligations under the 1989 revisions of the Montreal protocol that restrict ODCs. The amendments include measures such as a production phaseout schedule, a motor-vehicle air-conditioner recycling rule, a recycling and emission reduction program, a ban on nonessential products, and a labeling requirement.

ODC Phaseout Schedule

Title VI defines two chemical classes. Substances with a large ODP are Class I substances. Class I substances are further segregated into the five groups shown in Table 11.2 The phaseout schedule for each Class I chemical will depend upon the group to which it belongs. Currently, all Class II substances are HCFCs, which are relatively new solvents with low ODPs.

The amendments phase out production of TCA by 2002, which is 3 years earlier than the Montreal protocol deadline. However, subsequent production of TCA may be allowed until 2005 for medical devices and aviation safety purposes if no safe and effective substitutes are available. TCA has a relatively low ODP, but because it is used in extremely large volumes, it contributes significantly to atmospheric chlorine levels and ozone depletion.

To assure yearly compliance with the phaseout schedule of Class I substances, the U.S. EPA will issue allowances that will enable a company to produce or import a quantity of a regulated substance. The number of allowances issued for a chemical will depend on the base year production of the chemical and the phaseout year. For example, in 1997, CFC-113 can only be produced in quantities no greater than 15% of a company's 1986 (the baseline year for Class I substances, except TCA) produc-

Table 11.2. Class I Ozone-Depleting Substances According to the Clean Air Act Amendments

Group I	chlorofluorocarbon-11 (CFC-11)
	chlorofluorocarbon-12 (CFC-12)
	chlorofluorocarbon-113 (CFC-113)
	chlorofluorocarbon-114 (CFC-114)
	chlorofluorocarbon-115 (CFC-115)
Group II	halon-1211
	halon-1301
	halon-2402
Group III	chlorofluorocarbon-13 (CFC-13)
	chlorofluorocarbon-111 (CFC-111)
	chlorofluorocarbon-112 (CFC-112)
	chlorofluorocarbon-211 (CFC-211)
	chlorofluorocarbon-212 (CFC-212)
	chlorofluorocarbon-213 (CFC-213)
	chlorofluorocarbon-214 (CFC-214)
	chlorofluorocarbon-215 (CFC-215)
	chlorofluorocarbon-216 (CFC-216)
	chlorofluorocarbon-217 (CFC-217)
Group IV	carbon tetrachloride
Group V	methyl chloroform (TCA)

tion amount. However, this allowance system only pertains to companies that produced or imported the Class I substances and not to facilities that purchased the substance from industries in the United States. By restricting the production and importation of Class I chemicals, the U.S. EPA will implicitly reduce the use of the Class I substances. As the supply decreases, the price of regulated chemicals will increase, providing an incentive for developing and using viable alternatives.

The U.S. EPA considers HCFCs, which are Class II substances, as a temporary transitional substitute for the harmful CFCs. But because the HCFCs add chlorine to the stratosphere, the U.S. EPA plans to freeze production levels by 2015 and completely phase these chemicals out by 2020, if feasible. Although HCFCs currently may be considered a viable alternative to CFCs, replacements for these chemicals will be needed in the future. Developing substitutes to CFCs will be especially important for automobile air conditioners because these appliances are responsible for the largest CFC use in the United States. The U.S. EPA has developed rules to prevent unnecessary emissions during the servicing of automobile air conditioners.

Servicing of Motor Vehicle Air Conditioners

As of January 1, 1992, anyone who services automobile air conditioners must be certified by the U.S. EPA and must use recycling equipment that is certified by the U.S. EPA. Rather than developing a new set of certification regulations, the U.S. EPA will employ rules developed by trade groups. The U.S. EPA will approve recycling and recovery equipment that complies with standards that are at least as stringent as the standards developed by the Society of Automotive Engineers. Equipment previously certified by the Underwriters Laboratory (U.L.) will be considered "substantially identical" to approved equipment. The U.S. EPA will also approve technicians who are certified under programs established by the National Institute for Automotive Service Excellence (ASE), the Mobile Air Conditioning Society (MACS), or similar programs. Such programs properly train technicians to use approved mobile air-conditioning-service equipment. Similar regulations concerning the servicing of other appliances that contain CFCs will be developed by the U.S. EPA, as discussed below.

National Recycling and Emission Reduction Program

The U.S. EPA, under the National Recycling and Emission Reduction Program, will establish rules that will:

- Assure maximum recovery of Class I substances during the servicing of appliances or industrial-process refrigerators
- Forbid the deliberate venting of Class I or II substances into the atmosphere during the servicing of appliances
- Require that all Class I and II substances be removed from appliances before the appliances are disposed of
- Require that all appliances manufactured or sold that contain Class I or II substances in bulk be equipped with a device, such as a servicing aperture, to facilitate the recapture of the substances during service

The rule forbidding the deliberate venting of Class I or II substances during the servicing of appliances was effective on July 1, 1992. However, the other rules were not developed until approximately June 1992 and were not effective until 6 months after the rules had been established. To assure the maximum recovery of Class I substances during the servicing of appliances, the U.S. EPA again will require that individuals who service the appliances be certified by the U.S. EPA and use certified equipment. Currently, the U.S. EPA is planning to approve both the technician certification programs offered by the Refrigerator Service Engineering Society and the recycle and recovery equipment that has been approved by the Air Conditioning and Refrigeration Institute. Rules that will assure maximum recovery and emission reduction during the use and disposal of Class I and II substances, other than from appliances and industrial-process refrigerators, will be developed within 4 years of the enactment of the amendments.

Safe Alternatives Policy

Federal research programs will be established to identify safe alternatives to Class I and II substances. The alternatives may include chemical and product substitutes and alternative manufacturing processes. The list of approved safe alternatives and their unsafe counterparts will be available to the public. Once a safe alternative has been developed, use of its unsafe counterpart will be illegal.

Ban of Nonessential Products

As of November 15, 1992, selling nonessential items that release Class I substances into the environment, either during use or during the manufacturing process, was unlawful. In defining nonessential items, the U.S. EPA considers the item's purpose, the availability of substitutes, safety, health, and other issues. Some of the potentially nonessential items include halon fire extinguishers, CFC-propelled noise horns, and hair care products. Important products may be taken off the market in the future, but there will be replacements for them. Products that are considered essential may be subject to a labeling requirement.

Labeling Requirement

Products that contain ODCs and are introduced into interstate commerce will need to follow applicable labeling requirements. Containers of Class I or II chemicals and products that contain Class I chemicals will need to have a label stating that the container or product contains "a substance that harms public health and the environment by destroying ozone in the upper atmosphere." The products that contain Class II substances will be required to have a warning label only if the U.S. EPA has determined that a safe alternative exists for the product. In addition, products that are manufactured using Class I or II chemicals and have a U.S. EPA-approved safe alternative also will be required to have a similar label stating that the item was manufactured with such a chemical. Eventually, all products that are manufactured using Class I and II substances will need to fulfill the labeling requirement, regardless of whether a safe alternative exists.

Excise Taxes on CFCs and ODCs

To further discourage the continued use of controlled ODCs, the Omnibus Budget Reconciliation Act of 1989 imposed an excise tax on chemicals that deplete the ozone layer and on imported products containing or manufactured using these chemicals. This tax is imposed on an ODC when it is first used or sold by a manufacturer or importer. The manufacturer or importer is liable for the tax.

The excise tax was developed as an incentive for users to reduce their consumption of ODCs. Not only has the cost of using ODCs increased dramatically through excise taxes, but the cost of materials also has increased. The taxable ODCs per the Internal Revenue Service can be found in Table 11.3. The tax rate is determined by multiplying the weight in pounds of the ODC by the base tax rate for that year and by the ozone-depleting factor. After 1995, the base tax amounts for each pound of ODC will increase by 45 cents each year.

Floor Stocks Taxes

Any owner/operator holding title to any ODC or ODC-containing material is required to pay a floor stocks tax to the Internal Revenue Service (see IRS Publication 510, January 1991). This tax only applies if the holder has at least 400 pounds of ODCs on hand. In 1991, only post-1990 ODCs, halons, and rigid foam ODCs were counted in determining this 400-pound figure. In 1992, only post-1989 ozone-depleting chemicals were taken into account, and in 1993, all ozone-depleting chemicals were taken into account. In 1994, post-1990 ODCs (400 pounds or more), rigid

Table 11.3. Excise Tax Schedule

ODCs	Ozone-Depletion Factor	Rate of Tax ($/lb)			
		1991	1992	1993/94	1995
CFC-11	1.0	1.370	1.67	2.65	3.10
CFC-12	1.0	1.370	1.67	2.65	3.10
CFF-113	0.8	1.096	1.33	2.12	2.45
CFC-114	1.0	1.370	1.67	2.65	3.10
CFC-115	0.6	0.822	1.00	1.59	1.86
Halon 1211	3.0	0.2466	0.2505	0.2623	
Halon 1301	10.0	0.2466	0.2505	0.2650	
Halon 2402	6.0	0.2466	0.2505	0.2544	
CFC-13	1.0	1.370	1.67	2.65	3.10
CFC-111	1.0	1.370	1.67	2.65	3.10
CFC-112	1.0	1.370	1.67	2.65	3.10
CFC-211	1.0	1.370	1.67	2.65	3.10
CFC-212	1.0	1.370	1.67	2.65	3.10
CFC-213	1.0	1.370	1.67	2.65	3.10
CFC-214	1.0	1.370	1.67	2.65	3.10
CFC-215	1.0	1.370	1.67	2.65	3.10
CFC-216	1.0	1.370	1.67	2.65	3.10
CFC-217	1.0	1.370	1.67	2.65	3.10
Carbon Tetrachloride	1.1	1.507	1.83	2.91	3.41
Methyl Chloroform	0.1	0.137	0.16	0.26	0.31

foam ODCs (200 pounds or more), and halons (20 pounds or more) are taken into account.

If an owner/operator is liable for the tax, an inventory of the taxable ODCs must be prepared on January 1 of each applicable year. The floor stocks tax is due on June 30 of each year.

Most of the usage of CFC, halon, and certain chlorinated hydrocarbons has been curtailed significantly as a result of ozone depletion, global warming, worker health concerns, and the costs associated with managing a hazardous substance. Today's climate of production restrictions, high excise taxes, and public pressure is forcing industries to eliminate the use of these compounds.

APPROACHES TO SOLVENT WASTE REDUCTION

The four basic approaches to reducing or eliminating the adverse impacts of current solvents are the following:

- Modify equipment to reduce solvent emissions or waste production.
- Make operational improvements.
- Replace existing solvents with those that are more environmentally friendly.
- Recycle waste solvents.

Equipment Modifications

Many older vapor-degreasing units can be retrofitted to incorporate newer features designed to reduce emissions and minimize solvent usage. Typical equipment modifications are described below.

Add or improve cover. A good cover over open-top degreasers is perhaps the simplest yet most important device for reducing losses, especially during idle times. Reductions in solvent losses of 30% to 50% have been reported with proper use of covers. Sliding or rolltop covers are particularly beneficial because, unlike hinged covers, they do not cause turbulence when they are closed or opened.

Increase freeboard. As the ratio of freeboard height to width of tank (freeboard ratio) increases, solvent losses decrease. Various sources recommend upgrading equipment to a minimum freeboard ratio of 75% to 100%, and many new designs have freeboard ratios as high as 150%.[1]

Install refrigerated chillers. Solvent vapors are maintained at the desired level in the vapor degreaser by controlling the flow of coolant through primary condenser coils or in the jacket of the degreaser wall. To improve containment of vapors, additional coils are typically placed in the tank at the vapor level to control the upper boundary of the vapor zone. Either cooling water or a mechanically cooled refrigerant is used as the cooling medium in the coils.

Retrofit degreasers with a freeboard refrigerated chiller system. Although this appears to be a second set of condenser coils located slightly above the primary condenser coils, functionally it is different. Whereas primary condensers are used to produce the vapor cloud, refrigerated chilling coils impede diffusion of solvent vapors from the vapor zone into the work atmosphere. These chillers typically operate between 0° and -20°C. Manufacturers claim that 20% to 40% reductions in solvent loss can be achieved by adding freeboard chillers to an existing vapor degreaser.

Install recovery stills. Spent solvent is a valuable resource and should not be discarded. Solvent recovery stills should be used when feasible. It is typically worthwhile to compare a centralized distillation system with integral stills on individual degreasers. If distillation recovery is practiced, a regular testing program is critical to ensure that stabilizer, other additives, and impurities (contaminant) concentrations are maintained within acceptable ranges.

Install safety switches. Safety switches are preventive devices that reduce solvent usage and emissions during malfunctions. They have no effect on normal operations. The five main types of safety switches are (1) safety vapor thermostat, (2) condenser water flow switch and thermostat, (3) sump thermostat, (4) solvent level control, and (5) spray safety switch. The first four switches turn off the sump heat during malfunction; the fifth switch turns off solvent spray.

Add automated hoist. Installing a speed-control hoist can reduce losses caused when parts are loaded or withdrawn too rapidly. The vertical rate of entry or removal of parts should be limited to less than 11 ft/min.[1]

Modify spray wands. Handheld spray wands can be easily misused above the vapor zone by operators. Consideration should be given to switching to a fixed-spray system below the vapor zone.

Install a vapor recovery system. Three technologies are available for capturing and reusing the solvent vapors released by a degreasing operation. These technologies include (1) carbon adsorption with steam regeneration, steam condensation, and solvent/water separation, (2) reverse Brayton Cycle, and (3) chiller/condenser followed by solvent/water separation. The cost of these systems can range from $300,000 to $1,000,000 to treat a flow rate of 1,000 standard cubic feet per minute of solvent vapor.

Retrofit to alternative cleaning solvents. Existing vapor degreasing equipment can be retrofitted to use HCFC or an aqueous cleaner (explained later in this chapter). Modifications required to convert a chlorinated solvent degreaser (for example, TCA) to an HCFC degreaser are relatively minor. However, the cost of retrofitting will depend on which HCFC compound is chosen as a substitute. If it is not part of existing equipment, freeboard refrigeration typically will need to be added to control emissions and minimize solvent losses. Equipment modifications required to convert an existing vapor degreaser into an aqueous cleaner can be cumbersome and expensive. The cost of conversion is dependent on the work load; the type, shape, and size of parts being processed; and the type of aqueous cleaner selected.

Alternative Equipment

Hermetically sealed vapor degreasers were developed to minimize solvent usage, protect the global environment, and minimize worker exposure. These ultra-low-emission vapor degreasers incorporate both high-efficiency degreasing and solvent capture-and-recovery technologies. They can be conceptualized as a conventional vapor degreaser contained in an airtight enclosure. These units are equipped with automated controls, airlocks, and built-in vapor capture and recovery systems. Figure 11.3 shows a schematic of a hermetically sealed vapor degreaser. According to the manufacturers, these units can reduce solvent usage by 80% to 99% compared to conventional open-top degreasers equipped with covers and freeboard refrigerated chillers. Recent tests conducted by Battelle reported that hermetically sealed vapor degreasers had solvent losses of less than or equal to 1 g per hour and had maximum worker exposure of 3 ppm of solvent.[4]

Operational Improvements

Good operating practices are essential in any waste reduction program. Often overlooked, operational improvements can significantly reduce solvent usage and air emission losses. The following is a list of potential operational improvements:

- Modify exhaust fans, blowers, and ventilation ducts to minimize velocities across equipment while still providing adequate worker protection.
- Provide baffles and shields to protect the degreasers from air currents caused by exits and doorways.
- Limit operation to trained personnel.
- Generate adequate heat input to maintain a vapor blanket. This would prevent "work shock" or collapse of the vapor blanket.

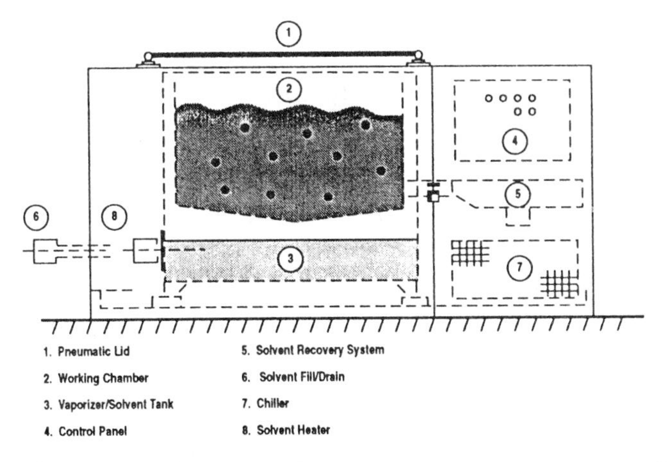

Figure 11.3. Schematic of a hermetically sealed vapor degreaser (Source: Durr Industries).

- Balance the heat input versus cooling capacity to maintain the vapor blanket on the center coil of the condenser.
- Use baskets with cross-sectional areas that are less than 50% of the cross-sectional area of the degreaser to avoid a piston effect when raising and lowering baskets in degreasers.
- Limit the vertical speed of work to less than 11 ft/min to avoid disturbing the vapor level.
- Turn on the cooling water or cooling fluid first during startup and turn it off last during shutdown.
- Avoid adding water or placing excessively wet parts into degreasers and maintain water separators. Water in a degreaser can cause corrosion of the equipment, spotting of the processed work, and increased solvent consumption.[1]
- Keep degreasers covered when not in use to avoid excessive loss of solvent.
- Avoid overloading the degreaser; overloading causes the vapor blanket to collapse.
- Maintain the stabilizer level by regular analysis and acid acceptance testing.
- Remove metal pieces, oil, fines, and sludge from the solvent sump regularly to prevent shortening the life of the solvent. Certain metal particulates (i.e., aluminum and zinc fines) can lead to the decomposition of TCA solvent if allowed to accumulate in the solvent sump. This accumulation can lead to rapid, violent reactions if the degreaser turns acidic. The reaction products are corrosive fumes and black, gummy residues of carbon and organic-metal compounds, which also are corrosive.
- Repair leaks and conduct routine equipment inspections to reduce solvent loss.
- Evaluate cleanliness requirements and the type of cleaner and process being employed. Control factors (storage and handling procedures) that contribute to surface contamination of the parts in order to reduce or eliminate cleaning requirements.
- Evaluate the need for the number and type of solvent cleaning operations at the facility. Consolidation of work loads is an easy first step toward waste reduction and lowering of emissions and solvent usage.

Table 11.4 lists several vendors who can supply state-of-the-art vapor degreasers, including hermetically sealed units (Durr Industries, Inc.) and equipment modifications.

Case Study 11.1: Pratt & Whitney's Vapor Degreaser Consolidation/Replacement Program, Phase 1

In 1988, Pratt & Whitney (P&W), an operating unit of United Technologies Corporation, began looking for pollution prevention strategies through source elimination and process modification. Their program, co-funded by the U.S. Air Force

Table 11.4. Vapor Degreaser and Solvent Cleaning Equipment Suppliers

Company	Address	Phone
Baron & Blakeslee	2001 N. Janice Avenue Melrose Park, IL 60160	(708) 338–0571
Durr Industries, Inc. Finishing Systems	40600 Plymouth Road P.O. Box 2129 Plymouth, MI 48170–4297	(313) 459–6800
Ultra-Kool, Inc.	P.O. Box 458/500 County Line Road Gilbertsville, PA 19525	(215) 367–2019

Industrial Modernization Incentives Program, was carried out in two phases: (1) eliminating redundant or unnecessary cleaning operations and (2) replacing required chlorinated solvent cleaning operations with safe alternatives. The first phase of the project is described below. The second phase of the project is described in Case Study 11.2.

Phase 1 focused on consolidation of solvent-cleaning operations. This phase was initiated with a data-gathering task involving the following questions:

- Where are the degreasers located?
- Who owns or is responsible for the degreasers?
- What solvent is being used? (PCE, TCA, or a CFC)
- What is being removed? (lubricants, oils, etc.)
- What type of parts are being cleaned? (large, small, intricate passages, material of construction, etc.)
- How often is the equipment used? (hours per day of operation)
- Why are the parts being cleaned? (between operations or before inspection)

Following data collection, informed decisions were made about where units and different solvent types were required. The low utilization of vapor degreasers, 10% to 30% of the workday, indicated that a number of vapor degreasers could be eliminated through consolidation. The data also showed that PCE, a high-temperature solvent, was being used to remove compounds that could easily be removed using lower-temperature solvents such as TCA. This finding led P&W to decide to begin using a single solvent, TCA, in the remaining degreasing operations.

Phase 1 was very successful to mid-1992. Since 1988, P&W has reduced the number of operating vapor degreasers from 161 to 74 while maintaining or increasing production. This cutback has contributed to a 60% reduction in toxic air emissions since 1988. In addition, P&W was very successful in converting to TCA as the standard solvent in vapor degreasers. Only two PCE degreasers currently remain in operation in all seven P&W facilities. It is P&W's belief that using a single solvent will greatly simplify the task of eliminating chlorinated solvents altogether from the manufacturing process.

ALTERNATIVE SOLVENTS

Chlorinated solvents were adopted by industry because of their ability to dissolve a wide range of organic contaminants. In addition, their low flammability and high vapor pressure allow them to evaporate at room temperature and leave a residue-free surface. However, concerns about the toxicity and environmental effects (ozone depletion in the upper atmosphere) of chlorinated solvents have led many companies to search for replacements.

Alternative chemicals marketed as substitutes to chlorinated solvents in cleaning operations generally fall into three categories: organic solvents, aqueous cleaners, and semi-aqueous cleaners.

Supercritical fluids, such as supercritical carbon dioxide, are also marketed as an alternative to chlorinated solvents, especially in precision cleaning applications. However, supercritical fluid equipment can be expensive and is currently limited to smaller-part applications for economic reasons.

Many of the alternative compounds exhibit zero or low potential for ozone depletion. However, uncertainty over the toxicity of some compounds (for example, d-Limonene); photochemical reactivity of others, which contribute to smog formation; and the safety hazards of flammable and combustible solvents limit their widespread use. Many of the substitutes are subject to regulatory restrictions.[5]

Alternative Organic Solvents

Several organic solvents are used as substitutes for chlorinated solvents in removing polar contaminants. In general, these compounds are classified as volatile organic compounds and are flammable. Alternative organic solvents include:

- ketones, such as methyl ethyl ketone (MEK) and methyl isobutyl ketone (MIBK)
- aromatics, such as benzene, toluene, xylenes, and m-Cresol
- aliphatics, such as mineral spirits, Stoddard Solvent, n-pentane, n-hexane, and n-heptane
- alcohols, such as methanol, ethanol, isopropanol, n-butyl and isobutyl alcohol, cyclocexanol, and ethylene glycol

Although these solvents can be effective in many cleaning applications, their acceptance is somewhat limited because of concerns about fire hazards, health and safety, and environmental effects. Despite these drawbacks, they are being used by many in the electronics industry.

Aqueous Cleaners

Aqueous cleaners are finding an increasing market as a replacement for organic solvents in cleaning operations. Although aqueous cleaners are not a universal replacement for chlorinated solvents, it is estimated that approximately 85% to 90% of all metal degreasing operations and nonsurface-mount electronics cleaning applications currently using chlorinated solvents could use aqueous cleaners.[5] Many companies have conducted extensive testing to determine the effectiveness of various aqueous cleaners with various equipment and operating conditions. Tests have been conducted on removal of soils (oils, grease, sediment, and dirt) from parts of many different shapes, sizes, and materials of construction. Some companies have switched to aqueous cleaning almost entirely; others are proceeding cautiously. In most cases, an extensive testing and pilot program is necessary to determine the type of cleaner and equipment that is best for replacing a solvent cleaning operation. Most aqueous cleaners are alkaline, with an operating pH between 9 and 11.[5] However, acidic cleaners and emulsion cleaners also are used. Table 11.5 lists suppliers of aqueous cleaning chemicals; Table 11.6 lists suppliers of cleaning equipment.

Alkaline Cleaners

Alkaline cleaners can be used to remove virtually all organic and inorganic contaminants that commonly are removed using chlorinated solvents. The active ingredients in alkaline cleaners are alkyl benzene sulfonates or other anionic or non-ionic

Table 11.5. Suppliers of Aqueous Cleaner Chemicals

Company	Address	Phone/*Fax*
Allied-Kelite Div., Witco Chemica Corp.	2701 Lake Street Melrose Park, IL 60160	(708) 450–7435 *(708) 450–7265*
Ardrox	16961 Knott Avenue LaMirada, CA 90638	(714) 739–2821 *(714) 670–6480*
DuBois Chemicals Div. Chemed Corp.	DuBois Tower Cincinnati, OH 45202	(513) 762–6000 *(800) 762–6601*
Eldorado Chemical Co., Inc.	Box 34837 San Antonio, TX 78265	(512) 653–2060 *(512) 653–0825*
Enthone-OMI	2779 El Presidio Long Beach, CA 90810	(310) 537–0288 *(310) 637–2595*
MacDermid, Inc.	245 Freight Street Waterbury, CT 06702	(203) 575–5700 *(203) 575–5630*
Oakite Products, Inc.	544 S. 6th Avenue City of Industry, CA 91746	(818) 968–1551 *(818) 968–1380*
Quaker Chemical	Elm and Lee Streets Conshohocken, PA 19428	(215) 832–4000 *(215) 832–4497*
Shipley Co., Inc.	2300 Washington Street Newton, MA 02162	(617) 969–5500 *(617) 332–2350*
Stuart Ironsides, Inc.	4346 Tacony Street Philadelphia, PA 19124	(215) 744–5859 *(215) 289–4417*
W.R. Grace & Company	55 Hayden Avenue Lexington, MA 02173	(617) 861–6600 *(617) 861–9066*
	2140 Davis Street San Leandro, CA 94577	(510) 568–3427 *(510) 562–2394*

Table 11.6. Suppliers of Cleaning Equipment

Company	Address	Phone/*Fax*
Baker Bros. Div., Systems Eng. & Mfg. Corp.	44 Campanelli Parkway Stoughton, MA 02072	(617) 344–1700 *(617) 341–1713*
Belco Industries, Inc.	115 E. Main Street Belding, MI 48809	(616) 794–0410 *(616) 794–3424*
Blue Wave Ultrasonics	960 S. Rolff Street Davenport, IA 52802	(800) 373–0144 *(319) 322–0144*
Branson Ultrasonics Corp.	41 Eagle Road Danbury, CT 06813–1961	(203) 796–0400 *(203) 796–0450*
Durr Industries, Inc.	40600 Plymouth Road P.O. Box 2129 Plymouth, MI 48170–4297	(313) 459–6800 *(313) 459–5837*
Finishing Equipment, Inc.	3640 Kennebee Drive St. Paul, MN 55122	(612) 452–1860 *(612) 452–9851*
Jensen Fabricating Engineers, Inc.	555 Wethersfield Road Berlin, CT 06037	(203) 828–6516 *(203) 828–0473*
Ransohoff Corp.	N. Fifth St. at Ford Boulevard Hamilton, OH 45011	(513) 863–5813 *(513) 863–8908*
ROTO-JET of America Co., Inc.	2819 San Fernando Boulevard Burbank, CA 91504	(818) 841–1520 *(818) 841–6448*
Turco Products Div.	Box 195 Marion, OH 43302	(614) 382–5172 *(614) 382–5343*

surfactants. Sodium salts of phosphates, carbonates, silicates, hydroxides, and zeolites aid in suspending soil and preventing redeposition. Silicate salts and polyphosphates are typical corrosion inhibitors added to the cleaners. Other additives may include anti-oxidants such as borate, stabilizers, and small amounts of water-miscible organic solvents such as glycol ether.[6]

Example. The Torrington Company manufactures automobile bearings. Oils are used in stamping and quenching operations in the manufacturing of the bearings. Previously, TCA was used in a vapor degreaser to remove the stamping and quenching oils. When the company expanded in 1982, they switched to a hot-water alkaline cleaner, using a washer for automated parts that is manufactured by Jenson Fabricating Engineers of Berlin, Connecticut. The custom-made washer, which includes a hot-air dryer, cost about $40,000 installed and was estimated to have had a payback period of 1 year, principally because of reduced solvent costs.[7]

Acidic Cleaners

Acidic cleaners are commonly used to remove rust and scale and also can be used to remove oxides, flux residues, corrosion products, and tarnish films. Acidic cleaners can be used to clean aluminum, which can be etched if cleaned with alkaline cleaners.[6]

Emulsion Cleaners

Emulsion cleaners consist of a solvent suspended (emulsified) in a water-based cleaning solution. These cleaners typically contain emulsifiable solvents such as alcohol, methylene chloride, or methyl chloroform.[6] Emulsion cleaners are used to remove organic contaminants on heavily soiled parts. Spent cleaning solutions can be difficult to treat or dispose of, and volatile emissions from these processes may need to be controlled.

Case Study 11.2: Pratt & Whitney's Vapor Degreaser Consolidation/Replacement Program, Phase 2

Following the accomplishments of the vapor degreaser consolidation program (Phase 1, described above in Case Study 11.1), the P&W team focused on the more difficult task of identifying alternatives for the solvent cleaning processes that remained. Phase 2 of the project consisted chiefly of laboratory testing of alternatives. This phase had several objectives; the most important are listed below along with the results.

- Develop a test procedure that could be used to evaluate alternative cleaners and establish a baseline level of cleanliness. A procedure was developed that relied on measured residual weight and water-break scores using test panels soiled with four different lubricants and oils.

- Perform the cleanliness test procedure (residual weight and water-break scores) on panels using PCE and TCA vapor degreasing to establish a baseline cleanliness level. All candidate cleaners were required to meet or exceed the cleanliness levels obtained using PCE and TCA on test panels.

- Apply the testing procedure to alternative cleaners. Twelve water-based and three nonchlorinated organic cleaners were tested. The test results were compared and ranked with the PCE and TCA test panel cleanliness scores. The following cleaners passed initial screening tests and are listed here for completeness; their listing does not constitute an endorsement by P&W or the author: Turco's Sprayeze LT, Oakite's 220 NP, and Modern Chemical's Blue Gold used as immersion cleaners; Turco's Sprayeze LT and W.R. Grace's Daraclean 283 used as spray-wash cleaners; and Stuart Ironside's KeyChem 013386 and Luscon's Voltkut 30GW nonchlorinated organic cleaners.

- Statistically analyze the data to determine the effect of the factors that were varied in the study. The study concluded that equipment selection (ultrasonic, spray, immersion, or agitation) was the most important factor in achieving adequate cleanliness levels, accounting for 45% of the total. The type of cleaner chosen accounted for only 12%, and cleaner concentration, only 6%. Cleaning time, cleaner temperature, and cleaner's soil loading each accounted for 3%. However, 28% of the effect on cleaning could not be accounted for in the six variables included in the study and was therefore unassigned.

Following the testing program, P&W began using these cleaners in the shop on actual parts and soil types. Initial studies indicate that the cleaners that were effective in the testing program are also effective in shop applications.

High-Pressure Hot-Water Washers

Hot-water washers can be used without solvents or detergents in some applications, such as cleaning oil and grease from engine compartments. Suppliers of high-pressure hot-water washers are listed in Table 11.7. A typical unit is shown in Figure 11.4.

Table 11.7. Suppliers of Self-Contained High-Pressure Hot-Water Washers

Name	Address	Phone
Hydro Systems Co.	3798 Round Bottom Road Cincinnati, OH 45244	(513) 271–8800
Sioux Steam Cleaner Co.	One Sioux Plaza Beresford, SD 57004	(605) 763–2776

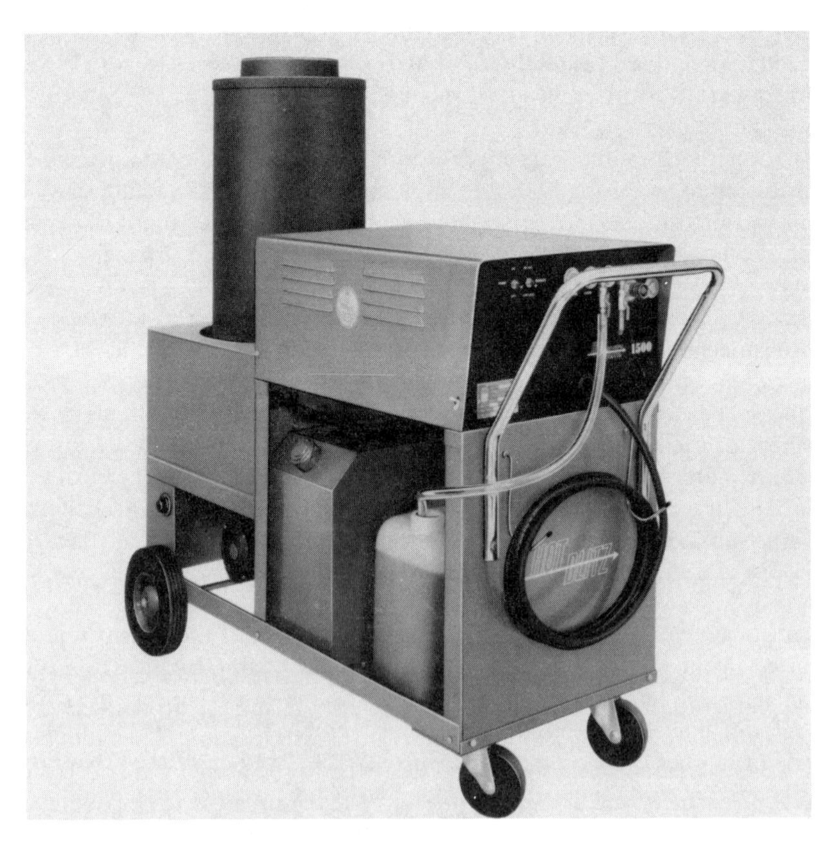

Figure 11.4. Hot-water washer (Photo courtesy of Hydro Systems Company, Cincinnati, Ohio).

Case Study 11.3: High-Pressure Hot-Water Washers for Vehicle Maintenance Cleaning

Cleaning and maintaining military vehicles can produce a waste that is difficult to treat. At Fort Polk, Louisiana, cleaning engine compartments with solvents, steam, and detergents resulted in an emulsified, frothy waste that had the consistency of chocolate mousse. The detergents tended to emulsify the oil in the condensed steam, making separation impossible in a simple oil/water separator. Solvents contaminated both the water and the oil, rendering both a hazardous waste. Disposal of this mixed waste cost $0.80 per gallon F.O.B. at the disposal site; in 1985, the cost was $84,000 plus shipping.

Similar problems were encountered at Fort Lewis, Washington. At this Army post, however, the vehicle maintenance racks were equipped with high-pressure (800 pounds per square inch [psi]) hot-water washers. Pressure washing at higher than 800 psi is not recommended because it can open grease fittings, causing water contamination of the engine grease. The self-contained, oil-fired, high-pressure hot-water washers cost less than $2,000 each. These units supply 3.5 gal/min of hot-water jets to a handheld wand that cuts the grease and oil from the engine compartments without the need for detergents or solvents. Replacing the old facilities (which used 30 gal/min of cold water and steam cleaners) with hot-water washers has reduced both water usage and maintenance.

Because the oil and grease were no longer emulsified, a simple oil/water separator was sufficient to treat this wastewater. In 1984, a total of 46,000 gallons of used oil was recovered and sold to a recycler for $10,800.

Substitute Refrigerants

Because of the planned phaseout of all ozone-depleting CFCs, refrigerant manufacturers are developing alternatives. The criteria for a viable alternative are that the replacement chemical must be compatible with the original refrigerant equipment and that the replacement is safe to use.

The best alternatives to CFCs used in refrigeration applications are HCFCs and HFCs. These products have little or no ODP and are nonflammable. Because HCFCs have a small ODP, their production is scheduled for phaseout by at least 2030. Development of HCFCs is continuing so that HCFC can be used as a transitional working fluid while research is conducted to determine a material with the desired thermodynamic properties and a zero ODP.

Table 11.8 shows possible replacements for some of the currently used refrigerants, and Table 11.9 shows the physical characteristics of these replacements.

HFC-125 is currently listed under the Toxic Substances Control Act (TSCA). HFC-125 is a non-ozone-depleting candidate replacement for CFC-502 in low-to mid-temperature refrigeration. Developmental quantities currently are available; however, application testing and lubricant development still are required before this material can be recommended for new equipment. HCFC-22 is another compound designed to replace CFC-502 in low-temperature refrigeration applications.

HFC-134a is a new alternative and is designed to replace CFC-12. It has a good refrigeration capacity and a relatively high critical temperature for high-temperature applications. However, for medium- and low-temperature applications, HFC-134a requires a high compression ratio, which reduces the refrigeration capacity of conventional reciprocating compressors (piston-type). Thus, the compressor must have

Table 11.8. Replacement Materials

Substitute Material	Manufacturer	ODP	Replaces	Application
HCFC-123	DuPont	.02	CFC-11	Centrifugal chillers
HCFC-124	DuPont	.02	CFC-11	Centrifugal chillers
HFC-134a, HCFC-22	DuPont	0 0.05	CFC-12	Centrifugal and reciprocating chillers; appliances, med. temp. refrigeration, and air conditioning
HFC-125, HCFC-22	DuPont	0 0.05	CFC-502	Retrofit in low temp. refrigeration units

a significantly larger displacement with HFC-134a than when CFC-12 is used to accomplish the same level of cooling. Furthermore, HFC-134a systems are expected to consume 10% to 15% more power than CFC-12 systems.

Both HFC-134a and HFC-125 have limiting characteristics when compared to existing refrigerants. Extensive testing is necessary to completely characterize their properties, and process engineering will be required to overcome their limitations. Research into the properties of these new refrigerants and how they affect power consumption is under way.

The greatest problem with HFC-134a is that its compatible lubricants degrade the insulation materials and some of the polymers used in many compressors. These lubricants are also less effective than current lubricants and may cause increased compressor wear. Thus, compressors may need to be redesigned before they can be used with HFC-134a.

Most properties of HFC-125 have not been well-characterized because of its unavailability, although its boiling point is known to be similar to that of CFC-502. One problem with HFC-125 is that it has a low critical temperature (150°F), above which it cannot be compressed to its liquid state at any pressure. Therefore, HFC-125 cannot be used in systems where condensing temperatures greater than 150°F might be encountered. In addition, it is anticipated that HFC-125 will have lubrication problems similar to those encountered with HFC-134a.

Some of the problems encountered with the new refrigerants may be solved with new compressor technology, such as the scroll compressor. The scroll compressor uses two mating scroll elements in a spiral progression, one that rotates while the other is stationary, to push the gas into pockets of diminishing volume. The high efficiency of the scroll compressor results in lower system discharge temperatures, thereby avoiding many lubricating oil breakdown problems associated with refrigerants such as CFC-22.

Table 11.9. Physical Characteristics of Refrigerants

Characteristics	CFC-11	CFC-12	HFC-134a	HCFC-12	HFC-125	HCFC-22
Boiling point	74.9F	−21.6F	−15.7F	81.7F	−55.3F	NA
Flammability	None	None	None	None	None	NA
TLV	1,000 ppm	1,000 ppm	1,000 ppm	1,000 ppm	1,000 ppm	NA
ODP	1.0	1.0	0.0	0.02	0.05	0.05

One additional feature can be used with the scroll compressor to reduce the discharge temperature. The scroll can be tapped in a variety of locations, allowing liquid or vapor at almost any pressure to be injected between the close of suction and the start of discharge without having to stage compression. Furthermore, injection of liquid for cooling after the close of suction does not decrease evaporator capacity.

Another advantage of the scroll compressor is that its volumetric efficiency (98% to 100%) is not affected by pressure ratio. Thus, it maintains a higher capacity at high-pressure ratios than a reciprocating compressor. For low-temperature applications, the scroll compressor's capacity is much higher than that of the reciprocating compressor.

Fire-Extinguishing Alternatives

Halons were commercialized as fire-extinguishing compounds in the late 1960s. Halons gained popularity because they were effective, electrically nonconductive, clean, and safe to humans. However, because of the concerns over ozone depletion, the high ODP of halons when compared to CFC-12, and the enactment of regulations concerning ODCs, halons are scheduled for production phaseout. Therefore, industry, insurance carriers, and government agencies have been working diligently to develop acceptable substitutes.

Table 11.10 lists the potentially viable alternatives for reducing or eliminating the use of halons for fire extinguishing. Before any of these alternatives are implemented, however, the following factors should be considered:

- ODP of alternatives (long-term availability)
- extinguishing requirements and capabilities
- economics
- worker exposure and health and safety issues

The alternatives are described in detail below.

Because of the planned phaseout of halons and the newly imposed excise taxes, chemical manufacturers are devoting a great deal of time to developing halon substi-

Table 11.10. Alternatives for Extinguishing Fires

Process	Current Compound	Short-Term Alternative	Long-Term Alternative
Flooding Agent	Halon 1301	Alternative Testing Method	Substitute extinguishing agent and possibly modify equipment
Portable and Flightline Extinguishing	Halon 1211	Replace with existing alternatives like carbon dioxide, water, dry chemical, or foam systems	Substitute extinguishing agent currently being developed

tutes. The following factors are typically used to evaluate potential halon substitutes:

- low toxicity
- zero or low ODP
- acceptable fire-extinguishing capability
- compatibility with existing fire-extinguishing systems

DuPont, Great Lakes Chemical, ICI Americas, North American Fire Guardian (NAFG), and 3M Company have developed alternative chemicals. The following is a summary of the materials currently under development.

Pentafluoroethane (FE-25)

FE-25 (CF_3CHF_2) is being developed by DuPont as a replacement for Halon 1301. It has a boiling point of -48.5°C (-55.3°F), a vapor pressure of 1,301 kPa at 25°C (190 psia at 77°F), and an ODP of zero. It has been shown to extinguish fires at a concentration of 7.6% by volume (cup burner test method). Toxicity testing of this chemical is under way. On the basis of other HFC data and limited animal inhalation studies, FE-25 is expected to have an extremely low order-of-inhalation toxicity. Some extended toxicity tests are being run by DuPont in collaboration with the Program for Alternative Fluorocarbon Testing (PAFT). Results of the extended testing were to be available in 1993.

FE-25 is considered compatible with standard construction materials such as steel, aluminum, and brass. It is also compatible with elastomers such as Buna S rubber, butyl rubber, and neoprene. The construction materials for each fire-extinguishing system should be reviewed for compatibility with FE-25. Manufacturers of the equipment will be able to provide test data for compatibility with their system.

FE-25 will be available for small-scale testing but will not be available on the market until all toxicological testing is complete. FE-25 is expected to cost up to two to three times more than Halon 1311.

Dichlorotrifluoroethane (FE-232)

FE-232 (CF_3CHCl_2) is currently being developed by DuPont as a replacement for Halon 1211 in portable fire extinguishers. It has a boiling point of 27.9°C (82.2°F), is a liquid at room temperature, and has an estimated ODP of 0.02. It has been shown to extinguish fires at a concentration of 7.5% by volume (cup burner method). Recently released data from a lifetime bioassay shows possible toxicity problems. Further analysis of these data is currently under way. On the basis of these results, DuPont recommends a workplace exposure limit of 10 ppm as an 8-hour, time-weighted, average. In addition, DuPont has recognized a provisional emergency exposure limit of 2,500 ppm for 1 minute. Long-term toxicity testing in conjunction with the PAFT is being conducted. Testing to date has indicated that FE-232 exhibits a low to moderate order of toxicity in acute, subacute, and subchronic testing. FE-232 has not been shown to be a mutagen or teratogen, and it is considered to be a moderate cardiac sensitizer. Two-year chronic inhalation exposure test results were to be available in 1992.

There are several compatibility problems with construction materials. FE-232 has high solvency power, and tests on elastomers have shown excessive swelling on Buna and butyl rubbers. Neoprene W and EPDM showed no change following immersion testing.

FE-232 is available in limited amounts for testing only. Increased production was expected in 1993, depending on remaining toxicity test results. The cost of FE-232 is expected to be similar to the cost of Halon 1211.

Trifluoroethane (FE-13)

FE-13 ($C_2H_3F_3$) currently is used in commercial production by DuPont and is being projected as a total flood agent, possibly for use in some normally occupied spaces. FE-13 currently is being used as a chemical intermediate, refrigerant, and etchant. It has a high vapor pressure, and, although it cannot be used in conventional halon systems, it may be suitable for use in hardware designed for carbon dioxide. It has a boiling point of -82°C (-115.7°F) and an ODP of zero. It has been shown to extinguish fires at a concentration of 13% by volume. Therefore, 1.75 times as many pounds of FE-13 as Halon 1301 are required to extinguish fires. FE-13's high vapor pressure shows its similarity to carbon dioxide; however, its greater liquid density and much lower toxicity and extinguishing concentration merit its consideration for use.

Toxicity testing of this chemical is under way and test data to date have demonstrated a low order of toxicity. Concentrations of up to 80% by volume do not produce cardiac sensitization in animals. FE-13 is not a mutagen in the Ames test.

Bromodifluoromethane (FM-100)

FM-100 (CHF_2Br) has been developed as a replacement for Halon 1301 and Halon 1211. In May 1990, Great Lakes Chemicals was granted a consent order by the U.S. EPA allowing use of FM-100 as a flooding agent for unoccupied space protection. Approval for general-use applications is pending a U.S. EPA review of additional toxicity data. Recent testing by the U.S. Navy, Ansul Fire Protection, and Fike Corporation shows FM-100 to be as effective or better than Halon 1211 and Halon 1301 for fire protection in a number of uses. The ODP of FM-100 is estimated at between 0.19 and 1.1, which may restrict its use after the year 2000.

Heptafluoropropane (FM-200)

FM-200 (C_3HF_7) is being developed by Great Lakes Chemicals as a replacement for Halon 1301. It has an ODP of zero and has been shown to extinguish fires at a concentration of 5.9 percent by volume. Full-scale fire testing has shown FM-200 to be an effective total flooding agent at concentrations ranging from 6 to 9%, depending on the fire size and type. Toxicity testing of this chemical is under way.

Bromotetrafluoroethane (124B1)

ICI Americas has announced the development of 124B1 (C_2HBrF_4)as a possible replacement for Halon 1211. 124B1 has a boiling point of 8.6°C and has been shown to extinguish fires at a concentration of 3.6% by volume. Toxicity testing on the chemical is under way. The chemical has an estimated ODP of 0.3, which may restrict its use after the year 2000.

NAF Series

North American Fire Guardian (NAFG) currently offers NAFI and NAFII extinguishing agents, which are based on the use of CFCs. Their availability will be restricted after the year 2000. NAFG is currently developing NAF S-III (a blend of three HCFCs and NAFGs detoxifying agent) as a replacement for Halon 1301. NAF S-III has a boiling point of -37.3°C and an ODP of 0.044. Total flooding tests of NAF S-III have been conducted in Italy. The product is available in Europe and is expected to be available soon in North America. NAFG is also developing the HCFC-based agents NAF I-A (portable/handheld extinguishers) and NAF II-A (fixed systems).

Perfluorocarbons

3M has announced its intention to develop perfluorocarbons, which are fully fluorinated chemicals currently used as heat-transfer fluids and electronic-testing fluids, as replacements for Halon 1301. 3M is also investigating the use of perfluorocarbons as replacements for Halon 1211. Perfluorocarbons have been shown to extinguish fires at concentrations ranging from 5.5 to 7.0 percent by volume, depending on the particular perfluorocarbon compound tested. Perfluorocarbons have an ODP of zero, and toxicity testing on these chemicals is under way.

Fluoroiodocarbons (FICs)

Fluoroiodocarbons were recently announced as a possible true "drop-in" replacement for CFCs, halons, and chlorinated solvents in many applications including cleaning, refrigeration, foam blowing, aerosol, and fire suppression. Two researchers announced in September 1993 that FICs are nonflammable, noncorrosive, non-ozone-depleting, non-global-warming, non-VOCs, and appear to be as effective or more so than CFCs, halons, and chlorinated solvents in many cases The more volatile FICs (e.g., trifluoromethyl iodide, CF_3I and perfluoro-n-propyl iodide, $CF_3CF_2CF_2I$) and their blends with HFCs show promise as refrigerants, aerosol propellants, and foam-blowing agents. For example, a blend of CF_3I and HFC-152a was successfully demonstrated as a high-performance drop-in replacement for R-12 in a domestic refrigerator. Moderately volatile FICs (e.g., perfluoro-n-butyl iodide, bp 67°C and perfluoro-n-hexyl iodide, bp 117°C) and their nonflammable blends with conventional solvents such as alcohols, esters, hydrocarbons, and ketones have shown excellent cleaning ability and materials compatibility in early tests. FICs may also aid in mitigating tropospheric ozone, a major contributor to urban smog. The compounds are currently planned to undergo rigorous performance, thermal stability, materials compatibility, and toxicity testing that is anticipated to take 2 to 3 years to complete. Full-scale commercialization is anticipated in 1996 under the trade name Ikon.

Propellants and Blowing Agents

Almost no CFCs are used currently in making foam. Several national foam-producing companies have switched from using CFC-12 as a blowing agent to using either carbon dioxide or water with various agents. Some of these alternative blowing agents are reported to work better and more efficiently than the CFCs. Typically, however, a more exotic chemistry is required to manufacture foams without CFCs. One promising, although relatively expensive replacement for foam insulation, is the use of vacuum panels. This alternative is being adopted in the next generation of high-efficiency refrigerators.

Additional Solvent Substitution Information

In response to the need for an easily accessible, central location for information on solvent substitutes, the Idaho National Engineering Laboratory (INEL) has developed the "Solvent Substitution Handbook."[8] Information on solvent substitutes is accessible electronically via the handbook.[9] INEL's program has consisted of identifying solvents (alternatives) that are not currently restricted by government regulations for use at DOD and DOE Defense Program maintenance facilities and private industry. In addition, the laboratory has compiled data on evaluations of cleaning performance, corrosivity, air emissions, and recycling and recovery possibilities for solvent alternatives. They also have tested compatibility of solvent alternatives with nonmetallic materials. If you would like more information on this useful tool, contact them at:

Idaho National Engineering Laboratory
P.O. Box 1625 MS2108
Idaho Falls, ID 83415-2208
(208) 526-7834

The U.S. EPA's Air and Energy Engineering Research Laboratory and the Research Triangle Institute have developed a *S*olvent *A*dditives *G*uid*e* (SAGE). This expert system uses a question-and-answer format to assist the user in identifying alternatives for existing solvent-cleaning chemistries and processes. Example questions include:

- What material is being cleaned?
- Does the material have a coating and, if so, is it to be removed?
- What is the size and shape of the part?
- What production volumes are required for the cleaning operation?
- What level of cleanliness is required?
- What types of soil are on the part to be cleaned?
- What cleaning chemistry is currently being used?
- What type of equipment is currently used for cleaning?

The software was to have been released by the end of February 1993. For additional information on the software's capability or on how to access the software, contact the hotline at (919) 541-0800. The electronic bulletin board number for accessing the software is (919) 541-5742.

REUSING WASTE SOLVENTS

Frequently, high-quality solvents are used once for precision cleaning and are then disposed of, either on a scheduled basis or because they do not meet stringent specifications. Such solvents could be reused, without treatment, for applications that do not require as high a standard of purity. Using an untreated waste produced in an operation with high purity standards for an operation with lower purity standards is called cascade reuse.

Example. At a national laboratory, Freon and isopropanol are used for periodically cleaning high-pressure liquified-gas lines. Once used, these solvents cannot be reused for their original purpose because of contamination by small amounts of water. However, these used solvents are suitable for general-purpose cleaning at the facility. The used solvents can be used directly, without treatment, because general cleaning does not require the high purity standards required for cleaning the gas lines.

Example. An aerospace missile manufacturer's procurement specifications require that isopropanol be provided at work stations in a specific-size squeeze bottle, with the solvent replaced on a set schedule. As a result of this requirement, unused solvent was disposed of when its expiration date was reached even though it was not contaminated. A program was instituted to collect this solvent to use for non-prohibited purposes.

RECYCLING WASTE SOLVENTS

Technologies for Cleaning Solvents

Several technologies are available for cleaning solvent so that it can be used again for its original purpose. One technology that can be applied at almost all industrial facilities is distillation. Solvent recovery using distillation can be implemented in five configurations: company-owned local recycling system, company-owned central recycling system, contract recycling, sale to recyclers, and manufacturer take-back. Other technologies that are commercially available and may be applicable for a specific process are centrifugation, filtration (Figure 11.5), ultrafiltration, reverse osmosis, and activated carbon adsorption.

Distillation consists of heating a solvent until it vaporizes and then condensing the vapor.[10] The condensed vapor is then reused. If the boiling point of the solvent is high (over 200°F), the distillation temperature can be reduced by using a vacuum to minimize the thermal decomposition of the solvent. Another technique used for solvents with a high boiling point is to inject steam into the solvent, forming an azeotropic mixture that has a lower boiling point. The condensate of water and

Figure 11.5. Small portable solvent filter system. Solvent is collected from a parts washer, filtered, and returned to washer. Approximately 500 gallons of Solvent 140 are reused each year in fleet operations at Hanford, WA due to this system (Photo courtesy of Westinghouse Hanford Company/BCSR, Richland, WA).

solvent is then separated using a gravity solvent/water separator. Typically, distillation reduces the waste to be disposed of to one-tenth to one-fifteenth of its original volume. Table 11.11 lists the solvents amenable to distillation, along with important physical parameters.

Most solvents are not pure chemicals, but are formulated mixtures of compounds. Proprietary inhibitors are added to commercial solvents to improve their properties or extend their service life. Three types of inhibitors—classified as antioxidants, metal stabilizers, and acid acceptors—are added to the commonly used chlorinated solvents TCE, TCA, and PCE. Antioxidants form stable compounds that suppress the decomposition reaction of unsaturated solvents and slow the propagation of auto-oxidation. Metal stabilizers inhibit solvent degradation that normally occurs in the presence of metals and their chlorides (e.g., aluminum [Al] and aluminum chloride [$AlCl_3$]). Inhibitors either react with the metal to form an insoluble deposit or complex the metal chloride to prevent degradation. Acid acceptors are either neutral (epoxide) or slightly basic (amine) compounds that neutralize the hydrochloric acid (HCl) produced in the breakdown of chlorinated solvents. If left unneutralized, HCl can further degrade the solvent and corrode degreaser equipment as well as the parts being degreased.

There is often a concern that distillation will alter or fail to recover these inhibitors and, thus, that recovered solvent will not provide adequate protection. Joshi et al.[11] studied the fate of inhibitors subjected to repeated use and recycle using distillation. They identified inhibitor compounds used in commercial chlorinated solvent mixtures (Tables 11.12 through 11.14). In tests of solvent reclamation by distillation of dirty solvent, 65% of the inhibitors were recovered with the clean solvent (Tables 11.15 through 11.17).

Table 11.11. Physical Properties of Commonly Used Solvents

Solvent	Atmospheric Boiling Pt (°F)	Azeotropic Boiling Pt (°F)	Density (lb/gal)
Aliphatic Hydrocarbon			
Hexane	157.0	142.9	5.51
Heptane	209.0	174.8	5.70
Stoddard	308–316.0	204.0	6.47
Aromatic Hydrocarbon			
Benzene	176.0	157.0	7.32
Toluene	232.0	185.0	7.20
Xylene	261–318.0	202.1	7.17
Chlorinated Hydrocarbon			
Trichloroethylene	189.0	163.8	12.2
Perchloroethylene	249.0	189.7	13.5
1,1,1-Trichloroethane	166.0	149.0	11.0
Methylene Chloride	104.0	101.2	11.07
Fluorocarbon			
Freon TF	117.6	112.0	13.06
Freon 112	199.0	166.0	13.69
Acetone	133.0	133.0	6.59
Methyl Ethyl Ketone	175.0	164.1	6.71
Methyl Isobutyl Ketone	241.0	190.2	6.67

Table 11.12. Additives/Impurities Identified in Trichloroethylene

Inhibitor	Formula	Molecular Weight	Boiling Point (°F)	Function
Butylene oxide	C_4H_8O	72.1	146	Acid acceptor
Ethyl acetate	$C_4H_8O_2$	88.1	171	Unknown
5,5-Dimethyl-2-hexene	C_8H_{16}	112.2	Unknown	Unknown, possibly anti-oxidant
Epichlorohydrin	C_3H_5OCl	92.5	242	Acid acceptor
N-Methylpyrrole	C_5H_7N	81.1	239	Antioxidant

Source: Adapted from Joshi.[11]

Table 11.13. Additives Identified in Perchloroethylene

Inhibitor	Formula	Molecular Weight	Boiling Point (°F)	Function
Cyclohexene oxide	$C_6H_{10}O$	98.2	269	Acid acceptor
Butoxymethyl oxirane	$C_7H_{14}O_2$	130.2	Unknown	Acid acceptor

Source: Adapted from Joshi.[11]

Table 11.14. Additives Identified in 1,1,1-Trichloroethane

Inhibitor	Formula	Molecular Weight	Boiling Point (°F)	Function
N-Methoxy-methanamine	C_2H_7NO	61.1	Unknown	Acid acceptor
Formaldehyde dimethyl hydrazone	$C_3H_8N_2$	72.1	Unknown	Aluminum stabilizer
1,4-Dioxane	$C_4H_8O_2$	88.1	214	Aluminum stabilizer

Source: Adapted from Joshi.[11]

Table 11.15. Inhibitor Concentrations of Reclaimed Trichloroethylene

	Inhibitor Concentration (wt fraction)			
Sample	Butylene Oxide ($\times 10^3$)	Epichlorohydrin ($\times 10^3$)	Ethyl Acetate ($\times 10^4$)	n-Methyl Pyrrole ($\times 10^4$)
New TCE	1.64	1.66	3.46	1.59
Spent TCE	0.685	1.69	2.85	2.18
TCE distillate	0.718	1.61	2.58	1.66
Carbon-adsorbed TCE	0.44	1.31	2.65	0.9

Source: Adapted from Joshi.[11]

Table 11.16. Inhibitor Concentrations of Reclaimed Perchloroethylene

	Inhibitor Concentration (wt fraction)	
Sample	Cyclohexene Oxide ($\times 10^3$)	Butoxymethyl Oxirane ($\times 10^3$)
New PCE	1.06	4.26
Used PCE	0.988	7.45
PCE distillate	0.968	5.42
Carbon-adsorbed PEC	0.091	5.40

Source: Adapted from Joshi.[11]

Table 11.17. Inhibitor Concentrations of Reclaimed Methylene Chloride

Sample	N-Methoxy-methanamine ($\times 10^4$)	Formaldehyde Dimethyl Hydrazone ($\times 10^3$)	1,4-Dioxane ($\times 10^3$)
New methylene chloride	8.92	5.78	17.2
Used methylene chloride	4.14	6.16	29.0
Methylene chloride distillate	4.60	7.22	19.6
Carbon-adsorbed methylene chloride	1.30	3.37	23.4

Source: Adapted from Joshi.[11]

Onsite Distillation Systems

Many commercially available stills are available with capacities ranging from 0.5 to 100 gal/hr. The smaller systems are self-contained, off-the-shelf units that can be installed in any sheltered area equipped with electrical power and cooling water (Figure 11.6). The larger units are generally more complex and require a supply of steam. The capital cost is generally about $5,000, plus $1,000/gal/hr capacity. For example, a 50-gal/hr still would cost about $55,000. Generally, the payback period for purchase of a still is between 6 months and 2 years. The normal lifetime of a still is about 20 years. Table 11.18 lists the major suppliers of solvent recycling equipment.

The operating costs of the distillation apparatus include labor, energy, cooling water, and maintenance parts. Normally, the largest component is labor. A moderately skilled operator is needed to tend the apparatus about 10% of the time during its operation.

For recycling to be effective, solvents should be segregated. If two or more solvents are mixed, an off-the-shelf still will often be unable to separate them, and a

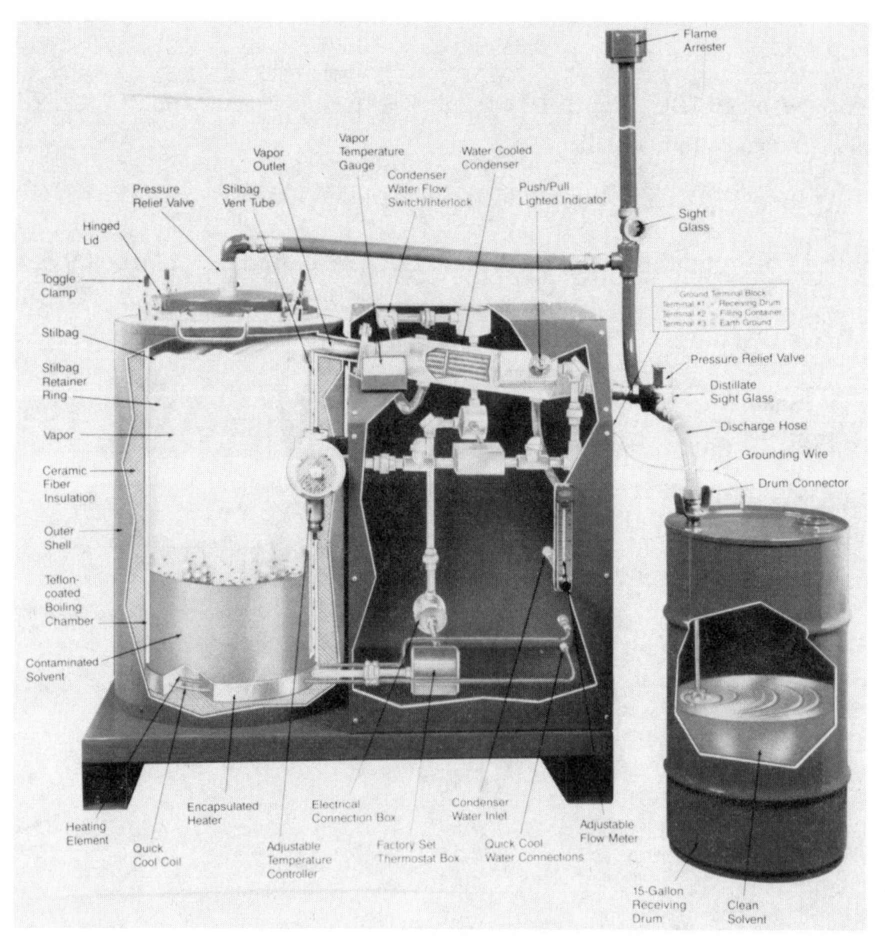

Figure 11.6. A typical atmospheric batch operated still (Photo courtesy of Finish Engineering, Erie, Pennsylvania).

Table 11.18. Suppliers of Solvent Recycling Equipment

Company	Address	Phone/*Fax*
Baron & Blakeslee	1500 W 16th Street Long Beach, CA 90813	(310) 491–1228 *(310) 491–1091*
Calfran International, Inc.	171 Shaker Road, Box 269 Springfield, MA 01101	(413) 732–3616 *(413) 732–9246*
Corpane Industries	250 Production Court Louisville, KY 40299	(502) 491–4433 *(502) 491–9944*
Detrex Chemical	4000 Town Center Suite 1100 Southfield, MI 48075	(313) 358–5800 *(313) 358–5803*
Durr Industries, Inc. Finishing Systems	40600 Plymouth Road P.O. Box 2129 Plymouth, MI 48170–4297	(313) 459–6800 *(313) 459–5837*
Environmental Associates, Inc.	460 SW Madison, Suite 13 Corvallis, OR 97333	(503) 758–7321 *(503) 754–1849*
Finish Thompson, Inc.	921 Greengarden Road Erie, PA 16501	(814) 455–4478 *(814) 455–8518*
Industrial Filter & Pump Mfg. Co.	5900 Ogden Avenue Cicero, IL 60650	(708) 656–7800 *(708) 656–7806*
Ionics, Inc.	65 Grove Street Watertown, MA 02172	(617) 926–2500 *(617) 926–4304*
Licon, Inc.	2442 Executive Plaza Pensacola, FL 32524	(904) 477–0334 *(904) 477–7234*
SRS Engineering Systems	711 Fox Wood Drive Oceanwood, CA 92057	(619) 722–8816 *(619) 722–8835*

much more expensive, customized unit will be required. Inability to enforce solvent segregation is often the major obstacle to solvent recycling.

Example. Annually, about 25,000 gallons of heptane are used to calibrate sensors for aircraft fluid flow at an aircraft maintenance facility. Heptane is used because its performance is more consistent than that of jet fuel. Used heptane is stored in an underground waste storage tank. A nonfractionating-batch atmospheric still was provided to the facility with instructions to find a use for it. This responsibility was assigned to a facilities engineer in addition to his regular duties. Local personnel were neither involved in the decision to recycle solvents nor in the selection of the type of still to be used. The engineer decided to use the still to reclaim heptane from the underground waste storage tank. In the still's first test, the recovered solvent failed to meet specifications for use as a calibrating fluid because of the presence of contamination that lowered its boiling and flash points. The contamination was traced to sinks that were connected to the waste storage tank into which solvents other than heptane were disposed of. Segregation of the heptane from other solvents could have alleviated the contamination problems. But, because the engineer was less than enthusiastic about the extra duty and because he believed he had proved that the "still did not work," he made no further efforts. The still was abandoned in place.

Alternatives to Onsite Distillation

If purchasing a still is unfeasible, other methods are available for distilling solvents. One method is to contract with a recycler to distill and return spent solvents. Another is to sell the solvents to a recycler. The best method to use depends on the availability of a local recycler, the type of solvent recycled, and the economics of onsite versus offsite recycling. Owning a still is usually preferable because of lower cost, better process control, and convenience, as well as elimination of the liability associated with giving an outside entity responsibility for handling one's hazardous waste.

One more option is to rent solvent cleaning equipment with a service contract to replace and recycle the solvent. Safety-Kleen Corporation supplies drums of solvent with self-contained sinks (Figure 11.7). A cleaning solvent is supplied to the sink from a reservoir in the drum, and the used solvent drains back to the drum. The drum of dirty solvent is replaced periodically with a drum of fresh solvent, and the dirty solvent is distilled at a central facility. On the basis of its success in supplying cleaning sinks, the company has started marketing a similar facility for cleaning spray-paint equipment (Figure 11.8).

With some specialized solvents, the manufacturer will reprocess the solvents at no

Figure 11.7. A rental solvent cleaning system (Photo courtesy of Safety-Kleen, Elgin, Illinois).

Figure 11.8. A rental paint spray equipment cleaner (Photo courtesy of Safety-Kleen, Elgin, Illinois).

charge or for a nominal fee. This recycling can be an advantageous alternative to disposal of used solvents.

MANAGING A SOLVENT RECYCLING PROGRAM

Companies can set up a centralized distillation facility for recovery or can install stills at points of use. Industrial facilities have been successful with both approaches. Regardless of where the distillation occurs, it is critical that waste solvents be properly handled and stored so that various solvents and impurities are not mixed.

Centralized Programs for Solvent Recycling

The main advantage of operating a large centralized facility is that capital costs can be recovered quickly because of economies of scale. A centralized facility can redistill large quantities of various types of solvents; there are, however, several disadvantages. Because many different types of solvents are recycled, great care must be taken with waste segregation and sample analysis. Another disadvantage is that solvents must be transported to and from the point of use. A centralized facility depends on an individual dedicated to initiating and supervising operation of the system and an enthusiastic staff dedicated to collecting, analyzing, recycling, and distributing the solvent.

Case Study 11.4: Centralized Solvent Recycle Program

Warner Robins AFB in Macon, Georgia, refurbishes airlift, fighter, bomber, utility, and remote-control aircraft as well as helicopters and missiles. The base repairs predominantly C-130 and C-141 transport planes and F-15 fighter jets.

In 1982, Robins AFB purchased a $48,000 batch atmospheric-pressure still manufactured by Finish Engineering Corporation. The still is used to reclaim trichloroethane, Freon-113, and isopropanol. In 1983, a total of 227 drums of chemicals were distilled for a savings of $81,000. O. H. Carstarphen, Solvent Reclamation Engineer, estimated that in fiscal year (FY) 1984 the recycling of those three chemicals saved the base $118,000 in virgin material and in costs for hazardous waste disposal. The cost to reclaim the used chemicals was only $13 per drum, whereas disposal of the chemicals and purchase of new materials would have cost from $250 to $500 per drum.

The still operated up to a temperature of 300°F in the pot and reclaimed organic fluids at a rate of up to 55 gal/hr. Freon and isopropanol were processed at a rate of approximately 50 gal/hr, and TCA was processed at a rate of 35 to 40 gal/hr. Recovery efficiency for isopropanol and Freon-113 was approximately 95%. The recovery efficiency for TCA was only 70% because the used solvent contained nonvolatile wax, dirt, and grease that were removed from metal parts during degreasing operations.

The Finish Engineering still was easy and inexpensive to operate and maintain. Some problems initially encountered with a feed pump when recycling Freon were solved. Even though reclaimed Freon did not meet specifications, it was used for initial cleaning. Virgin material was then used for final assembly cleaning operations.

Annually, 584 drums of degreasing solvents were used by the Directorate of Maintenance. Because it was the predominant solvent used at Robins, approximately 175 drums of TCA per year were reclaimed for reuse in vapor-degreasing tanks located in the plating shop. Laboratory tests of the reclaimed TCA indicated that the material met military specifications. The Directorate of Maintenance estimated that between 1982 and 1985, recovery of waste TCA saved approximately $79,000.

In the past, isopropanol, which is used for cleaning electronic parts, had been discarded when the solution became contaminated with oil and dirt. Isopropanol was reclaimed by the organic fluid recovery system, resulting in a savings of $16,200 in FY 1983 and $18,500 in FY 1984. A 5-micron filter was installed in the discharge line for removing fine metal particles that were carried over with the alcohol vapors. The reclaimed alcohol had a purity of 99.8%.

Recycling at Robins was successful because personnel segregated wastes and kept excessive water and other impurities out of the waste slop cans and drums. Segregation of the waste liquids is necessary to maintain the usefulness of recovered organic fluids. For example, two common paint thinners, MEK and toluene, could easily be mixed together in the waste slop drums in the painting shop. However, if this were to occur, the mixture could not effectively be separated by single-stage batch distillation because the boiling points of the two thinners are similar.

At Robins AFB, management was strongly committed to the organic fluid recovery operation, as demonstrated by the facilities and personnel dedicated to the operation of the system. The Chemical Control Group, consisting of 10 people, collected waste chemicals at 30 different areas. These covered collection areas had controlled access and were on diked concrete pads. The areas were used to dispense fresh solvents from drums and to collect waste solvents in separate, labeled drums. Site managers were responsible for segregating wastes at the different sites.

The Chemical Control Group was also responsible for performing the following tasks: sampling all drums; redistilling Freon, TCA, and isopropanol wastes; and transporting the reclaimed materials back to their source. In addition, analytical chemists were required to perform two analyses for each drum of waste. First, as each drum was received, the contents were analyzed to confirm the labeling. Then, after each distillation run, the recovered solvent was analyzed to ensure that it met appropriate specifications.

One additional management tool implemented at Robins AFB that helped the reclamation program succeed was educating base personnel about hazardous wastes. The Directorate of Maintenance developed a course titled "Storage, Handling, and Disposal of Industrial Chemicals," which is attended by all personnel who store, handle, use, or dispose of industrial chemicals. The scope of this training includes industrial materials terminology, personnel protective equipment, hazard identification systems, emergency procedures, and industrial waste collection and disposal.

Case Study 11.5: Centralized Solvent Recycle Program

At Tyndall AFB in Panama City, Florida, solvents are used in the general maintenance of jet aircraft and motor vehicles. In 1981, the Air Force Engineering and Services Laboratory initiated a research project at Tyndall to determine if Stoddard Solvent could be economically recycled on the base. The Air Force estimated that approximately 13,000 gal of Stoddard Solvent were being used per year at 19 different shops, making it the most widely used solvent at Tyndall in 1981.

A vacuum still, manufactured by Gardner Machinery of Charlotte, North Carolina, was purchased to reclaim solvents. The system had a rated capacity of 200 to 225 gal of solvent per hour and was designed to process Stoddard Solvent, naphtha, mineral spirits, and petroleum spirits. The solvent recovery system cost approximately $50,000 to purchase and install. The savings dropped from $3.72 per gal of solvent recovered in 1982 to $1.44 per gal in 1983, primarily as a result of a dramatic drop in the price of fresh Stoddard Solvent from $4.51 per gal to $1.92 per gal over the same period.[12] Only 4,500 gal of Stoddard Solvent were reclaimed, for a savings of approximately $7,000.

There were several reasons for the lack of success of the program. The financial incentive was reduced because of the drop in price of virgin material. Collecting, transporting, and storing the waste Stoddard Solvent generated in the numerous small shops was difficult. Inadequate involvement and commitment of the operational personnel also may have contributed to the limited success of the collection system because the concept had been developed by an outside group and was implemented as a research project. In addition, management's commitment to the success of the project was not as evident as at Warner Robins AFB.

Of the 19 shops that used Stoddard Solvent in 1981, only the tire shop actively collected and stored waste solvent for recycle. This shop used two 300-gal dip tanks that contained Stoddard Solvent. The cleaning solution was used to remove carbon, grease, and grit from aircraft wheel bearings. Every 4 months, the spent Stoddard Solvent was discharged into ten 55-gal drums. The waste solvent in the drums was then pumped to the still holding tank for recycling.

From 1981 to 1984, the still was operated nine times, or approximately 1 day every

4 months. An average of 506 gal of solvent were recycled at a recovery efficiency of 97% during each of the nine runs. Samples of the recycled solvent were analyzed and generally failed to meet specifications because of an undetected internal leak in the still and a buildup of iron oxide in the system during periods of nonuse.

Because the recycled solvent did not meet specifications, it could not be accepted by the base supply department for distribution and reuse. Most of the recycled solvent was, however, reused in the tire shop, which did not require solvent that met the specifications. Some of the solvent bypassed the supply department and was sent directly to users who expressed an interest in the free material. Although maintenance personnel at the tire shop were pleased with the quality of the recycled solvent, they noticed that the recycled material took longer to dry than fresh solvent.

Operation of the still was discontinued because of its limited use, failure of recycled product to meet specifications, and resultant poor economic performance. In 1985, the still was given to Warner Robins AFB in Macon, Georgia, to supplement its existing still.

Case Study 11.6: Distillation to Recover Solvent and DI Water

An electronics manufacturer developed an improved method of cleaning soldered semiconductor substrates. The process included cleaning of the semiconductor materials with N-methyl-2-pyrrolidone (NMP), a flux removal agent, followed by rinsing in deionized (DI) water. This process resulted in the production of a mixture of NMP and DI water. Engineering consultants from CH2M HILL developed, tested, and prepared a conceptual design for a distillation system to recover NMP and DI water from this mixture.

In the cleaning process, the NMP solvent becomes contaminated with flux solids, tin, and lead. In addition, the solvent removes minute quantities of oil and grease from the mechanical equipment. Water is introduced into the NMP bath by splashover from the subsequent DI rinse step. The semiconductor substrates carry NMP (and its contaminants) over into the DI rinse. The characterizations of the contaminated NMP and DI waste streams are shown in Table 11.19.

The goal was to develop a feasible process for purifying these waste streams; to recover an NMP solvent contaminated with less than 500 ppm of water, 5 ppm of flux, and trace amounts of oil; and to recover a DI stream with less than 500 ppm of NMP.

Purification processes considered, and their advantages and disadvantages, are listed in Tables 11.20 and 11.21. Laboratory testing was performed to investigate the technical feasibility of the distillation, ion exchange, filtration, and evaporation

Table 11.19. Composition and Production of Waste NMP and DI Water

Parameter	NMP	DI Water
Production (gal/min)	45	15
NMP concentration	99%	1%
Water	1%	99%
Flux	100 ppm	Trace
Metals	Trace	Trace
Oil	20 ppm	Trace

Table 11.20. Evaluation of Potential NMP Purification Processes

Unit Process	Advantages	Disadvantages
Filtration	• Removes solids • Simpler regeneration	• Potential plugging • No removal of water, solubles, or isopropyl alcohol
Precipitation	• Removes metals and solids	• Chemical additions
Ultrafiltration/ reverse osmosis	• Removes dissolved solids	• Membrane durability • Fractional split of components
Crystallization/ cryogenic	• Likely to render high purity product	• Minimal existing data base • Delicate process
Molecular sieves	• Removes water	• Only applicable to trace removal • Not resistant to upsets • Difficult regeneration and operation • Media durability unknown • Isopropyl alcohol may adsorb
Ion exchange	• Removes organic acids • Potential elimination of high boilers • DI water operators understand system • Unattended regeneration	• Resin durability • Complex regeneration • Significant regeneration wastes • Doubtful metals removal • Generates water
Carbon adsorption		• Not selective removal of NMP
Distillation	• Separates NMP from water, metals, and flux • Flexible system • Resistant to upsets	• Operator training • Long startup time
Evaporation	• Removes water	• Partial impurity split
Gas stripping	• Removes water; gas dryness is limiting • Simpler regeneration	• NMP entrainment
Solvent extraction	• Removes water	• Not a binary system • Solvent selection

alternatives for NMP recovery. Laboratory tests were also performed for DI water recovery using distillation and carbon adsorption.

In the laboratory tests, distillation was shown to be a technically feasible alternative to accomplish the desired recovery and purity of both DI water and NMP. A vacuum distillation system was recommended (Figure 11.9). In this design, both

Table 11.21. Evaluation of Potential DI Water Purification Processes

Unit Process	Advantages	Disadvantages
Carbon adsorption	• Removes NMP • NMP is recoverable	• Regeneration hysteresis • Carbon durability
Solvent extraction	• Removes NMP	• Further processing required to recover NMP
Ultrafiltration/ reverse osmosis		• Separation unknown • Membrane durability unknown
Distillation	• Removes water from NMP	• Heat of vaporization of water
Molecular sieves		• Water concentration too high

NMP and DI water are recovered as distilled condensate. One advantage of this design is that the presence of impurities (such as iron or suspended solids) is minimized in the recovered products.

The estimated capital and operating costs for the design basis system (42 gal/min of NMP and 15 gal/min of DI water) were estimated to be $4,700,000 and $1,700,000, respectively (1981 dollars). Detailed breakdowns of these estimates are provided as Tables 11.22 and 11.23, based on the costs presented in Table 11.24.

Local Facilities for Solvent Recycling

Local solvent stills are sometimes preferable because they allow the waste generator to have total control over the recycling operation. Because only a few types of solvents are redistilled at the small facilities, laboratory analysis of waste solvents is often not required. Local facilities are also advantageous in that they eliminate labor-intensive transportation and segregation. However, use of decentralized facilities requires training of more personnel than would be required if using a centralized facility, and local facilities must be convinced to adopt solvent recovery as part of their routine.

Case Study 11.7: Recovering Painting Solvents with Local Stills

Cleaning operations in the paint shop at Norfolk Naval Shipyard (NSY) generate approximately 15 gal/day of numerous waste solvents including mineral spirits, ketones, and epoxy thinners containing paint pigments. Historically, Norfolk NSY disposed of the waste mineral spirits and other waste organic fluids at a reported cost of $7.80/gal.

Norfolk NSY now uses a nonfractionating batch still, Model LS-15V, manufactured by Finish Engineering, Erie, Pennsylvania. This model is designed to recover 15 gal of solvent per shift of operation (i.e., one full charge of the still pot). The system employs an electrically heated pot with a residue collection pan, a water-cooled shell and tube condenser, a reclaimed solvent collection tank, and an electric vacuum pump. The system is designed to recover organic fluids with boiling points in the range of 100 to 320°F without using the vacuum system. The vacuum system, which produces a vacuum of 25 in. of mercury during operation, is designed to recover organic fluids with atmospheric pressure boiling points up to 500°F.

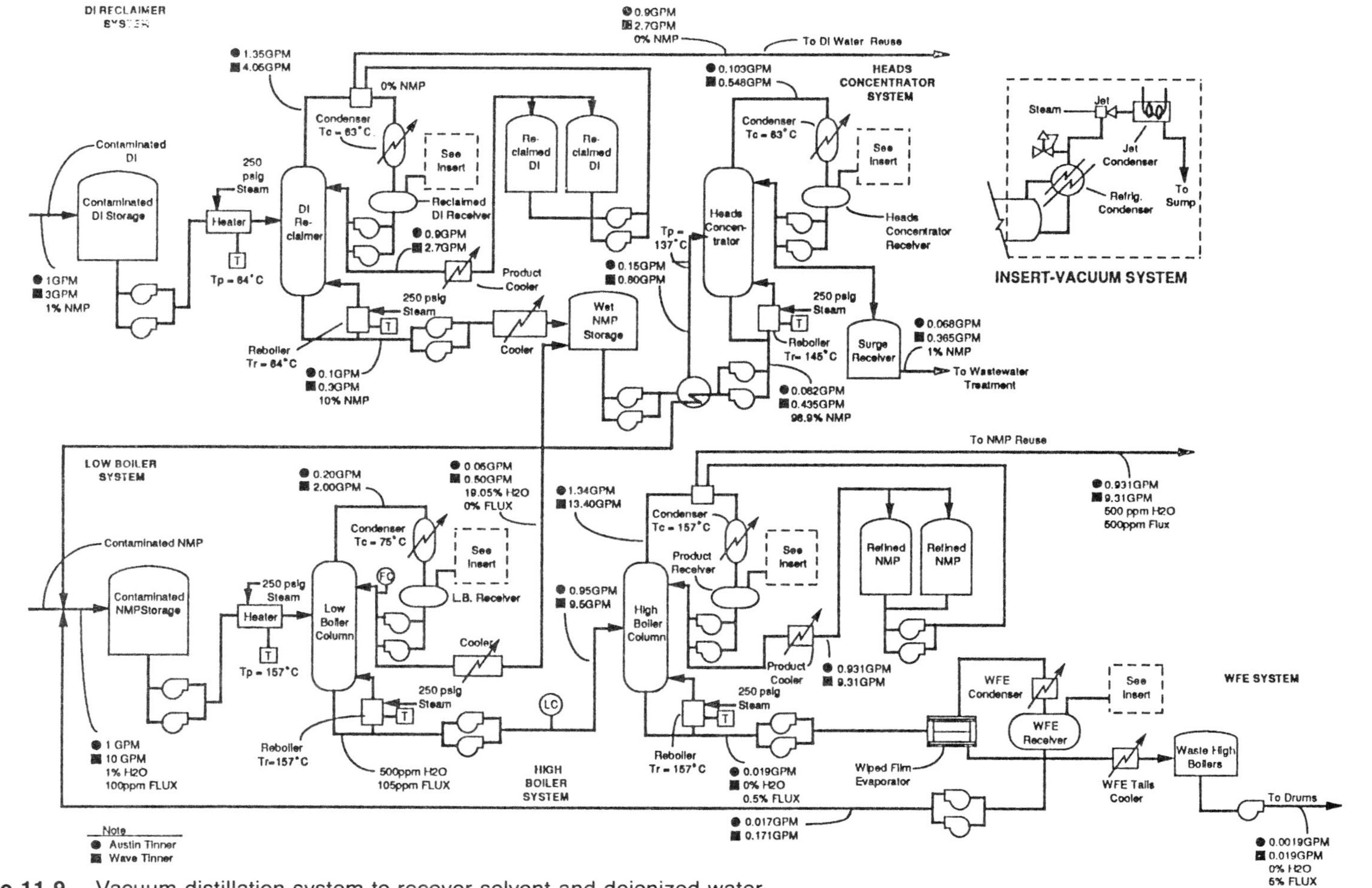

Figure 11.9. Vacuum distillation system to recover solvent and deionized water.

Table 11.22. Capital Cost Estimate for Vacuum Distillation

Cost Category/ System[a]	Storage Tanks and Pumps	DI Reclaimer	Heads Concentrator	Low Boiler	High Boiler	WFE	Total
Equipment	$ 93,800	$456,200	$196,600	$245,100	$455,000	$160,000	$1,606,700
Installation	18,800	463,700	61,100	144,600	283,100	49,700	1,021,000
Structural	10,300	50,200	21,600	27,000	50,100	17,600	176,800
Mechanical	44,100	214,400	92,400	115,200	213,900	75,200	755,200
Instrumentation	6,600	31,900	13,800	17,200	31,900	11,200	112,600
Electrical	3,800	18,200	7,900	9,800	18,200	6,400	64,300
Subtotal	$177,400	$1,234,600	$393,400	$568,700	$1,052,200	$320,100	$3,746,400
25% Contingency	44,400	308,700	98,400	142,200	263,100	80,000	936,900
Total	**$221,800**	**$1,543,300**	**$491,800**	**$710,900**	**$1,315,300**	**$400,100**	**$4,683,200**

[a]Total project cost including site development, utility supply, industrial buildings, taxes, and engineering, legal, and administrative fees will be greater than Process Module Total. No allocation for inflation is included.

Table 11.23. Annual Operating Cost Estimate for Vacuum Distillation

Cost Category/ System[a]	DI Reclaimer	Heads Concentrator	Low Boiler	High Boiler	WFE	Total
Energy (steam)	$355,800	$13,600	$142,400	$230,700	$3,200	$745,700
Cooling water	74,800	9,600	12,500	52,600	400	149,900
Maintenance	45,600	19,700	24,500	45,500	16,000	151,300
Waste Disposal losses and solvent losses (1/10%)	—	63,400	140,000	134,200	270,600	608,200
Total	**$476,200**	**$106,300**	**$319,400**	**$463,000**	**$290,200**	**$1,655,100**

[a]Labor costs not included.

Table 11.24. Basis for Operating Cost Estimate

Energy Costs[a]	
Electricity	\$0.04/kWh
Steam, 250 psig	\$4.00/1,000 lb_m
City water, 70°F	\$0.18/1,000 gal
Material Costs	
Carbon[b]	
Ion exchange resin[c]	\$1.16/lb_m
	\$205/ft^3
Disposal Costs[a]	
Waste solvent drumming	\$55/55-gal drum
Wastewater treatment[d]	\$0.20/lb of BOD
Chemical Costs[d]	
NMP	\$1.15/lb_m
Caustic, 10%	\$300/ton
Isopropanol	\$2.50/gal
DI water	\$5.00/1,000 gal
Liquid nitrogen	\$0.22/gal

Note: Based on 24 hr/day, 320 day/yr, 7,680 hr/yr.
[a]Cost data supplied by client.
[b]Barnebey-Cheney Type PC.
[c]Rohm & Haas Amberlyst A-27.
[d]Cost data supplied by CH2M HILL.

The system produces a solid residue in the still pot's residue collection pan. The collection pan is then removed, and the residue is emptied into a container for disposal. In 1984, the cost of the system (uninstalled) was approximately $9,000. The cost of the same system without the vacuum system option was $5,000.

On the first day of system operation with the vacuum accessory, preparation for startup took only 15 min; the system was started by pressing only one button, and then it ran unattended. On the startup day, mineral spirits were distilled under vacuum. Dry paint solids remained in the collection pan after the cycle was completed and were easily removed for disposal. The system recovered approximately 13 gal of solvent from 15 gal of waste solvent, for an 85% recovery.

The system has also been used successfully without the vacuum system to recover organic fluids with boiling points below 320°F. When the still was operating at atmospheric pressure, more than 50% of the waste solvent was recovered at a cost of about $0.05/gal.

This solvent recovery operation has three key elements that combined to make it a success: personal dedication of a production representative, technical innovation and ease of operation, and physical location near the waste generation site. Jake Coulter, the paint shop foreman, has been the champion of this solvent recovery operation. He wanted it to work, and it appears to have been a great success.

Case Study 11.8: Vapor Degreasers with Integral Stills

Anniston Army Depot reconditions used tanks and other armored vehicles. Reconditioning consists of completely disassembling the tanks and dismantling their components. Paint, rust, and dirt are removed from these components prior to remanufacturing. Paint is removed by sand blasting or is stripped by using organic solvents or alkaline strippers. Grease and oil are removed using solvent vapor degreasers followed by alkaline cleaners.

Approximately 15 to 20 TCE vapor degreasers are being used at Anniston Army Depot. All are equipped with stills for recovering TCE from the solvent-oil mixture for reuse in the degreasers. Most stills at Anniston are manufactured by Detrex Corporation. They run continuously when the vapor degreasers are in operation, normally 8 hr/day, 5 days/wk. Dirty solvent is fed from a degreaser boiling sump through a water separator to the recovery still. The steam-heated stills have the capacity to recycle 20 gal/hr of TCE.

Anniston Army Depot has reported no problems in operating and maintaining the distillation units. Twice a year (during shutdown) the vapor degreasers and stills are taken out of service for cleaning and general maintenance. Vapor degreaser TCE baths have never had to be dumped during normal operation or shutdown. Some TCE is lost through drag-out, evaporation, and disposal in waste still bottoms.

Still bottoms, typically containing 11 to 17% TCE as well as oil, grease, and dirt, are automatically discharged to waste holding drums. This hazardous waste is sent to a commercial contractor for treatment. Anniston investigated the cost-effectiveness of recovering TCE from still bottoms and determined that the still bottoms would have to contain 40% TCE before it would be economical to recover additional solvent. Table 11.12 lists several vendors who can supply solvent recycling equipment.

REFERENCES

1. *Manual on Vapor Degreasing*. 3rd Edition. Compiled by ASTM Subcommittee D26.02 on Vapor Degreasing. ASTM Manual Series: MNL 2. Revision of Special Technical Publication (STP) 310A, June 1989.
2. Cheng, S.C., et al. "Alternative Treatment of Organic Solvents and Sludges From Metal Finishing Operations," U. S. Environmental Protection Agency, EPA-600/2-83-094, September 1983.
3. Burgess, W.A. *Recognition of Health Hazards in Industry: A Review of Materials and Processes*. (New York: John Wiley & Sons, 1981).
4. Gavaskar, A., R. Olsenbuttel, L. Hernon-Kenny, J. Jones, M. Salem, and J. R. Becker. *Onsite Solvent Recovery Study. Draft Technology Evaluation Report*. Battelle Memorial Institute for Risk Reduction, Engineering Laboratory, Office of Research and Development, Columbus, OH, prepared for the U.S. Environmental Protection Agency, Cincinnati, OH, Contract No. 68/CO 0003, Work Assignment 0-06, 1992.
5. Locklin, J.M. "Alternative Technologies for Environmental Compliance," presented at The First Annual International Workshop on Solvent Substitution, December 4-7, 1990, in Phoenix, AZ, sponsored by the U.S. Department of Energy, Office of Technology Development, Environmental Restoration, and Waste Management and the U.S. Air Force Engineering & Services Center.
6. D'Riuz, C.D. *Aqueous Cleaning as an Alternative to CFC and Chlorinated Solvent-Based Cleaning*. (Park Ridge, NJ: Noyes Publication, 1991).

7. Kohl, J., P. Moses, and B. Triplett. "Managing and Recycling Solvents," Industrial Extension Service, School of Engineering, North Carolina State University, Raleigh, NC, December 1984.
8. Chavez, A.A., "Solvent Substitution Handbook," presented at The Second Annual International Workshop on Solvent Substitution, December 10–13, 1991, Phoenix, AZ, sponsored by the U.S. Department of Energy, Office of Technology Development and the Air Force Civil Engineering Support Agency.
9. DOE/DOD Solvent Utilization Handbook Information Sheet. Idaho National Engineering Laboratory, Idaho Falls, ID.
10. *Vacuum Still Operation Manual.* (Charlotte, NC: Gardner Machinery Corporation).
11. Joshi, S. B., et al. "Methods for Monitoring Solvent Conditions and Maximizing Its Utilization," presented at the 8th ASTM Symposium on Hazardous and Industrial Solid Waste Testing and Disposal, Clearwater, FL, November 12–13, 1987.
12. Tapio, G.E. "A Limited Test of Solvent Reclamation at an Air Force Base," AFESC/RDV, Tyndall Air Force Base, Panama City, FL, 1984.

GENERAL REFERENCES

Isooka, Y., Y. Imamura, and Y. Sakamoto. "Recovery and Reuse of Organic Solvent Solutions," *Metal Finishing*, June 1984.

Johnson, J.C., et al. "Metal Cleaning by Vapor Degreasing," *Metal Finishing*, September 1983.

CHAPTER 12

Metal Plating and Surface Finishing

*Kevin Klink, George Cushnie, Peter Gallerani, and Thomas Higgins**

DESCRIPTIONS OF METAL-FINISHING OPERATIONS AND WASTES

Metal-finishing operations involve preparing and finishing metal parts for final use. Metal-finishing operations can include cleaning, degreasing, pickling (acidic removal of surface oxides), electroplating and electroless metal plating, etching, passivation, phosphating, chemical electropolishing, chemical milling, and anodizing conversion coating. These processes involve applying a functional, protective, or decorative coating to a metal part (or in circuit board manufacture, adding a metallic coating to a plastic substrate) to add value to that product. Cleaning processes are used to prepare parts for coating or other processing and to improve the adhesion of a coating to the surface.

Metal-finishing operations produce waste streams containing acids and bases, toxic heavy metals, and solvents and oils. Table 12.1 shows a list of waste components typically found in metal-finishing operations. Metals and solvents are the principal components regulated under wastewater treatment and hazardous waste regulations.

Chrome Plating

Chrome plating is performed for one or more of the following reasons: to enhance corrosion resistance, to provide a bright metallic appearance, or to impart improved mechanical properties (hardness, lubricity) to the underlying base metal. Plating for corrosion protection or appearance is usually referred to as "decorative chrome plating" and generally involves adding a thin (50 millionths of an inch) coating to a part. This process can be accomplished in only a few minutes. Many parts are plated

*Mr. Klink is a senior process engineer for CH2M HILL's Corvallis, Oregon, office and may be reached at (503) 752-4271; Mr. Cushnie is vice president of CAI Engineering in Oakton, Virginia, and may be reached at (703) 254-0039; Mr. Gallerani is president of Integrated Technologies in Cheshire, Connecticut, and may be reached at (203) 272-6014; and Dr. Higgins is a vice president in CH2M HILL's Reston, Virginia, office and may be reached at (703) 471-1441.

Table 12.1. Metal-Finishing Operations and Typical Wastes

Source	Waste
Degreasing	Solvents, oils, still bottoms, metal chips
Cleaning	Alkalis, metals, chips and salts, chelates, solvents, oil and grease
Pickling	Acids, metals, chromates
Metal plating	Acids, metals, cyanide, alkalis, chelates, chromates, filters
Etching	Metals, acids, chelates
Conversion coating	Chromates, phosphates, metals

in a single tank, often using automated equipment, with large volumes of plating solution being "dragged out" of the plating bath each day on plated parts.

Plating to improve mechanical properties is usually referred to as "hard-chrome plating." Hard-chrome plating often involves adding tens of thousandths of an inch thickness of plate, a process that can take more than 24 hr to complete. As a result, drag-out rates for hard-chrome plating are significantly less than the rates for decorative plating in which parts are removed from the plating tanks after a few minutes of plating.

Nickel Plating

Nickel is an attractive and ductile metal with many engineering and decorative applications. It is easily buffed to a high luster, exhibits good resistance to corrosion and abrasion, and is readily machined. Because of its combined physical and chemical properties, nickel is the most popular and useful metallic coating. Thin nickel coatings are used mainly for corrosion protection, for improving the ability to braze or solder difficult materials, for decorative purposes, and as an undercoating for other metal deposits that subsequently are plated. Heavy nickel coatings are used primarily for combined corrosion and wear resistance, salvage of worn or corroded parts, and electroforming.

The mechanical properties of the nickel deposit and its effect on the base material (e.g., tensile strength, internal stress, ductility, and hardness) are influenced by the chemical composition and the operation of the plating bath. In the aerospace industry, where low-stress deposits are a requirement, nickel sulfamate plating baths are used extensively. Watts nickel plating and duplex nickel plating are used as undercoatings for decorative chromium.

Cadmium and Zinc Plating

Sacrificial cadmium and zinc coatings are normally applied to protect the base metal, typically iron or steel. A thin surface-coating is usually applied for corrosion protection, for improvement of wear or erosion resistance, for reduction of friction, or for decorative purposes. Cadmium is significantly more expensive and toxic than zinc; therefore, it is used as a protective coating only in those circumstances requiring its special properties. Zinc and cadmium are normally coated with a chromate-conversion coating.

Cadmium is selected over zinc as a protective coating for the following reasons: (1) it is more easily soldered than zinc; (2) its corrosion products do not swell and are not bulky (unlike the "white rust" formed by zinc) and hence do not interfere with functional moving parts; and (3) cadmium is somewhat superior to zinc in corrosion protection in marine (salt) environments. Parts that are to be cadmium or zinc plated typically are cleaned of grease, dust, oil, and rust by undergoing solvent vapor degreasing, or alkaline cleaning and acid pickling. After a part is cleaned, it is plated, possibly baked to relieve hydrogen embrittlement, and then chromate-conversion coated.

Copper, Gold, Silver, and Tin Plating

Copper has excellent conductive properties for electricity and is plated on plastic in the manufacture of printed circuit boards. Gold, silver, and tin are plated on electrical contacts and high-value circuits because of their superior physical properties, such as higher conductance and inertness. Gold and silver also are plated on products for aesthetic appeal.

Aluminum Finishing

Aluminum and aluminum alloys have various types of finishes applied to enhance appearance, improve functional properties of the surfaces, or both. Aluminum products are widely used in manufacturing segments of industry including aerospace, electronic equipment, motor vehicles, machinery, furniture, and household appliances.

Although aluminum and aluminum alloys can be electroplated with chromium, nickel, cadmium, copper, tin, zinc, gold, or silver, aluminum requires more preparation than other base metals (e.g., steel, copper alloys) require for electroplating.[1] Finishing aluminum by anodizing, conversion coating, or painting is more common than electroplating. The most common aluminum finishing method is anodizing, which is the conversion of the aluminum surface to aluminum oxide. Anodizing increases corrosion resistance and paint adhesion and improves appearance. It is performed electrolytically in solutions containing chromic acid, sulfuric acid, sulfuric/oxalic acid, sulfuric/boric acid, or phosphoric acid.

Conversion coating of aluminum is performed for corrosion retardation, improvement of adhesion for organic coatings (e.g., paint), wear resistance, and decorative purposes. Conversion coatings are applied by spray or immersion processes, without electricity. The most common conversion coating is chromate, which is widely applied by aerospace and aircraft industries because it provides maximum resistance to corrosion.

Waste products from aluminum finishing include spent process solutions and rinse waters. Process solutions include prefinishing (desmut and deoxidize solutions usually containing nitric or chromic acids) and post-finishing (dichromate and hot-water seals). Mostly, these solutions are discarded when they build up intolerable levels of contaminants such as dissolved aluminum or trivalent chromium. Bath maintenance methods such as membrane electrodialysis, ion exchange, and diffusion dialysis can be applied to extend bath life. Rinse waters are generally treated, although ion exchange is being more widely applied in recent years for recovery of constituents such as chromium.

Electroless Nickel Plating

In electroless nickel (EN) plating, nickel ions in solution are reduced to nickel metal at the surface of a catalytic substrate. The most commonly used reducing agent is sodium hypophosphite. The nickel metal deposits evenly on prepared catalytic substrate surfaces; deposit thicknesses of 10 mils and greater are common. Electroless nickel deposits are an alloy of nickel and phosphorus and deposit properties are very much a function of the percentage of phosphorus in the deposit. After the initial substrate surface is coated, nickel plates on the substrate autocatalytically.

For electroless nickel plating, the metal ion reducing agent, complexing agent, and stabilizer concentrations need to be controlled closely to maintain proper plating quality. Trace organic contamination, heavy metals, and other impurities can result in poor plating quality and decrease bath life and even result in spontaneous plate out of the bath. An important side reaction in EN plating is the oxidation of hypophosphate to orthophosphate, which becomes a significant contaminant as the concentration increases. Bath maintenance practices are currently being developed and basically involve ion exchange, electrodialysis, or lime precipitation.

Manufacture of Printed Circuit Boards

Production of printed circuit (wiring) boards involves the plating and selective etching of flat circuits of copper supported on a nonconductive sheet of plastic. Production begins with a sheet of plastic laminated with a thin layer of copper foil. Holes are drilled through the board using an automated drilling machine. The holes are used to mount electronic components on the board and to provide a conductive circuit from one layer of the board to another.

Following drilling, the board is scrubbed to remove fine copper particles left by the drill. The rinse water from a scrubber unit can be a significant source of copper waste. In the scrubber, the copper is in a particulate form and can be removed by use of filters or a centrifuge. Equipment is available to remove this copper particulate, allowing recycle of the rinse water to the scrubber. However, once mixed with other waste streams, the copper can dissolve and contribute to the dissolved copper load on a wastewater treatment plant.

After being scrubbed, the board is cleaned and etched to promote good adhesion and then is plated with an additional layer of copper. Since the holes are not conductive, electroless copper plating is employed to provide a thin continuous conductive layer over the surface of the board and through the holes. Electroless copper plating involves using chelating agents to keep the copper in solution at an alkaline pH. Plating depletes the metal and reduces the alkalinity of the electroless bath. Copper sulfate and caustic are added (usually automatically) as solutions, resulting in a "growth" in the volume of the plating solution. This growth is a significant source of copper-bearing wastewater in the circuit board industry.

Treatment of this waste stream (and the rinse water from electroless plating) is complicated by the presence of chelating agents, making simple hydroxide precipitation ineffective. Iron salts can be added to break the chelate, but only at the cost of producing a significant volume of sludge. Ion exchange is used to strip the copper from the chelating agent, typically by using a chelating ion exchange resin. Regeneration of the ion exchange resin with sulfuric acid produces a concentrated copper sulfate solution without the chelate. This regenerant can then be either treated by hydroxide precipitation, producing a hazardous waste sludge, or concentrated to produce a useful product.

Growth from electroless copper plating is typically too concentrated in copper to treat directly by ion exchange. Different methods have been employed to reduce the concentration of copper sufficiently either to discharge the effluent directly to a sewer or to treat with ion exchange. One method is reported by Hewlett-Packard in which growth is replenished with formaldehyde and caustic soda to enhance its autocatalytic plating tendency, and then mixed with carbon granules, on which the copper plates in a form suitable for reclaiming.[2]

Following electroless plating, copper is electroplated on the board to its final thickness, and a layer of tin-lead solder is plated over the copper. A photoresist material is then applied to the board and exposed by photoimaging a circuit design. After developing and stripping a selected portion of the photoresist, the exposed portion of the tin-lead plate is etched to reveal the copper in areas between the final desired circuit pattern.

The exposed copper is then removed by etching to reveal the circuit pattern in the remaining copper. Ammonia-based solutions are most widely used for etching. Use of ammonia complicates waste treatment and makes recovery of copper difficult.

An alternative to ammonia etching is the use of etching solution containing sulfuric acid or hydrogen peroxide. The latter solution is continuously replenished by adding concentrated peroxide and acid as the copper concentration increases to about 80 g/L. At this concentration, the solution is cooled to precipitate copper sulfate. After replenishing with peroxide and acid, the etching solution is reused. Disadvantages of the sulfuric acid-peroxide etching solution are that it is relatively slow when compared with ammonia, and controlling temperature can be difficult.

Preparation for Metal Finishing

Preparation for metal plating produces waste streams from several sources. Precleaning and removing surface oxides (pickling) in process baths result in metal-bearing acidic and caustic waste streams. Each operation is typically followed by a rinsing operation that produces a dilute waste stream. Spent or contaminated process baths constitute a concentrated, intermittent waste source.

The metal-finishing process itself produces several waste streams (Figure 12.1). Following each step in the process, parts are rinsed to remove finishing solutions that adhered to the parts (drag-out). Most plating operations use single-overflow rinse tanks that operate at flow rates of from 2 to 8 gal/min. Rinse water is typically the predominant wastewater at plating facilities. Additional discharges of waste include materials used in cleanup of spills; aerosol spray from operations such as chromium plating, which is exhausted to the atmosphere or removed by wet scrubbers; and discarded process solutions.

In metal-finishing operations, wastewaters are typically produced from many separate plating or finishing processes. The result can be production of a mixed wastewater that contains several metals and chelating agents. Conventional metal hydroxide precipitation is unworkable for treating these chelated wastewaters.

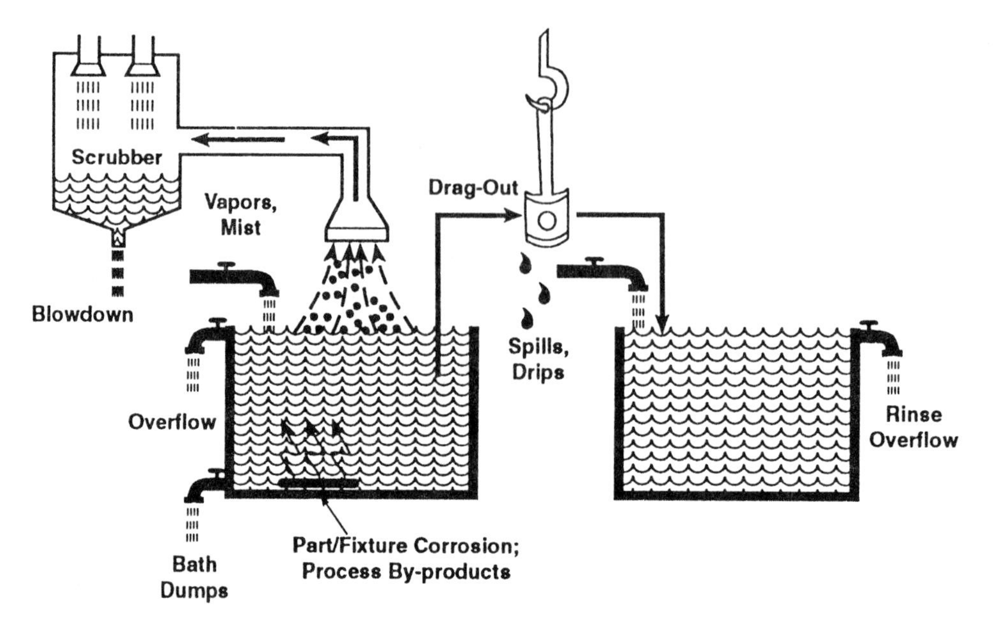

Figure 12.1. Sources of waste from metal-finishing processes.

TECHNIQUES FOR REDUCING GENERATION OF PLATING WASTES

To successfully implement pollution prevention techniques, metal-finishing processes must be thoroughly understood and critical process parameters identified. Some critical parameters are:

- process concentration and physical limits
- process contamination limits
- rinse water quality requirements
- process drag-out and drag-in

Some process parameters are critical (i.e., a change would affect product characteristics in a negative way) and some parameters are not critical (i.e., changes can be made within some range that will not affect product quality). A thorough understanding of the process allows changes to be made that can substantially and directly reduce waste or improve recovery opportunities.

Several process modifications have been proposed for reducing, at their source, waste generation from metal plating. These modifications include:

- improving housekeeping and maintenance practices
- reducing drag-out
- modifying rinsing
- modifying process operations/controls
- reducing or eliminating tank dumping
- recovering metals from rinse waters
- substituting less hazardous materials

An overview of each modification is presented in this section.

Table 12.2 lists some major vendors that provide pollution prevention equipment for the metal-plating and surface-finishing industry.

Improving Housekeeping and Maintenance Practices

Changes in housekeeping and maintenance practices can be made rapidly with little capital investment. Successfully implemented, these changes can increase production rates, improve product quality, and improve workplace safety, as well as reduce generation of waste. Reducing generation of waste can yield significant savings in raw material usage and waste treatment. The following list of housekeeping and maintenance practices, although not all-inclusive, could save plating shops thousands of dollars a year:

- Repair all leaking tanks, pumps, valves, and pipe fittings.
- Inspect tanks and tank liners periodically to avoid failures that may necessitate bath dumps. Inspect steam coils and heat exchangers to prevent accidental contamination or leakage of steam condensate and cooling water into the plating bath.
- Replace or repair equipment before leaks or spills occur.
- Install alarms on plating tanks to warn of high levels to avoid accidental bath dumps.
- Maintain plating racks and anodes to prevent contamination of baths. Remove racks and anodes when baths are not in use.
- Minimize the volume of water used during cleanup.
- Train plating personnel to understand the importance of minimizing bath contamination and wastewater discharge.
- Thoroughly clean and rinse parts before plating to minimize contamination of plating baths. Areas of parts that are not to be plated should be masked or stopped off with tape or wax to limit contamination from corrosion in these areas. Parts should be removed from the bath when not being plated. Regularly check tanks and remove fallen parts.
- Remove anodes from tanks when plating tank is not in use. In baths where erosion of the anodes provides replacement metal, dissolution of anodes during periods of nonuse can result in a buildup of metal to a concentration higher than acceptable. This buildup often results in the need to dispose of a portion of a bath to reduce the metal concentration.
- Provide for segregated collection of spills.
- Use in-tank pump/filtration systems.
- Use seal-less pumps.
- Use proper materials to provide for corrosion protection.

Table 12.2. Waste Reduction Equipment Vendors for Metal-Finishing Applications

Company	Address/Phone/*Fax*	Technology
Ag-Met Equipment Corporation	600 Greenwich Avenue West Warrick, RI 02883 (401) 823–5690	Electrolytic recovery
Aqualogic, Inc.	30 Devine Street North Haven, CT 06473 (203) 248–8959 *(203) 288–4308*	Ion exchange Atmospheric evaporators
Asahi Glass America, Inc.	1185 Avenue of Americas 30th Floor New York, NY 10036 (212) 704–3155 *(212) 764–3384*	Diffusion dialysis Membrane electrodialysis
Baker Bros., Division of Systems Engineering & Mfg. Corp.	44 Campanelli Parkway Stoughton, MA 02072 (617) 344–1700 *(617) 341–1713*	Diffusion dialysis Electrodialysis Electrolytic recovery Ion exchange
Baker, Company, M.E.	25 Wheeler Street Cambridge, MA 02138 (617) 547–5460 *(617) 354–7396*	Ion exchange
BEWT Recovery Technologies, Inc.	1380 Hopkins Street Unit 11 Whitby, Ontario Canada L1N 2C3 (416) 668–8100 *(416) 430–7667*	Electrolytic recovery
Bio-Recovery Systems, Inc.	2001 Copper Avenue Las Cruces, NM 88005 (505) 523–0405 *(505) 523–1638*	Ion exchange
Calfran International, Inc.	171 Shaker Road Box 269 Springfield, MA 01101 (413) 525–4957 *(413) 732–9246*	Evaporators/stills
ConRec, Inc.	140 Waterman Avenue Centredale, RI 02911 (401) 231–3770 *(401) 231–3360*	Electrolytic recovery
Corning, Inc.	Big Flats Plant Big Flats, NY 14814 (607) 974–7201 *(607) 974–7203*	Evaporators/stills

Table 12.2. Waste Reduction Equipment Vendors for Metal-Finishing Applications (Continued)

Company	Address/Phone/*Fax*	Technology
Covofinish Company, Inc.	P.O. Box 145 N. Scituate, RI 02857 (401) 568–9191 *(401) 568–9196*	Electrolytic recovery
Eco-Tec	925 Brock Road, South Toronto, Ontario L1W 2X9 (416) 831–3400 *(416) 831–3409*	Acid retardation Electrowinning Electrolytic recovery Ion exchange Crystallization
Eltech International Corporation	625 East Street Fairport Harbor, OH 44077 (216) 357–4080 *(216) 357–4077*	Electrolytic recovery
Graver	2720 U.S. Highway 22 Union, NJ 07083 (908) 964–2400 *(908) 964–7770*	Reverse osmosis/ Ultrafiltration
Ionic Industries International, Ltd.	829 Highams Court Woodbridge, VA 22192 (703) 491–1190 *(703) 491–8831*	Atmospheric evaporators Ion exchange Membrane electrodialysis
Ionpure Technologies, Corporation	10 Technology Drive Lowell, MA 01851 (508) 934–9349 *(508) 441–6025*	Reverse osmosis/ Ultrafiltration
Ionsep Corp.	P.O. Box 258 Rockland, DE 19732 (302) 798–7402 *(302) 798–7425*	Membrane electrodialysis
Integrated Technologies, Inc.	1781 Highland Avenue Cheshire, CT 06410 (203) 272–6014 *(203) 272–2804*	Software
Kinetico, Inc.	P.O. Box 193 10845 Kinsman Road Newbury, OH 44065 (216) 564–9111 *(216) 338–8694*	Electrolytic recovery Electrowinning Ion exchange Reverse osmosis Ultrafiltration
Koch Membrane Systems, Inc.	850 Main Street Wilmington, MA 01887 (508) 657–4250 *(508) 657–5208*	Ultrafiltration

Table 12.2. Waste Reduction Equipment Vendors for Metal-Finishing Applications (Continued)

Company	Address/Phone/*Fax*	Technology
Licon, Inc.	P.O. Box 10717 Pensacola, FL 32504 (904) 434–5088 *(904) 438–2040*	Evaporators/stills
Memtek Corporation	28 Cook Street Billerica, MA 01821 (508) 667–2828 *(508) 667–1731*	Electrolytic recovery Electrowinning Ion exchange Reverse osmosis Microfiltration
Napco, Inc.	Napco Drive Box 26 Terryville, CT 06786 (203) 589–7800 *(203) 589–7304*	Atmospheric evaporators Ion exchange
Osmonics	5951 Clearwater Drive Minnetonka, MN 55343 (612) 933–2277 *(612) 933–0141*	Reverse osmosis Ultrafiltration
Penfield Liquid Treatment Systems	8 West Street Plantsville, CT 06479 (203) 621–9141 *(203) 621–2380*	Reverse osmosis Ultrafiltration Ion exchange
Poly Products	P.O. Box 151 Atwood, CA 92601 (714) 538–0701 *(714) 538–0691*	Atmospheric evaporators
Prosys corporation	187 Billerica Road Chelmsford, MA 01824 (508) 250–4940 *(508) 250–4977*	Microfiltration
RFE Industries, Inc.	19 Crows Mill Road Keasbey, NJ 08832 (908) 738–5200 *(908) 738–5319*	Electrolytic recovery Ion exchange
Techmatic, Inc.	133 Lyle Lane Nashville, TN 37210 (615) 256–1416 *(615) 242–7908*	Atmospheric evaporators
Trionetics, Inc.	2021 Midway Drive Twinsburg, OH 44087 (216) 425–2846 *(216) 425–9704*	Electrolytic recovery
U.S. Filter Corporation	181 Thorn Hill Road Warrendale, PA 15086 (412) 772–0044 *(412) 772–1360*	Electrolytic recovery Ion exchange

Reducing Drag-Out

Reducing drag-out (formation of film on part surface) from plating baths saves bath replacement costs and reduces waste disposal costs. To evaluate the effectiveness of drag-out reduction, existing drag-out must be quantified. For example, the drag-out from tanks using barrel plating is usually 10 times greater than that from plating the same area of parts using rack plating. The amount of pollutants contributed by drag-out is a function of such factors as the design of the racks or barrels carrying the parts to be plated, the shape of the parts, plating procedures (including part withdrawal technique and rate), and several interrelated parameters of the process solution, including concentration of chemicals, temperature, viscosity, and surface tension. Table 12.3 provides factors that can be used to estimate drag-out rates for different shapes of parts and different plating solutions.

Many devices and procedures can be used successfully to reduce drag-out by altering viscosity, chemical concentration, surface tension, velocity of withdrawal, or temperature. Most drag-out reduction methods are inexpensive to implement and are repaid promptly through savings in plating chemicals. Additional savings worth many times the cost of the changes will be realized in pollution control savings. A more favorable rate of return is realized by implementing drag-out reduction techniques at plating lines for decorative chrome, cadmium, and zinc. In those lines, plating times are relatively short and drag-out is significantly greater than in hard-chrome plating.

Table 12.3. Estimate of Drag-Out Presented in the Literature

A. Average Drag-Out Losses (Soderberg)

Nature of Drainage	Drag-Out Rate (gal/1,000 ft^2) Vertical	Horizontal	Cup Shapes
Well drained	0.4	0.8	8.0
Poorly drained	2.0	–	–
Very poorly drained	4.0	10.0	24.0

B. Drag-Out Amounts (Hogaboom)

Solution Type	Drag-Out Rate (gal/1,000 ft^2) Flat Surfaces	Contoured Surfaces
Brass	0.95	3.3
Cadmium	1.00	3.1
Chromium (33 oz/gal)	1.18	3.0
Chromium (52 oz/gal)	4.53	11.9[a]
Copper Cyanide	0.91	3.2
Watts Nickel	1.00	3.8
Silver	1.20	3.2
Stannate Tin	0.83	1.6
Acid Zinc	1.30	3.5
Cyanide Zinc	1.20	3.8

[a]It is of interest to note the effect of increased viscosity. There is roughly a threefold increase in drag-out volume with less than double the concentration of solution.

Controlling Plating Solutions

Drag-out can be reduced by decreasing either bath viscosity or surface tension. Viscosity can be decreased by reducing the chemical concentration of the bath or by increasing the temperature of the bath. Surface tension can be decreased by either adding wetting agents or increasing the temperature of the bath. These modifications will improve the drainage of plating solutions back into plating baths and/or reduce the concentration of metal in the drag-out.

For years, wetting agents have been used in process solutions to aid in the plating process. These substances are used, for instance, in bright-nickel plating to promote disengagement of hydrogen bubbles at the cathode. They also can be used as an aid to reduce drag-out. A wetting agent is a substance, usually a surfactant, that reduces the surface tension of a liquid, causing it to spread more readily on a solid surface. A typical plating bath solution has a surface tension close to that of pure water at room temperature, about 0.0050 lb/ft. The addition of very small amounts of surfactants can reduce surface tension considerably, to as little as 0.0017 to 0.0024 lb/ft.

Kushner estimates that the use of wetting agents will reduce drag-out loss by as much as 50%.[1] He recommends the use of nonionic wetting agents that are not harmed by electrodialysis in the plating bath.

Positioning Work on Racks

The metal finisher's primary consideration in the positioning of workpieces on a rack ("racking") is proper exposure of the parts to the anodes for optimal coverage and uniform thickness of the electrodeposit. Drainage and rinsability are also important considerations in racking. A cup-like depression in a part or rack can literally bail gallons of plating solution from a bath. Significant drag-out reduction can be accomplished if platers carefully rack and remove parts to minimize entrapment of bath materials on surfaces and in cavities (Figure 12.2). Damage to the workpiece surface can be caused by insufficient or inefficient rinsing, and succeeding process solutions can be contaminated by chemicals carried over from the previous bath.

Several rules apply to racking work to minimize drag-out. The basic principle, however, is that every object can be positioned in at least one way that will minimize drag-out. This position could be determined by experiment, but experimentation may not be justifiable unless a significant number of similar items are to be plated; therefore, it may be advisable to follow these suggestions:[2]

1. Tilt solid objects with plane or single-curved surfaces so that drainage is consolidated; that is, turn the part so that the clinging fluid will flow together and off the part by the quickest route.

2. Rack parts to minimize the average depth to which the parts are lowered into a solution to decrease the film thickness of the drag-out.

3. If possible, avoid racking parts directly over one another to reduce the drainage path of the plating solution.

4. Avoid table-like flat horizontal surfaces by tipping the part, but not if doing so causes formation of solution "pockets."

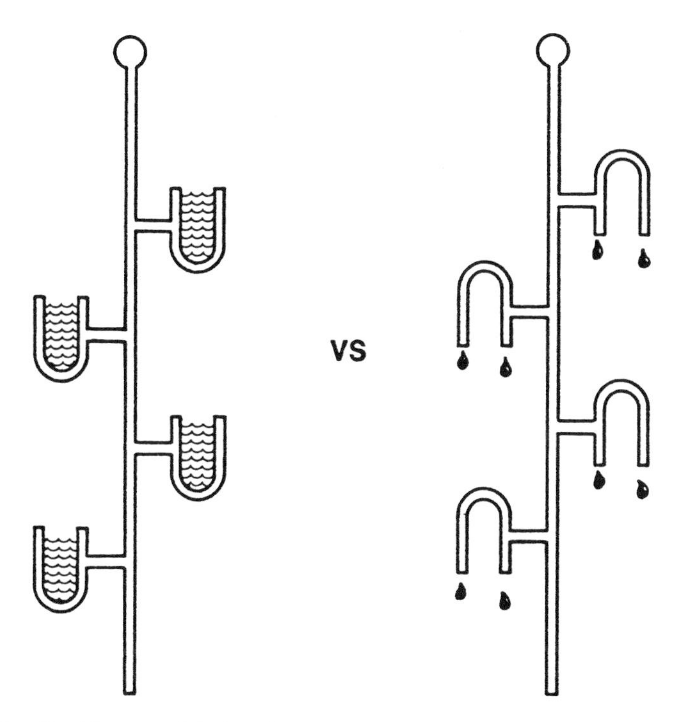

Figure 12.2. Racking to minimize drag-out.

5. Orient parts to minimize surface area in contact with the liquid surface on leaving the plating solution.

Workpiece Withdrawal

The velocity with which work is withdrawn from the process tank has a marked effect on drag-out volume. The faster an item is pulled out of the tank, the thicker the drag-out layer will be. The effect is so dramatic that Kushner suggests that most of the time available for withdrawing and draining the item should be used for withdrawal.

On automated plating systems, the velocity of withdrawal of work from the process tank usually can be controlled by adjusting vertical hoist speed. Drainage time can be extended by using limit switches or adjusting the timer that controls the horizontal movement of the hoist. In either case, there is no substitute for operator training. If the metal-finishing cycle is operated by hand, however, the withdrawal velocity is less controllable. The best control method is to place a bar or rail above the process tank on which the rack can be suspended for drainage while its predecessor is removed from the rail and transported to the next phase of the finishing cycle.

The withdrawal motion also affects drag-out. When a rack is jerked from a process solution, surface tension forces do not have time to pull solution from the part, and a much larger volume of liquid will cling to the surface than when the part is slowly and uniformly removed from the bath. An automatic machine that provides smooth, gradual withdrawal usually will drag out significantly less solution per item than will manually operated equipment.

It is impossible to accurately predict the drag-out reduction to be achieved by slowing withdrawal speed or by adopting a smooth withdrawal motion. Savings may be expected, but the volume will be determined by the specific application.

Draining time over the tank can be limited by the tendency of the plated object to spot if the plating solution dries on the surface. A fog spray that uses water from the first rinse can be very effective in keeping the surface from drying, accelerating the draining process, and extending the time available for draining.

Maintenance and Design of Racks and Barrels

Industrywide, maintenance of racks, fixtures, and rack coatings has been poor. Transport of chemicals from one process to another inside loose rack coatings is not uncommon. Chromium-bearing solutions, for example, appear in plant effluent in spite of treatment systems designed to handle the normal chromium discharge sources. These solutions have been traced to rinse tanks and process solutions that are located some distance from the chromium discharge points. The chromium-bearing solutions can reach these remote areas by way of loose rack coatings. Increased attention to rack maintenance not only will eliminate this potential hazard but also will contribute to a welcome reduction in the number of workpieces rejected because of poor contact.

Drain Boards, Drip Bars, and Drip Tanks

Drag-out can be captured by the use of drain boards, drip bars, and drip tanks and can be returned to the bath (Figure 12.3). These simple devices save chemicals, reduce rinse requirements, and prevent unnecessary floor washing.[3]

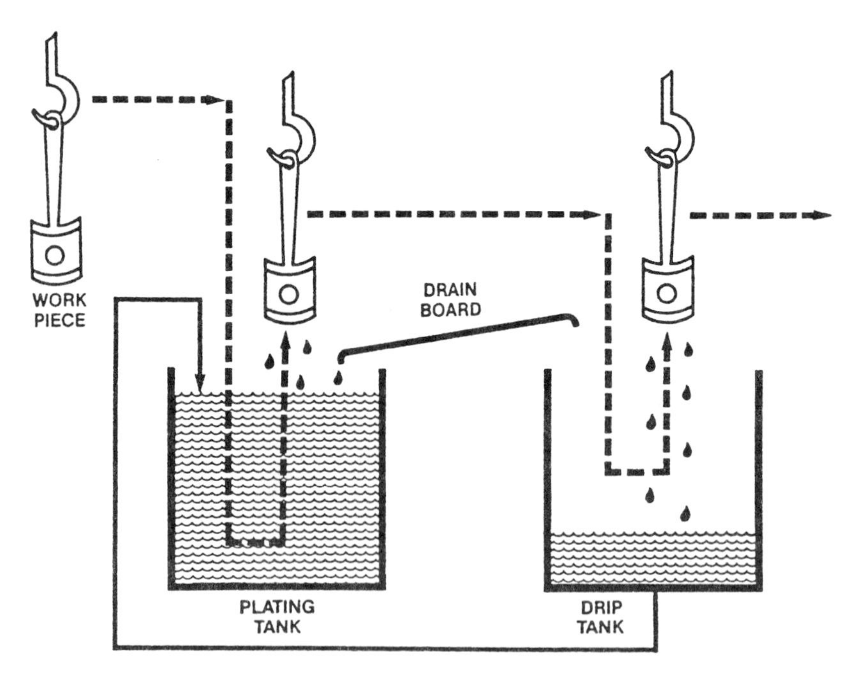

Figure 12.3. Drag-out recovery devices.

Miscellaneous Techniques for Reducing Drag-Out

Parts should be designed to promote drainage (Figure 12.4). Air knives can be used to knock plating films off parts and back into process tanks. This technique is particularly effective in removing ambient-temperature solutions from plated parts.

Modifying Rinsing

Rinsing is used to remove residual drag-out from parts and racks. Rinse water must be sufficiently clean to reduce the concentration of these chemicals in a reasonable period of time. Most platers and surface finishers employ continuously flowing single-tank rinses to wash out contaminants. Rinse flows are typically controlled manually and left running continuously. Flows are usually set to eliminate "color" in the water, rather than using a chemical analysis.

Decreasing rinse water flow may not reduce the amount of toxic metals to be disposed of, but it can reduce the volume of liquid waste that must be processed in industrial wastewater treatment plants. However, concentrations of metals would increase, possibly impairing treatability. Thus, decreasing rinse water flow may not appreciably reduce costs of wastewater treatment.

If rinse flow rates are reduced sufficiently, it is possible to use rinse water to make up for evaporative losses in the plating tanks, resulting in metal recovery and reduced waste discharge. Reducing flows can also increase the efficiencies of metal recovery by using concentration processes such as evaporation, ion exchange, and reverse osmosis (discussed in a later section of this chapter).

The following are descriptions of techniques that improve rinse efficiency.

Rinsing Over Plating Tanks

Rinsing over the plating tanks effectively removes drag-out from parts. A plated part is held over the plating tank and sprayed with rinse water. More than 75% of the plating chemicals drain back to the plating bath. This modification is best suited for parts that are plated at elevated temperatures so that evaporation rates are high, thus making space available for the rinse water added to the tank.[4]

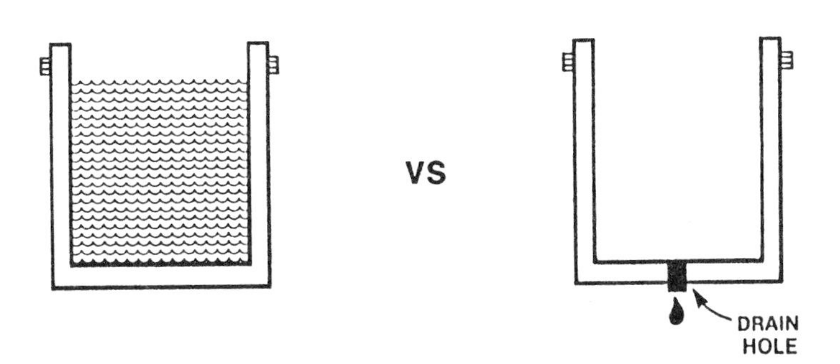

Figure 12.4. Part design to minimize drag-out.

Example: At NASA's Marshall Space Flight Center, the plater simply rinsed plated parts over the plating bath with a handheld hose, recovering the chemical and replacing water lost to evaporation in the bath. This simple but effective technique recovered virtually all of the drag-out and reduced the concentration of plating metals in the rinse water to less than 0.1 mg/L.

Spray Rinse

Spray and fog rinses are techniques for using rinse water more efficiently. If evaporation is sufficient, drainage can be directed back into the process tank or into a drag-out tank.

Spray rinsing typically involves the use of a dedicated spray rinse tank in which two to four rows of high-velocity spray-jet nozzles are mounted. Spray nozzles are available in a wide range of flow rates and spray patterns. Generally, the sprays are operated at the water pressure in the municipal service line. However, in some cases, water pressure is increased by pumping to improve the rinsing action. Typically, spray rinses use $^1/_8$ to $^1/_4$ of the amount of water that would be used for a single-stage overflow rinse.[1] However, the savings in water use depend heavily on the part configuration, as discussed below. Other factors influencing the efficiency of spray rinsing are the arrangement of nozzles to workpieces, water pressure, specific flow rate, spray time, and mechanical design of the delivery system. As an example, water can be wasted through drainage of an improperly designed spray manifold after part rinsing has ceased.

The most effective application of spray rinsing is for flat-surfaced parts where the spray can directly strike the entire surface of the part. Spray rinsing is ineffective for rinsing parts with recessed and inaccessible surfaces. In addition, it cannot be used for small racked parts that could be displaced from the rack by the force of the spray or with plating barrels that prevent the spray from adequately reaching the parts.

Fog rinsing is a specialized type of spray rinsing used at the exit point of process tanks. A fine fog is sprayed on the work, diluting the drag-out film and causing plating solution to run back into the tank. Fog rinsing is applied when process operating temperatures are high enough to produce a high evaporation rate and allow replacement water to be added to the process in this manner. Fog rinsing prevents dry-on patterns by wetting and cooling the workpieces, but it may preclude the use of a drag-out tank as a recovery option because the fog rinse makes up some of the liquid lost to evaporation. This limitation can be overcome by using rinse water for fog rinsing. Potential ventilation problems associated with this configuration must be closely evaluated. For fog rinsing to be effective, work must be withdrawn slowly from the process tank.

Drag-Out Tank

Still rinse tanks (also called "dead rinse," "stagnant rinse," and "recovery rinse") (Figure 12.5) can be used before using rinse tanks with flowing clean water. Water from the drag-out tank or still rinse tank can be returned to the bath to make up for evaporation losses. Increasing plating bath temperatures to increase evaporation may be justified.

The drag-out tank is a rinse tank(s) that can be assumed to be filled initially with pure water. As the plating line is operated, the drag-out rinse tank remains relatively

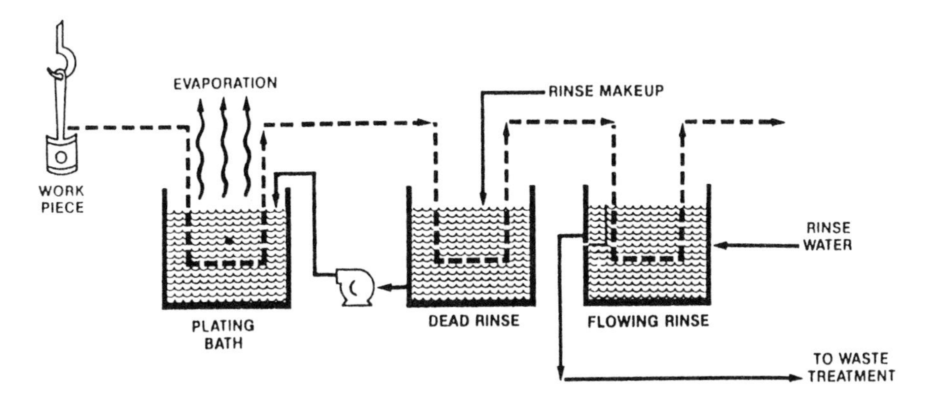

Figure 12.5. Dead rinse for recovery of plating chemicals.

stagnant; the salt concentration increases as more work passes through the rinse tank. Agitation must be used to aid the rinsing process because there is minimal water flow within the tank to cause turbulence. The presence of a wetting agent is helpful. After a period of operation, the diluted plating solution in the drag-out tank can be used to replenish evaporation losses to the plating bath. If sufficient evaporation has occurred, then a portion of the drag-out tank solution can be added directly to the plating bath. Evaporation usually will be sufficient in baths that are operated at elevated temperatures, such as those for nickel plating.

As a general rule, the use of a drag-out tank will reduce chemical losses by 50% or more. The recovery rate is derived from the ratio of evaporation to drag-out. The efficiency of the drag-out tank arrangement can be increased significantly by adding a second drag-out tank. Use of a two-stage drag-out system usually reduces drag-out losses by 70% or more. Multiple countercurrent drag-out tanks can be used to completely close the loop and return 100% of drag-out.

Low-temperature baths have minimum surface evaporation and their temperatures cannot be increased without degrading heat-sensitive additives. Recently, new additives, which are not as readily degraded by heat, have been developed for many of these plating baths. These additives might make operation of the plating bath at higher temperatures possible, facilitating drag-out recovery by recycle techniques. Usually the value of the recovered chemicals is much higher than the increased energy cost of operating the bath at a higher temperature. Additional benefit is derived from the decreased viscosity and drag-out of a plating solution operated at an elevated temperature.

Example: At a small plating shop at Williams AFB, the plater kept a plastic bucket of distilled water next to each plating bath. After plating, he would dip each rack of parts into the bucket before final rinsing in a flowing rinse tank. Then, before starting work each morning, he would empty the bucket from the previous day into the plating tank, and refill the bucket with distilled water. This simple low-cost system recovered most of the plating chemicals, which otherwise would have been discharged to the treatment plants.

Flow Control

Water supply control valves can be used to reduce wastewater flows to a minimum. These inexpensive devices (approximately $30 each) accurately regulate the feed rate of fresh water despite variations in line pressure. These valves can usually be set to regulate flow within a 1/2-gal/min range. One difficulty with limiting flow to a rinse tank is that water agitation is essential to good rinsing; a high flow of water often provides useful agitation in the rinse tank.

Conductivity Control

Conductivity controllers (Figure 12.6) can be used to operate rinse water control valves automatically, thereby reducing demands on plating personnel. Conductivity control operates on the principle that clean water has a lower conductivity than water contaminated with plating solutions. A conductivity probe, controller, and valve reportedly can cost less than $1,000 to purchase and install; such systems have been installed in many plating shops.[5] Unfortunately, these units have not performed well in most installations because of the fragility of the probe and the need for frequent calibration and cleaning. Selecting the optimum conductivity setpoints can be difficult. Plating personnel frequently override or disconnect conductivity controllers because of dissatisfaction with their operation. Alternatively, microswitches can perform similar functions, as described in the example that follows.

Example: An aerospace company manufactures experimental circuit boards at its New England facility. Because of the experimental nature of the operation, only a small number of circuit boards are manufactured on a full-size plating line. When conductivity controllers were installed on all of the rinse tanks, the platers discovered that the controllers could not be set accurately enough to turn on the rinse flow each time a rack was lowered into the rinse tank, which was unacceptable since they relied on fresh-water flow to provide mixing. To remedy the problem, conductivity

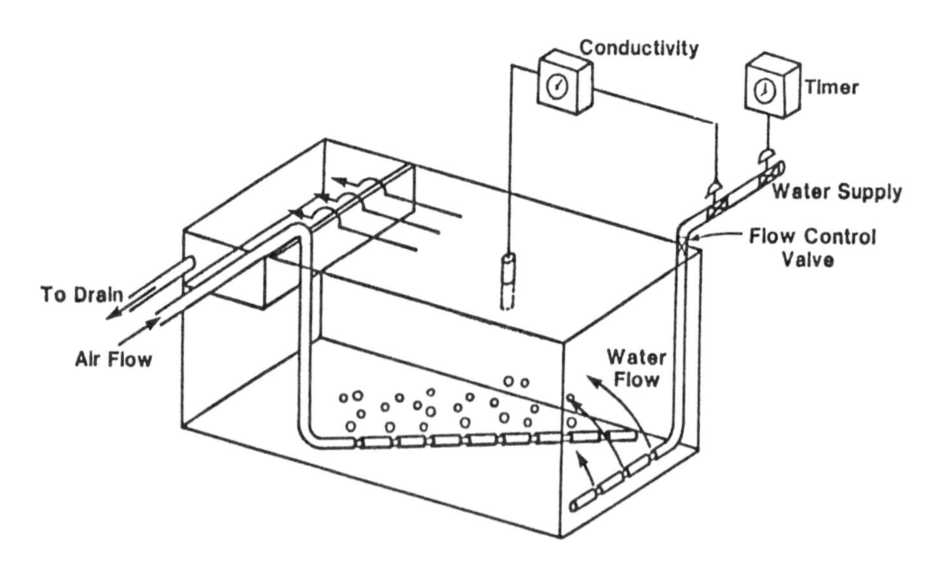

Figure 12.6. Use of controls and air agitation to reduce rinse water usage.

controllers are being replaced by micro-switches on the rack holders. These switches will activate valves to provide rinsing whenever a rack is lowered into the tank. In addition, aeration is being added to improve mixing, thus increasing the efficiency of rinsing even at reduced water flows.

Improved Mixing in Rinse Tank

Improved mixing in the rinse tank can increase the efficiency of water use (Figure 12.6) and, therefore, allow a reduced flow of water that normally would not be acceptable. A submerged influent water line evenly distributes fresh water through the tank and creates a rolling action. Most improvements in rinse tank mixing are accomplished by aeration. Existing facilities can be retrofitted with aeration mixing using low-pressure blowers and inexpensive plastic (polyvinyl chloride [PVC]) piping.

Timers

Timer controls can be added that will reduce the daily volume of rinse water while providing an adequate flow to maintain rinse water quality and assure adequate mixing when rinsing is required. With a timer, the plater does not have to remember to shut off the valve after rinsing (Figure 12.6). For reasonably uniform plating operations, timers can be used to operate rinse valves on a preset cycle. For intermittent plating operations, timers can be installed that can be initiated manually, providing adequate flow when needed for mixing and an adequate period of flow for reducing the concentration of contaminants for the next batch. For automated systems, rinsing can be controlled by the automatic sequencer.

Example: For a California company, CH2M HILL recommended that manually activated timers be installed on the rinse tanks. Pushing a convenient button at the front of each tank will provide up to 1 hr of flow to the rinse tank. Equipment costs for switch, timer, valve, and controls total approximately $400 per tank.

Cascade Rinsing

In cascade rinsing, overflow from one rinse tank is used as the water supply for another compatible rinsing operation. For example, rinse water from an acid dip tank can be cascaded to an alkaline cleaner rinse tank. Interconnecting rinsing tanks can complicate operations, but the savings often exceed the additional operation cost.

Countercurrent Rinsing

In countercurrent rinsing, parts are sequentially immersed in a series of tanks, countercurrent to the rinse flow (Figure 12.7). Countercurrent multiple rinse tanks can reduce rinse flows by more than 95% compared to single overflow rinses. Optimum countercurrent rinsing usually employs three tanks operating in series.

The rinse rate needed for adequate cleaning is governed by a logarithmic equation that depends on the concentration of plating chemicals in the drag-out, the concen-

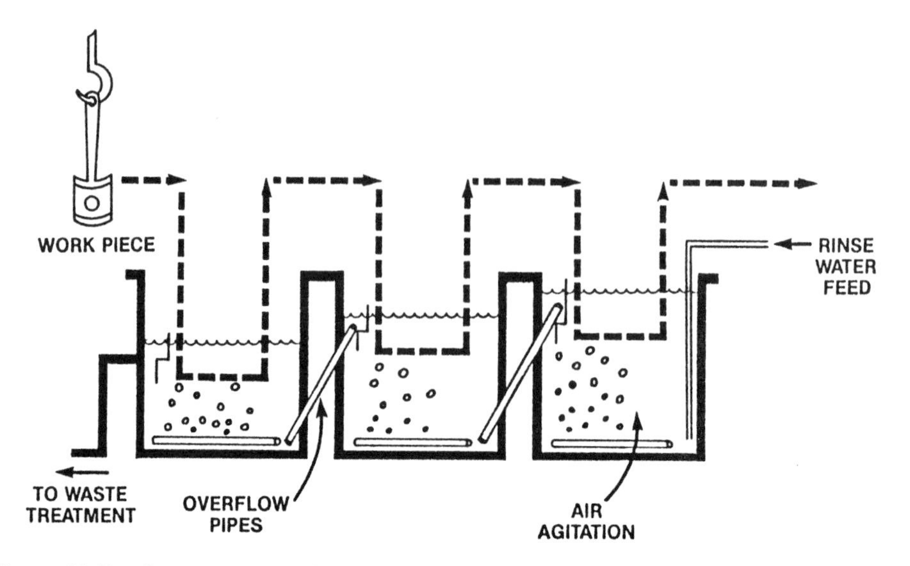

Figure 12.7. Countercurrent rinsing.

tration of plating chemicals that can be tolerated in the final rinse tank before resulting in poor plating, and the number of countercurrent rinse tanks. The mathematical rinsing models are based on complete rinsing (i.e., removal of all drag-out from the part/fixture) and complete mixing (i.e., homogeneous rinse water). These conditions are not achieved or even approached unless there is sufficient residence time and agitation in the rinse tank.

The most commonly applied countercurrent rinsing model follows:

$$R = (Cp/Cn)^{1/n}$$

and

$$F = DR$$

where Cp = concentration in process solution
Cn = required concentration in last rinse tank
n = number of rinse tanks
R = the rinse ratio
F = rinse water flow rate
D = process drag-out

This model does not predict required rinse rates accurately when the value of R falls below 10. In addition, complete rinsing will not be achieved unless there is sufficient residence time and agitation in the rinse tanks.

Example: As an example using the above equation, a typical Watts-type nickel-plating solution contains 270,000 mg/L of total dissolved solids, and the final rinse must contain no more than 37 mg/L of dissolved solids. The ratio of Cp/Cn is 7,300, and approximately 7,300 gal of rinse water are required for each gallon of process solution drag-in with a single-tank rinse system. By installing a two-stage rinse system, water requirements are reduced to 86 gal of water per gallon of process solution drag-in. The rinse water consumption is reduced by 99%. The mass flow of pollutants exiting the rinse system remains constant.

Many plating facilities do not include countercurrent rinsing because the required additional space is not available. Another consideration is the additional production time, because parts must be rinsed in more than one tank. Where space is available, the cost of additional rinse tanks can range from approximately $1,000 to $10,000 per tank, depending on size, shape, and construction materials.

It is sometimes difficult to provide adequate mixing in countercurrent tanks because, when compared with conventional single-tank rinse, flows are considerably reduced. Aeration or other means of providing sufficient mixing is required.

Example: At one facility, countercurrent rinsing was considerably less efficient than is normally expected. Aeration to provide mixing caused a backflow of rinse water to adjacent rinse tanks, resulting in considerably lower water savings than predicted. When the tanks could not be balanced to provide adequate mixing without backflow, countercurrent rinsing was abandoned.

Countercurrent rinse systems can be retrofitted in existing tanks by adding baffles, weirs, pipes, and pumps. Savings vary considerably as a result of differences in costs of raw water and wastewater treatment. At many facilities, the payback period can be as short as 1 year. When plating solution is recovered, further savings can be realized by returning the most concentrated rinse water to the plating bath to make up for evaporative losses. Similar savings can be accomplished by employing a dead or still rinse, followed by a flowing rinse. The contents of the still rinse are periodically returned to the plating bath to recover the plating chemicals.

Modifying Process Operations and Controls

The following modifications to plating operations and controls are recommended to reduce hazardous waste generation:

- Use level controls and alarms to eliminate accidental tank overfills.

- Automate process lines, where practical, to provide consistent and improved control of drag-in and drag-out, to improve the effectiveness of rinsing, and to provide for more consistent high-quality metal-finishing processing.

- Enclose process lines, where practical (in conjunction with process-line automation and proper ventilation design), to minimize airborne contaminants in process solution tanks and rinse tanks.

- Automate chemical monitoring and replenishment systems to significantly extend process bath life and minimize wastes. For example, electroless nickel baths have a limited lifespan because when they consume solution components they are self-contaminating. Automatic monitoring and chemical replenishment systems can significantly extend the life of these baths.

- Install mist eliminators to capture and return mist lost from plating solution (e.g., hard-chrome plating solutions).

Recovering Metals from Rinse Waters

Evaporation, ion exchange (IX), reverse osmosis (RO), and electrodialysis (ED) have been used to concentrate chemicals from rinse waters for recovery. These processes reconcentrate plating solutions from rinse water, sometimes producing a relatively pure water, that can be reused for rinsing. Both general and site-specific factors must be evaluated to determine the recovery process best suited for a particular plating operation. Factors include the type of metal being plated, drag-out rates, rinse water concentrations and flows, space and staffing resources, availability of utilities (such as steam or electricity), and costs for water supply, wastewater treatment, and sludge disposal.

Evaporation

Evaporation is the oldest method used to recover plating chemicals from rinse streams. In this process, rinse water is boiled off to concentrate the solution sufficiently to return it to the plating bath. The steam can be condensed and reused for rinsing. Evaporators can be operated under a vacuum to lower the boiling temperature, thus reducing energy consumption and preventing thermal degradation of plating additives.

The degree of concentration required of the evaporator can be reduced by increasing the evaporation rate from plating baths. Raising the operating temperature can significantly increase the evaporation rate but only at the expense of added heating costs. Use of air agitation in a plating tank also can increase the evaporation rate.

Because of their high energy use, evaporators are most cost-effective in concentrating rinse waters that are to be returned to hot baths, such as those used in chromium plating, where high evaporation rates reduce the concentration required.

Evaporative recovery also has been used successfully for ambient temperature nickel baths and various metal cyanide baths. The capital and operating costs of an evaporator can be reduced by using countercurrent rinsing to produce a low-volume, concentrated rinse stream. One study estimated that chrome-plating shops at Naval shipyards could save $17,000 a year (1983 dollars) by using a countercurrent rinse system in conjunction with evaporative recovery (Figure 12.8).[6] The payback period was estimated to be less than 1 year. A typical atmospheric evaporator is shown as Figure 12.9.

The two primary types of evaporators are atmospheric and vacuum. The basic difference between the two is that the atmospheric evaporator depends on air flow, temperature, and to some extent relative humidity and operates at atmospheric pressure, whereas the vacuum evaporator is a vacuum distilling device that can operate effectively at relatively low temperatures by operating at low pressures. Both, however, are relatively expensive in terms of energy requirements and are thus less practical than other techniques when large volumes of low concentration waters are to be handled.

To achieve a high initial starting concentration and the low volume required for cost-effective evaporator use, counter-flow rinsing or recovery rinsing are almost always integrated into the recovery scheme with the evaporative equipment.

Advantages of evaporative systems in general are that very high recovery rates can be achieved and, once the system is in place, no additional reagents are required for the process. The principal disadvantages are the relatively high operating costs of the simple atmospheric systems and high procurement costs of the more complicated

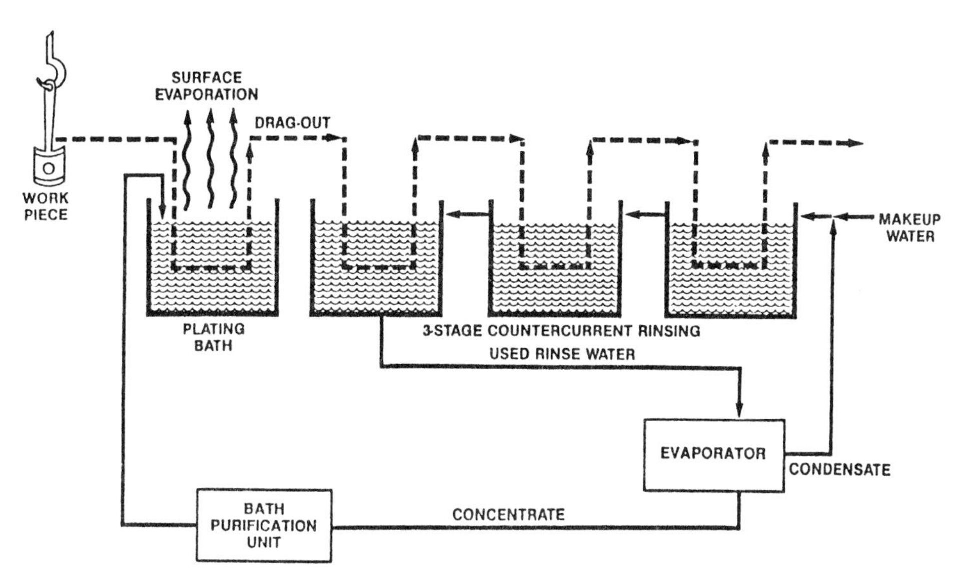

Figure 12.8. Evaporative metal recovery.

Figure 12.9. Atmospheric evaporator (Photo courtesy of Poly Products Corporation, Atwood, California).

vacuum systems. In general, atmospheric evaporators require the least maintenance of all recovery technologies.

Atmospheric Evaporators. Atmospheric evaporators essentially consist of a pump to move the rinse water, a blower to move the air, a heat source, an evaporation chamber in which the water and air can be mixed, and a mist eliminator to remove any entrained liquid from the exit air stream. The evaporation chamber is usually filled with a packing material to provide a large air-to-water interface area. In operation, the water is heated and introduced into the evaporation compartment. Air is then blown through the compartment to evaporate some of the water, and then vented from the chamber. In some systems, the air passing through the evaporative chamber is preheated to increase its capacity for water vapor. Because of their simplicity, atmospheric evaporators are relatively inexpensive to build and maintain.

Two basic strategies are used with atmospheric evaporators to achieve chemical recovery. In the first, solution from a heated plating tank is fed to and concentrated by the evaporator and returned to the plating tank. This approach is used to increase the quantity of recovery rinse water that can be transferred to the plating tank. In a second approach, recovery rinse water is fed to the evaporator and the concentrated liquid is transferred to the plating tank. With this second approach, the recovery rinse water often is fed to the evaporator from a heated transfer tank, and the higher temperature of the solution increases the evaporation rate of the system.

Atmospheric evaporation systems have two disadvantages that could tend to outweigh the advantages of low initial procurement costs. First, energy must be available to evaporate the necessary amount of water and heat the air in the chamber sufficiently to hold the evaporated water. Providing this energy can be expensive, although systems are available that can use heat sources inherent to the basic process to reduce the requirement somewhat. Second, large volumes of air are vented by atmospheric evaporators and may require a permit or pose a pollution hazard.

Generally, energy costs limit the use of atmospheric evaporators to applications where the required evaporation rate is 50 gal per hour (gph) or less. Typical commercial units have evaporation rates of 10 to 30 gph, depending on the size of the unit and operating conditions (e.g., solution temperature). It is not uncommon to use multiple atmospheric evaporators in parallel.

Vacuum Evaporators. Vacuum evaporators operate on the principle that the rate of vaporization is directly related to the level of the vacuum and the temperature of the water. In operation, heated water is introduced into the vacuum chamber. Because of the vacuum, the boiling point of the water is reduced, and the resultant vapor is removed from the chamber and can either be discharged or condensed for return to the basic process (as distilled water). Because of their relative complexity, vacuum systems are more expensive to construct and maintain than atmospheric systems.

Several types of vacuum evaporators are used in the plating industry: rising film, flash type, and submerged tube. Generally, each consists of a boiling chamber that is under a vacuum, a liquid/vapor separator, and a condensing system. Site-specific conditions and the mode of operation influence the selection of one system over another.

Two techniques have been applied successfully for reducing energy (i.e., steam) demand for evaporation; both involve reusing the heat value contained in the vapor from the separator. The most common technique is to use a multiple-effect evaporator. Essentially, these are vacuum evaporators operated in series, each with a different boiling point and each operated at a different vacuum pressure. The solution to be concentrated is fed into the boiling chamber of the first effect, and external heat is introduced to volatilize the water. The water vapor is then condensed at a different vacuum level, and the energy is used to heat the subsequent vacuum chamber. Therefore, the same energy is used several times in multiple stages.

The second technique is to use a mechanical (or vapor recompression) compressor. Water is evaporated in a boiler operating under a vacuum. The vapor enters the suction of a compressor, where its temperature and pressure are increased. The hot vapor is then passed through a heat exchanger where it condenses, transferring heat to the water on the vacuum side of the heat exchanger. Thus the latent heat of evaporation, normally lost to the condenser, is recycled by the compressor, providing a temperature difference across the heat exchanger. The energy required is only that used for increasing the pressure to provide the temperature difference.

There are a number of advantages of vacuum systems. One is that they are essentially independent of the requirement to heat and move large volumes of air, thus reducing the air pollution problem, at least as compared to atmospheric systems. Further, they are operated at relatively low temperatures, which could be important in systems that handle temperature-sensitive products. In addition, vacuum systems are advantageous with alkaline cyanide solutions that build up carbonates naturally; atmospheric evaporators aerate the solution and accelerate the buildup of carbonates.

Ion Exchange

Ion exchange uses charged sites on a solid matrix (resin) to selectively remove either positively charged ions (cations) or negatively charged ions (anions) from the solution. Ions removed from the solution are replaced by an equivalent charge of ions displaced from the resin, hence the name ion exchange. Exchanged rinse water is normally recycled.

Following saturation of the exchange sites, ion exchange resins are usually regenerated by passing an acid (cation exchange) or base (anion exchange) through them, producing a concentrated metal solution that can be recycled.

In metal plating operations, anionic exchange resins have been used to recover chromic acid from rinse waters, typically exchanging hydroxide ions for the negatively charged chromic acid anions (Figure 12.10). Anionic resins also have been used to recover cyanide and metal cyanide complexes. Cationic exchange resins have been used to recover metal cations. An IX system typically consists of a wastewater storage tank, prefilters, cation or anion exchanger vessels, and caustic or acid regeneration equipment.

In general, IX systems are suitable for chemical recovery applications where the rinse water is relatively dilute and where a relatively low degree of concentration is required for recycling the concentrate. Recovery of plating chemicals from acid-copper, acid-zinc, nickel, tin, cobalt, and chromium plating baths has been demonstrated commercially (Figures 12.11 and 12.12). The process has also been used to recover spent acid-cleaning solutions and to purify plating solutions for longer service life.

A U.S. EPA study estimated that an IX system being operated 5,000 hr/yr would pay for itself in 5.2 years.[7] IX recovery systems are not cost-effective, however, when drag-out rates are low. According to a U.S. EPA study, a favorable payback

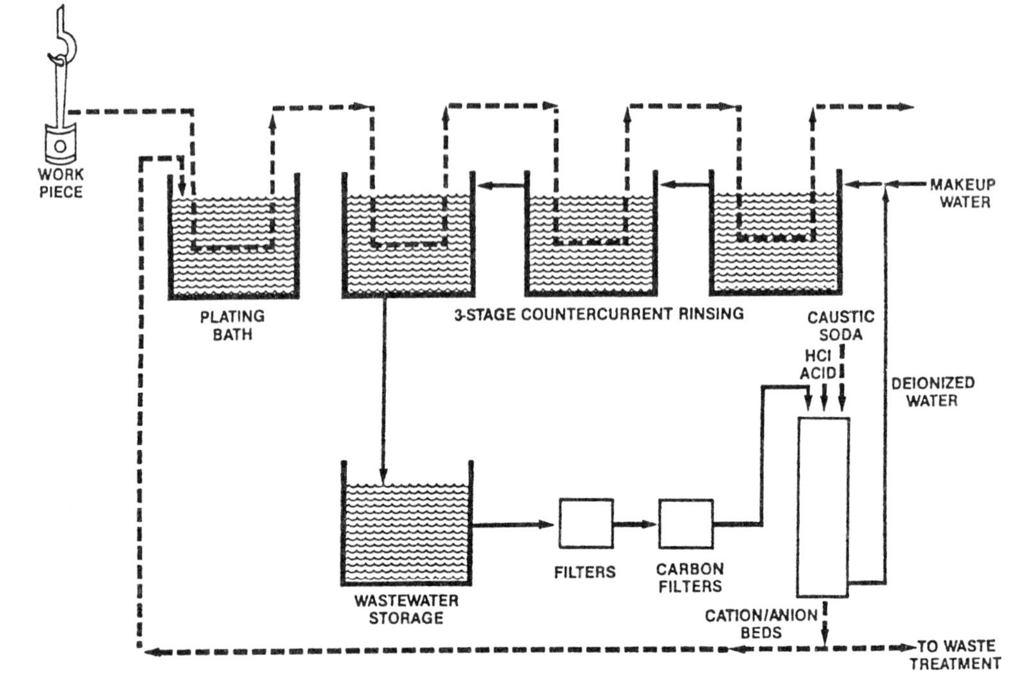

Figure 12.10. Ion exchange chromic acid recovery.

period of 2.8 years was estimated for chromic acid recovery from rinse water where the chromic acid drag-out rate is 3 lb/hr.[8] For drag-out rates significantly lower (e.g., those used in hard-chrome plating), an IX recovery system is not normally cost-effective. IX may also be uneconomical where wastewater treatment and sludge disposal costs are minimal.

IX has been most successful when recovering copper from rinse water, but problems have been encountered in concentrating mixed-metal solutions. By using the ion exchanged water for rinsing, fresh water consumption can be reduced by 90%. However, waste regenerant brine can be difficult and expensive to treat and dispose of. Often, the environmental and economic benefits of reduced water consumption can be offset by an increased use of treatment chemicals.[9]

Reverse Osmosis

Reverse osmosis is a demineralization process in which water is separated from dissolved metal salts by forcing the water through a semipermeable membrane at high pressures (400 to 800 pounds per square inch gauge [psig]). The basic components of an RO unit are a membrane, a membrane support structure, a containment vessel, and a high-pressure pump (Figure 12.13). A typical RO recovery process is shown in Figure 12.14. Rinse water must be filtered to prevent fouling the membranes by solid particles. RO units can concentrate most divalent metals (e.g., Ni, Cu, Cd, Zn) from rinse waters to a 10 to 20% solution. The concentrated solution is fed back to the plating bath to make up for plating and drag-out losses. Activated carbon adsorption is commonly used to remove organic contaminants. The cleaned rinse water can then be reused.

Figure 12.11. Ion exchange system for recovering chromic acid (Photo courtesy of Eco-Tec, Pickering, Ontario).

Figure 12.12. Ion exchange system used to recycle chromium-bearing wastewater (Photo courtesy of Napco, Inc., Terryville, CT).

Figure 12.13. Reverse osmosis equipment (Photo courtesy of Osmonics, Minnetonka, Minnesota).

For an example RO system installation, in terms of savings associated with plating chemicals, wastewater treatment, and sludge disposal, the payback period was reportedly 4.3 years.[7]

According to a U.S. EPA study, the main plating application of RO has been for concentrating rinse water from slightly acidic nickel-plating baths using cellulose acetate membranes.[10] Since 1970, more than 150 RO systems have been installed for

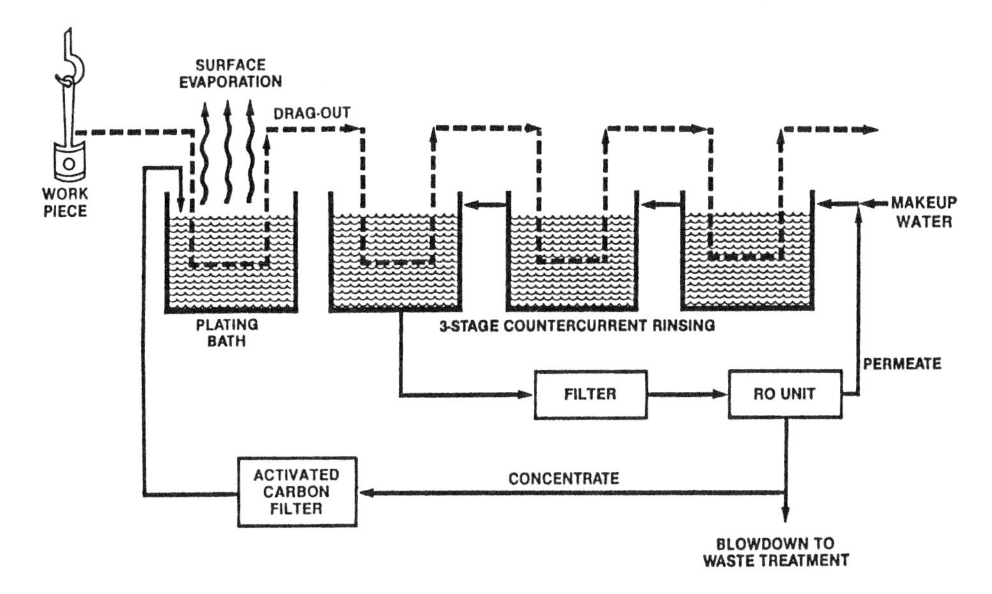

Figure 12.14. Reverse osmosis metal recovery system.

nickel-plating baths. Recovery efficiencies have been reported to be from 90% to 95%, with membranes lasting from 1 to 3 years.[11]

More than 100 RO systems have been installed for recovering copper sulfate, copper cyanide, zinc sulfate, brass cyanide, and hexavalent chromium. RO use for these solutions is limited, since RO membranes are attacked by solutions with a high-oxidation potential (e.g., chromic acid) or extremes of pH (less than 2.5 or greater than 11). The use of RO for nonnickel baths is expected to increase because of the anticipated development of membranes that can withstand corrosive and oxidizing environments. RO use is limited to moderately concentrated rinse waters. For this reason, it is often coupled with a small evaporator when used to concentrate rinse waters from ambient temperature baths, such as copper and zinc sulfate.

Example: A U.S. EPA study evaluated the use of RO and evaporation for recovering zinc cyanide from rinse water.[12] To reach an adequate concentration for reuse in the ambient temperature plating bath, an evaporator was required to supplement the RO system. In 1981, capital costs for the RO system and evaporator were $25,000 and $40,000, respectively, for a total cost of $65,000. The operating cost of the complete system was $12,000 per year. A savings of $10,000 per year in wastewater treatment, water, and makeup chemical costs were insufficient to offset operating and capital recovery costs. Another U.S. EPA study[13] demonstrated that RO could be effectively used to recover copper cyanide from rinse water for recycling in a plating bath. However, because of low rinse-water concentrations, short membrane lives, and low wastewater disposal costs, this process was found not to be cost-effective.

In summary, RO has been shown to be cost-effective in concentrating nickel in rinse water for reuse in nickel-plating baths. However, for ambient temperature plating baths, RO must be supplemented with evaporators to concentrate the metals in rinse water to plating bath strength. The cost-effectiveness of an RO metal-recovery system depends on production rate, type and concentration of constituents in the rinse water, costs of fresh water supply and wastewater disposal, and expected useful life of the RO membrane used. Process and operating uncertainties associated with membrane processes that can significantly affect their cost-effectiveness include membrane fouling, bath chemical balance, wastewater generation, and operation and maintenance requirements.

Electrodialysis

Electrodialysis concentrates or separates ionic species in a water solution through use of an electric field and semipermeable ion-selective membranes. Applying an electrical potential across a solution causes migration of cations toward the negative electrode and migration of anions toward the positive electrode. ED units are packed with alternating cation and anion membranes. Cation membranes pass only cations, such as copper, nickel, and zinc, whereas anion membranes pass only anions, such as sulfates, chlorides, or cyanides. Alternating cells of concentrated and dilute solutions are formed between the cation and anion membranes. Packaged ED units contain from 10 to 100 cells.

ED has been used to recover cationic metals from plating rinse water. In a typical application, as depicted in Figure 12.15, rinse water from a stagnant or dead rinse (i.e., no inflow or outflow) tank is continuously fed to an ED unit and concentrated

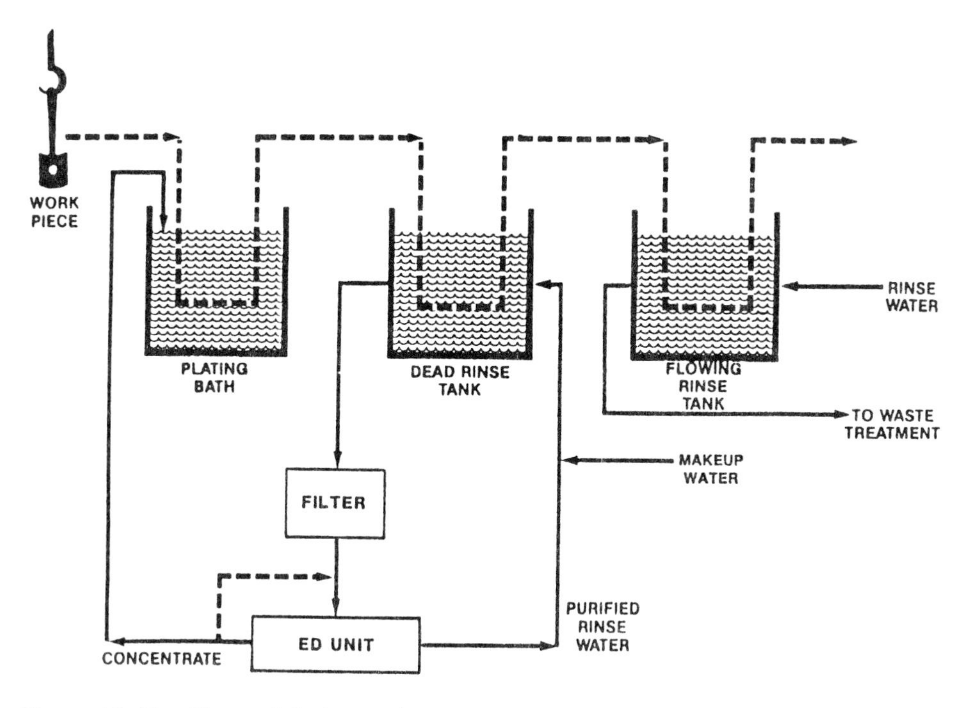

Figure 12.15. Electrodialysis metal recovery system.

by a factor of 10:1. The concentrate is then returned to the plating bath. The water in the dilute cells is combined with makeup water and returned to the dead rinse tank.

Unlike IX and RO, the maximum concentration of an ED unit is limited only by the solubility of the ions in solution. Therefore, ED generally can produce a more concentrated solution than IX or RO, eliminating the need for an evaporative concentrator when used with ambient-temperature plating baths. ED units are reportedly easy and economical to operate, require little space, and operate continuously without requiring regeneration.[4]

One disadvantage of ED and RO is that all ionic species are removed. Therefore, ionic impurities—organic brighteners, wetting agents, and other compounds that accumulate in the dead rinse tank—are returned to the plating bath along with the recovered metal. As a result, plating baths must be periodically treated to remove impurities, and the dead rinse tanks must occasionally be disposed of. In addition, if the applied voltage exceeds the hydrogen electrode potential, water will be converted to gaseous hydrogen and hydroxide ions. The subsequent increase in pH can cause precipitation of metal hydroxides that can foul the membranes.[14]

Example: A Navy study estimated a payback period of less than 1 year for an ED recovery system for a cadmium cyanide plating bath operating 4,000 hours per year at drag-out rates of 1.3 lb/day of Cd and 5.1 lb/day of Cn. This evaluation did not include the costs of removing impurities from the baths or of maintaining the ED units and replacing the membrane modules.[4]

Example: A U.S. EPA study evaluated the use of ED for recovering nickel from rinse water.[15] The ED unit was able to recover 95% of the nickel salts from the rinse water and return the concentrated solution to a Watts-type nickel-plating tank. The study estimated that $16,000 per year could be saved by employing ED in a nickel-plating line that operated 4,000 hours per year. The cost estimate considered only savings in chemical usage, wastewater treatment, and sludge disposal and did not consider the cost of operating and maintaining the ED system.

The following case study illustrates the use of process modifications and innovative rinsing to reduce and recover plating solution from rinse water and the use of ED to remove contaminants from plating solution and recycle the solution.

Case Study 12.1: Recovery of Chromium by Innovative Rinsing Techniques

The most common electroplating process found at Naval Aviation Depots (NADEPs), Naval Shipyards (NSYs), and Naval Air Stations (NASs) is hard-chrome plating. Typical hard-chrome plating methods used at Naval facilities for more than 20 years include:

- Areas on worn parts that do not require a chrome buildup are masked with wax, aluminum foil, lacquer, or tape.
- After masking, the parts are fastened to racks and suspended in the plating bath. These racks are then secured by C-clamps to the cathode bus bar, which provides physical support for the part and completes the electrical circuit.
- Heavy lead anode bars are then hung from the anode bus bar and positioned around the racked part. Since the lead anodes are 8 feet long and weigh more than 50 lb each, they cannot be easily removed by one person, so they are often left in the plating solution when not in use. As a result, the anodes slowly become passive and ineffective.
- After plating, parts must be rinsed to remove plating solution that is dragged out of the bath. Continuous-flow rinse tanks are generally used to clean plated parts.

The NADEP at Pensacola uses hard-chrome plating for remanufacturing parts from H-3 and H-53 helicopters and A-4 jet aircraft.

At the Pensacola NADEP, these rinse flows ranged from 3 to 12 gal/min per rinse tank for approximately 24 hours per day, 260 days per year of operation. This resulted in substantial costs for supply of fresh water and for treatment of wastewater.

Plating baths become contaminated with metal ions leached from parts, plating tanks, racks, and anodes, and from conversion of hexavalent to trivalent chromium. These contaminants can blemish a plated surface, reducing plating efficiency and quality. Because of a buildup of impurities, plating baths at Pensacola had been dumped about every 2 years. Treating plating wastewater and replacing the plating solution with new material was expensive.

To mitigate these problems, the Naval Civil Engineering Laboratory (NCEL) at Port Hueneme, California, adapted an innovative chrome-plating system for use at

Navy plating shops.[4] The "new" plating process uses a technology that was developed more than 50 years ago in the Cleveland area, hence it is called the "Cleveland process" or the "Reversible Rack 2 Bus Bar System." The Navy laboratory converted three of the seven plating baths at Pensacola to the Cleveland process to demonstrate this technology. Although the plating method varies considerably from conventional procedures, the system greatly improved plating efficiencies, and the end product meets all military specifications.

Modifications from the standard Navy practice of hard-chromium plating were as follows:

1. Changes in the process that improve plating quality and speed and reduce rejects and rework
2. Use of a recirculating spray rinse system (Figure 12.16)
3. Operation at higher temperatures (140°F vs 130°F) that permits better drag-out recovery
4. Use of a continuous system of bath purification to remove contaminating cations from the plating solution (Figure 12.17)

Photographs of the components of this innovative chrome-plating system are shown as Figures 12.18 through 12.22.

To reduce the amount of rinse water used, a prototype spray rinse system was installed in an existing rinse tank (Figure 12.16). A foot-activated pump recirculates rinse water through eight high-velocity spray nozzles located around the perimeter of the rinse tank. Clean rinse water is also available from a handheld sprayer. After repeated use, some of the rinse water is pumped through a cloth filter into the plating tank to replace water lost to evaporation (Figure 12.17). When compared to

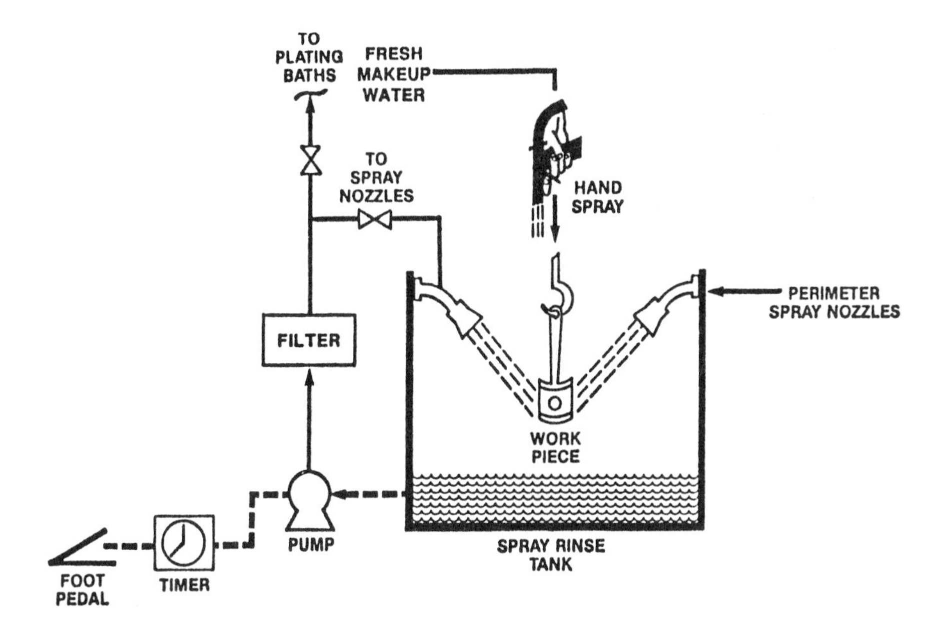

Figure 12.16. Recirculating spray rinse system.

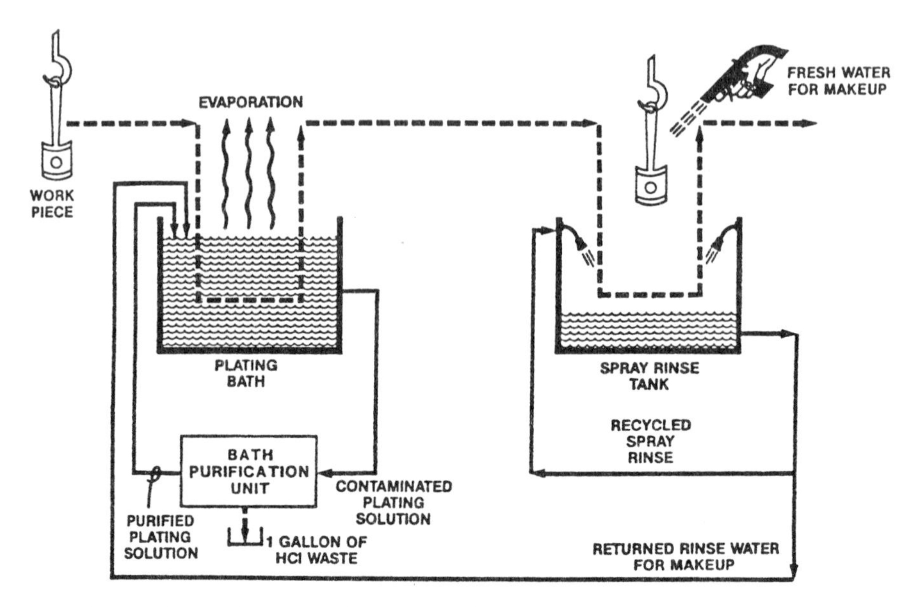

Figure 12.17. "Zero-discharge" chrome-plating system.

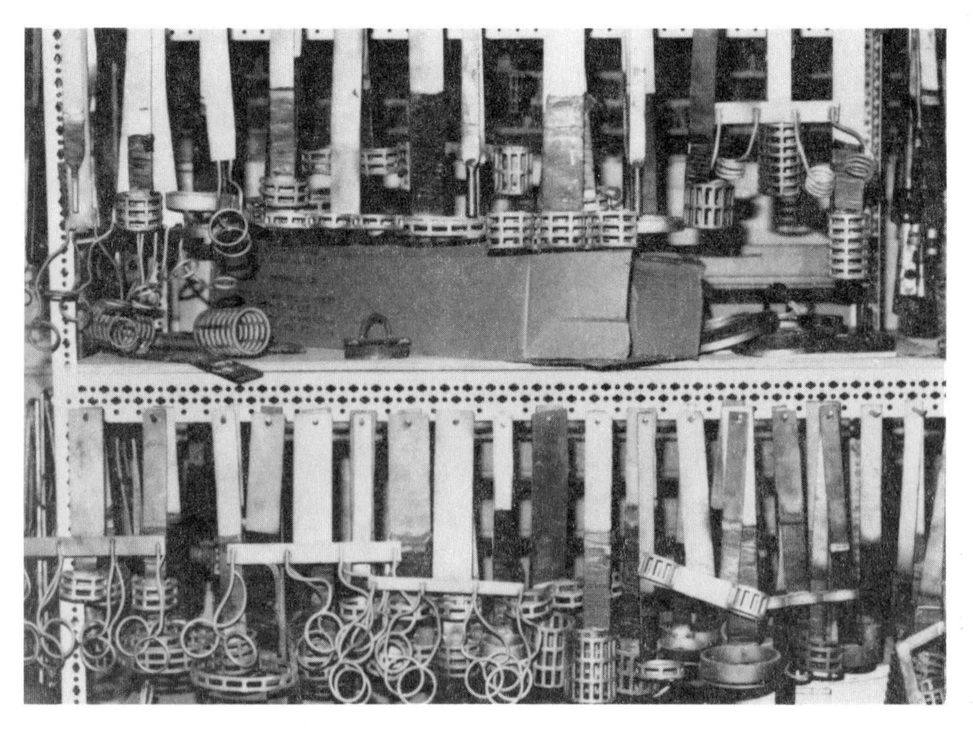

Figure 12.18. Assortment of conforming anodes.

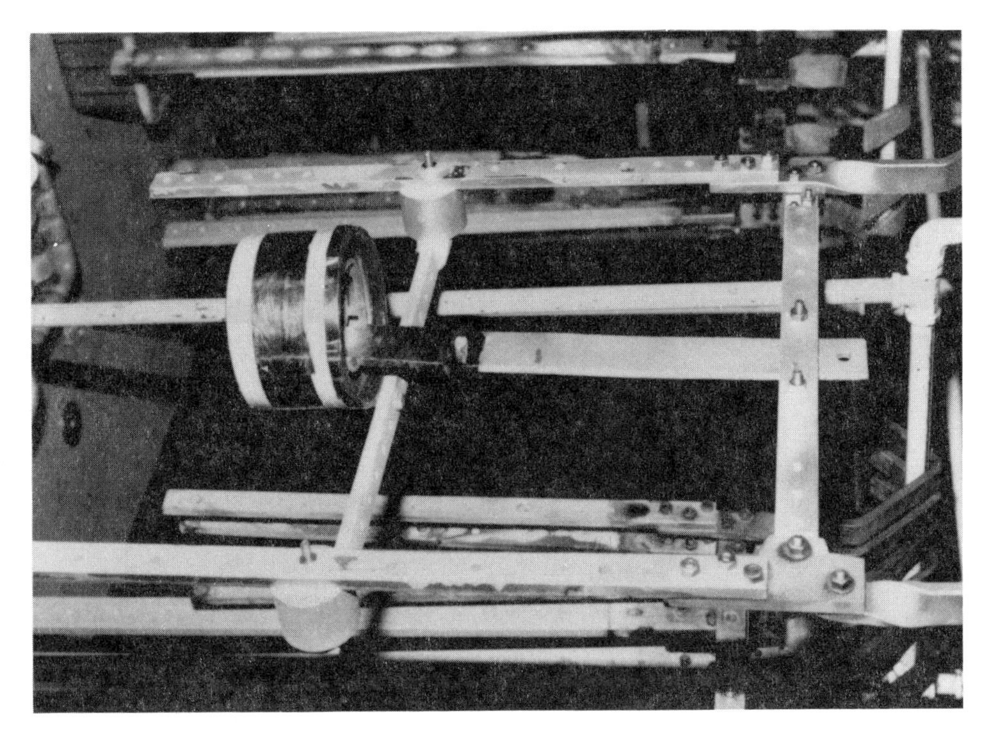

Figure 12.19. Reversible rack with conforming anode.

Figure 12.20. Chrome-plating bath with two bus bar reversible rack.

Figure 12.21. Spray rinse of plated part and reversible rack.

Figure 12.22. Hand spray rinse.

conventional chrome plating, operation of the plating bath at a higher temperature results in higher evaporation and increased plating rates. Because of these changes, less rinse water is produced than is needed to make up for evaporation losses in the plating bath. The result is a closed-loop plating system.

Without drag-out to help remove contaminants from the plating bath, a cleanup process was desirable to reduce the need for bath dumps. Testing at the facility demonstrated that ED technology is cost-effective for maintaining hard-chrome plating bath purity for the closed-loop system.

PLATING BATH PURIFICATION

Maintaining the purity of the plating solution can significantly reduce bath dumps. Periodic removal of impurities from plating baths is necessary to maintain closed-loop systems. For example:

- Metal impurities such as iron, copper, aluminum, nickel, zinc, and trivalent chromium need to be removed from chromium-plating baths.
- Organics and metal impurities such as copper, zinc, lead, and iron need to be removed from nickel baths.
- Carbonates and metal impurities need to be removed from cyanide baths.

Table 12.4 provides an overview of purification technologies for plating bath and some example applications. Table 12.5 provides an example of plating solutions to be maintained by particulate filtration and carbon purification.

Table 12.6 provides an example of solutions potentially amenable to purification by membrane electrodialysis. Present-worth cost evaluations for various ED systems for hard-chrome plating and chromic acid anodizing showed payback times of 3 to 6 years. These evaluations estimated costs for system purchase and installation, catholyte replacement, membrane replacement, and anode replacement. Savings are the result of lower costs for chemicals and reduced waste disposal.

On the basis of technology and cost evaluations, a membrane electrodialysis system (a simple ED configuration) is recommended for solution purification for hard-chrome plating. The process comprises a membrane electrochemical cell designed for immersion in a process liquor; a rectifier; a process liquor containing metal salts; a catholyte solution; and a pump to flow the catholyte solution through the cell. The membrane separates the process liquor from the catholyte solution and lets metal cations go from the process liquor through the membrane (which filters out the metals) into the catholyte solution and keeps anions in the process liquor. The metal cations are continuously converted to hydroxides in the catholyte solution, and the anions are continuously converted to acids in the process liquor. The hydroxides of multivalent metals (cadmium, zinc, iron, copper, aluminum, calcium, etc.) are substantially insoluble in the catholyte and can be removed for use. The process is unique in that salts of multivalent metal cations can be converted. There is essentially no electrodeposition of metals. The process can be operated at reproducible capacities for months. The capacity of the process is varied by current (Figure 12.23). Another ED system is shown in Figure 12.24.

Table 12.4. Overview of Plating Solution Purification and Recovery Technologies

Potentially Applicable Technology	Solution Purification	Example Application
Filtration	Remove suspended solids.	See Table 12.5.
Carbon adsorption	Remove organic contaminant.	See Table 12.5.
Membrane electrodialysis	Remove metal contaminants; return solution concentrate.	Removes iron, aluminum, copper, other tramp metals from hard-chrome plating baths. Caustic etch regeneration. Acid regeneration.
Diffusion dialysis	Metal contaminant removal.	Purification recovery of acid solutions.
Electrowinning	Remove metals that build up to excessive concentration levels.	Cadmium removal/ recovery from cadmium strip solutions.
	Recover metal from spent process solutions.	Copper recovery from copper pickling and milling solutions, including sulfuric acid, cupric chloride, and ammonium chloride solutions.
Dummy plating	Remove metal impurities.	Nickel- and chrome-plating solutions. Reoxidize trivalent chromium to hexavalent chromium.
Ultrafiltration and microfiltration	Remove suspended dirt and oil and high molecular weight molecules.	Purification of alkaline cleaners.
Ion exchange	Remove tramp metals.	Regeneration of spent chromic acid anodize or dilute hard-chrome plating solutions.
Clarification of recirculation streams	Remove precipitates.	Phosphating baths.
Crystallization	Remove metal salts.	Remove ferrous sulfate from sulfuric acid steel-etch solutions.
Hydrolysis and crystallization	Remove excess carbonates.	Remove ferrous sulfate from sulfuric acid steel etch solutions.

Table 12.4. Overview of Plating Solution Purification and Recovery Technologies (Continued)

Potentially Applicable Technology	Solution Purification	Example Application
	Remove dissolved aluminum from etching solutions.	Solutions formulated with sodium cyanide, including some copper plating, copper strike, cadmium plating, and silver rhenium plating solutions.
		Caustic etch regeneration.
Acid sorption and desorption	Sorb acid on resin; desorb purified acid with water.	Acid pickling baths.
Decant and remove sludge	Separate plating solution from sludge and allow for return/reuse.	Tin plating.

Table 12.5. Applications of Filtration and Carbon Purification to Metal-Finishing Solutions

Solution Description	Particulate Filtration	Carbon Purification
Chromium strip	✓	
Cadmium strip	✓	
Chrome plate	✓	
Electroless nickel	✓	
Anodize seal	✓	
Red dye	✓	
Green dye	✓	
Blue dye	✓	
Black dye	✓	
Sulfuric acid anodize	✓	
Hard anodize	✓	
Phosphoric acid anodize	✓	
Chromic acid anodize	✓	
Zincate	✓	
Copper strike	✓	✓
Copper plate	✓	✓
Nickel plate	✓	✓
Nickel strike	✓	✓
Cadmium plate	✓	✓
Magnesium phosphate	✓	
Silver plate	✓	✓
Gold plate	✓	✓
Ag/Rh plate	✓	✓
Silver strip	✓	✓
Rhodium plate	✓	✓
Tin plate	✓	✓

Table 12.6. Potential ED Systems for Plating-Solution Baths

Solution Name	Bath Constituents	ED System[a]
Sulfuric acid etch	Sulfuric acid Glycerine	2-C or 3-C
Phosphoric acid etch	Phosphoric acid	3-C
Nitric acid etch	Nitric acid	2-C or 3-C
Rust remover	Sodium hydroxide	Special
Chromic acid bright dip	Chromic acid	2-C
Conversion coat	Iridite	3-C
Chromic acid dip	Chromic acid	2-C
Nitric acid strip	Nitric acid	2-C or 3-C
Deoxidizer	Sodium dichromate Sulfuric acid	2-C
Alodine 1200	Chromic acid Potassium ferrocyanide Barium silicofluoride	3-C
Chromic acid anodize	Chromic acid	2-C
Sulfuric acid anodize	Sulfuric acid	2-C or 3-C
Nitric acid etch	Nitric acid	2-C or 3-C
Aluminum alloy etch	Turco 9H (mostly NaOH)	Special
Chrome pickle treatment	Sodium dichromate Nitric acid	2-C
Dichromate	Sodium dichromate Calcium fluoride	2-C
Phosphoric chrome strip	Phosphoric acid	2-C or 3-C
Alkaline chrome strip	Sodium hydroxide	3-C
Nital etch 1	Nitric acid	3-C
Nital etch 2	Sodium hydroxide	Special
Chromium plating	Chromic acid Sulfate	2-C
E-Ni strip	Sodium hydroxide Sodium cyanide Meta-nitro benzene sulfanate	3-C[b]
E-Ni tank strip	Nitric acid	2-C or 3-C

[a]Data developed in cooperation with IONSEP Corporation. The designations "2-C" and "3-C" refer to 2-compartment and 3-compartment cells as explained in the description of this technology. Where both 2-C and 3-C are listed, additional data are needed to select the proper system. The "special" designation indicates that the application requires a more complex arrangement, including ED and non-ED components.
[b]Requires nickel electrowinning followed by ED reformulation.

SUBSTITUTING LESS HAZARDOUS MATERIALS

In recent years, efforts to find substitutions for plating solution have focused primarily on reducing or eliminating the use of cyanides, hexavalent chromium, cadmium, and chlorinated and ozone-depleting solvents. Waste reduction for solvent cleaning and degreasing is discussed in a separate chapter in this book.

Past efforts have focused on treatability issues; the new paradigm requires a focus on recycling. Substitutes are often more difficult to recycle.

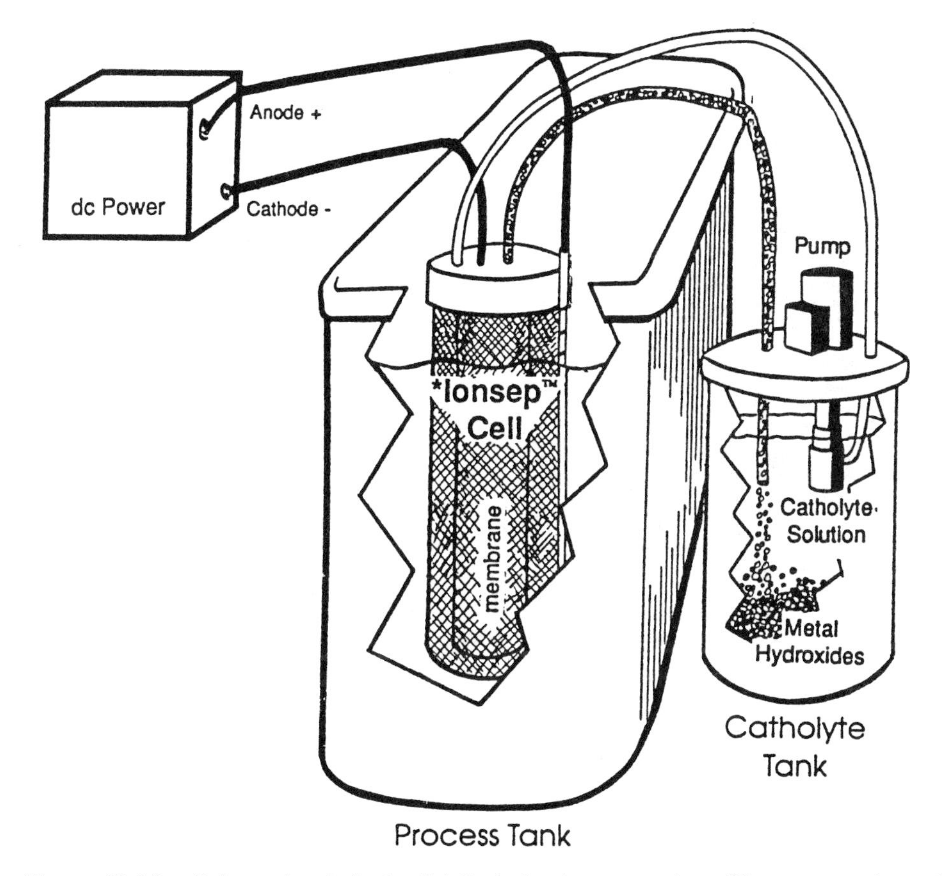

Figure 12.23. Schematic of electrodialytic bath cleanup system (Diagram courtesy of Ionsep Corporation, Inc., Rockland, Delaware).

Noncyanide Baths

Noncyanide baths have the advantages of improving workplace safety, as well as reducing the costs of disposing of rinse water. Traditionally, cadmium, zinc, brass, and precious metals have almost universally been plated from alkaline cyanide baths because of the superior plate produced from stable metal cyanide complexes. Unfortunately, cyanide baths are costly and dangerous to operate, and the wastes generated are difficult and expensive to treat.

In the late 1960s and early 1970s, extensive research was conducted to develop noncyanide zinc electroplating baths. As a result, several alternative zinc baths were developed. Alternatives include low-cyanide baths; noncyanide alkaline baths; neutral ammonium chloride and potassium baths; and a number of acidic baths containing sulfate, chloride, and fluoborate ions.[4]

Low-cyanide baths contain approximately 20% of the cyanide found in conventional cyanide baths and have similar operating characteristics. However, process control is more difficult, and cyanide treatment is still required.

Neutral chloride baths use ammonium or potassium salts to form a zinc complex. These baths usually require the addition of proprietary brighteners and chelating

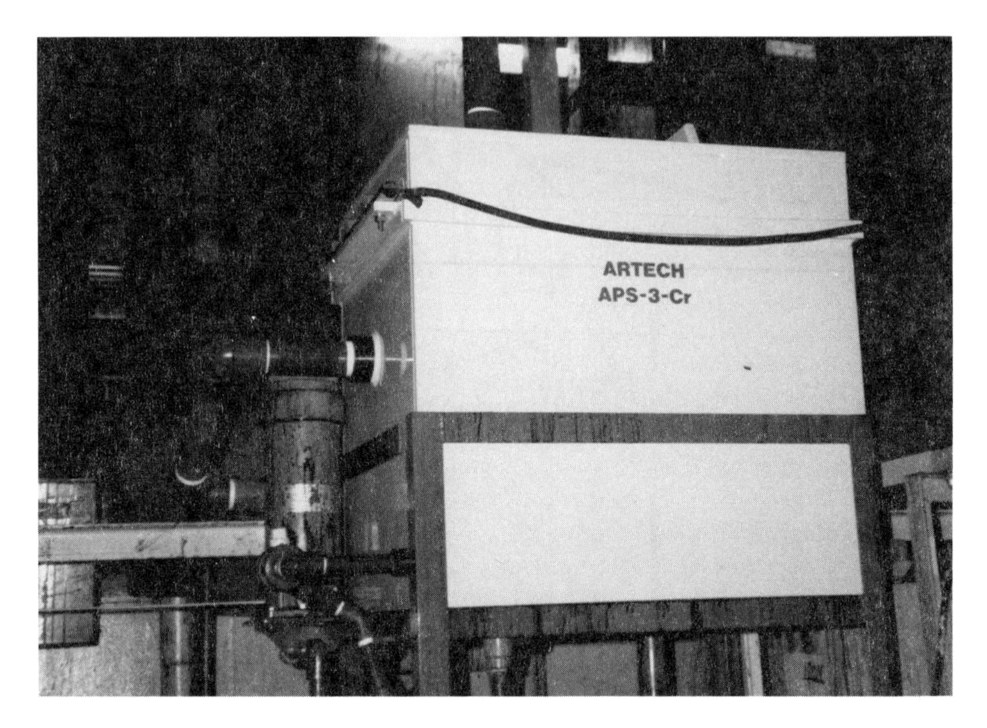

Figure 12.24. Electrodialysis system for hard-chrome plating bath purification (Photo courtesy of I^3, Inc., Woodbridge, Virginia).

agents that form zinc complexes. Unfortunately, zinc complexes can be difficult to remove in subsequent waste treatment.

Acid sulfate, chloride, and fluoborate baths have become the most popular noncyanide zinc baths. With the recent development of new additives, acid zinc baths are capable of producing bright deposits that are competitive with alkaline cyanide baths for general plating applications.[4]

Less effort has been expended in developing noncyanide cadmium baths because the volume of cadmium plating is only 5% to 10% that of zinc plating. However, because of increased environmental and safety concerns with operating and disposing of cadmium cyanide baths, alternative proprietary acidic cadmium baths similar to zinc baths have recently been developed to replace cyanide baths.

Most of these acidic baths consist of cadmium oxide, sulfuric acid, distilled water, and anionic compounds. Because many old tanks for alkaline cadmium cyanide plating are made of bare steel, conversion to these acidic baths may require that the existing tanks be refurbished or replaced. Thus, material substitution may require a considerable capital expenditure. However, the savings in eliminating cyanide treatment can make the modification economically attractive.[16]

Parts being plated in noncyanide cadmium baths may require more thorough cleaning before plating than parts being plated in cyanide baths. The noncyanide cadmium baths reportedly have less throwing power and lower cathode efficiency than cyanide baths. Despite the disadvantages, however, some platers prefer the new noncyanide plating baths because of the reduced complexity of waste treatment. Some platers have reported that drag-out from noncyanide cadmium baths is less than from cyanide baths.

Noncyanide zinc and cadmium baths usually cost more than cyanide baths. However, to properly evaluate the cost-effectiveness of the material substitution, facilities must also consider the following factors: cost of new corrosion-resistant equipment, difference in labor and chemical costs, change in production rate, and savings realized by eliminating cyanide treatment and costs of new chelate-breaking waste treatment.

Example: In 1983, the Charleston NSY successfully switched from alkaline cyanide baths to an acidic noncyanide solution and eliminated the cyanide oxidation process from the waste treatment plant.[17]

Example: The plating shop of Air Force Plant 6 in Marietta, Georgia, operated by Lockheed-Georgia Corporation, switched from an alkaline cyanide cadmium bath to a proprietary acidic noncyanide cadmium bath. Lockheed found that product quality was improved by switching from the alkaline cyanide baths to the acidic noncyanide cadmium baths; however, more careful process control was required. The new plating solution, costing approximately $3 per gallon, is more expensive than the old formulation; however, reduced costs for waste treatment resulted in a net savings, because the wastewater treatment plant no longer had to operate the alkaline chlorination treatment process for cyanide. The material substitution was implemented primarily to reduce the safety hazards associated with operation and disposal of cyanide baths. Improved quality and decreased costs have ensured the permanent adoption of the process modification.

Trivalent Chromium Plating

Trivalent chromium baths have been used in place of conventional solutions of hexavalent chromium in decorative applications. It is unnecessary to add sodium bisulfite or other reducing agents in waste treatment of trivalent chromium rinse water for conversion of hexavalent to trivalent chromium before precipitation. Trivalent solutions are also less concentrated (22 g/L versus 150 g/L for hexavalent solutions), thus reducing the amount of chromium drag-out. Consequently, sludge produced from trivalent baths is about one-seventh of the volume from hexavalent baths and is much less toxic.[18]

The main disadvantage of trivalent solutions is that they cost two to three times more than hexavalent solutions. Some researchers have reported that higher production rates and lower rejection rates can be realized with trivalent chromium plating solutions; however, the main advantage is the lower cost of wastewater treatment and sludge disposal. Before a plating shop converts to trivalent chrome solutions, a detailed study should be performed to determine if the projected savings in waste treatment exceed the increased operating costs. Closed-loop operation of hexavalent chromium baths is proven and is less difficult than converting to trivalent chromium baths.

Vacuum Deposition of Cadmium

Vacuum deposition of cadmium was developed as an alternative to electroplating. Problems with electroplating arise from cadmium cyanide baths because of the toxicities of cadmium and cyanide. Switching to noncyanide plating baths (discussed previously) removes one of these problems. Use of vacuum deposition of cadmium also eliminates the need for cyanide.

Vacuum deposition of cadmium is a line-of-sight process, making it difficult to provide a uniform deposit on an irregularly shaped part. Parts need to be rotated at intervals during processing for more uniform coverage. The deposit does not adhere to the base metal as well as it adheres using conventional cadmium plating. In addition, occupational and environmental hazards can result from the evacuation of cadmium vapors and condensed aerosols. The vacuum exhaust must be carefully filtered to prevent cadmium vapors and condensed aerosols from escaping to the work environment.

Ion Vapor Deposition of Aluminum

Ion vapor deposition (IVD) of aluminum can substitute for cadmium plating for corrosion protection. It was developed because of the many hazards inherent in working with cadmium and the increasingly stringent requirements being placed on disposal of wastes that contain even traces of cadmium. Aluminum coating is a logical replacement for cadmium in providing corrosion protection because aluminum is anodic to steel and provides galvanic protection similar to that afforded by cadmium. In addition, aluminum's corrosion products are not bulky or unsightly, and aluminum is less expensive than cadmium and zinc on a volume basis. Moreover, aluminum-coated parts can be used at temperatures up to 925°F, compared with a maximum of 450°F for cadmium.

These features have resulted in considerable interest in the possibility of aluminum plating, with many attempts to develop a successful method. However, the electrode potential of aluminum is too negative for it to be successfully plated from a water bath.[19] Aluminum has been deposited on steel by hot dipping or by using a metal spray system. These methods do not provide the thin, uniform coating required on aircraft parts, nor do these coatings adhere to substrates as strongly as plated cadmium.

As a logical extension of vacuum deposition of cadmium, IVD of aluminum was developed by McDonnell Douglas Corporation as a substitute for cadmium plating on steel aircraft parts.[20-22] The IVD (Ivadizer) system (Figure 12.25) consists of a vacuum chamber, a resistance-heating aluminum-vaporization system, and a high-voltage system to ionize the aluminum and to impart a negative charge to the parts. The charge causes aluminum ions to electrodeposit on the parts. Air in the vacuum chamber is replaced by a low-pressure inert gas. Aluminum vapor ions interact with the inert gas to coat the parts uniformly. Without the ionization and interaction with the inert gas, IVD would be restricted to line-of-sight coating as in the vacuum deposition of cadmium.

Compared to cadmium, advantages of IVD of aluminum include a higher useful temperature, improved throwing power, and better adhesion of the aluminum coating. Safer working conditions are another advantage of IVD of aluminum. Furthermore, parts plated with cadmium require baking to prevent hydrogen embrittlement; if oven temperatures are not carefully controlled, parts must be scrapped.

Example: Personnel at the North Island NADEP have been using IVD of aluminum for about 11 years, having procured one of the first commercially available systems. After many problems with this developmental model, NADEP procured a

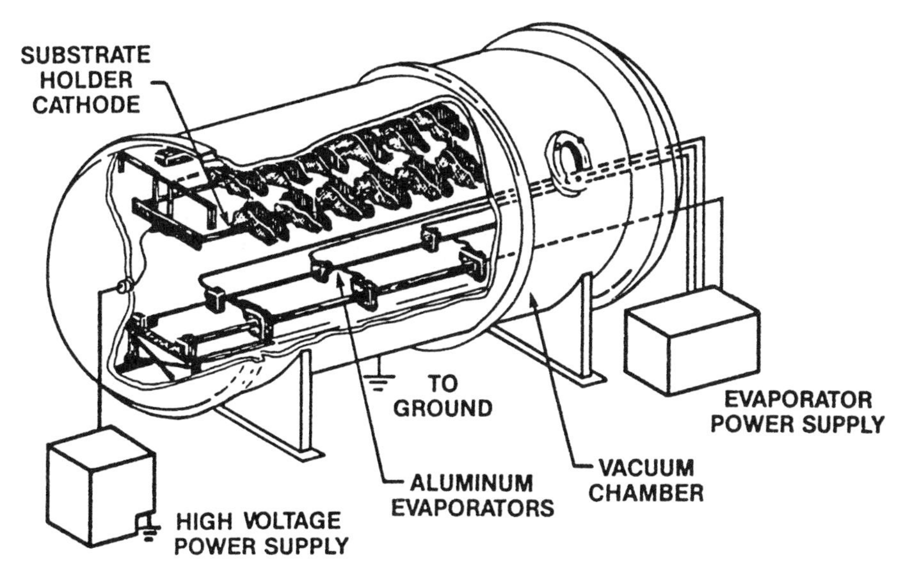

Figure 12.25. Equipment for ion vapor deposition of aluminum.

more recent, improved model. However, numerous problems persist in adapting the process to existing plating operations.

Problems developed with the newer IVD system for two principal reasons. First, the equipment was installed in the open plating shop, where ambient gases contaminated the vacuum chamber. Second, the IVD system is complicated to operate because many operating variables need to be adjusted to produce a good coating. Personnel from the plating facility were assigned to operate this complicated system without adequate skills, training, or incentives.

Production personnel oppose complete conversion to IVD because the process is more complex and requires more labor and skill than cadmium plating. For this reason, parts are evaluated individually for conversion to IVD aluminum coating. Most parts are still either electroplated from cadmium cyanide baths or vacuum-deposited with cadmium. The limited use of IVD has aggravated problems with the system at the depot, because extensive use is needed to, as one facilities engineer put it, "work the bugs out of the system." This limited use of the system and a general lack of cost information make it impossible to evaluate the economics of the process.

In summary, the technology appears to have considerable potential. When the process is performed correctly, the coating is as protective as cadmium coating. From an environmental standpoint, widespread adoption to replace cadmium plating would eliminate a significant source of hazardous waste. However, unless these systems are made easier to operate and maintain, unless they are located in cleaner facilities than plating shops, and unless they are supported by skilled and well-trained operators, it is unlikely that IVD of aluminum will displace cadmium plating at NADEPs.

Miscellaneous Substitutions

Electroless Nickel Plating

Electroless nickel plating was developed in 1946 to coat a substrate without using an outside source of electrical current. Electroless nickel plating uses the substrate to catalyze a chemical reduction reaction. However, because of the expense of the chemical reducing agents, electroless plating is not cost-effective in applications where conventional electroplating can be used.

Most nickel plating is done in an acidic (pH between 1.5 and 4.5), elevated-temperature (between 110°F and 150°F) Watts bath that contains nickel sulfate, nickel chloride, and boric acid. An electrical current causes the nickel to be plated on the substrate.

The two main advantages of electroless nickel plating are that (1) throwing and covering power is essentially perfect and (2) deposits provide greater protection of the substrate because they are less porous.[5] In addition, the nickel concentrations of electroless baths are approximately one-eleventh those of conventional Watts nickel baths. Therefore, drag-out quantities and sludge production from an electroless bath are much less than from a conventional bath.

Alternative Chemistry for Nickel Strike Solution

Switching the nickel strike solution to an alternative chemistry, nickel sulfamate strike, may have some operational advantages over the Wood's nickel strike chemistry currently used. Nickel sulfamate strike (NSS) has been used by other plating shops in the U.S. as an alternative to Wood's nickel strike for the following reasons:

- NSS is operated at temperatures between 100 and 130°F, which results in significant evaporation. This allows for evaporative recovery of countercurrent rinse water.
- NSS has substantially better "throwing power" than Wood's nickel strike.
- NSS chemistry is compatible with sulfamate nickel plating and can be used without rinsing prior to sulfamate nickel plate.

PLANNING AND DESIGNING METAL-FINISHING FACILITY RENOVATIONS

The hazardous waste reduction measures discussed in the previous section can be implemented for specific facility operations, or for more comprehensive renovations of metal-finishing facilities. Planning for comprehensive facility renovation involves a range of interrelated technical issues and provides opportunities for achieving overall operational improvements, increasing cost-efficiency, and reducing environmental liabilities. These issues and benefits are discussed below.

In addition to waste reduction goals, planning for comprehensive facility renovation and design includes consideration of:

- space and layout constraints and optimization
- construction phasing (funding or operational constraints)

- operations efficiency and reliability
- maintenance or improvement of product quality
- environmental permit requirements
- reuse of existing equipment
- capital and O&M costs

Planning and designing a comprehensive facility renovation, including waste reduction, involves participation and coordination of multidisciplinary process engineering, facilities engineering, regulatory, and industrial health and safety team members. The planning and design team should be linked through a project organization that allows for effective project management and communications and exchange of interrelated information between team members. A general description of primary responsibilities for each of these team members is summarized below:

- **Process engineer:** Define production process requirements, size and configure production processes to efficiently meet production requirements with minimal waste generation. Size and configure ancillary waste management systems. Perform dynamic process simulations to confirm that the design can meet the production and waste reduction requirements. Estimate multimedia pollution source emissions. Coordinate with other planning and design team members. Perform cost-effective system optimization. Provide instrumentation and control expertise. Provide corrosion engineering expertise. Prepare design plans and specifications, and estimate costs.

- **Facilities engineer:** Coordinate delineation of facilities support requirements for production processes and hazardous waste reduction systems with the process engineering team members. Plan systems. Prepare design plans and specifications for facility civil, architectural, structural, mechanical, heating, ventilation, air conditioning, and electrical systems.

- **Regulatory expert:** Coordinate with other team members and regulatory agencies to determine general facility and environmental regulatory and permit requirements. Coordinate preparation and submittal of permits and other required submittals to regulatory agencies. Environmental regulatory requirements include wastewater, solid and hazardous waste, and air permits and requirements; hazardous materials inventory and use requirements; pollution prevention and waste minimization requirements; and environmental impact analyses and risk assessments.

- **Industrial health and safety expert:** Review design plans and specifications for health and safety systems, health and safety requirements during construction and systems operation, and emergency preparedness and response measures.

The process engineering planning and design work defines the basic system process requirements that form the basis for work by the other planning and design team members. A general approach for the process engineering work to develop production process operations and hazardous waste reduction and management systems for renovating metal-finishing operations is outlined below.

1. Define individual process requirements: Includes identifying all processes and associated pre-processing and post-processing requirements. Block diagrams are created to define the sequence of specific requirements for each unique production process. Figure 12.26 shows an example process block diagram for chromium electroplating on aluminum alloys. Table 12.7 shows a sample data collection form to help define electroplating production process requirements and sizing.

2. Create draft process layout: Includes grouping processes (e.g., specific electroplating lines) and allowing for work-flow clearances, maintenance and operations access; placement of waste reduction and management systems (e.g., ED systems, filters, IX systems); and utilities and ancillary process support equipment. Figure 12.27 shows an example draft layout of the plating process. Figure 12.28 shows a cross-sectional diagram of this layout.

3. Perform dynamic production process simulation: Determine throughput capabilities and define tank and hoist requirements. Figure 12.29 shows an example screen from a dynamic process simulation for a plating line.

4. Refine process layout, update process simulation runs: These steps should be repeated until a sufficiently optimized system design is defined to meet the desired production processing needs.

After determining production process sizing and layout requirements, process facility support systems and waste reduction and waste management requirements can be developed.

The following case study summaries present experience from various phases of planning and implementing facility renovations for metal-finishing shops.

Case Study 12.2: Department of Defense Plating Shop Renovation Design

The existing DOD plating shop uses 20-year-old technology, has a significant portion of out-of-service tanks and equipment, has corrosion problems from spills of plating chemicals onto rectifiers and other equipment located in the basement, and generates substantial quantities of hazardous waste and contaminated wastewater.

Bath contaminants are contributed to the metal-finishing solutions by airborne particulates falling into the tanks; contaminants in makeup water and chemicals; drag-in of foreign chemicals and solids; decomposition of bath constituents and formation of unwanted compounds; corrosion of base metals and deposits; and corrosion of racks, bus bars, and electrodes.

CH2M HILL, with subcontractors Integrated Technologies, Inc., and CAI Engineering, prepared a renovation predesign for the plating shop in 1991. This plan included comprehensive metal-finishing process and building service modifications. Table 12.8 presents key features of the shop renovation and their expected benefits to shop operations.

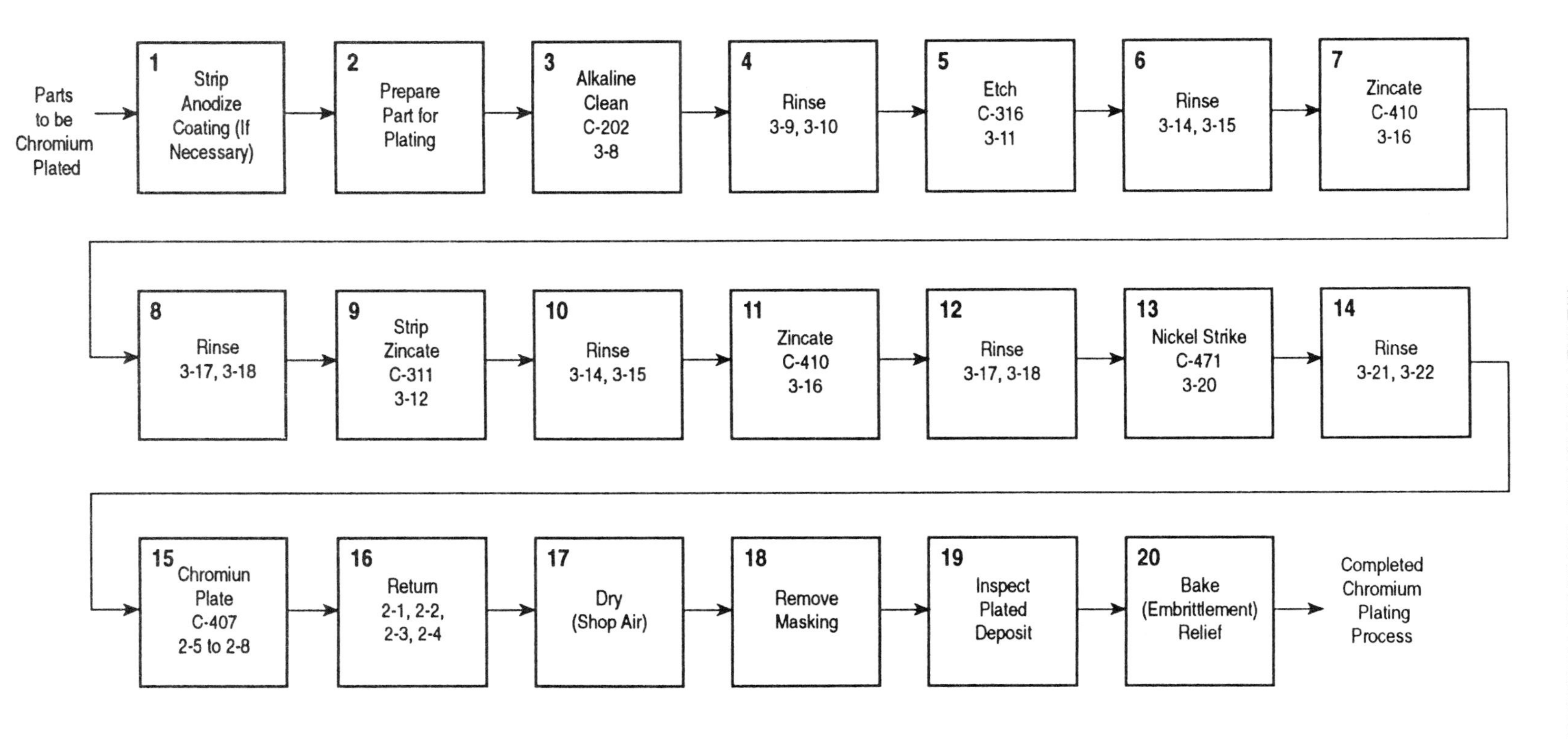

Figure 12.26. Process diagram for chromium deposition on aluminum alloys.

Table 12.7. Sample Data Collection Form

Date: ____________________

Name: ____________________

Part No. ____________________ Description:

End Item Part No. ____________________ Description:

Part system name: ____________________

Number of parts in lot: ____________________

Part Description:

Base metals: ____________

Part dimensions (l × w × h): ______ in. × ______ in. × ______ in.

Surface area: Total ______ in.² Plated only ______ in.²

Process Description:

Preprocess steps

☐ Vapor Degrease ☐ Glass Bead Blast ☐ Other (list)______
☐ Masking ☐ Stress Relief ______

Plating Processes (list in order used*)

	Plating Thickness (mils)	Rectifier Setting (ASI)	Plating Time (min.)
1. ________	________	________	________
2. ________	________	________	________
3. ________	________	________	________

Post Plating Processes

☐ Remove Masking ☐ Corrosion Prevention
☐ Baking ☐ Other (list) ____________________
____

Labor (minutes)

Preplate __________ Plating __________ Post Plate __________

Rework Factor ______

Plater(s) __________
____
____
____

*Use process diagram number, if applicable.

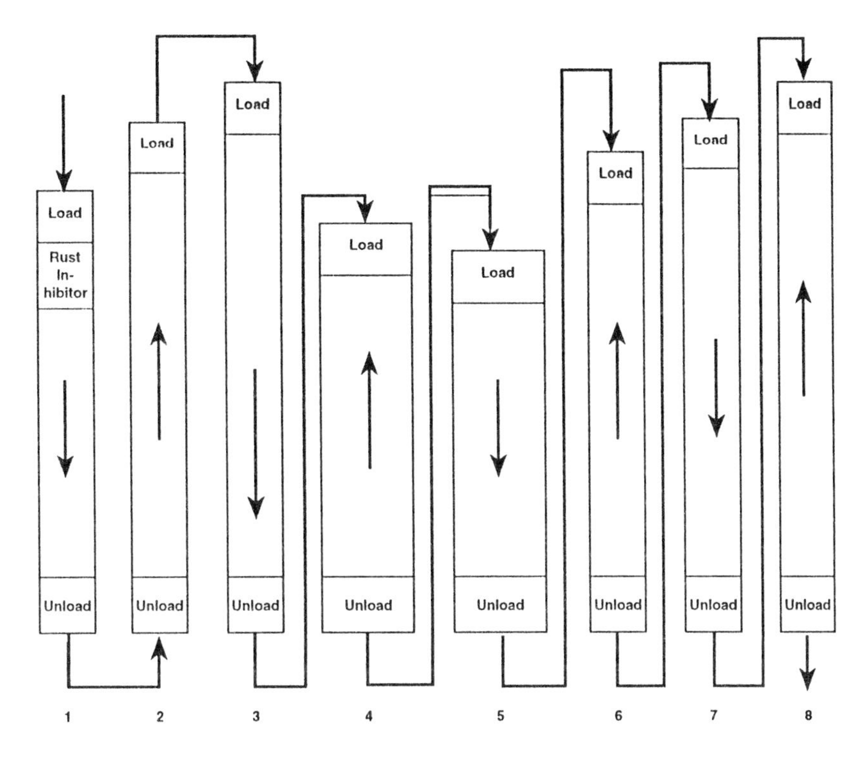

Figure 12.27. Example layout of plating process.

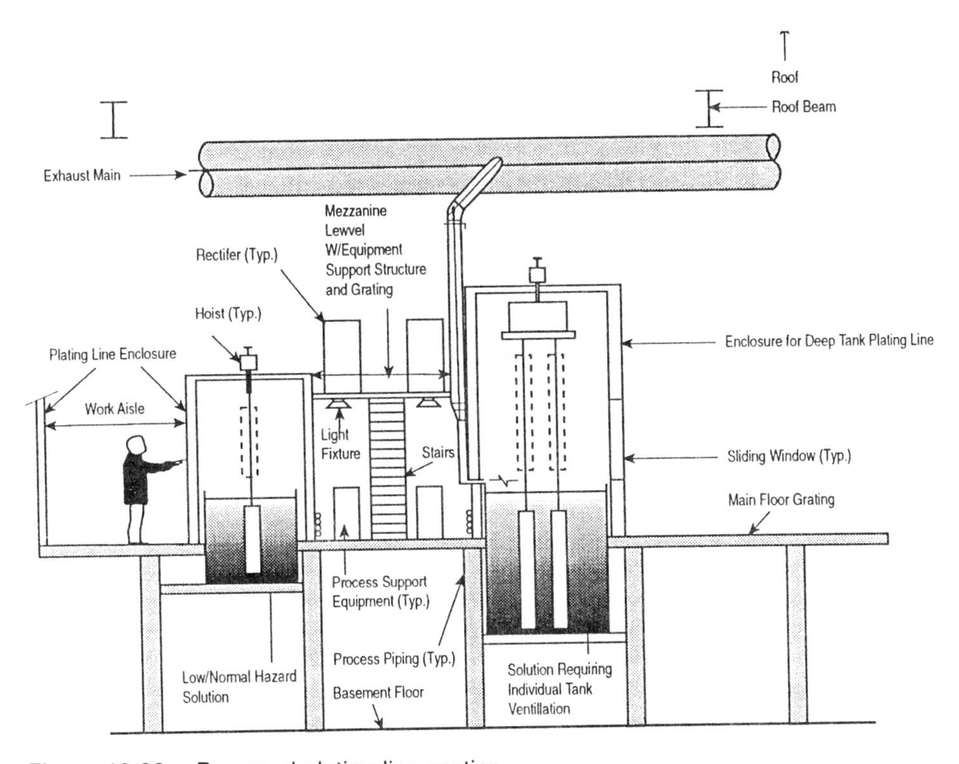

Figure 12.28. Proposed plating line section.

Process simulation of workload demonstrates production capabilities of new layout.

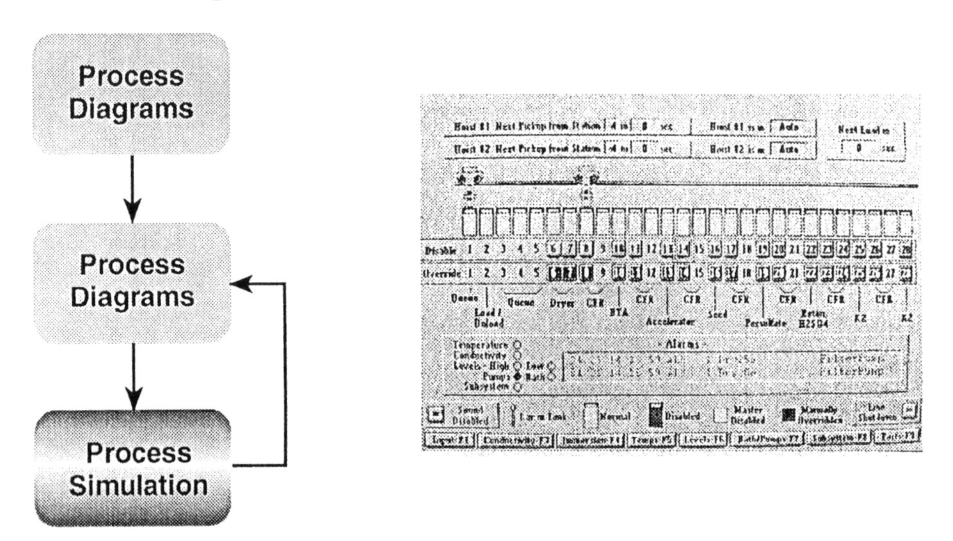

Figure 12.29. Process simulation software screen (Courtesy of Geotronics).

Case Study 12.3: Pratt & Whitney's Facility "Zero-Discharge" Program

In 1986, Pratt & Whitney's (P&W) facility in North Haven, Connecticut, began to plan conceptually for a "zero-discharge" metal-finishing capability.[23] At that time, P&W discharged about 1 million gal/day of treated wastewater, of which approximately 400,000 gal/day were generated by metal-finishing operations. Plating lines were mostly 1960s vintage. It was readily apparent that such a broad goal would require an evolutionary implementation program. Companywide, P&W has committed ". . . to reduce emissions of hazardous waste by 40% and toxic air by 50% by the year 1994," said Bob Daniel, Chairman and CEO, United Technologies Corp., (UTC) parent company of P&W.

P&W are involved in a corporate-wide waste reduction program that encompasses the following:

- Identify waste reduction opportunities that encompass good operating practices and proven technologies.
- Work with the chemical processing industry to investigate emerging waste reduction technologies that can be piloted and evaluated.
- Determine new hazardous waste reduction areas for research that can meet specific corporate needs.

Table 12.8. Renovation Design Features and Anticipated Benefits for Post-Renovation Plating Shop Operations

Renovation Design Features	Benefits
Layout • Tanks in process lines • Tanks positioned perpendicular to aisle ways and adjacent to mezzanines • Mezzanines used for support equipment (e.g., rectifiers)	• Maximizes use of space • Allows for easy access for operation and maintenance • Reduces equipment corrosion and worker hazards in basement • Maximizes good work flow
Automation • Automated hoists for all new plating lines • A central supervisory control system, control consoles for each of the plating lines, and input/output cards that interface with field devices	• Increases process efficiency and work quality • Provides for proper use of multiple rinse systems, proper withdrawal rates from process solutions, and dwell times above process tanks to minimize drag-out generation and losses.
Enclosures/Ventilation Improvements • Enclosures housing the process tanks and hoist systems • Mist elimination • Improved ventilation efficiency	• Reduces emissions • Provides for enhanced worker exposure protection • Provides bath protection from airborne contaminants • Reduces total ventilation rate
Rinse Improvements • Multiple countercurrent rinse systems for many of the processes • Strict use of DI water • Automated rinse controls • Improved agitation	• Provides direct drag-out recovery and return to the process bath • Reduces water use by 75 to 90% • Reduces contaminants from the water supply • Improves rinsing efficiency
Solution Maintenance Systems • Filtration • Carbon adsorption • Refrigerated carbonate removal • Electrodialysis • Electrowinning • Automatic solution analysis and chemical addition for electroless nickel	• Minimizes the buildup of bath contaminants • Extends solution life, resulting in reduced cost for disposal of hazardous waste replacement of solution
Construction Phasing • Planned renovation in six phases	• Minimizes shop operation downtime during construction
Corrosion Engineering • Service specific materials of construction for all wet process equipment and areas where spills could occur	• Extends equipment life. • Prevents waste due to leaks and equipment corrosion

The North Haven facility has taken a jump ahead of other P&W and UTC divisions and has already "closed the loop" on key processes including: Wood's nickel strike; sulfamate nickel plating; hard chromium; cadmium plating; chromating; and cadmium, chromium, and nickel stripping. The contribution of metal finishing to the total waste stream volume has been reduced from 40% to approximately 5%.

P&W's accomplishments required extensive conceptual planning, many visits to other plating facilities, attendance at many industry conferences such as the American Electroplating and Surface Finishing/U.S. EPA conference, and extensive assistance from consultants. However, the expected result is a return on investment of 100% in less than 2 years, and enhanced production capacity and product quality. Through waste reduction efforts, raw material costs have been reduced by about 80% and water usage by 95%.

Transportation and disposal costs and associated liabilities have been reduced by the same magnitude as a result of decreased sludge production and decreased shipments of concentrated solution wastes to P&W's East Hartford facility for treatment.

Implementation

Pratt & Whitney's zero-discharge program has been organized around the following implementation hierarchy:

In **Phase One**, good operating practices were defined. The following is a summary of P&W's Phase One efforts:

- defining minimum water quality standards; all water now used at North Haven is either deionized (critical) or softened (noncritical)
- using countercurrent rinses to reduce water usage
- using continuous process purification, as opposed to batch purification, to maintain consistent process quality; this includes dummy plating and carbon and particulate filtration
- using online process monitors to control solution additions
- optimizing process solutions to control drag-out (for example, reducing concentration and increasing temperature)
- optimizing preplate rinsing to control drag-in of contaminants

- installing automatic level controls on all heated processes
- training operators to understand proper rinsing and work transfer techniques to reduce drag-out and drag-in
- treating small concentrated batches, which is preferable to treating high-volume dilute waste streams

In **Phase Two**, procedural changes were implemented based upon established good operating practices.

In **Phase Three**, closed-loop technology was verified on a single process. For this phase, P&W chose an existing nickel-plating process encompassing a Woods nickel strike and four sulfamate nickel-plating tanks. Figures 12.30 and 12.31 show the before and after process schematics, respectively. P&W wanted to evaluate implementation factors, product quality, and operator acceptance and training. A modularized approach has allowed the fine tuning of designs, and disruption of production capabilities has been minimized. Operator acceptance has been very good, despite some initial skepticism. Product quality has improved. Implementation was completed in August 1989 and followed with an extensive study completed in October 1990.

In **Phase Four**, good operating practices and closed-loop technologies were incorporated in the design of planned and appropriated new plating lines. New plating lines encompassing nickel and chromium plating; cadmium, chromium, and nickel stripping; and titanium descaling were already on the drawing board. Initial plans were revised to incorporate the following changes:

- countercurrent rinses
- ion exchange: nickel strike, nickel, cadmium, and chromium stripping
- atmospheric evaporation: hard chromium, sulfamate nickel
- deionized water in all critical rinses and softened water in all noncritical rinses and noncritical evaporation makeup

In **Phase Five**, new plating lines were installed. This was completed in October 1990. Figures 12.32 and 12.33 show views of the new plating lines.

In **Phase Six**, the remaining existing processes were renovated, including cadmium cyanide plating and chromating.

Future Plans

Pratt & Whitney has outlined several goals for future waste reduction projects:

- implementation of continuous process-purification technologies to include the following processes: chromium and cadmium plating, acids, alkaline cleaners, chromating, and stripping
- implementation of closed-loop technologies on preplate rinses (acids and alkaline cleaners)
- implementation of online process monitoring

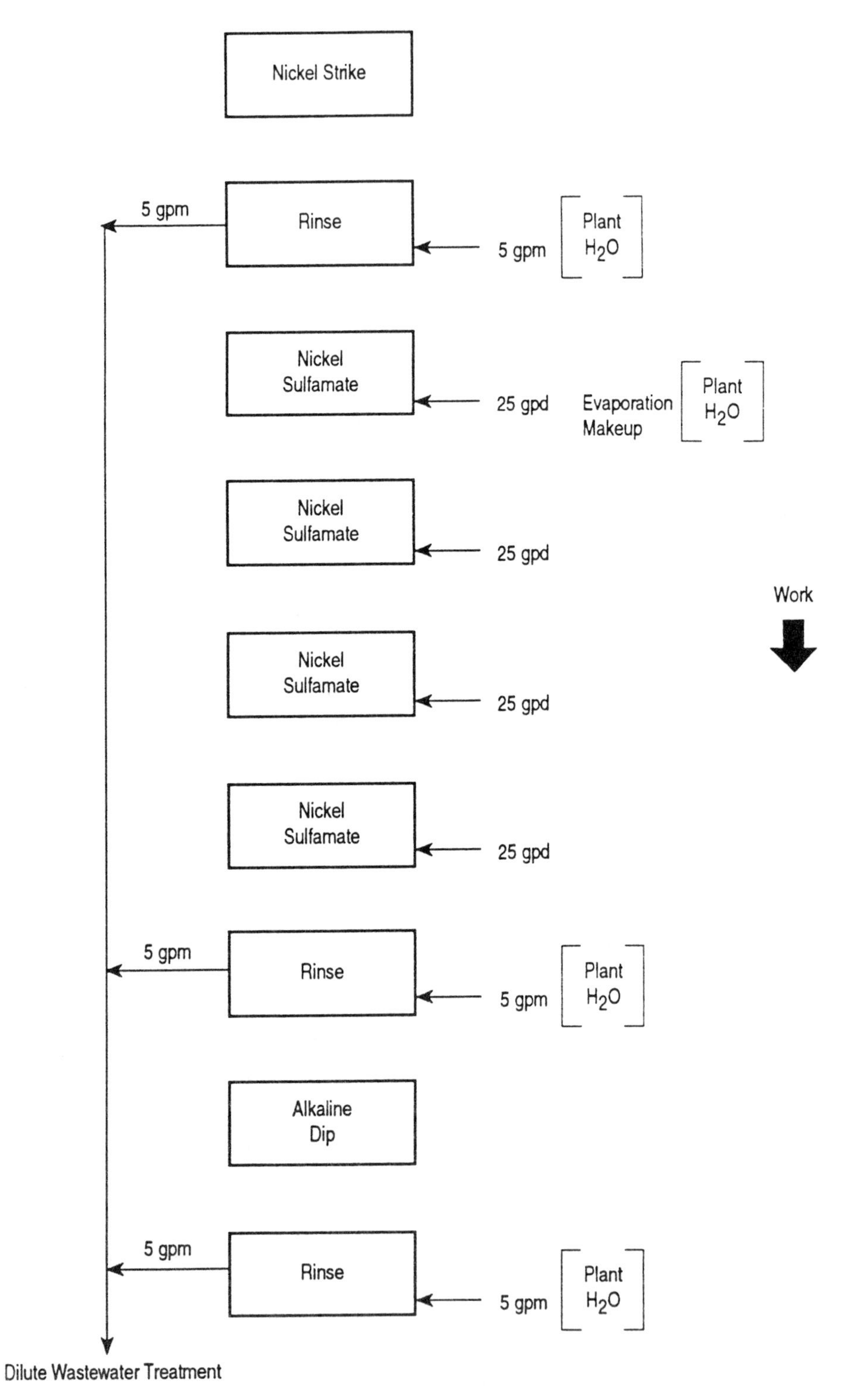

Figure 12.30. Nickel plate schematic of current operations (Courtesy of Pratt & Whitney Aircraft).

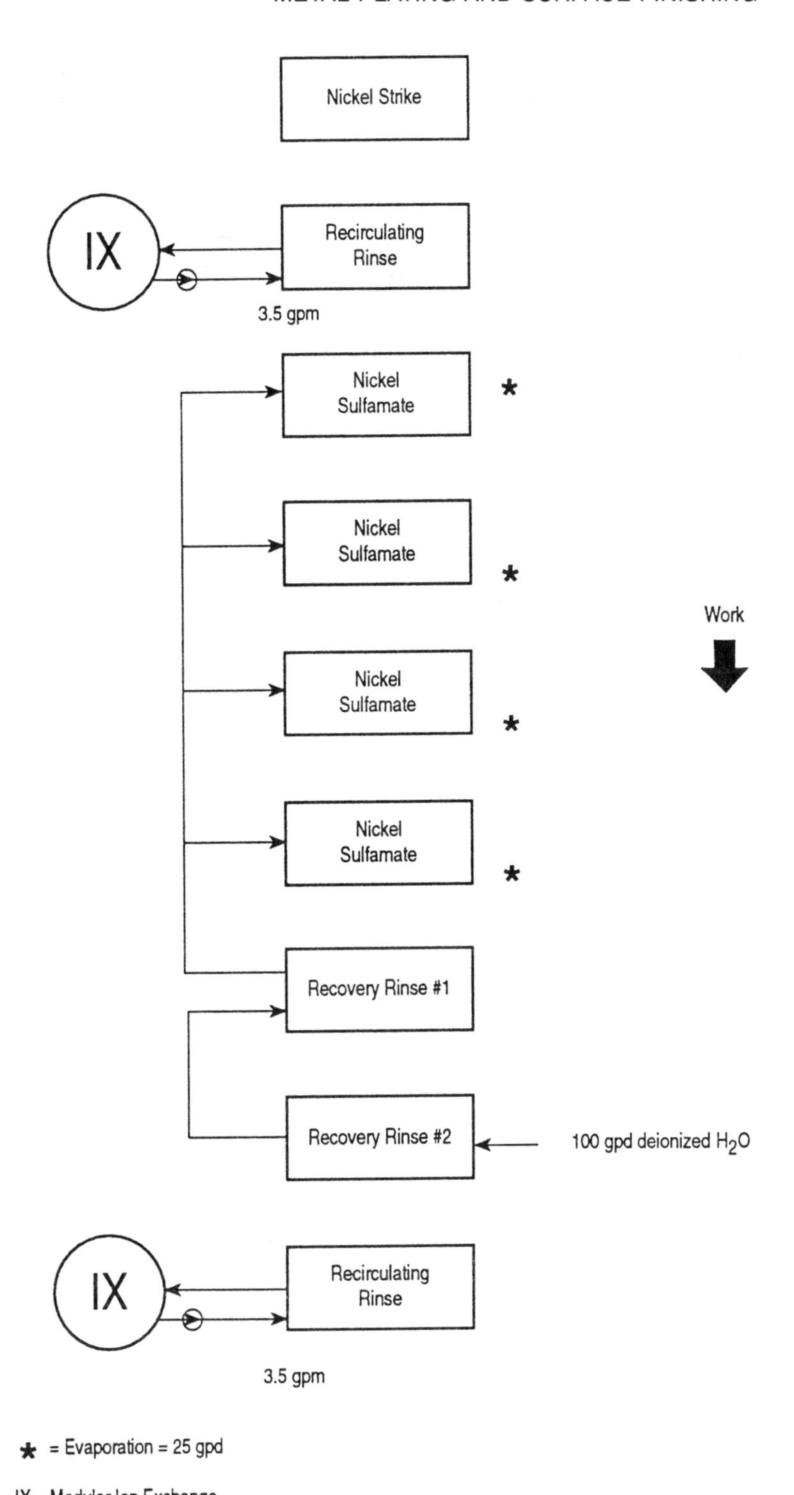

Figure 12.31. Nickel plate schematic of optimized operations.

Figure 12.32. Pratt & Whitney's new North Haven nickel-plating and titanium-cleaning line.

Figure 12.33. Endview of Pratt & Whitney's new nickel-plating line showing modular ion exchange regeneration system.

Case Study 12.4: New Plating Facility Design for B.F. Goodrich Aerospace, Simmonds Precision Aircraft Systems

CH2M HILL and Integrated Technologies, Inc., provided Simmonds Precision Aircraft Systems (SPAS) with conceptual and detailed design services for the construction of a new plating facility. SPAS requirements include nickel electroforming on aluminum mandrels, caustic etchout of mandrels, chromic and sulfuric anodizing, chromate conversion coating, electroless nickel plating and miscellaneous other wet processes. The existing facility generated approximately 30,000 gallons per day (gpd) of wastewater and 25 gpd of concentrated waste for offsite treatment and disposal, with a substantial volume of work sent to outside sources for offsite plating. The new facility is expected to generate less than 3,000 gpd of wastewater and 15 gpd of concentrated waste and sludge for offsite treatment and disposal, while substantially reducing out-sourced work.

These reductions in waste generation are achieved with minimal dependence on capital-intensive up-the-pipe technology and maximum use of techniques such as countercurrent rinsing, natural evaporation, and recovery rinsing. The design team used a three-step assessment process to verify and optimize design factors:

1. Develop a baseline characterization
   - drag-out estimation matrix, material, and flow-balance chemistry
   - baseline simulation and verification using Plato's Process Planner®

Plato's Process Planner® is a software system for calculation and process simulation that was developed for organization, configuration, and optimization of metal-finishing facility processes and outfalls. Plato simulates evaporative-recovery techniques, counter-current rinsing, and conventional (hydroxide precipitation) end-of-pipe waste treatment, and generates economic summaries for comparison of alternatives. Plato generates load data required for ion exchange, reverse osmosis, and other technology sizing. Plato also models wastewater treatment influent loading and flows, estimates sludge reduction, and is very useful for segregation planning. Plato is a useful tool for metal-finishing professionals who are charged with optimizing metal-finishing processes.[24]

2. Determine new facility requirements
   - good operating practices
   - evaluation of chemistry substitutions
   - consolidation and elimination of redundant or obsolete processes
   - estimation of up-the-pipe and end-of-pipe "treatment" technologies
3. Perform process/facility optimization
   - use of Plato to model new facility requirements and design concepts

Plato's Process Planner® was used to model multiple "what if" rinse-water management scenarios for each process. Sensitivity analyses were performed on the rinse-water recovery systems by modeling mean, peak, and maximum work loads. Initial rough sizing estimates were fine tuned, providing a "leaner" design recom-

PROCESS SIMULATION REPORT

Facility: SIMMONDS PRECISION, VERGENNES, VERMONT
Line : Line 1
Process : Chromic Anodize

Outfall:

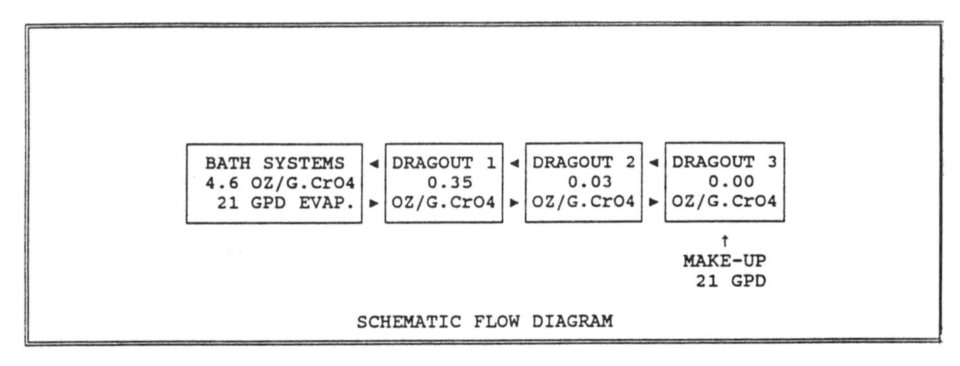

BATH SYSTEMS

Operating Hours :	16.0 Hours/Day	Operate Days:	5.0 Days/Week
Bath Temperature :	100 Deg.F.	Bath Surface:	32 Sq.Ft
Evaporation Total:	21 GPD	Bath Dragout:	1.60 GPD
Dragout Recovery :	99.96%	Evap/Dragout:	13.1
Last Dragout Conc:	14 mg/l.		

RINSING LOSSES

Total Bath Chemical Losses: 0 Grams/Day
Chromate = 0

REVENUE IMPACT

Bath Loss :(GPD)	0.0	($/GAL.)	$3.73	($/YEAR)	$0
Make Water :(GPD)	21	($/MGAL)	$5.00	($/YEAR)	$26
Rinse Water:(GPD)	0	($/MGAL)	$5.00	($/YEAR)	$0
Sewer Usage:(GPD)	0	($/MGAL)	$0.00	($/YEAR)	$0
Evaporation:(GPD)	21	($/GAL.)	$0.10	($/YEAR)	$525
Treatment :(MGPD)	0.00	($/MGAL)	$0.00	($/YEAR)	$0
Haul Sludge:(GPD)	0.0	($/GAL.)	$0.00	($/YEAR)	$0
Total Process Losses and Waste Costs				($/YEAR)	$551

Figure 12.34. Plato's Process Planner® simulation report for chromic anodize process.

mendation. As an example, SPAS's chromic acid anodizing process used a spray rinse, followed by an immersion rinse. Both rinses were shared with sulfuric anodizing. These processes generated approximately 3,000 gpd of wastewater. Initial rough estimates anticipated the need for an atmospheric evaporator with a three-stage countercurrent recovery rinse for closed-loop operation. The sulfuric anodizing process would require a three-stage countercurrent rinse with a 5-gpm discharge. Process modeling showed that natural bath evaporation was sufficient for the chromic process and that a two-stage rinse with 2-gpm discharge was sufficient for the sulfuric process. Figure 12.34 presents a Plato's Process Simulation Report for the chromic anodize process. Figure 12.35 presents a Plato's Process Simulation Report for the sulfuric anodize process.

A "zero-discharge option" was evaluated, but proved to be economically impractical because of higher operating costs.

PROCESS SIMULATION REPORT

Facility: SIMMONDS PRECISION, VERGENNES, VERMONT
Line : Line 1
Process : Sulfuric Anodize
Outfall: acid/alkalie

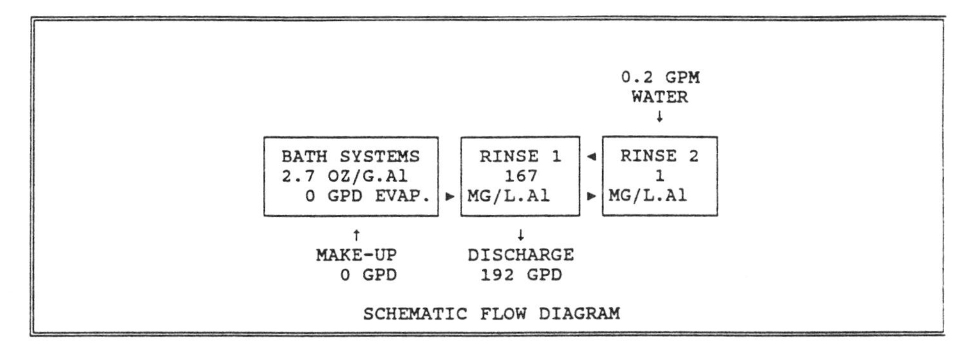

BATH SYSTEMS

Operating Hours :	16.0 Hours/Day	Operate Days:	5.0 Days/Week
Bath Temperature :	70 Deg.F.	Bath Surface:	32 Sq.Ft
Evaporation Total:	0 GPD	Bath Dragout:	1.60 GPD
Dragout Recovery :	0.00%	Evap/Dragout:	0.0

RINSING LOSSES

Total Bath Chemical Losses:		1,271 Grams/Day
Aluminum	= 121	
Sulfate	= 1,150	

REVENUE IMPACT

Bath Loss :(GPD)	1.6	($/GAL.)	$0.40	($/YEAR)	$160
Make Water :(GPD)	0	($/MGAL)	$5.00	($/YEAR)	$0
Rinse Water:(GPD)	192	($/MGAL)	$1.00	($/YEAR)	$48
Sewer Usage:(GPD)	192	($/MGAL)	$1.00	($/YEAR)	$48
Evaporation:(GPD)	0	($/GAL.)	$0.10	($/YEAR)	$0
Treatment :(MGPD)	0.19	($/MGAL)	$10.00	($/YEAR)	$480
Haul Sludge:(GPD)	0.1	($/GAL.)	$4.00	($/YEAR)	$140
Total Process Losses and Waste Costs				($/YEAR)	$876

Figure 12.35. Plato's Process Planner® simulation report for sulfuric anodize process.

REFERENCES

1. Kushner, J. *Water and Waste Control for the Plating Shop.* (Cincinnati, OH: Gardner Publications, Inc., 1976).
2. Johnnie, S.T. "Waste Reduction in the Hewlett-Packard, Colorado Springs Division, Printed Circuit Board Manufacturing Shop," *Hazardous Waste Hazardous Materials* 4(9) (1987).
3. "Control and Treatment Alternatives for the Metal Finishing Industry, In-Plant Changes," U.S. Environmental Protection Agency, EPA 625/8-82-008, January 1982.
4. Cushnie, G.C. "Navy Electroplating Pollution Control Technology Manual," written for Naval Civil Engineering Laboratory, Port Hueneme, CA, Report No. 84.019, February 1984.
5. "Environmental Pollution Control Alternatives: Economics of Wastewater Treatment-Alternatives for the Electroplating Industry," U.S. Environmental Protection Agency, EPA 625/5-79-016, June 1979.

6. Moore, Gardner & Associates. "Naval Shipyards Industrial Process and Waste Management Investigation," prepared for Naval Facilities Engineering Command, Contract No. N00025-80-C-0015, July 1983.
7. Mouchahoir, G.E., and M.A. Muradaz. "Clean Technologies in Industrial Sectors of Metal Finishing, Non-Ferrous Metals, and High Volume Organic Chemicals," U.S. Environmental Protection Agency, EPA 68-01-5-21, June 1981.
8. "Summary Report-Control and Treatment Technology for Metal Finishing Industry-Ion Exchange," developed by the Industrial Environmental Research Laboratory, EPA 625/8-81-007, June 1981.
9. "Water Pollution Abatement Technology Capabilities and Costs: Metal Finishing Industry," Lancy Laboratories, NTIS PB-24 808, October 1975.
10. McNulty, K.J., and P.R. Hoover. "Evaluation of Reverse Osmosis Membranes for Treatment of Electroplating Rinsewater," EPA-600/2-80-084, NTIS PB80-202385, May 1980.
11. Cartwright, P.S. "An Update on Reverse Osmosis for Metal Finishing," *Plating and Surface Finishing*, April 1984.
12. McNulty, K.J., and J.W. Kubarewicz. "Demonstration of Zinc Cyanide Recovery Using Reverse Osmosis and Evaporation," EPA-600/2-81-132, NTIS PB-231243, July 1981.
13. McNulty, K.J., et al. "Reverse Osmosis Field Test: Treatment of Copper Cyanide Rinse Waters," EPA-600/2-77-170, NTIS PB-272473, 1979.
14. Eisenmann, J.L. "Membrane Processes for Metal Recovery from Electroplating Rinse Water," Second Conference on Advanced Pollution Control for the Metal Finishing Industry, EPA-600/8-79-014, June 1979, pp. 99-105.
15. Eisenmann, J.L. "Nickel Recovery from Electroplating Rinsewaters by Electrodialysis," NTIS PB81-227209, EPA-600/2-81-130, July 1981.
16. Jorczyk, E.R. "A New Non-Cyanide Cadmium Electroplating Bath. Water Pollution Abatement Technology Capabilities and Costs: Metal Finishing Industry," prepared for National Commission on Water Quality, NTIS PB-248 808, October 1975.
17. Cushnie, G.C. "Initiation Decision Report-Treatment of Electroplating Wastes," prepared for Naval Civil Engineering Laboratory, TM No. 54-83-20CR, October 1983.
18. Garner, H.R. "Meeting the Regs: How Trivalent Helps," *Products Finishing* 47(12) (September 1983).
19. Lowenheim, F.A. *Electroplating*. (New York: McGraw-Hill, 1978).
20. Muehlberger, D.E. "Ion Vapor Deposition of Aluminum: More Than Just a Cadmium Substitute," *Plating and Surface Finishing* 24 (November 1983).
21. Fannin, E.R. "Ion Vapor Deposited Aluminum Coatings," Proceedings of the Workshop on Alternatives for Cadmium Electroplating in Metal Finishing, EPA-560/2-79-003, 1979, p. 68.
22. Steube, K.E. "Fabrication and Optimization of an Aluminum Ion Vapor Deposition System," Technical Report AFML-TR-78-132, Wright-Patterson Air Force Base, OH, Air Force Materials Laboratory, June 1978.
23. Gallerani, P., and R. McCarvill. "Waste Minimization and Pollution Prevention at Pratt & Whitney Aircraft," *Plating and Surface Finishing*, March 1991.
24. Walton, C.W. "Software Review: Plato's Process Planner,"® *Plating and Surface Finishing*, December 1992.

GENERAL REFERENCES

Cushnie, G., and W. Anderson. "Removal of Metal Cations from Chromium Plating Solutions," presented at 10th AESF/EPA Conference on Environmental Control for the Metal Finishing Industry, January 23-25, 1989.

Drake, D., and P. Gallerani. "Wastewater Management for the Metal Finishing Industry in the 21st Century," presented at AESF/EPA Conference on Environmental Control for the Metal Finishing Industry, January 1993.

Integrated Technologies, Inc. "Waste Minimization and Pollution Prevention: Metal Finishing, A Self Audit Manual," prepared for The Connecticut Hazardous Waste Management Service, September 1990.

Karrs, S., D.M. Buckley, and F.A. Steward. "Ion Exchange for Metal Recovery: A Discussion of Trade-Offs," *Plating and Surface Finishing*, April 1986.

Metal Finishing, *Metal Finishing Guidebook and Directory Issue* (Hackensack, NJ: Metal Finishing, 1992).

Rosenblum, J., and M.J. Naser. "Heavy Metals Waste Minimization: Practice and Pitfalls," *Plating and Surface Finishing*, April 1991.

Steward, F.A., and W.J. McLay. "Waste Minimization Alternate Recovery Technologies," in *Metal Finishing Guidebook and Directory* (1986 ed.). Reprint from Lancy International, Inc., Zelienople, PA.

Metals Handbook, Ninth Edition, Vol. 5: Surface Cleaning, Finishing, and Coating. (Metals Park, OH: American Society for Metals, 1982).

CHAPTER 13

Painting and Coating

*John Soebbing and Thomas Higgins**

DESCRIPTIONS OF PAINTING PROCESSES AND WASTES

Solvent-based paints are applied to surfaces of parts, components, and assembled products for corrosion protection, surface protection, identification, and aesthetic appeal. Most painting is performed by conventional liquid spray technology.[1] In spray painting, the paint is mixed with a carrier, usually an organic solvent, and is applied to the surface with an air-pressurized sprayer (Figure 13.1). Spray painting is usually done in a horizontal or downdraft booth.

Painting processes generate two significant sources of wastes: paint sludge and waste solvents. The largest volume of waste generated from painting is paint sludge produced in air emission control equipment. During typical spray painting, 50% of the paint is deposited on the surface being painted; the other 50%, called overspray, is sprayed into the air and removed by ventilation equipment.[2] As the paint dries, the solvent evaporates to the air. Air exhausted from a paint booth is often passed through a water scrubber that separates the paint from the air before discharge to the atmosphere. The scrubber water is normally recycled, and paint solids are concentrated in the scrubber sump. When the sump fills with paint sludge, it is removed and put in drums for disposal.

The solvents used to clean painting equipment are the second largest source of solid waste generated in painting processes. Most paints are organic- and solvent-based, so they require organic solvents for cleanup. The type of solvent used varies with the paint. Some of the more common solvents are methyl ethyl ketone; xylene; 1,1,1-trichloroethane; toluene; butyl acetate; ethylene glycol; monoethyl acetate; and alcohol. In addition, paint-stripping solvents (as described in Chapter 14) may be used to remove hardened paint from surfaces. These wastes are classified as hazardous because of their general flammable and toxic properties.

Many painting processes generate varied types of hazardous waste. The common feature of almost all of these wastes is that their hazardous characteristics are derived both from the paint constituents (heavy metals and solvents) and from the

*Mr. Soebbing is a corrosion engineer in CH2M HILL's Denver, Colorado, office and may be reached at (303) 771-0900. Dr. Higgins is a vice president in CH2M HILL's Reston, Virginia, office and may be reached at (703) 471-1441.

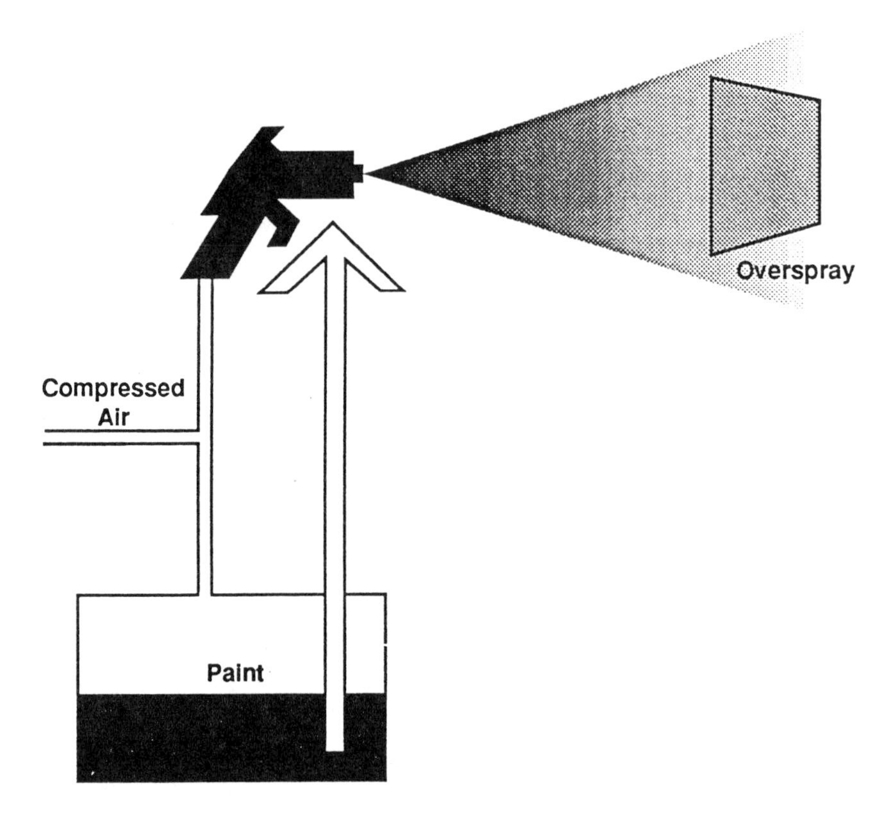

Figure 13.1. Air-atomized spray painting.

solvents used in cleanup (toxic and flammable organics). For example, when dry-scrubber paint booths are used instead of water scrubbers, the filter material can become contaminated with the paint and require disposal as a hazardous waste. However, if water-based paints are used, solvents are not needed for cleanup, thus reducing the volume of hazardous waste generated.

Another waste product from painting is volatile organic compounds (VOCs) emitted to the air. Federal VOC limits for paint are 420 g/L for paints that cure below 90°C and 360 g/L for paints that cure above 90°C.[2] Some state air pollution control agencies are setting strict VOC content limits for paint. For example, the South Coast Air Quality Management District in California has set a 300 g/L VOC limit for general air-dried paints used for coating metal parts and products in fabricating and painting shops. Local regulatory agencies also control VOCs by setting total permissible discharge limits from facilities. These limits include point sources and fugitive sources. The U.S. EPA is required to develop limits for toxic air emissions; these limits will likely have an impact on both the types of solvents used in paint and those used in cleanup.

Some facilities house both paint-stripping and painting operations in the same area. Compared to the disposal of wastes from stripping processes, disposal of paint waste is less problematic because much less hazardous waste is generated and only a small amount of wastewater is produced.

ALTERNATIVE PAINTING TECHNIQUES

Alternatives to conventional solvent-based spray painting can reduce wastes and pollutants. These alternatives require an integrated approach in which paint materials and techniques are improved to reduce or eliminate hazardous materials. For example, modified painting techniques can minimize the amount of paint waste that ultimately must be disposed of as hazardous waste. Companies also can formulate paints that minimize or eliminate carrier and cleanup solvents, both of which contribute to hazardous waste and air pollution. In addition, substitute solvents may be used to minimize air pollution and to produce less-toxic hazardous waste. The transfer efficiencies of various alternatives are listed in Table 13.1.

More-promising developments in the area of painting modifications and substitutions are summarized in the following pages. Suppliers of painting equipment are listed in Table 13.2.

POWDER COATING

Powder coating, also called "dry-powder painting," is one of the major advances in paint application. This technique is based on depositing specially formulated thermoplastic, or thermosetting, heat-fusible powders on metallic substrates. No solvents are used; therefore, the system eliminates the pollution and safety problems associated with solvent-based paints. Powder coating produces little or no VOC emissions, greatly reduces cleanup solvents, and eliminates use of paint primers and thinners. In addition, no waste (old) paint has to be disposed of.[5]

Along with the environmental advantages offered by dry-powder painting, the process provides technical, production, and cost benefits. Productivity is increased because, without solvents, the coating can be cured immediately after application. Because curing is thermoactivated, curing times are short. One technical advantage is that special coating materials such as nylon, which cannot be applied by conventional solvent-based systems because of the lack of appropriate solvents, can be applied by dry-powder painting. In addition, complex surfaces are more evenly coated in dry-powder systems. For some applications, a single coating can replace the multiple-coating applications used in conventional spray painting. Dry-powder

Table 13.1. Transfer Efficiency of Various Coating Methods

Coating Method	Transfer Efficiency
Air-Atomized Spraying	30 to 40%
Air-Electrostatic Spraying	60 to 70%
Airless and Air-Assisted Electrostatic Spraying	70 to 75%
Airless Electrostatic Spraying	70 to 80%
Airless Spraying	50 to 60%
Dip Coating	95 to 100%
Electro Coating	95 to 100%
Electrostatic, High-Speed Rotating	80 to 90%
Low-Pressure Air	70 to 90%
Powder Coating	95 to 100%

Source: Adapted from *Product Finishing Directory*, Gardner Publications, Cincinnati, OH, October 1992.

Table 13.2. Painting Equipment Suppliers

Type	Company	Address	Phone/*Fax*
E, F	Belco Industries, Inc.	115 E. Main Street Belding, MI	(616) 794–0410 *(616) 794–3424*
A, B, C, D, F, G, H, I, J	Binks Manufacturing Co.	9201 Belmont Avenue Franklin Park, IL 60131	(708) 671–3000 *(708) 671–6489*
E, F	Cincinnati Industrial Machine Co.	3280 Hageman Street Cincinnati, OH 45262	(513) 769–0700 *(513) 769–0697*
B, E, I	Conforming Matrix Corp.	830 New York Avenue Toledo, OH 43611	(419) 729–3777 *(419) 729–3779*
E, F	Despatch Industries	P.O. Box 1320 Minneapolis, MN 55440	(612) 781–5363 *(612) 781–5353*
A, B, C, D, G, H, I, J	DeVilbiss	1724 Indian Wood Circle Maumee, OH 43537	(419) 891–8200 *(419) 891–8205*
E, F	Durr Industries, Inc.	40600 Plymouth Road Plymouth, MI 48170	(313) 459–6800 *(313) 459–5837*
E, F	Finishing Group	P.O. Box 7567 Ocala, FL 32672	(904) 821–1005 *(904) 821–0833*
E, F	General Fabrications Corp.	7777 Milan Road Sandusky, OH 44870	(419) 625–6055 *(419) 625–7843*
E, F	GLA Finishing Systems	38830 Taylor Parkway North Ridgeville, OH 44039	(216) 327–3323 *(216) 327–4498*
A, B, C, D, G, H, J	Graco Inc.	P.O. Box 1441 Minneapolis, MN 55440	(612) 623–6743 *(612) 623–6893*
E, F	Koch Sons Inc.	10 S. Eleventh Avenue Evansville, IN 47744	(812) 465–9600 *(812) 465–9724*
A, B, C, G, I, J	Kremlin, Inc.	P.O. Box 1219 Addison, IL 60101	(708) 543–1177 *(708) 543–1201*
A, B, C, G, H, J	Nordson Corp.	555 Jackson Street Amherst, OH 44001	(800) 626–8303 *(216) 985–1471*
A, B, C, H, I	Wagner Spray Tech Corp.	1770 Fernbrook Lane Minneapolis, MN 55447	(612) 553–7000 *(612) 553–7288*

Type Code:
A: Air-Assisted Airless
B: Air-Atomized Spray
C: Airless Spray
D: Centrifugal
E: Dip Coating
F: Electro Coating
G: Electrostatic, High-Speed Rotating
H: Electrostatic Spray
I: Low-Pressure Air
J: Hot Spray

Source: Adapted from *Product Finishing Director*, Gardner Publications, Cincinnati, OH, October 1992.

techniques are also readily adaptable to current production methods and are easy to learn.

The one major limitation with the most commonly used dry-powder painting techniques is that the items to be painted must be able to withstand the typical curing temperatures of 350°F for 30 minutes.[6] Aluminum alloys cannot be subjected to these conditions without significant loss of strength.

Dry-powder painting techniques that are commercially available include electrostatic dry-powder painting, fluidized bed method, and plasma spraying. A description of each follows. Suppliers of powder-coating equipment are listed in Table 13.3. Suppliers of powder-coating materials are listed in Table 13.4.

Electrostatic Dry-Powder Painting

Electrostatic dry-powder painting is the most widely used powder coating technique. With this method, dry powder is metered into a compressed-air-driven spray gun and is sprayed at the prepared surface (Figures 13.2 and 13.3). An electrode in the spray gun ionizes the air and powder suspension using direct current, and the dry-powder particles then become charged. The surface to be coated is given the opposite charge, and the powder is electrostatically attracted to the surface. As the charged powder builds up, the coating thickness is limited by the loss of attraction of the powder to the surface, resulting in a uniform thickness, even on complex shapes. The coating is then fused to the surface and cured in conventional ovens.

In commercial applications, a ventilation system collects the powder overspray and the powder is separated by an air-filter system and recycled, thus eliminating the need to dispose of scrubber water associated with solvent-based paints. A powder usage efficiency of 90 to 99% is possible. Table 13.5 compares annual operating costs of a dry-powder system with a conventional solvent painting system at an Air Force facility. The comparison is based on O&M costs required to coat 12 million ft^2 of parts with a 1-mil-thick polyester coating.[5]

Case Study 13.1: Dry-Powder Painting at Hughes Aircraft Company

Developmental Program Description. At Air Force Plant 44 in Tucson, Arizona, operated by the Missile Systems Group of Hughes Aircraft Company, electrostatic dry-powder painting is being used in a developmental program for painting missile parts. Selected for the initial application of paint to the inside fuselage section of the Phoenix missile, dry-powder painting was chosen over other conventional paint systems because of enhanced surface protection, better coverage, and reduced solvent emissions.

The results of the developmental program have been very successful. In addition to satisfactorily achieving the initial goals, the developmental program for dry-powder painting has shown that this technique provides additional significant benefits, including reduced hazardous waste, elimination of wastewater, fewer work hours, less paint use, and lower overall cost per square foot of painted surface. The developmental program is continuing, and in-house implementation is being evaluated.

Table 13.3. Suppliers of Powder-Coating Equipment

Type	Company	Address	Phone/*Fax*
ES	ACS Engineered Systems	741 Hurlingame Avenue Redwood City, CA 94063	(415) 368–1704 *(415) 368–1706*
ES, FB	Advanced Power Coatings, Inc.	620 Stone Hill Road Denver, PA 17517	(215) 484–4789 *(215) 484–0272*
ES, FB, FE	Binks Manufacturing Co.	9201 Belmont Avenue Franklin Park, IL 60131	(708) 671–3000 *(708) 671–6489*
FB	Erie Powdercoating Technology	227 Hathaway Street Girard, PA 16417	(814) 774–8238 *(814) 774–9372*
FB, FE	Fusion Coatings, Inc.	Road #1, Route 422 Robesonia, PA 19551	(215) 693–5886 *(215) 693–6802*
ES	GEMA-Volstatic	3939 W. 56th Street Indianapolis, IN 46208	(317) 298–5001 *(317) 298–5881*
FB	GLA Finishing Systems	38830 Taylor Parkway North Ridgeville, OH 44039	(216) 327–3323 *(216) 327–4498*
ES, FB, FE	Industrial Coating Systems, Inc.	374½ W. Chicago Road Coldwater, MI 49036	(517) 369–2529 *(517) 369–2830*
ES, PS	Iontech	1050 W. Mehring Way Cincinnati, OH 45203	(513) 421–0008 *(513) 421–0089*
FB, FE	Koch Sons, Inc.	10 S. Eleventh Avenue Evansville, IN 47744	(812) 465–9600 *(812) 465–9724*
PS	Metco, Division of Perkin Elmer	1101 Prospect Avenue Westbury, NY 11590	(516) 334–1300 *(516) 338–2414*
FB, FE	Moco Thermal Industries	First Oven Place Romulus, MI 48174	(313) 728–6800 *(313) 728–1927*
PS	Nordson Corp.	1150 Nordson Drive Amherst, OH 44001	(800) 626–8303 *(216) 985–1471*
ES, FB, FE	Powder Equipment Partsland	100 Wellington Drive Stamford, CT 06903	(203) 329–8588 *(203) 329–8876*
ES, FB, FE	Reclaim	140 Joey Drive Elk Grove Village, IL 60007	(708) 640–0111 *(708) 640–0152*
ES, FB	Rockwell Co., Inc.	208 Eliot Street Fairfield, CT 06430	(203) 259–1621 *(203) 259–8304*

ES: Electrostatic Spray FE: Fluidized-Bed Electrostatic
FB: Fluidized Bed PS: Plasma Spray

Source: Adapted from *Product Finishing Directory*, Gardner Publications, Cincinnati, OH, October 1992.

Table 13.4. Suppliers of Powder-Coating Materials

Type	Company	Address	Phone/*Fax*
A, B, C, F, G, K, L	Armitage & Co.	545 National Drive Gallatin, TN 37066	(615) 452–6556 *(615) 452–4147*
B, C, G, K	Beaver Paint Co.	710 Beaver Road Girard, PA 16417	(814) 774–3177 *(814) 774–2521*
A, B, C, D, E, F, G, H, I, J, K, L	Cardinal Industrial Finishes	901 Stimson Avenue City of Industry, CA 91745	(818) 336–3345 *(818) 336–0410*
B, C, G, K	Elpaco Coatings Corp.	P.O. Box 447 Elkhart, IN 46515	(219) 295–3991 *(219) 293–0497*
A, B, C, E, G, K	Evodex Powder Coatings	90 Carson Road Birmingham, AL 35215	(205) 854–5486 *(205) 854–2566*
A, B, C, G, K	EVTECH (A Kodak Co.)	9103 Forsyth Park Drive Charlotte, NC 28241	(704) 588–2112 *(704) 588–2280*
B, C, G, K	Farboil Co.	8200 Fisher Road Baltimore, MD 21222	(410) 477–8200 *(410) 477–8995*
A, B, C, G	Ferro Corp.	4150 E. 56th Street Cleveland, OH 44105	(216) 641–8580 *(216) 696–6958*
B, C, G, K	Fuller Co.	3200 Labore Road Vadnais Heights, MN 55110	(612) 481–9558 *(612) 481–9047*
A, B, C, E, G, K	Glidden Co.	925 Euclid Avenue Cleveland, OH 44115	(216) 344–8000 *(216) 344–8744*
B, C, G, K	Herberts Powder Coatings, Inc.	4150 Lyman Drive Hilliard, OH 43026	(614) 771–7881 *(614) 771–7901*
B, C, G, K	International Paint Powder Coatings, Inc.	P.O. Box 924224 Houston, TX 77292	(713) 682–1711 *(713) 682–0065*
A, C, G, K, L	Lilly Industries, Inc.	P.O. Box 946 Indianapolis, IN 46206	(317) 634–8512 *(317) 687–6741*
A, B, C, E, F, G, K, L	Morton International	P.O. Box 1–5240 Reading, PA 19612	(215) 775–6600 *(215) 775–6691*
B, C, G, K	O'Brien Powder Products, Inc.	6800 Genaro Houston, TX 77041	(713) 939–4000 *(713) 939–4029*
B, C, G, K	Porter Powder Coatings	400 S. 13th Street Louisville, KY 40203	(502) 588–9200 *(502) 588–9671*
A, B, C, G, K	Pratt & Lambert, Inc.	40 Sonwil Industrial Park Buffalo, NY 14225	(716) 683–6831 *(716) 683–6204*
A, B, C, G	Tiger Drylac	9605 Arrow Street, Suite S Rancho Cucamonga, CA 91730	(714) 980–7977 *(714) 980–8461*
B, C, G, K	Valspar Corp.	95 Quaker Oats Drive Jackson, TN 38301	(901) 424–9200 *(901) 424–4746*

A: Acrylic
B: Epoxy
C: Epoxy, Polyester
D: Ethylene Vinyl Acetate
E: Fluorocarbon
F: Nylon
G: Polyester
H: Polyethylene
I: Polypropylene
J: Polyphenylene Sulfide
K: Urethane
L: Vinyl

Source: Adapted from *Product Finishing Directory*, Gardner Publications, Cincinnati, OH, October 1992.

Figure 13.2. Electrostatic powder coating equipment (Courtesy of Nordson, Amherst, Ohio).

Industrial Process Description. Most painted parts used in the fabrication of missiles are painted using the solvent-based wet-spray technique. Paint is applied in spray booths, where overspray is collected in a conventional air-ventilation system equipped with a recirculating water curtain scrubber, which removes the overspray from the exhaust air. The scrubber wastewater containing the overspray is treated in the facility's central wastewater treatment system, where the overspray ultimately becomes part of the treatment plant's wet sludge, which is a hazardous waste. Waste solvents and paint/solvent waste also are generated from mixing operations, cleaning operations, empty containers, and waste materials; all are hazardous wastes and must be disposed of as such.

Alternative painting technologies were evaluated for painting the interior surface of the fuselage section of the Phoenix missile, an area previously left unpainted. Coating this area was desirable for enhancing corrosion protection from the salt environment on aircraft carriers and from sulfur dioxide in jet engine exhaust. The area is small, approximately 9 ft^2, and unit production is nominal, approximately 50

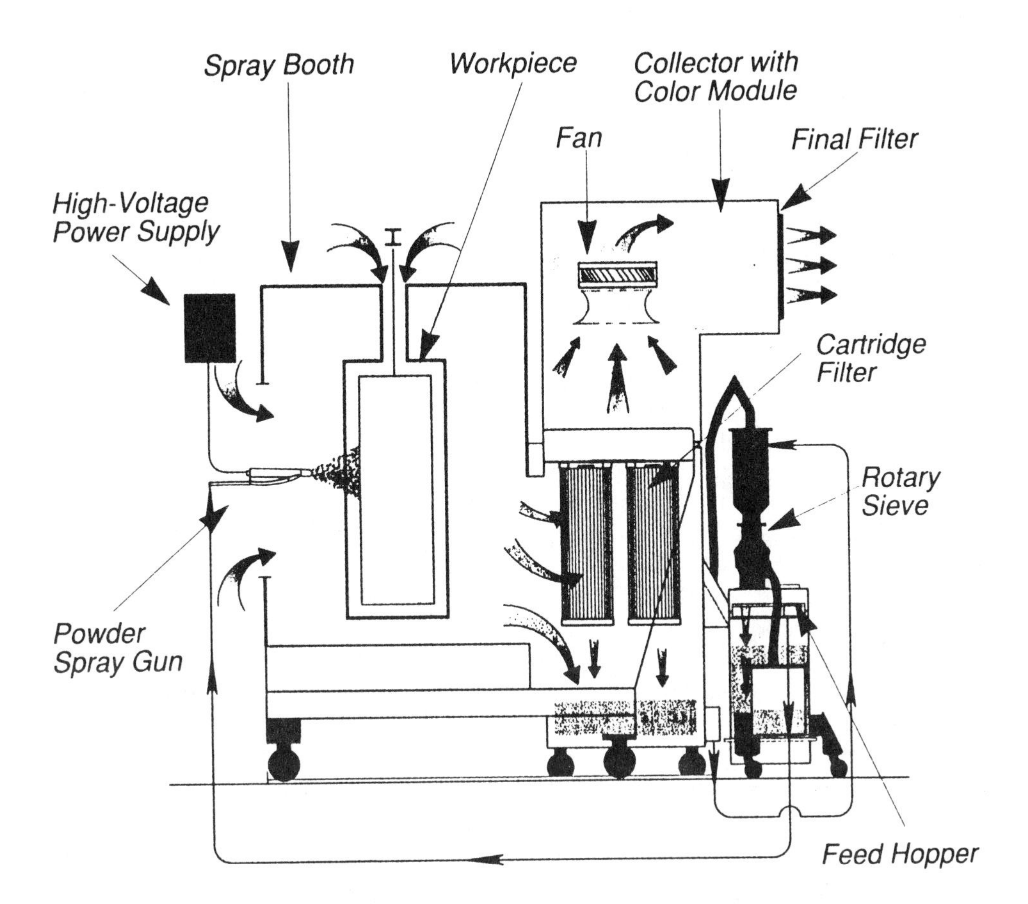

Figure 13.3. Electrostatic powder coating schematic (Courtesy of Nordson, Amherst, Ohio).

Table 13.5. Cost Comparison of Solvent Painting and Powder Painting

Item	Conventional Solvent	Dry Powder
Material	$333,600	$242,400
Labor and cleanup	132,100	75,600
Maintenance	18,000	10,000
Energy	29,100	15,700
Hazardous waste disposal	10,800	1,100
Total annual cost	$523,600	$344,800
Cost per square foot	$ 0.044	$ 0.029

per month. The requirements provided the opportunity to test alternative painting technologies on a developmental scale.

Process Modification Description. The paint system is a polyester and epoxy powder coating that is electrostatically applied and fusion bonded. Paint materials are Type I, thermosetting polyester epoxy powder base coating, or Type II, thermosetting epoxy powder base coating, and Class 1, nonzinc-filled polyester or epoxy powder base coating, or Class 2, zinc-filled polyester or epoxy base coating. The paint system standards include Mil-C-5541 (Chemical Films and Chemical Film Materials for Aluminum and Aluminum Alloys) and Mil-C-5624 (Turbine, Fuel, Aviation, Grades JP-4 and JP-5). Material vendors are Polymer Corporation, Reading, Pennsylvania, for Type II, Classes 1 and 2, and Ferro Corporation, Cleveland, Ohio, for Type I, Class 1.

In the developmental program, a local vendor is being used to apply the paint. The vendor is using Solids Spray 90XC manual powder-coating equipment manufactured by Volstatic, Inc. The equipment provides consistent coating thickness, even on complex surfaces. This portable unit has a storage drum that holds 45 lb of powder. The powder is fluidized and delivered through a venturi gun applicator. Total air consumption is minimal, at 6 standard cubic feet per minute (scfm), with good delivery rates for dry powder of up to 1 lb/min. Constant or variable voltage control provides the electrostatic charge to the powdered particles, which electrostatically bind to the surface being coated. The coated part is fusion-cured in conventional ovens. For this specific application, curing temperatures are between 325 and 375°F, as compared to the 180°F required by conventional solvent-based paint.

Comparison of Electrostatic Dry-Powder Painting versus Conventional Painting. Technical and economic advantages of the electrostatic dry-powder painting process over conventional solvent-based painting are (1) a one-third reduction in curing time, saving both energy and labor and (2) a reduction in the number of coats per unit from two to one, saving material cost and labor. The material and labor cost savings are estimated to be $1.05 per square foot of coated surface. The cost for the electrostatic painting equipment is relatively low; the unit used in the developmental program costs approximately $3,500.

Implementation of the electrostatic dry-powder paint system requires minimal facility changes. At the Hughes Tucson plant, the portable powder-coating equipment is used in existing conventional wet-spray booths. Estimated personnel training time is 2 weeks. If the use of dry-powder painting is expanded, the water scrubber system used in wet spraying booths could be replaced with a dry-powder overspray collector, which would eliminate the wastewater from these scrubbers.

Hazardous waste production is minimized using dry-powder painting. The coating is dry; therefore, the empty powder containers are free of residual material and can be disposed of as normal refuse. In addition, because the material is dry, solvent use for cleanup is much reduced, and solvent use for mixing paint formulas is eliminated. The number of paint types needed also may be reduced because dry-powder paint can be used for multiple applications; thus, waste from partially used containers and stored material with limited shelf-life will be reduced. Using a dry-powder overspray collector will minimize the volume of hazardous waste generated by overspray and eliminate the wastewater that would need treatment.

The process modification was successful because it improved the production rate and quality, decreased staff requirements, and, consequently, decreased costs. As in previously discussed case studies, an improvement in production had been the primary motivation for implementing the process modification. The subsequent reduction in hazardous waste generation became a secondary benefit.

Fluidized-Bed Powder Coating

Fluidized-bed powder coating typically is used to apply relatively thick coatings (10 to 60 mils) to small objects.[7] In the fluidized-bed method, a dense cloud is created by passing air through a powder container to create a suspension of powder that behaves like a fluid. The part to be coated is preheated and immersed in the fluidized powder, and the powder fuses to the part. The coated part is then cured in a conventional oven.

Example. At Norfolk NSY in Virginia, the repair shop was constantly replacing the heavy-duty springs used for the doors of assault landing craft. While in service, these springs were exposed to salt water and beach sand. The combination of corrosive and abrasive environments was too much for conventional paint coatings. The paint shop foreman jury-rigged a fluidized-bed powder-coating unit out of a 55-gal drum and a blower. The springs were heated in a shop oven, then immersed in a fluidized bed of powdered nylon to coat. The resulting finish was of such a high quality that the foreman never saw powder-coated springs returned to the shop for recoating.

Plasma-Spray Powder Coating

Plasma-spray powder coating is relatively new, and applications in industrial, transportation, and marine environments are still being identified. Dry thermoplastic powder is fed into an extremely hot (5,000 to 15,000°F) gas stream, where the hot gas melts the plastic and forms a plasma of gas and plastic. The residence time of the powder in the gas is kept very short to prevent material decomposition. The plasma stream is sprayed onto the prepared substrate where a dense, pore-free coating forms as the paint material condenses.

The advantage of this system over other dry-powder techniques is that the coating is applied and cured in one step, eliminating the need for subsequent heat treatments. Since the substrate surface temperature does not exceed 185°F, this coating system can be used on substrates that are heat sensitive. For example, tests have shown that 7075-T78 aluminum alloy was not affected when painted by the plasma technique, but a 10% loss of tensile strength occurred with a curing temperature of 255°F.[8] The plasma technique also can be used for items too large to be cured in conventional ovens. However, personnel protection would be required because of the high temperatures of the spray.

WATERBORNE COATINGS

Waterborne coatings are used extensively in industry and on a limited basis by the military. In waterborne coatings (as the term suggests), the carrier is a water-based solvent rather than an organic solvent.

Fewer hazardous pollutants are generated when using waterborne paints than when using solvent-based paints. The most significant decrease is in VOC emissions, which are almost eliminated in waterborne painting because the volume of solvents used is reduced and the solvents used comply with air pollution control regulations. In addition, no solvents are needed for paint thinning, and the use of solvents for cleanup is greatly reduced. Wastewaters generated from waterborne painting contain fewer toxic organics because of the limited solvents in the paint.

In industry, waterborne paints are normally used where surfaces require only moderate protection and where decorative requirements are most important. Waterbornes are extensively used for decorative or protective coatings on metallic surfaces, as well as for nonmetallic surfaces such as hardboard, wood cabinetry, and plastics.[9]

There are several key disadvantages of waterborne paints. First, the surface must be completely free of oil-type films or the paint will not adhere well. Second, waterborne coatings require longer drying times or even oven drying in cold or humid weather; this requirement may result in significant expense to outfit a facility for waterborne paints.[10] In addition, automotive industry studies indicate that the transfer efficiency of water-based paints is significantly less than solvent-based paints unless sophisticated application techniques are used.

Positive results with water-based paints have been achieved in applications where conditions were acceptable.

Case Study 13.2: Use of Waterborne Primer on Aircraft

At the NADEP in Pensacola, waterborne primers are being tested with the goal of using them in place of existing solvent-based primers. The waterborne primer selected is a water-reducible, amine-cured, epoxy primer manufactured by Deft Chemical Industry, Inc. This waterborne paint contains some solvent, but less than solvent-based primers. The paint's volatile fraction contains approximately 80% water and 20% solvent. Use of waterborne primer not only reduces solvent emissions and wastewater discharge, but also allows effective cleanup with soap and hot water.

Deft water-reducible coatings are supplied in two components—a pigmented resin solution containing corrosion inhibitors and a clean, unpigmented curing agent solution. The two components, which are packaged in a 1-gal kit, must be carefully mixed with 3 gal of deionized or distilled water. The function of the water is solely to control the paint's viscosity during application. After the paint is mixed, the material is catalyzed and is ready for application. The catalyzed coating must be used immediately, since it has a pot life of only six hours. After mixing, the primer can be applied using conventional spray guns. After the primer is applied, the water evaporates, leaving the nonvolatile coating and some of the unevaporated solvent. At this stage of drying, the film is similar to a high-solids coating without water. The solvent then evaporates and leaves the low molecular weight-pigmented material. Reportedly, the final film is physically and chemically identical to an analogous film deposited by a solvent-based coating.

Painters at Pensacola initially tried to spray paint entire H-53 helicopters with water-based primers, relying on solvent-based primers for touch-up work only. However, because of the presence of ingrained oil, the primer frequently would not dry properly or adhere adequately to the porous surface, thus making use of water-based primers infeasible for these helicopter surfaces. At first, Pensacola personnel tried to clean helicopter surfaces with Freon and alcohol, but the oils remained entrapped on the surfaces. Because approximately 50% of all painting performed at the NADEP is the overspray of whole aircraft, water-based primers only have the potential to partially replace solvent-based primers. Use of the water-based primer on parts, however, proved to be effective.

In the past, approximately 20% of parts painted with solvent-based primers were rejected and had to be repainted. This rejection rate has been reduced to 2% with the new waterborne primer. However, there are also disadvantages to the waterborne primers. They are slower to dry than solvent-based paints, some painting supervisors believe that these paints do not provide the same overall corrosion protection, and the paint's inability to adhere to oily surfaces is well documented. Nevertheless, personnel at Pensacola NADEP have found that in most of their applications, water-based primers are superior overall because they are easy to apply, they decrease overspray, they lower the rejection rate, and they make cleanup easier.

Case Study 13.3: Lockheed's Experiences with Water-Based Paints

In 1960, a paint line was installed at Lockheed for coating aircraft parts to protect them from scratching and corroding during aircraft assembly. From 1960 to 1983, Fabrifilm, a solvent-based coating, was used to protect aircraft surfaces. Water-based coatings were tested in 1983 and are currently being used along with Fabrifilm coatings. However, water-based coatings are used only to provide in-house protection of aircraft surfaces during assembly and are removed before final aircraft painting.

Lockheed also tested water-based primers to determine if they could replace solvent-based primers. The company was hesitant to make the change, believing that the useful life of water-based primers was shorter than that of solvent-based primers. Whereas water-based primers meet military specifications for a useful life of 500 hours, solvent-based primers can last up to 2,500 hours. Therefore, Lockheed reportedly does not intend to make the change on a permanent basis, regardless of the quantities of hazardous waste produced, unless the performance of the water-based primer can equal or exceed the performance of solvent-based primers. In addition, Lockheed personnel expressed the belief that solvent-based paints are lighter for the same thickness than water-based paints, less expensive, easier to apply, easier to remove for inspection, and more durable. Solvent-based primers also dry more rapidly than water-based primers. If Lockheed were to make the change, ovens would have to be installed to hasten the drying of painted aircraft parts.

HIGH-SOLIDS COATINGS

High-solids solvent-based coatings, which are similar in composition to conventional solvent-based coatings, are becoming more widely used in industrial applications. These coatings contain about 50% to 95% solids and, compared to solvent-based coatings, use lower molecular weight paint resins with highly reactive sites to aid in coating polymerization. The finished coat is comparable to typical solvent-based coatings.

The general opinion of industry experts is that high-solids solvent-based coatings will become the "standard" to replace regular solvent-based coatings. The major advantage will be the capability to comply with the more stringent VOC emission limits while using the same basic paints, equipment, and application techniques.[11]

High-solids coatings require special spray equipment for application because of their high viscosity. Because less solvent is used, less is available to wet metallic surfaces that are contaminated with oils; therefore, surface preparation for removal of oils is more critical. In addition, spray application can be wasteful because there is a tendency to apply too much coating to achieve a "wet" appearance similar to that obtained with normal solvent coatings.

Improving the efficiency of paint application reduces waste generation in two ways. Reduced use of paint results in less solvent to evaporate from the final paint film. Reduced overspray means less paint sludge to be removed from a water-wall scrubber or less paint to be eventually stripped off the walls or floor of a paint booth. Techniques for improving painting efficiency are discussed below.

ELECTROSTATIC PAINTING

In theory, wet electrostatic painting is similar to depositing dry-powder coatings by electrostatic attraction. It differs in that some solvent is used as a thinner (the solvent content is lower, however, than in conventional spray painting). Overspray is minimized, if not eliminated, resulting in hazardous waste reduction. Wet electrostatic painting is widely used for painting aircraft parts and other small, complex, nonaluminum metallic articles. Concern exists, however, about the potential safety hazard of imparting high voltage to an aircraft that may still contain fuel vapors.

A variation of electrostatic painting is the spinning disc or bell type (Figure 13.4). These high-speed discs atomize the paint finer than air-atomization and direct more paint to the target. They are particularly efficient at applying difficult-to-disperse high-solids paints.

Electrocoating

Electrocoating is similar to metal plating and is commonly used in automotive body coating. In this process, metallic parts or other electrically conductive parts are dipped into a solution that contains specially formulated ionized paint. The action of an electric current induces the paint ions to deposit on the part. The paint formulations are a special class of waterborne nonvolatile organic compounds. Hazardous waste production is minimal, and VOC air emissions are almost eliminated.

One constraint inherent in this process is the requirement for dip tanks, which limits the size of items that can be painted. A more important disadvantage is that the system can be used to apply only one coat (either a prime coat or a single finish coat) because the electrocoated surface prevents further electrodeposition.[8]

AIRLESS AND AIR-ASSISTED SPRAY PAINTING

Airless spray painting can be used for most applications where air spray is used. Airless sprayers have 20% to 30% better transfer efficiency than air spray, resulting in less overspray waste and lower VOC emissions (Figure 13.5). The disadvantages

Figure 13.4. Electrostatic spinning disc paint atomizer (Photo courtesy of DeVilbiss Company, Toledo, Ohio).

are a coarser finish and a higher paint flow rate, requiring a higher degree of control of paint application.

Air-assisted spray painting systems combine the best characteristics of both air and airless spraying.[12] They use an airless fluid-spray tip to atomize the coating into a fan pattern at moderate pressures; a second low-pressure air stream is injected just after the nozzle to improve atomization and spray pattern. This new system is reported to provide finer control, less overspray, and a higher transfer efficiency than airless spraying.

AUTOMATED PAINTING

Automated conveyor systems can be designed and operated to maximize painting efficiency while minimizing the generation of hazardous waste.

Example. In 1980, the Lockheed-Georgia Company installed a modern conveyor system in an automated paint and process line used for painting small aircraft parts at its C-5 assembly plant in Georgia. Painters spray parts as they move along the conveyor system. The conveyor allows parts to be plated, painted twice, and oven-cured, if necessary, all without being touched. The system has improved product quality because impurities from handling are eliminated; in addition, operators can concentrate on improving painting technique, thus reducing overspray and excess paint use. Overall, the system is more efficient and produces less hazardous waste from overspray and cleanup.

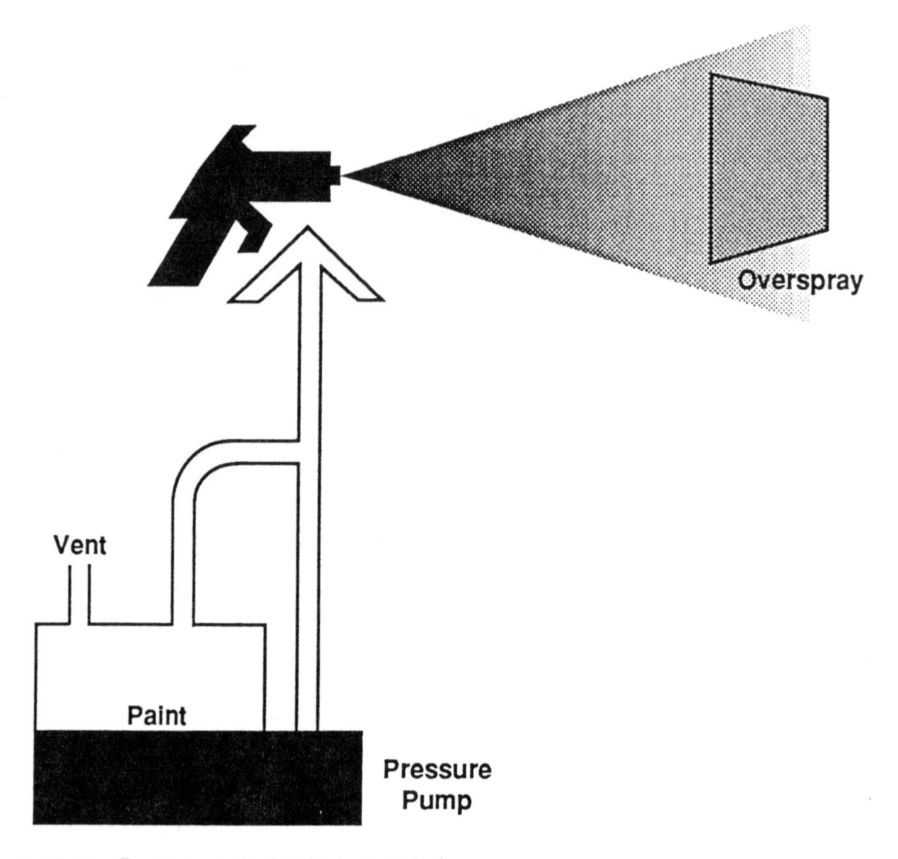

Figure 13.5. Pressure atomized spray painting.

In conjunction with the new conveyor system, Lockheed has been retraining operators and inspectors to help them determine the proper paint thickness. The primary reason for the training program is to reduce aircraft weight and paint material cost. If the training program is successful, the quantities of waste solvents and paint sludge also should be reduced substantially.

Robotic control in painting applications produces tremendous potential for reducing hazardous waste. Not only can overspray and spills be reduced, but the higher temperatures required for application of low-solvent and no-solvent formulations can be sustained without human discomfort. Available robotic systems are best suited for industries that mass produce products or components, such as the automotive industry.

Example. At Lockheed-Georgia, a robotic painting system was installed to reduce paint overspray and improve product quality and efficiency. The robot had the capability to paint an 8-ft by 6-ft rectangular area and could be used for both normal spray painting and electrostatic painting. Its use was discontinued, however, because of difficulty in spraying a wide variety of irregularly shaped aircraft parts.

REFERENCES

1. Schmitt, G.F., Jr. "U.S. Air Force Organic Coatings Practices for Aircraft Protection," *Metal Finishing*, November 1981.
2. Brewer, G.E.F. "Compliance Solvents for Formulation and Thinning of Spray Paints," *Metal Finishing*, January 1984.
3. "Calculations of Painting Wasteloads Associated with Metal Finishing," U.S. Environmental Protection Agency, June 1980.
4. "Update: Regulations on Coating Operations for Heavy Duty Industrial Maintenance Painting," *Journal of Protective Coatings & Linings*, February 1991.
5. Cole, G.E., Jr. "VOC Emission Reductions and Other Benefits Achieved by Major Powder Coating Operations," Paper 84-38.1, Air Pollution Control Association, June 25, 1984.
6. Bowden, C.C. "Powder Coatings on Aluminum Substrates," *Metal Finishing*, February 1983.
7. Minuti, D.V., and M.J. Devine. "Innovative Application of Materials for Aircraft Wear Prevention," Naval Air Development Center, Warminster, PA.
8. "Electrocoating Today," *Products Finishing,* February 1983.
9. Albers, R. "Waterborne Has Solvent Borne Properties," *Industrial Finishing,* April 1984.
10. Higdon, M. "Hyster's Experiences: Air Dry, Water-Borne Primers," *Products Finishing*, April 1981.
11. Nickerson, R.S. "Applying High-Solids Coatings," *Products Finishing*, November 1981.
12. Sinclair, R. "Air-Assisted Airless Spray Painting," *Products Finishing*, February 1984.

GENERAL REFERENCES

Anderson, C.C. "High Solids Automotive Topcoats," presented at 78th Annual Meeting of Air Pollution Control Association, June 16–21, 1985.

Cole, G.E., Jr. "Powder Coating: 1984," *Products Finishing*, January 1984.

Kikendall, T.R. "Converting from Conventional to Compliance Coating Systems," *Industrial Finishing*, June 1982.

Schmitt, G.F., Jr. "U.S. Air Force Organic Coatings Practices for Aircraft Protection," *Metal Finishing*, November 1981.

Shaffer, P.D. "When to Use Airless Electrostatic Spray," *Products Finishing,* February 1984.

CHAPTER 14

Removal of Paint and Coatings

*Ty Daniel and Thomas Higgins**

DESCRIPTIONS OF PAINT-STRIPPING PROCESSES AND WASTES

Paint stripping is the process of removing paint and paint-type coatings from surfaces, usually as a preparation for inspection, dismantling, reconditioning, or repainting. Traditionally, industry has applied solvents and/or solvent-chemical mixtures to the surface to physically destroy either the paint coating itself or the coating's ability to stick to the surface. When this process is complete, the paint/solvent residue is removed from the surface, usually by pressurized water wash or scraping, or both. In many instances, the solvent stripper must be reapplied to remove multiple coats or paint that is particularly resistant.

The wastes generated in the stripping process are a significant source of pollutants. These wastes include solvent/paint residue, which can be collected separately, and waste wash water, which contains solids and dissolved chemicals from paints and solvents. Collected solvent/paint residue is typically put in drums and transported to a licensed hazardous waste disposal site. Waste wash water requires treatment in an industrial wastewater treatment plant to remove the paint-stripping solvents (usually phenolic or methylene chloride components) and metals contained in the paint or removed from the metal under the coating.

Strip baths are also used for removing paint. In this method, parts to be stripped are immersed in tanks of stripping solvent. After the solvent dissolves the paint, the stripped parts are removed from the tank and washed with water. The stripping baths are replaced periodically, generally once or twice a year. The solvent/paint liquid and sludge from the bath are then disposed of at a hazardous waste disposal site. The wash water is discharged to an industrial waste treatment plant.

The hazardous and toxic characteristics of the wastes generated at stripping facilities vary considerably. Various paints contain different hazardous constituents (e.g., chromium, lead) that affect the degree of hazard and the disposal method used. In addition, because paint-stripper solvents are formulated from many different com-

*Mr. Daniel is a chemical engineer in CH2M HILL's Portland, Oregon, office and may be reached at (503) 235-5000. Dr. Higgins is a vice president in CH2M HILL's Reston, Virginia, office and may be reached at (703) 471-1441.

pounds, the mixture of wastes can vary significantly, affecting the hazardous and toxic characteristics of the waste and the choice of disposal methods.

The concentrated waste from stripping baths and surface scraping contains mostly solvent and paint residues having hazardous characteristics. However, the degree of toxicity of wastewater from washing varies depending on the type of paint and solvent, the amount of solvent, and the volume of wash water used. Table 14.1 presents typical concentration ranges of paint-stripping wastewater.[1]

Another significant source of pollutants is solvent emissions that are discharged into the atmosphere. When solvent is exposed to air, some of the solvent evaporates. To prevent hazardous working conditions, solvent-stripping areas are ventilated with large volumes of fresh air, which reduce harmful levels of solvent vapor. Improved and upgraded facilities incorporate improved cross-draft ventilation. Ventilated air is normally discharged to the outside, where solvent vapors are diluted and dispersed. For additional protection, workers can wear respiratory apparatus that minimize inhalation of organic vapors.

Air emissions generated by solvent stripping are difficult to quantify. Emissions generally are expected to include VOCs found in the solvent itself, mainly methylene chloride and phenol compounds. Little information on the subject is available however, primarily because these emissions have only recently been regulated by the U.S. EPA and by state and local agencies. Until now, the need to quantify these emissions to comply with Occupational Safety and Health Administration (OSHA) requirements, which specifically concern on-the-job safety of workers, has been minimal. The U.S. EPA is developing standards for toxic air emissions, and these limits will affect solvent-stripping operations.

MODIFICATIONS TO CONVENTIONAL PAINT STRIPPING

Several techniques for reducing waste from paint stripping have been demonstrated or are practiced by the military and industry.[1] These techniques are generally nontechnical, labor-intensive methods that reduce the volume of hazardous liquid and wash water.

Example. At Norfolk NADEP, paper is placed on the floor of the paint-stripping hangar to collect the loosened paint and the spent stripper solution. This

Table 14.1. Paint-Stripping Wastewater Characteristics

Parameter	Range
pH, unit	6.2–8.0[a]
Phenols, mg/L	17.7–45.2
Methylene chloride, mg/L	3.8–219.2
Chromium (hexavalent), mg/L	0.10–1.12
Total chromium, mg/L	0.164–1.187
Cadmium, mg/L	0.024–1.09
Lead, (mg/L)	0.001–0.01

[a]Caustic strippers may exceed pH 10.

water-free technique has eliminated the high volume of solvent-laden wastewater normally produced in such a facility. The reduced volume of waste is incinerated.

Example. Reuse of stripping solvent has been investigated at Hill AFB as a means of reducing waste generation. In laboratory testing, filters were used to extract paint particles from collected solvent/paint residues. In theory, the filtered solvent stripper could be reused. Initial tests showed some loss of stripping characteristics, which could probably be remedied by adding makeup chemicals. The major problem of collecting the solvent/paint residue without using either water or significant hand labor was not solved. A full-scale solvent reuse system could save $60,000 per month at Hill AFB if a cost-effective method was found for collecting the solvent/paint residue.

Industry has attempted to reduce wastes by using labor-intensive methods to collect solvent/paint residue in concentrated form, thereby minimizing the volume of hazardous waste and wash water generated.

Example. Pan American Airlines at John F. Kennedy Airport in New York used aluminum troughs taped to the sides of the aircraft to collect waste solvent/paint residue and to convey it directly into 55-gal drums. This technique minimized washwater use, thus decreasing the waste volume. Plastic troughs and sheets beneath aircraft have also been used to collect stripping waste and to minimize the use of washwater for cleanup. Another practice is the use of manual squeegees to remove the maximum amount of stripper before washing.

Two other methods of reducing onsite generation of waste are the use of paint-stripping contractors (who merely move the problem from one place to another) and the use of easily removed decals of paint instead.

The current practices to reduce waste production in conventional operations for chemical paint stripping generally have limited benefits, although in certain applications the practices may prove effective. The major problems associated with chemical strippers have not, for the most part, been solved. Solvent air emissions remain at the same levels because solvent use is basically the same. Lower volumes of more concentrated solvent wastes are produced, but the total amount is still considerable and is probably even more hazardous to handle because it is more highly concentrated. In addition, washwater is still required for final surface washing, and although the concentrations of contaminants are lower, the wastewater still must be treated to meet local discharge limits.

ALTERNATIVES TO CONVENTIONAL PAINT-STRIPPING TECHNIQUES

Several alternatives to traditional techniques are available for chemical paint stripping. These alternatives, which reduce the generation of hazardous waste, require new equipment and facilities. The techniques include plastic-media blasting (PMB) paint stripping, wet-media stripping, laser-paint stripping, flashlamp stripping, dry-ice blasting, water-jet stripping, wheat-starch blasting, biodegradation, cavitating water-jet stripping, salt-bath paint stripping, burn-off systems, and hot-caustic stripping. The more promising improvements in paint stripping are summarized below. Vendors for paint stripping equipment and suppliers are listed in Table 14.2.

Table 14.2. Paint-Stripping Equipment Suppliers

Type	Company	Address	Phone/*Fax*
F	AC Molding Compounds	So. Cherry Street Wallingford, CT 06492	800–523–2262 *203–284–4306*
A	Advanced Curing Systems, Inc.	3701 So. Ashland Avenue Chicago, IL 60609	312–247–3600 *312–247–3069*
B	Air Products & Chemicals, Inc.	7201 Hamilton Boulevard Allentown, PA 18195	215–481–2349 *215–481–2556*
A, D, E	Ajax Electric	60 Tomilson Road Huntington Valley, PA 19006	215–947–8500 *215–947–6757*
C	Aqua-Tool Inc.	Rt 130 N/Cranbury S. River Road Cranbury, NJ 08512	609–655–4443 *609–655–2012*
A	Armature Coil Equipment, Inc.	4725 Manufacturing Road Cleveland, OH 44135	800–255–1241 *216–267–4361*
C, F	Automated Blasting Systems, Inc.	46 Schweir Road South Windsor, CT 06074	203–528–5525 *203–528–8368*
A	Bayco	4350 Pell Drive Sacramento, CA 95838	916–929–8061 *916–929–5041*
A	Belco Industries, Inc.	115 E. Main Street Belding, MI 48809	616–794–0410 *616–794–3424*
C	Black Co. Inc.	24075 Hoover Road Warren, MI 48089	313–756–6670 *313–756–7182*
A	Blu-Surf, Inc.	P.O. Box 190 Parma, MI 49469	517–531–3346 *517–531–3589*
A	Brown Engineering	550 S. Monroe Street Seattle, WA 98108	800–426–6384 *206–763–2546*
C	Centri-Spray	39001 Schoolcraft Road Livonia, MI 48150	313–464–6100 *313–464–1845*
F	Clemco Industries Corp.	One Cable Car Dr. Washington, MO 63090	800–239–0300 *314–239–0788*
F	Custom Metal Fabricators, Inc./Vacublast	P.O. Box 286 Herington, KS 67449	800–255–7910 *913–258–2584*
A	Dukes Industries, Inc.	2237 Waterworks Drive Toledo, OH 43609	800–543–3853 *419–389–0304*
F	DuPont Co.	1007 Market Street Wilmington, DE 19898	800–572–1568 *800–872–3448*
C	Durr Industries	40600 Plymouth Road Plymouth, MI 48170	313–459–6800 *313–459–5837*
A	F Systems, Inc.	400 Industrial Drive Lynn, IN 47355	317–874–2531 *317–874–1199*

Table 14.2. Paint-Stripping Equipment Suppliers (Continued)

Type	Company	Address	Phone/*Fax*
A	Finishing Group	P.O. Box 7567 Ocala, FL 32672	904–821–1005 *904–821–0833*
A	GLA Finishing Systems	38830 Taylor Parkway North Ridgeville, OH	800–551–3097 *216–327–4498*
A	Greene Inc.	710 Myrtle Avenue Boontown, NJ 07005	201–335–1630 *201–335–8118*
E	Hones, Inc.	607 Albany Avenue North Amityville, NY 11701	516–842–8886 *516–842–9300*
C	Hydro-Blast	501 S. E. Columbia Shores Boulevard Vancouver, WA 98661	800–332–1590 *206–696–5948*
F	Inventive Machine Corp.	P.O. Box 369 Bolivar, OH	800–325–1074 *216–874–2933*
F	Jet Wheelblast Equipment	401 Miles Drive Adrian, MI 49221	517–263–0502 *517–263–0038*
C	K.E.W. Cleaning Systems	130 E. St. Charles Road, Suite B Carol Stream, IL 60188	800–942–1690 *708–690–2789*
F	Kleiber & Schultz	2017 New Highway Farmingdale, NY 11735	516–293–6688 *516–293–1856*
A	Koch Sons, Inc.	10 S. Eleventh Street Evansville, IN 47744	812–465–9600 *812–465–9724*
E	Kolene Corp.	12890 Westwood Avenue Detroit, MI 48223	800–521–4182 *313–273–5207*
F	Larry Hess & Associates	P.O. Box 1554 Salisbury, NC 28145	800–535–2612 *704–638–9311*
F	Maxi-Blast, Inc.	630 E. Bronson South Bend, IN 46601	800–535–3874 *219–234–0792*
F	Metal Dimensions	4720 District Boulevard Vernon, CA 90058	213–582–1955 *213–582–8138*
A	Moco Thermal Industries	First Oven Place Romulus, MI 48174	313–728–6800 *313–728–1927*
A	Mote Ltd.	214 Newkirk Road Richmond Hill Ontario, Canada L4C 3G7	416–884–5250 *416–737–5188*
F	MPC Industries	638 Maryvale Pike, S.W. Knoxville, TN 37920	615–573–5411 *615–693–6007*
C	NLB Corp.	29830 Beck Road Wixom, MI 48393	313–624–5555 *313–624–0908*
F	Pangborn Corp.	P.O. Box 380 Hagerstown, MD 21741	800–638–3046 *301–739–3500*

Table 14.2. Paint-Stripping Equipment Suppliers (Continued)

Type	Company	Address	Phone/*Fax*
A	Pollution Control Products	2677 Freewood Drive Dallas, TX 75220	214–358–1539 *214–358–3379*
F	Potters Industries	20 Waterville Boulevard Parsippany, NJ 07054	201–299–2900 *201–335–9350*
D	Procedyne Corp.	11 Industrial Drive New Brunswick, NJ 08901	908–249–8347 *908–249–7220*
C, F	Progressive Technologies	4695 Danvers, S.E. Grand Rapids, MI 49512	616–957–0871 *616–957–3484*
C	Serfilco, Ltd.	1777 Shermer Road Northbrook, IL 60062	800–323–5431 *708–559–1995*
D	Service Tectonics, Inc.	2827 Treat Street Adrian, MI 49221	517–263–0758 *517–263–4145*
C	Spraying Systems, Co.	North Ave at Schmale Road Wheaton, IL 60189	708–665–5000 *708–260–0842*
F	Stripping Technologies, Inc.	2949 E. Elvira Road Tucson, AZ 85706	800–999–0501 *602–741–9200*
C	Tally Cleaning Systems	P.O. Box 1305 Attleboro Falls, MA 02763	508–695–1007 *508–695–6335*
D	Techne Inc.	3700 Brunswick Parkway Princeton, NJ 08540	800–225–9243 *609–987–8177*
A	Therma-Tron-X Inc.	1155 S. Neenah Avenue Sturgeon Bay, WI 54235	414–743–6568 *414–743–5486*
A	Torrid Ovens, Inc.	P.O. Box A Buffalo, NY 14225	716–852–0882 *716–678–6258*
E	Upton Industries, Inc.	30345 Groesbeck Highway Roseville, MI 48066	800–541–1204 *313–771–8970*
F	US Technology Corp.	79 Connecticut Mills Avenue Danielson, CT 06239	800–634–9185 *203–779–1403*
A	Viking Spray Booth	4350 Pell Drive Sacramento, CA 95838	916–924–8061 *916–924–5041*
C	Wagner Spray Tech Corp.	1770 Fernbrook Lane Minneapolis, MN 55447	800–292–4637 *612–553–7288*

Type Code:
A: Burn-off
B: Cryogenic
C: High-Pressure Water Spray
D: Hot-Fluidized Bed
E: Molten Salt Bath
F: Plastic-Media Blast

Alternative Blasting Media

Conventional sand blasting, abrasive-media blasting, and glass-bead blasting have been extensively used for decades to remove paint and rust from metal surfaces. These removal techniques cannot be used in many applications, however, because the abrasive media can damage aluminum or fiberglass surfaces and small or delicate steel parts. Silica dust generated from sand and glass blasting can also cause silicosis, a respiratory ailment among workers. Softer dry media (e.g., walnut shells, and rice hulls) have been used with limited success for various paint-stripping operations where sand and glass could not be used. The "soft media" blasting method has received considerable attention for both military and industrial applications. These natural, soft materials are reasonably effective but are difficult, if not impossible, to recycle and, during storage, are susceptible to bacterial growth, which is reported to cause respiratory infections among workers.

Research into the use of substitutes for abrasive blasting media by the Navy resulted in a change in the type of abrasive blasting media used for removing paint from ships.

Example. The United States Navy identified the need for an abrasive blasting media that would be less harmful to health and the environment than traditional blasting media. This need has arisen as a result of concerns about the use of silica-based abrasive media that cause silicosis among workers, and about environmental contamination from airborne particulate matter, noise, and trace metals. Because of these concerns, the Navy has adopted a military specification that limits total metals and soluble metals content of blasting media. The specification addresses these concerns while maintaining performance standards for the media. This specification motivated the Navy and manufacturers of blast media to work together to produce a new, safer media.

Plastic-Media Blasting

A new media has been developed and is manufactured in commercial quantities for blast-stripping painted surfaces without damaging the undersurface. The new material has many advantages over other materials, including engineered abrasive characteristics; it is recyclable, durable, and nonhazardous. The material is constructed of soft plastic formed into rough-edged granular media. Old paint is dislodged using the new media and conventional sand-blasting equipment. The resulting dry waste of pulverized paint and plastic media has significantly less volume and is more readily disposed of than the wastewater produced in solvent-based paint stripping.

United States Plastics and Chemical Company was the first manufacturer of the plastic media. The plastic is available in three material hardness grades (Polyextra, Polyplus, and Type 3) and six grain-size sieve distributions (12–16, 16–20, 20–30, 30–40, 40–60, and 60–80). The plastic is used in a wide variety of applications for stripping coatings from substrate materials. Conventional sand-blasting equipment can be used, although hoses and nozzles are usually modified to account for the plastic's lower abrasion of equipment (Figure 14.1).

Paint stripping by PMB is the most promising alternative to conventional solvent stripping. It has been successfully demonstrated for aircraft renovation at Hill AFB,[2,3] Pensacola NADEP,[4,5] Republic Airlines, United Airlines, and Boeing Ver-

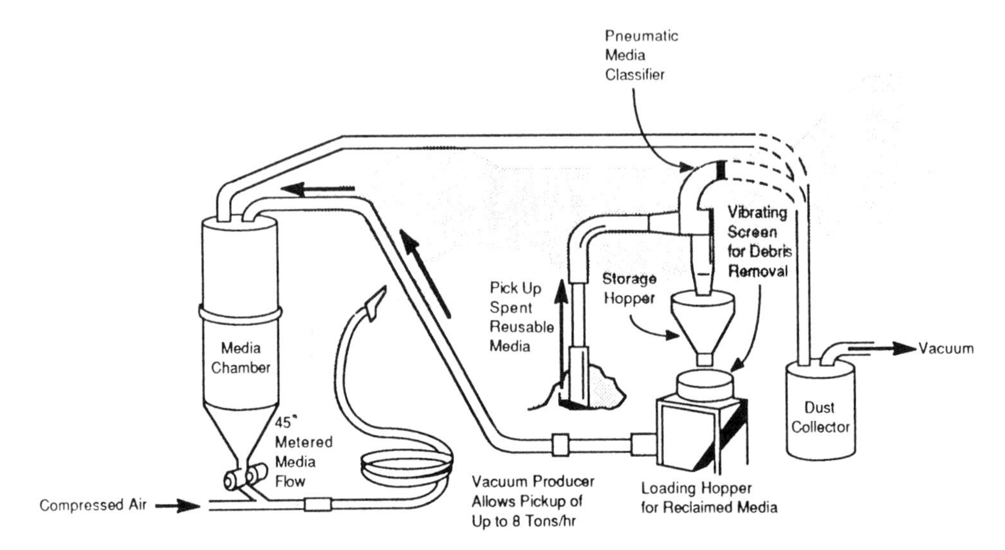

Figure 14.1. Schematic of a self-contained plastic-media blasting paint-stripping unit.

tol. Many other industrial facilities are considering the process because of its highly successful demonstrations and testing as well as its cost-effectiveness.[5-7]

By carefully controlling media size and process conditions, the plastic media can be separated from the loosened paint particles and recycled. PMB completely eliminates the generation of wet hazardous waste (solvents and paint sludge in water); however, PMB does produce a small volume of dry waste, which can be classified as hazardous if its metals content exceeds regulatory limits.

Through a lease program, PMB that is no longer effective as stripping media can be recycled. Composite Leasing Corporation of Minocqua, Wisconsin, offers a program that eliminates the hazardous waste generated by a PMB user. This program also has the advantage of reducing the cost of final disposal of the PMB. Under this lease program, the PMB is leased to the user. When the useful life of the PMB is expended, the material is returned to Composite Leasing. The attrited plastic material is then pyrolized to yield methylmethacrylate (MMA) monomer. From MMA, acrylic sheets are produced. The entire PMB waste stream is therefore effectively recycled without generating hazardous waste.

The two key parameters for successful use of PMB are hardness and reusability. First, the plastic media must be harder than the paint, but softer than the surface underneath the paint coat. Second, the media must be durable enough to be reused repeatedly to minimize the amount of residue that must be disposed of.

With some very hard paints (such as epoxy and urethane paints), presoftening with a solvent (such as methylene chloride) has been used before PMB stripping. However, recent test data provided by the media supplier indicate that, with modifications to media selection and application methods, even very hard paint can be successfully removed without presoftening.

The PMB technique has been effective in stripping and removing a variety of coatings from a number of substrate surfaces. However, extreme care must be exercised when removing coatings from composite surfaces, thin-skinned aluminum, and other fragile materials. In some instances, the number of times that an aircraft substrate can be stripped using PMB is being limited. Boeing has specified that its aircraft may only be stripped twice using PMB over the service life of the aircraft.

In particular, composite fibers have sometimes unravelled when composite surfaces that did not have a resin-rich surface were blasted. In some instances, using excessive pressure or holding the nozzle too close has resulted in surface damage. Even though the PMB process is relatively simple, considerations such as these make it imperative that operators receive adequate training.

The blasting action of the PMB technique helps to stress-relieve surfaces when removing paint from titanium, stainless steel, alclad, and anodized aluminum. Alclad aluminum surfaces have a sandblasted appearance after blasting because the aluminum cladding is softer than the plastic media. This soft cladding is shifted, but not removed; in fact, after blasting its surface is much better for repainting.

Many additional applications will be developed as testing continues. Table 14.3 lists some of the coatings and substrates successfully stripped with plastic media.[8]

The potential savings associated with PMB stripping are substantial. The potential savings for the Department of Defense in labor, chemicals, and waste treatment or disposal amount to more than $100 million annually.[6] Savings resulting from decreased energy costs and costs for compliance with future environmental regulations coupled with increased productivity are likely to be equally significant. The preliminary cost estimate presented in Table 14.4 illustrates the potential savings.

The plastic media paint-removal process is so simple and efficient that it lends itself to a wide variety of uses. The most notable feature of the process, however, is its elimination of solvent wastes. The only waste from this system is dry fine plastic dust and paint particles that contain trace heavy metals from the paint. This waste is easily contained within sealed drums and can be safely transported for disposal or storage. No liquid waste is generated. Because the air system can be self-contained and dust removal facilities provided, air pollution can be eliminated. The plastic media are recycled and reused with little degradation. Energy, materials, labor, and product efficiency all cost significantly less than in conventional operations for solvent stripping. However, to protect workers from occupational hazards (dust and noise), hearing protection, goggles or masks, and filtered air for breathing must be provided.

Table 14.3. Applications for Paint Stripping by PMB

Coatings	Substrates	Applications
Polyurethane	Aluminum	Aircraft fuselage
Epoxy polyamide	Alclad aluminum	Components
Acrylic lacquer	Anodized aluminum	Ship bilges
Enamel	Steel	Vehicle bodies
Fluorocarbons	Magnesium	Boat hulls
Metallic spray	Anodized magnesium	Engine components
Koropon primer	Titanium	Truck wheels
Rain erosion	Carbon graphite	Propeller blades
Fuel sealants	Fiberglass (except	Molds
Structural adhesive	Radomes and Kevlar)	Heat exchangers
Corrosion buildup	Honeycomb	Alloy fuel tanks
Lubricants		
Polysulfide sealants		
Carbon buildup		

Table 14.4. Estimated Savings from Adopting PMB at DOD Facilities

Item	Solvent/Chemical Stripping	Plastic Media Stripping
Labor and material		
Work-hours	3,360,000 hr	1,426,000 hr
Solvents/chemicals	7,000,000 gal	0
Wash water	100,000,000 gal	0
Wastes	107,000,000 gal	500,000 lb dry
Operating costs		
Work-hours	$136,516,800	$ 67,698,380
Material supplies	30,960,000	4,440,000
Waste treatment and disposal	8,000,000	1,500,000
Total operating costs	$175,476,800	$ 73,598,380
ANNUAL COST SAVINGS		$101,878,420

Case Study 14.1: PMB Paint Stripping at Hill AFB

Hill AFB in Ogden, Utah, has been the lead facility in developing and implementing PMB. The development of this process modification demonstrates the key elements necessary for successfully implementing a pollution prevention project. The process itself is elegant in its simplicity—conventional sand-blasting equipment was adapted to provide media recovery and classification of the reused plastic. More than 700 aircraft have been stripped since 1985.

Initially, Bob Roberts, the champion of the process, was motivated to replace the existing wet-solvent process because it was environmentally objectionable and an occupational hazard. After extensive testing on aircraft components to demonstrate PMB's effectiveness and safety, personnel at Hill AFB completely stripped an F-4 fighter plane in July 1984. This test demonstrated that in addition to being less hazardous, the process was much less labor-intensive than solvent stripping. The aircraft was completely stripped in 40 work-hours versus 340 work-hours required for wet-solvent stripping. Moreover, greater control with PMB, compared with wet-solvent stripping and sanding, resulted in reduced damage to underlying surfaces.

The environmental and labor advantages of PMB led the Air Force to have a full-sized hangar constructed specifically for PMB (Figure 14.2). The hangar incorporates a live floor-vacuum system to provide ventilation and dust removal and a separation system for media recovery and reuse. The hangar was funded under a Productivity Enhancement Capital Investment fund, which allowed the demonstration facility to be built within one year rather than being built according to the standard military construction schedule. Photographs of the facility are shown as Figures 14.3 through 14.12.

Based on initial prototype testing at Hill AFB, estimated savings from using plastic media compared with using solvent for paint stripping are summarized in Table 14.5. The quantity and cost-saving estimates are based on stripping 100 F-4 aircraft annually. J. D. Christensen of Hill AFB assisted with the update of these estimates, which are based on the following assumptions.

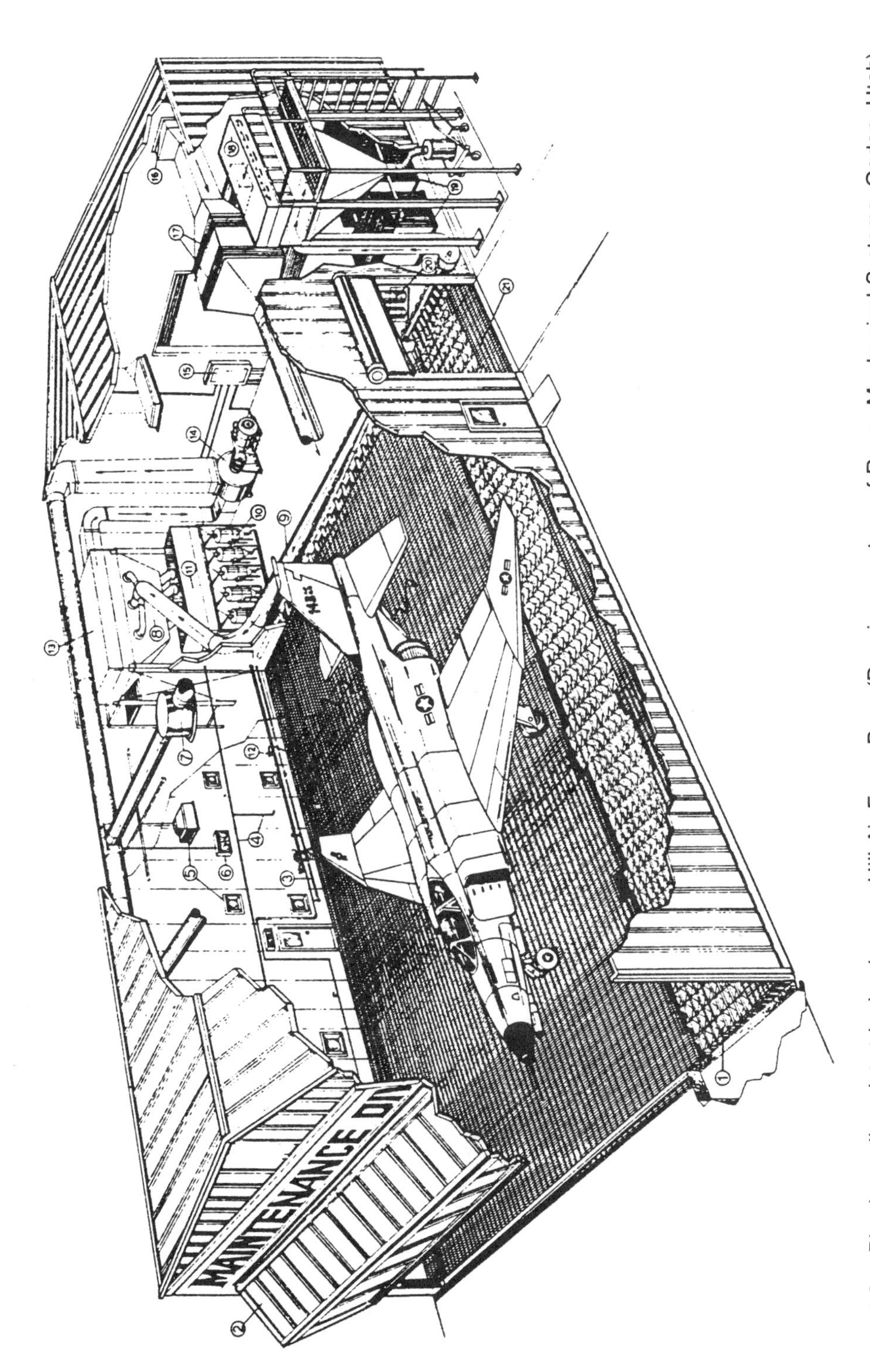

Figure 14.2. Plastic-media paint-stripping hangar, Hill Air Force Base (Drawing courtesy of Royce Mechanical Systems, Ogden, Utah).

Figure 14.3. Plastic-media blasting paint-stripping facility at Hill Air Force Base.

Figure 14.4. Hill Air Force Base plastic-media blasting paint-stripping equipment.

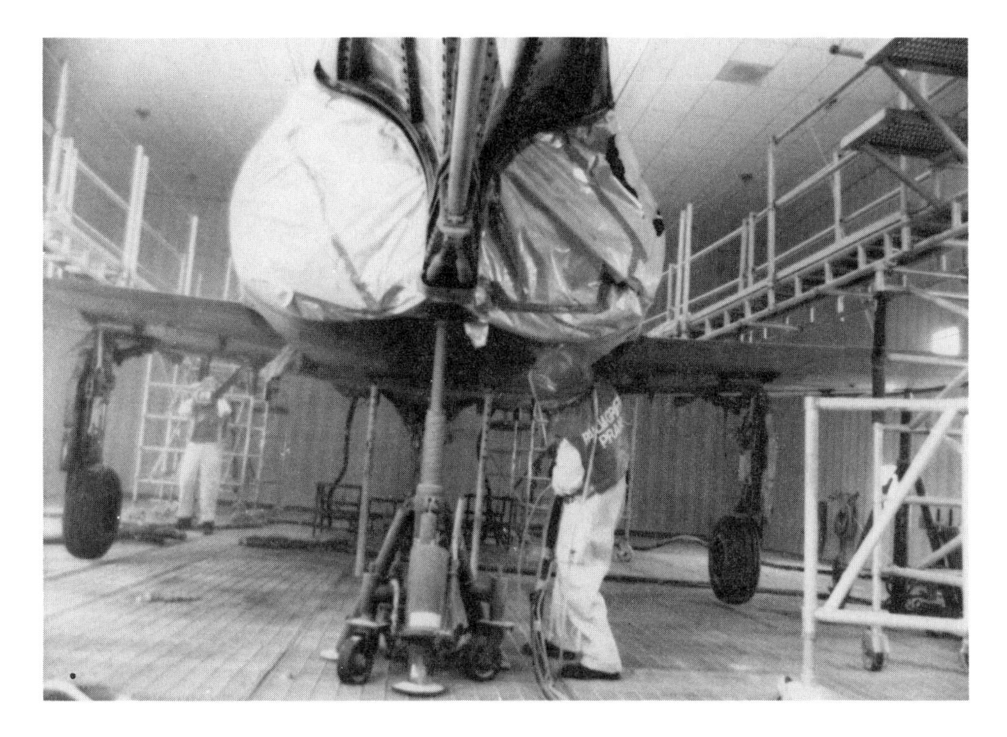

Figure 14.5. Plastic-media blasting paint stripping of F-4 aircraft (floor-level view).

Figure 14.6. Plastic-media blasting paint stripping of F-4 aircraft (view from above).

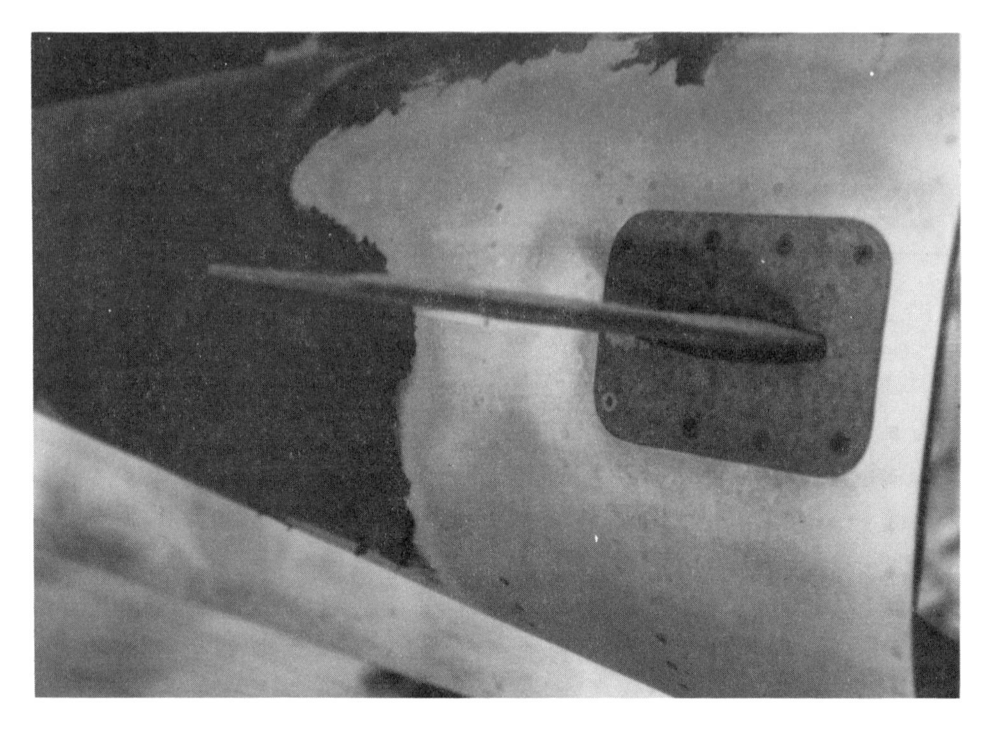

Figure 14.7. Plastic-media blasting paint stripping of anodized aluminum surface.

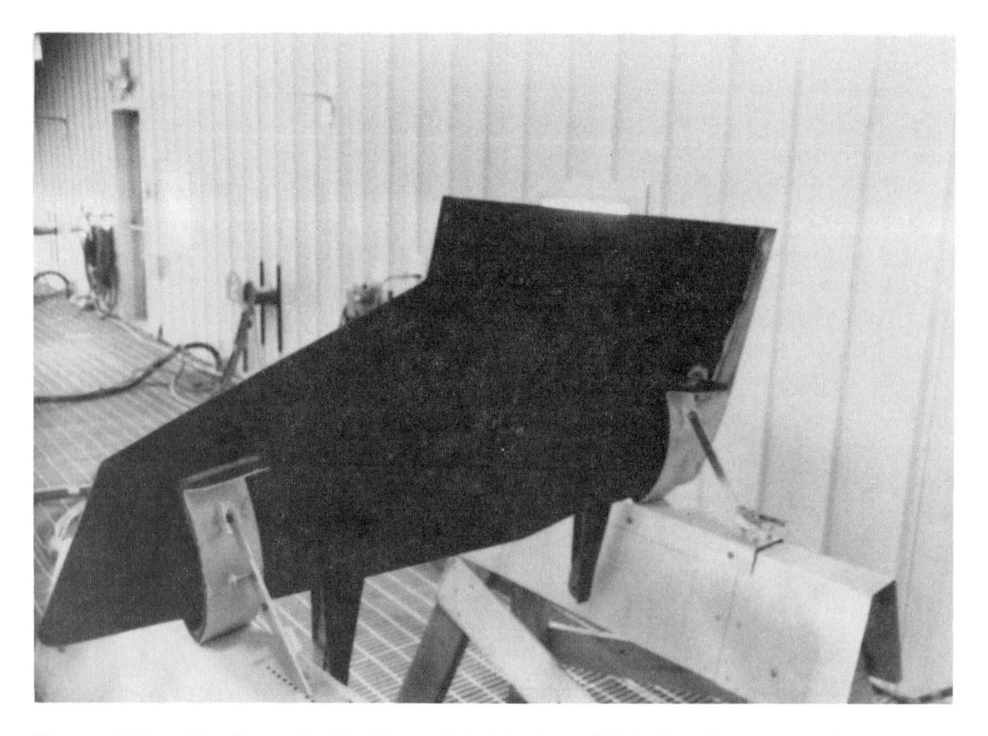

Figure 14.8. Plastic-media blasting paint stripping of F-4 aircraft component.

Figure 14.9. F-4 aircraft component after stripping.

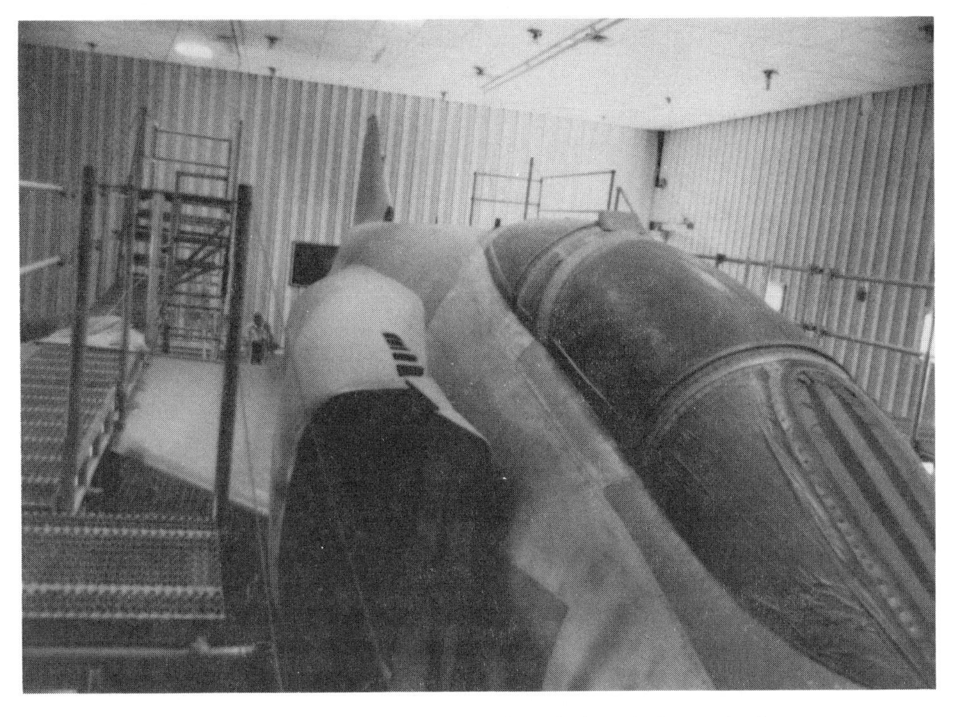

Figure 14.10. Plastic-media blasting paint-stripped aircraft (front view).

Figure 14.11. Plastic-media blasting paint-stripped aircraft (view from above).

Figure 14.12. F-4 aircraft stripped by conventional solvent.

Table 14.5. Estimated Savings from Adopting PMB at Hill AFB

Item	Potential Savings	Annual Cost Savings
Hazardous waste	Generates 1/100 the waste sludge, which requires hazardous waste disposal	$ 207,000
Wastewater pollution	Eliminates generation of 210,000 gal/day of wastewater, which must be treated in on-base waste treatment plant before discharge to the city municipal treatment plant	$ 526,375
Materials	Eliminates the use of chemical solvents and requires minimal use of plastic media to make up for worn-out media	$352,000
Labor	Requires 1/2 the labor	$540,000
Energy	Required 1/10 the energy	$ 224,000
Flow days	Provides increased flow-day use of aircraft	$1,050,000
TOTAL ANNUAL SAVINGS FOR 100 F-4 AIRCRAFT		$2,373,000

Hazardous Waste. Hill AFB's existing wastewater treatment plant produced approximately 3,000 tons per year of sludge (10% solid weight), which is hazardous. The sludge was transported by truck to California, where it was disposed of at a licensed hazardous waste disposal site for a total cost of $200/ton.

Containing solvent and paint residue, the wastewater generated from solvent stripping F-4 aircraft was approximately 35% of the total sludge produced from Hill AFB. Therefore, total sludge contributed by solvent stripping was estimated at 1,050 tons. The only hazardous waste produced by plastic-media stripping is the dry stripped-paint residue, which is estimated at 120 lb per aircraft, and the dry spent plastic media, which amounts to 200 lb per aircraft—a total of only 16 tons of hazardous waste per year. Thus, the estimated annual reduction in hazardous waste products was 1,034 tons, a 98% reduction, and a savings of $207,000.

Wastewater Pollution. The changeover from solvent stripping to PMB stripping at Hill AFB eliminated the need for washwater for the stripping process. Of the 600,000 gal per day of wastewater treated in the on-base industrial waste plant, 35% (210,000 gal/day) was generated by solvent-stripping operations. Approximately 20,000 to 30,000 gal of water was used to wash off solvents paint residue for each stripper application. Several applications of stripper were normally required per aircraft. Water was also used for washing floors and for general area maintenance, further contributing to the wastewater flow.

The annual cost of treatment chemicals at the industrial waste plant was $912,500. A 35% reduction in wastewater should correspond to a proportionate reduction in the use of treatment chemicals, or annual savings of $319,375. Additional annual savings in O&M expenses for the industrial waste plant (i.e., labor and equipment repair and replacement) were estimated at $207,000. Thus, the estimated annual chemical and O&M cost savings total $526,375.

Materials. Solvent stripping of an F-4 aircraft requires 468 gal of chemical stripper at a cost of $11.40/gal and 12 rolls of aluminum masking tape at a cost of $7.30/roll. The total cost per aircraft is $5,422, or $542,200 annually (for 100 aircraft). Plastic media is recycled, but losses occur as a result of abrasion. Media loss is estimated to be 1,100 lb per aircraft. At a cost of $1.73/lb, the cost per aircraft is $1,900, or $190,000 for 100 aircraft. Savings in material costs amount to $3,519 per aircraft or approximately $352,000 in annual savings for 100 aircraft.

Labor. One of the most significant advantages that media stripping has over solvent stripping is its lower labor costs. Solvent stripping an F-4 aircraft requires 341 work-hours per aircraft; at a labor rate of $33.56/hr, the cost per aircraft is $11,444 ($1,144,000 annual cost for 100 aircraft). Plastic-media stripping is estimated as requiring 180 work-hours per aircraft; at the same $33.56/hr labor rate, the cost per aircraft is $6,040 ($604,000 annual cost). Estimated annual labor savings amount to approximately $540,000. Typical prototype production comparisons for F-4 aircraft components and estimated production comparisons for other aircraft and equipment are shown in Table 14.6.

Energy. Two components make up most of the energy used for paint stripping: the energy required to maintain the building interior at the required temperature, and the energy required to operate the stripping equipment's electrical motors. Solvent-stripping operations require significant energy to heat the building because the building interior air must be maintained at 72°F ± 2°F for proper solvent action on the paint, and large fresh air flows are required to ventilate the solvent vapor emissions. These annual heating and ventilating costs are $201,600 (based on 507,000 cubic feet per minute of fresh air, average annual temperature 51°F, 16

Table 14.6. Production Comparisons of PMB vs Solvent Stripping at Hill AFB

Item	Solvent Strip Time	Plastic Media Strip Time
F-4 Component:		
Rudder	3 hr 36 min	15.6 min
Inboard leading edge flap	2 hr 48 min	21.6 min
Spoiler	40 min	14.4 min
Outboard leading edge flap	2 hr 48 min	18.6 min
Aileron	6 hr 28 min	32.4 min
Wingfold	8 hr 45 min	54.1 min
Stabilator	9 hr 49 min	55.2 min
Aircraft and equipment:		
F-4 (prototype)	342 hr	39 hr
F-100 (museum aircraft)	290 hr	25 hr
P-8 pumper (fire truck)	52 hr (sanding)	4 hr
D-50 Pickup (compact)	40 hr (sanding)	1 hr 20 min
1/2 Ton Pickup (full size)	60 hr (sanding)	1 hr 55 min

hr/day and 260 days/year building use, steam cost $5.59/million Btus). Mechanical equipment used in solvent stripping also requires significant amounts of electrical energy to operate.

The annual electrical energy cost is $49,634 (based on 320-hp motor, 16 hr/day at 260 days/year equipment use, $.05/kWh electrical energy cost). Plastic-media-stripping operations require much less electrical energy for heating and electrical equipment operation. Almost no building heating is required because the waste heat generated from the air compressor equipment used in blasting plastic media generates sufficient heat to warm the building during cold weather. And, because no solvents are emitted into the air (which would have required large fresh air flows), no significant amount of energy is required to heat fresh air; heated air can be recirculated throughout the building. In addition, electrical energy requirements for plastic-media-stripping equipment are much lower than those for solvent-stripping equipment. Plastic-media equipment requires only $27,305 in annual electric energy costs (based on 340-hp motor, 8.25 hr/day for 260 days/year, $0.051 kWh electrical energy cost). Therefore, plastic-media-stripping operations are estimated to save $224,000 in annual energy costs.

Flow Days. Plastic-media stripping decreases the overall time needed to renovate aircraft for use; therefore aircraft use is increased. Plastic-media stripping requires 3 flow days, compared to 5 to 9 flow days to complete the solvent-stripping process. On the basis of Air Force cost and planning factors (AFR 173–3), the flow-day efficiency cost savings amount to $1,050,000 annually.

The new plastic-media-stripping facility (also called "blast booth") at Hill AFB, Building 223, is a full-scale facility specifically constructed for F-4 aircraft maintenance. The major components of the facility are shown in Figure 14.2. Royce Mechanical Systems (Ogden, Utah) provided fast-track component design and facility construction. The facility includes a steel prefabricated insulated building (45 feet × 75 feet × 25 feet high) and all process and support mechanical and electrical equipment. The total facility cost, including equipment and labor, was $647,389. The payback period for the facility was approximately 3 months.

The PMB process at Hill AFB is in the process of being modified to incorporate robotics to control the PMB stripping process. A robotics project is now in the startup phase. Two robots have been installed in a blast booth and nine F-16s have been stripped on a nonproduction basis. Upon completion of verification testing, automated stripping of aircraft is expected to begin on a production basis.

In conjunction with the automation of the PMB stripping process, Hill AFB has entered into an agreement with Composite Leasing Corporation. Under this lease program, Hill AFB will return PMB media that has been attrited, through use, to a size that is no longer efficient for stripping process. The PMB will then be recycled into acrylic sheets.

Benzyl Alcohol Paint Stripping

Paint strippers formulated with benzyl alcohol are being tested. These paint strippers currently can be formulated with compounds that are not on the Toxics Reporting Inventory (TRI) list.

N-Methyl Pyrrolidone Paint Stripping

N-methyl pyrrolidone (NMP) has generally been recognized by the industry as an alternative to methylene-chloride-based paint strippers. When used alone, NMP's paint-stripping performance is similar to that of methylene chloride. However, NMP is less toxic and less volatile than methylene chloride. The reduced volatility also contributes to NMP's lower flammability and a reduction in worker exposure. The primary disadvantage of NMP as a paint stripper is its high cost. Generally, blending of NMP with other industrial solvents has resulted in reduced performance as a paint stripper; however, research of NMP use continues.[12]

Laser-Paint Stripping

In the Air Force's testing of lasers for removing paint, research has been directed at the development of a pulsed carbon dioxide (CO_2) laser system. The pulsed laser was chosen because it minimizes energy consumption; CO_2 was selected because its wave length is readily absorbed by paint. Actual pilot-scale tests showed that paint was completely removed from test surfaces. The system, in which a full-scale operational installation would be based on a robot-operated pulsed CO_2 laser, is still in the experimental stage.

Although this alternative appears to be technically feasible, it has several drawbacks. Before it can be considered for widespread use, system reliability, effects of the lasers on aircraft substrate and components (electronics, sensors, etc.), generation of air pollutants, and other factors need to be researched extensively. In addition, careful engineering control must be exercised during laser-paint stripping; for example, special consideration must be given to worker eye protection. It may take several years for this technology to be commercially available. In addition, by one estimate, the initial capital outlay for an aircraft-size facility with an automated laser system would be at least $10 million, which is 10 times as expensive as a comparable PMB facility.

Flashlamp Stripping

Flashlamp stripping is similar to stripping with lasers, but high-energy quartz lamps are used for vaporizing paint. The Air Force is conducting research and development on this process. Unlike laser stripping, flashlamp stripping has been proven not to harm aircraft electronics. However, flashlamps are difficult to use and operators must have extensive training. Unresolved questions about this technique include the potential for damage to various substrates and generation of toxic air pollutants. In Navy tests, this method failed to remove barnacles from the bottom of ships, and the equipment produced loud, annoying "bangs" when operating.

Dry-Ice (CO_2) Blasting

The Lockheed Company investigated dry-ice blasting for removing aircraft paint.[9] Dry ice or carbon dioxide pellets were used as a blasting medium. Dry-ice blasting can remove a broad range of coatings, sealants, and adhesives. The attractiveness of this technology is that dry-ice pellets vaporize after being used and the only waste product is the dry paint chips. Research is needed, however, on potential damage to surfaces, effectiveness of paint removal, intrusion into seams, and operation costs. One disadvantage of this technique is that carbon dioxide from the dry ice displaces oxygen in a room, necessitating the use of a contained air supply when blasting. Fog produced from cool humid air is also a problem.

Other applications of this technology involve combining dry-ice blasting with chemical softeners. Tinker AFB has examined using benzyl alcohol as a paint softener followed by dry-ice blasting. This two-step technique has enhanced paint-removal rates and has been effective for removing coatings that dry-ice blasting alone would not remove. However, because of the use of benzyl alcohol, this process generates VOCs.

Another application of dry-ice blasting is with flashlamp stripping. The equipment is bulky so it is being adapted to robotic operation.

Cryogenic Coating Removal

This method operates on the principle that organic coatings become brittle and tend to debond from substrate metals at low temperature because of differential thermal contraction of the coating and the substrate metals. Small cabinet-size equipment based on cryogenics is commercially available. Liquid nitrogen is sprayed on the coating to lower the surface temperature to −100°F, and plastic-media blasting is used to break off the frozen paint.[10] This system is not suitable for large-scale operations.

High-Pressure Water-Jet Blasting

Both the Air Force and the Navy investigated water-jet blasting for removing paint. To remove paint, this process uses pulsed or continuous water-jet blasting produced by high-pressure pumping (10,000–35,000 psi). As with the other systems discussed, the use of a water jet is technically feasible; however, questions need to be resolved about the system's control and reliability, potential damage to soft surfaces, ability to remove a wide range of coatings, and worker safety. High-pressure water-jet blasting is currently used by the automotive industry to remove paint buildup from the floor gratings of paint booths.

Medium-Pressure Water-Jet Blasting

Medium-pressure water-jet blasting combines the use of alcohol-based paint strippers with water blasting. The chemical-based stripper is applied in a similar manner to methylene chloride-based paint strippers. Once the coating has been allowed to soften, a medium-pressure water jet (less than 10,000 psi) is used to strip the coating from the aircraft. Lufthansa Airlines of Germany is the leader in developing this paint-stripping technology.

Wheat-Starch Blasting

Wheat-starch blasting is similar to PMB paint stripping; instead of plastic, wheat starch crystals are the blasting media. Further research and testing is needed to develop application methods. Using this process, wheat starch crystals can either be recycled as blasting media, or disposed of by incineration or biodegradation. This process requires that dust collection equipment be used to recover the wheat starch crystals. In comparison to PMB, paint-stripping rates for this process are slower.

Example. In 1991, the Beech Aircraft Company began using wheat-starch blasting for production stripping of aircraft components.[13] These components include thin unsupported aluminum and magnesium skins. Also in 1991, Field Aircraft obtained approval to use wheat-starch blasting to remove paint from aircraft frames on a production basis.

Sodium Bicarbonate Stripping

Sodium bicarbonate has been investigated as a potential method for stripping paint from aircraft. A slurry of sodium bicarbonate is applied to the aircraft at medium to high pressures. The process is relatively new and little quantitative information is available about its effects on metal and on composite aircraft substrates.

Biodegradation

Enzymatic biodegradation is being developed as a paint-stripping technique by Technical Research Associates, Inc. under a Small Business Innovative Research grant. This research has indicated that biodegradation may be used alone or with other paint-stripping technologies for removing paint. However, biodegradation may be more suitable for treating paint waste generated by other paint-stripping techniques. In laboratory tests, biodegradation has removed various types of coatings. This process has slower stripping rates than other techniques.

Cavitating Water-Jet Stripping

This process uses the phenomenon known as cavitation, or the release of dissolved gases from a liquid. Using specially designed nozzles, bubbles are carried to the metal surface to be stripped. The bubbles implode on the metal surface, enhancing the erosive forces of the liquid. The resulting waste stream contains paint solids and water. The cavitating water-jet process has been investigated by the U.S. Navy for removing paint from ships, but may have applications for removing paint from other types of equipment.[11]

Salt-Bath Paint Stripping

Equipment is commercially available to strip paints in molten salt baths operating at a temperature of 900°F (Figure 14.13).[10] This method is used by automotive and appliance manufacturers. In this process, items to be stripped (generally steel) are immersed in the molten salt bath (a mixture of sodium hydroxide, sodium or potassium nitrate, sodium chloride, and catalysts), which destroys the paint. This process cannot be used on parts or equipment constructed of aluminum, nonmetallics, and alloys, because of the adverse effects of heat on these materials.

Figure 14.13. Molten salt paint-stripping equipment (Photo courtesy of Kolene Corporation, Detroit, Michigan).

Burn-Off Systems

High-temperature flames, ovens, and fluidized beds are commercially used to literally burn paint off. This method requires the control of potentially toxic gaseous by-products produced during the oxidation of the coating. This technology is limited to steel parts.[10]

Hot-Caustic Stripping

Hot-caustic-solution stripping is practiced commercially, and equipment is readily available. Hot caustic baths, typically at temperatures over 200°F, are very effective in removing caustic-sensitive paints. Applications are limited, however, because many coatings, such as epoxies, are both caustic and heat-resistant. The technique is only useful for steel parts because the caustic corrodes many materials, including aluminum.[10]

Acid Strippers

Acid-based paint strippers have been used by commercial and military maintenance facilities for removing aircraft coatings. This type of paint stripper is generally considered to be the least favored alternative for paint removal. Two primary reasons are cited: worker safety and degradation of aircraft components. The hazards to workers are obvious. In addition, acid strippers can cause hydrogen embrittlement of high-strength steel. Even the most cautious operators have difficulty controlling the amount of damage done to an aircraft substrate.

REFERENCES

1. Law, A.L., and N.J. Olah. "Initiation Decision Report: Aircraft Paint Stripping Waste Treatment System," Naval Facilities Engineering Command, Technical Memorandum, TM 71-85-06, Naval Civil Engineering Laboratory, Port Hueneme, CA (December 1984).
2. Roberts, R.A. "Plastic Bead Blast Paint Removal," Aircraft Division, Directorate of Maintenance, Hill Air Force Base, Ogden, UT, March 1, 1985.
3. Roberts, R.A. "Mechanical Paint Removal System: Special Report on Plastic Impact Cleaning Media," Hill Air Force Base, Ogden, UT, July 31, 1984.
4. "Preliminary Economic Analysis of Paint Stripping Using Plastic Impact Media at NARF Pensacola," NARF Pensacola, FL, March 28, 1984.
5. "Coating Removal via Plastic Media Blasting," Materials Engineering Division, NAVAIR, Engineering Support Office, Naval Air Rework Facility, Pensacola, FL, July 18, 1984.
6. Boubel, R.W. "Evaluation of Dry vs. Wet Paint Stripping," Memorandum for the Record to Peter S. Daley, Office of the Secretary of Defense, August 1, 1984.
7. "Status Report on Plastic Media Paint Stripping (F-45 Paint Strip)," Engineering Report 002-85. NAVAIR WORK FAC NESO, North Island, San Diego, CA, March 1985.
8. "Synopsis of Testing Performed by U.S. Military Facilities Evaluating U.S. Plastic and Chemical Corporation Plastic Abrasives," U.S. Plastic and Chemical Corporation, September 30, 1984.
9. Schmitt, G.F., Jr. "U.S. Air Force Organic Coatings Practices for Aircraft Protection," *Metal Finishing*, November 1981.
10. Mazia, J. "Paint Removal (Stripping Organic Coatings)," *Metal Finishing Guidebook Directory*, 1985.
11. Adema, C.M., and Smith, G.D. "Development of Cavitating Water Jet Paint Removal System," Proceedings of the 15th Environmental Symposium, Long Beach, CA, April 1987.
12. Sullivan, C.J. "NMP Formulations for Stripping in the OEM Market Sector," *Reducing Risk in Paint Stripping*, Proceedings of an International Conference, Washington, D.C., February 1991.
13. Pauli, R., and C. Drake. "Dry Stripping—Seven Years Later," 1992 DOD/Industry Advanced Coatings Removal Conference, Orlando, FL, May 1992.

GENERAL REFERENCES

Whinney, C. "Blast Finishing—Part 1," *Metal Finishing*, November 1983.
Whinney, C. "Blast Finishing—Part 2," *Metal Finishing*, December 1983.
Whinney, C. "Blast Finishing," *Metal Finishing*, January 1985.

CHAPTER 15

Motor Oil and Antifreeze

*Scott MacEwen**

DESCRIPTION OF WASTE-GENERATING OPERATIONS

Used motor oil and antifreeze are generated during gasoline and diesel engine repair and maintenance operations. Motor oil and antifreeze generally become wastes when they no longer meet the specifications required to perform their intended purposes (lubrication or temperature and corrosion protection). Disposal and replacement with new material is rarely based on measuring the performance of these fluids; rather it is based on an engine running time or mileage standard. The volume of waste oil and antifreeze generated at any one shop is typically small; however, more generators produce these types of wastes than any other type of hazardous waste.

Motor oil or antifreeze that is no longer being used for its intended purposes is often referred to as "off-specification" or "used" rather than "waste," to differentiate it from hazardous waste. Although used motor oil and antifreeze are not in themselves regulated as hazardous wastes, contaminants such as heavy metals, which enter this waste stream through normal use, or chlorinated solvents, which enter this waste stream through poor handling practices, can necessitate classifying the entire waste stream as hazardous. Even nonhazardous waste oil and antifreeze have come under increasingly stringent state and federal regulation, which is likely to make handling these wastes more and more costly. Regulations that affect classification and handling of used oil and antifreeze are discussed in later sections.

Motor Oil

Motor oil is composed of two major components: a base stock, which is typically 100% petroleum, consists of a wide range of long- and short-chain hydrocarbons along with additives for cleaning, preventing corrosion, and reducing wear. Synthetic motor oils use man-made polymers rather than petroleum as a base stock.

Motor oil provides lubrication for the internal workings of gasoline- and diesel-powered engines used in vehicles such as motorcycles, automobiles, trucks, trains,

*Mr. MacEwen is an environmental engineer in CH2M HILL's Reston, Virginia, office and may be reached at (703) 471-1441.

boats, and airplanes, and in stationary equipment such as generators. Under normal use, motor oil picks up contaminants that degrade the performance of the additives that are needed to protect the engine. Water and fuel can accumulate in the oil through condensation or leaks. Dirt, sediment, paint chips, and metals are picked up through engine wear, and polymers break down to form sludge. Other contaminants such as solvents, PCBs, antifreeze, and solid debris may be mixed with the used oil after it has been removed from the engine.

It should be kept in mind that the oil base stock itself is usually not worn out or degraded by use. It is usually the additives that become depleted and can no longer provide the necessary protection. The oil is then drained from the engine and replaced. Other sources of used oil include leaks, spills, and unused product that exceeds its dated shelf-life.

The rated life of motor oil is usually specified in engine miles or operating hours. For instance, most automobile manufacturers recommend that motor oil be changed every 3,000 to 5,000 miles. Construction equipment and power generators may require a change every 200 to 300 operating hours. The U.S. Department of Defense operates an oil analysis program where oil samples are actually collected from individual vehicles and tested to determine when changes should be made. This program typically pertains only to equipment such as ships, tanks, and other specialized vehicles that require large volumes of oil and where substandard oil may damage expensive machinery.

Used oil is generated by any vehicle or equipment maintenance operation, from backyard mechanics to commercial or government motor pools. Quantities of used oil generated per vehicle range from 6 qt for a standard 6-cylinder mid-size automobile to 12 gal for a large diesel-engine snowplow to 55 gal for an M-1 tank. Likewise, quantities by generator can vary from several hundred gallons a year from the corner gas station to tens of thousands of gallons per year from a quick-lube operation and several hundred thousand gallons per year from a large U.S. Department of Defense motor pool.

A common rule of thumb used by vehicle fleet managers to estimate used oil generation is that 50% of all virgin oil used will be lost during service through leaks, filter holdups, piston blowby, and spills during handling operations. The U.S. Department of Defense estimated in 1984 that it generated 4 million gal of used oil.[1]

It is estimated that the United States currently generates approximately 1.4 billion gal of used oil annually. Where does this used oil wind up? According to a 1991 study by a coalition of industry and environmental groups, 56% is blended with virgin oil and burned as fuel, 5% is re-refined and reused as lubricating oil, 3% is used in various other industrial processes, and 36% is dumped in landfills or storm sewers.[2]

Antifreeze

Antifreeze concentrate is mostly ethylene glycol to which additives such as corrosion inhibitors, dyes, and foam suppressants are added. It is usually mixed with tap or distilled water in a 50:50 mixture when it is used in the cooling systems of gasoline and diesel engines. The mixture protects the engines from temperature extremes. To be effective, antifreeze must have a low freezing point and a high boiling point. It also must have low corrosivity (a particular problem with aluminum engines), little tendency for foaming, a relatively high pH, and low total dissolved solids (TDS).

During normal use, antifreeze picks up contaminants, which break down the additives and affect the physical and chemical properties of the fluid. Common contaminants include oils, which affect the pH and cooling and freezing properties of the antifreeze and promote foaming; sulfates and chlorides, which increase corrosivity; and metals, paint, dirt, and other solids that cause engine wear.

When the properties of antifreeze pass outside acceptable limits, antifreeze is drained and replaced. As with motor oil, antifreeze replacement frequency is usually based on a time or mileage limit. Automobile manufacturers typically recommend when to change antifreeze, usually in the beginning of winter when antifreeze must provide maximum protection from freezing.

Generators of waste antifreeze are the same as those that produce used oil. They include backyard mechanics; gas stations; auto repair shops; car and truck fleet maintenance garages; locomotive, aircraft, and ship maintenance facilities; and maintenance facilities for power generators.

ENVIRONMENTAL REGULATIONS

Regulations for Used Motor Oil

The U.S. EPA specifically exempts used motor oil from regulation as a hazardous waste. This exemption is significant because without it most used oil would be considered a characteristic hazardous waste because it contains concentrations of lead and other hazardous constituents.

Although used motor oil is not currently regulated as a hazardous waste, federal regulations have established limits for discharging oil to publicly owned treatment works (POTWs) and to navigable waterways. Federal regulations also mandate management standards for used oil generators, collection centers, transporters, marketers, processors, re-refiners, and burners.

In the Used Oil Recycling Act of 1980, Congress recognized the value of used oil as a resource and the magnitude of the threat that it posed if disposed of or reused improperly. This act mandated U.S. EPA to issue regulations for recycling and reusing oil. In addition, the act provided states with technical assistance to implement these regulations, and with financial backing to establish used-oil recycling and reuse programs.

The most recent federal regulations concerning management standards for used oil are outlined in 40 CFR 279. These regulations, which went into effect March 8, 1993, pertain to recycling used oil and burning it as a fuel. These regulations also specify which used oils are regulated for recycling and burning as fuel. The used oil specifications are presented in Table 15.1. Used oil that does not meet these specifications is considered off-specification oil and is subject to these regulations. Waste oil that meets these specifications is subject to minimal requirements only. It should be noted that waste oil containing greater than 1,000 ppm total halogens is considered a hazardous waste even though it meets the specification, unless it can be shown by analyses that listed wastes and hazardous constituents account for less than 100 ppm of this total.[3]

In addition, waste oil that is mixed with a listed hazardous waste must be treated as a hazardous waste whether or not it meets the specifications listed in Table 15.1. Regulations that affect specific waste reduction practices will be discussed in more detail in the sections that follow.

Table 15.1. Specifications of Used Oil[a]

Constituent/Property	Allowable Level
Arsenic	5 ppm maximum
Cadmium	2 ppm maximum
Chromium	10 ppm maximum
Lead	100 ppm maximum
Flash Point	100°F minimum
Total Halogens	4,000 ppm maximum

[a]Used oil exceeding these specifications is regulated by 40 CFR 279 when burned for energy recovery.

Regulations for Used Antifreeze

Federal regulations do not specifically cover used antifreeze. Used antifreeze is not a listed hazardous waste and typically does not exceed any of the limits for a characteristic hazardous waste; however, it may become a hazardous waste and be regulated as such if it is mixed with a hazardous waste. In some instances antifreeze can, under normal conditions, pick up contaminants such as metals, which make it a hazardous waste.

Dilute ethylene glycol can be treated biologically by acclimated bacteria. Under some circumstances, antifreeze may be discharged to local POTWs through the sanitary sewer system or be treated in an onsite system. However, some characteristics of waste antifreeze, such as high BOD and TDS, and potential contaminants, including oils, phenols, aldehydes, ketones, and acids, are usually addressed by facility discharge limits that may restrict this type of disposal.

NPDES regulations effectively ban direct discharge of untreated used antifreeze to surface water.

TECHNIQUES FOR REDUCING AND REUSING USED MOTOR OIL

Effective waste reduction and reuse techniques for motor oils can range from simple implementation of proper handling and storage procedures to complex processes for re-refining. The general categories of practices addressed are:

- improving motor oil handling and housekeeping practices
- extending motor oil product life
- segregating used motor oil from other waste
- removing contaminants from used motor oil
- burning used motor oil as a fuel
- re-refining used motor oil

This chapter focuses on those techniques that are available and practical for a typical engine maintenance shop. Examples of vendors that furnish equipment and technologies for implementing some of these techniques are listed in Table 15.2.

Table 15.2. Technology Vendors: Used Motor Oil

Company	Address	Telephone
Reprocessing/Re-refining		
Safety-Kleen	777 Big Timber Road Elgin, Illinois 60123	800/669–5840
Evergreen Oil Company Evergreen Holdings, Inc.	18001 Cowan, Suites C&D Irvine, California 92714	714/757–7770
Space Heaters		
Black Gold® Heater Robert Sun Company	240 Great Circle Road Suite 344 Nashville, Tennessee 37228	800/351–0643
Clean Burn, Inc.	83 South Groffdale Road Leola, Pennsylvania 17540	800/531–0183
By-Pass Oil Filters		
Gulf Coast Filters, Inc.	P.O. Box 2787 Gulfport, Mississippi 39505	601/832–1663
Oil Change Equipment		
Fast Lube Oil Change System (FLOCS), Aeroquip Division	300 Southeast Avenue Jackson, Michigan 49203	517/787–8121
Oil Filter Cutting Machines		
CTEC, Inc.	P.O. Box 1768 Salt Lake City, Utah 84110	801/973–7977
Oil Filter Recyclers		
American Waste	P.O. Box 306 Maywood, Illinois 60153	708/681–3999
Petroleum Recycling Corporation	2651 Walnut Avenue Signal Hill, California 90806	310/595–7431

Improving Motor Oil Handling and Housekeeping Practices

Unusable virgin motor oil and waste generated from spills or leaks of unused motor oil are just as costly to dispose of as used oil. However, these waste streams can be minimized and even eliminated by improving purchasing, inventory, handling, and storage practices.

Virgin motor oil can become unusable if its shelf life is exceeded, engine manufacturers' specifications change, it is stored so that contaminants get into it, or it is exposed to extreme temperatures that affect the solubility of additives.

Purchasing and inventory practices should be established to minimize the number of grades of oil used at a facility and should require stock rotation, with material being used on a first-in, first-out basis. Although motor oil conceivably has an indefinite shelf life if stored properly, a conservative shelf life of 2 years has been adopted by some refiners.

Purchasing and inventory personnel should also anticipate changes in American Petroleum Institute (API) motor oil classifications and vehicle manufacturers' specifications. API motor oil classifications have typically been shown to change every 8 to 12 years (the most recent change was scheduled for 1994). As API classifications change, so do vehicle manufacturers' specifications on vehicles built on or after that year. This is particularly significant for operations such as rental car fleets where the fleet can change completely in a relatively short time. If these changes are not anticipated and accounted for during purchasing, a shop could get stuck with and have to dispose of outdated and theoretically unusable oil.

Virgin motor oil should be stored in closed containers where it can be kept clean and dry because water and dirt will affect its usefulness. Oil should not be stored in areas where temperatures vary drastically because this may cause additives to separate and settle. Oil storage and dispensing areas should be equipped with spill prevention or containment features such as drip pans, containment berms, and sumps to prevent releases to the environment and to maximize the potential to recover spilled or leaked oil.

Extending Motor Oil Product Life

The volume of used motor oil generated from engine maintenance activities can be reduced by extending the period of time between motor oil changes. Oil change recommendations made by engine manufacturers are typically based on conservative assumptions about operating conditions. Under certain conditions it is possible and economical to extend the life of engine oil without risking damage to the engine. The two methods for increasing motor oil life discussed here are motor oil testing and bypass filtration.

It should be noted that the amount of time between oil changes recommended by manufacturers is only a recommendation and is usually not a requirement of vehicle warranties. Most warranties state only that the oil must meet certain specifications. As discussed below, this requirement can be met with periodic testing and documentation.

Motor Oil Testing

In motor oil testing programs, samples of used motor oil are analyzed to determine if the oil still meets the specifications necessary to protect the engine. Loss of protection is generally a result of contamination rather than a chemical change in the oil. Analyses would typically be performed for water, metals, coolant, fuel, and other contaminants.

For economical reasons, motor oil testing programs are generally limited to use on vehicles and equipment with engines that require large volumes of oil (more than 10 gal) or that are run infrequently and under a variety of conditions. Samples of motor oil are collected and tested on an engine-by-engine basis at regular intervals. If results show that the oil does not meet certain standard specifications, it is replaced. Otherwise, it is left in the engine until the next scheduled test.

The major constraint associated with a motor oil testing program is the need for a laboratory with a quick turnaround time on analytical results. This usually means an onsite laboratory, which is not economically feasible except for the largest maintenance facilities.

Example. The U.S. Army, Air Force, and Navy all operate oil analysis programs designed to extend motor oil life and provide engine protection for most of their larger and more-expensive equipment, such as tanks, personnel carriers, construction equipment, ships, and generators. A single oil change on these engines can generate between 10 and 100 gal of used motor oil.

Under the Army program, oil samples are collected every 25 to 50 hours of vehicle operation and sent to one of the program's laboratories. The laboratory performs a spectrophotometric test for metals, a viscosity test to evaluate contamination with coolant and fuel, a hot plate splatter test to determine water content, and a blotter test to check for a variety of other contaminants. If the motor oil is found to be within standard specifications, it is left in the engine for another 25 to 50 operating hours, then retested.[4]

On-Board Bypass Filtration

Most automobile, truck, and other engines are factory equipped with a full-flow oil filtration system. In this type of system, all of the oil that lubricates the engine first passes through an oil filter. This filter must be quite porous in order to pass the required flow rate and, therefore, is designed to remove only relatively large particles (greater than 40 microns) that could seriously damage an engine. Other contaminants that could degrade the oil's protective properties, create sludge, and cause engine wear, such as metals, microscopic dirt and carbon particles, and water, pass through readily.

A bypass filter system, consisting of a much less porous element, slowly filters only a portion of the oil flow (usually less than 1/2 gpm compared with 4 to 5 gpm that is typical for a full-flow filter). Oil is withdrawn from the bottom of the crankcase, passed through the bypass filter, and returned to the crankcase. Some bypass filter systems remove not only solids to the submicron level but also moisture.

Oil analyses have shown that a properly serviced bypass oil filter system used in tandem with the manufacturer-installed full-flow filter system can maintain motor oil in a condition where it would never need to be replaced. These systems have also been shown to prolong engine life and to reduce the need for major engine repairs. Bypass filtration appears to be particularly cost-effective on large expensive engines, such as diesel trucks, construction equipment, boats, and ships.

One bypass filter system manufactured by Gulf Coast Filters, Inc., of Gulfport, Mississippi, uses inexpensive paper-towel and tissue-paper rolls as filter elements. These filters have been shown to remove particulate contaminants down to 1 micron and reduce moisture to less than 40 ppm. The filter canister can be mounted directly on the engine block. Filter changes are recommended every 5,000 to 10,000 miles. Changing takes 10 minutes, one or two rolls of paper towels, and several quarts of makeup oil, and does not require tools. Gulf Coast paper-towel filters also can be mounted on portable trailers and used to filter diesel fuel, as shown in Figure 15.1, or used to filter antifreeze.

Case Study 15.1: Bypass Oil Filtration

Dairy Fresh Corporation of Hattiesburg, Mississippi, operates a fleet of 40 trucks and cars, which in 1978 required almost 14 man-hours per day to maintain. Oil changes were performed every 5,000 to 10,000 miles, generating several thousand

Figure 15.1. The U.S. Army National Guard at Camp Shelby, MS, uses a portable 12-unit paper-towel filter for cleaning diesel fuel before placing it in an M1 tank (Photo courtesy of Gulf Coast Filters, Gulf Port, Mississippi).

gallons of used oil per year. A typical engine life was between 65,000 and 75,000 miles. Two years after installing bypass filters (Gulf Coast paper-towel filters) on all of their vehicles, mechanics' time was down to 8 hours per day. Oil changes were no longer necessary—oil was simply sampled and analyzed to verify its condition. The average life of an engine increased to between 130,000 and 160,000 miles, and problems such as sticking valves, which were prevalent in the past, no longer occurred.[5]

Segregating Used Motor Oil from Other Waste

The cost of reclaiming, recycling, or disposing of used motor oil depends on the types and concentrations of contaminants in the oil. Used oil that meets specifications for metals and halogens (shown in Table 15.1), and contains only low concentrations of water and sediment may be sold to a reclaimer for upward of 10 to 20 cents per gallon; however, disposing of oil that is mixed with a listed hazardous waste may cost the generator several dollars per gallon.

Waste disposal is minimized and resale income is maximized by keeping used oil as clean as possible. Most of the contaminants that affect reuse and disposal costs are picked up through inadvertent mixing after removal from an engine. This can be prevented by implementing relatively simple waste segregation practices.

The typical sources of outside contaminants found in used motor oil are wastes associated with other operations common to the automobile maintenance and repair industry. These wastes include hydraulic fluids, automatic transmission fluids

(ATF), antifreeze, synthetic lubricating oils, spent degreasing solvents, and spent paint solvents. An effective segregation program should provide separate accumulation and storage facilities for each type of waste. Used motor oil, hydraulic fluid, and ATF can be consolidated, whereas radiator fluid, spent solvents, and most used synthetic lubricants should be segregated.

Used oil should be segregated from the moment that oil is removed from a vehicle until it is removed from the site or reprocessed on the site. The first opportunity for outside contaminants to enter the used oil waste stream is during crankcase draining. Simple techniques can promote segregation during crankcase draining, transport, and storage. For small-volume shops, dedicated buckets or oil pans should be used to drain oil from vehicles. For large-volume shops, several quick-drain technologies on the market not only promote segregation, but also reduce labor costs.

The Fast-Lube Oil-Change System (FLOCS) is a vacuum-operated system that can drain oil directly from the bottom of the engine crankcase into a dedicated storage tank or drum. Tests conducted by the Army have shown that FLOCS is capable of draining the oil from an engine in approximately one-tenth the time required to drain it by gravity. Use of the system also eliminates the need for placing the vehicle on a lift. There is an initial expense, however, because a quick-disconnect fitting must be installed on the crankcase of each vehicle. Because of this requirement, the technology is only appropriate for fleet maintenance.[1]

A second quick-drain device is the vacuum dipstick drain system. This system, like FLOCS, provides waste segregation and reduces labor, with the added advantage that individual vehicles need not be modified. A major drawback is that, because it drains oil from the top, it does not remove sludge from the bottom of the crankcase.[1]

Dedicated storage containers should be used to accumulate used oil before reclamation or recycling. They should be clearly labeled, and employees should be informed of handling procedures. Storage containers should be covered to keep out water and debris, and should be locked, if necessary, to prevent their inadvertent use for disposal of other wastes.

Removing Contaminants from Used Motor Oil

Once water and sediment have contaminated the used oil, they can be removed to some degree using simple and commonly available technologies. Removing sediment and water reduces the volume of oil to be disposed of, and more importantly, improves the quality of the used oil and thus its resale value.

Oil/water/sediment separation is typically based on physical techniques that take advantage of the differing densities of water, oil, and solids. A basic separator consists of a tank equipped with baffles and strainers in which heavy solids and water settle to the bottom and oil rises to the top and is skimmed.

A wide variety of oil/water separators are on the market. They can be designed to operate in either batch or continuous mode. The type and size of separator that is best for a specific situation will depend on the volume of the waste stream to be treated and its characteristics (the water-to-oil ratio, the amount of sediment, and the type of oil).

Example. At one vehicle maintenance facility, oil/water separation is performed in batch mode by placing the used oil in an aboveground tank, allowing it to sit for awhile, and then draining the settled water through a valve in the bottom of the tank. The valve is closed when oil begins to drain from the valve. Although this technique removes some water, it can also pollute the environment.[1]

Example. Fort Carson Army Base uses a simple and innovative method for removing solids from used motor oil. As used oil is generated during oil changes, it is placed in a covered, aboveground, bathtub-like sump. The bottom of the sump is sloped to one end. At the low end, a 3-in.-diameter (vertical) pipe extends approximately 10 in. from the bottom. This pipe drains to a 1,000-gal underground storage tank. This design allows heavy solids and debris to settle out while oil overflows to the tank through the drainpipe.[1]

Burning Used Motor Oil as a Fuel

Burning used oil as a fuel is by far the most widely used method for ultimate disposal of this waste stream. It has been estimated that 56%, or 780 million gal, of the approximately 1.4 billion gal of used oil generated annually is burned as fuel.[2] Oil generators have several options for burning oil as fuel. The applicability and cost-effectiveness of each option depends on the volume and chemical composition of the waste oil, the size and type of onsite and offsite facilities available, and the local and state regulations that apply. Options for burning used oil include:

- onsite burning in space heaters
- onsite burning in boilers or furnaces
- offsite burning in boilers or furnaces

The federal regulations that address burning nonhazardous used oil as fuel are contained in 40 CFR 279 *Standards for the Management of Used Oil.* In these regulations, the primary factors that affect the conditions under which a used oil may be burned are the concentrations of contaminants in the oil and whether the oil is being burned at the site where it is generated or at an offsite location.

Different technical, procedural, and reporting requirements apply, depending on whether a used oil is classified as on- or off-specification. An oil is off-specification if it exceeds any of the criteria listed in Table 15.1. Used oils classified as hazardous wastes also may be burned as fuel in some instances. These wastes are regulated by 40 CFR 266 Subpart H—*Hazardous Waste Burned in Boilers and Industrial Furnaces.*

Regulations also vary, depending on whether the used oil is burned on the site by the generator, sold directly to an offsite burner, or sold to a burner through a marketer. Figures 15.2, 15.3, and 15.4 present flow charts showing how these regulations pertain to generators, marketers, and burners of on- and off-specification used oil and oil classified as hazardous waste. In general, only minimal requirements apply to generators of on-specification used oil who burn it on the site. More-stringent regulations apply to off-specification used oil and any used oil burned off the site. Hazardous waste treatment, storage and disposal (TSD) facility regulations may apply if the used oil is hazardous.

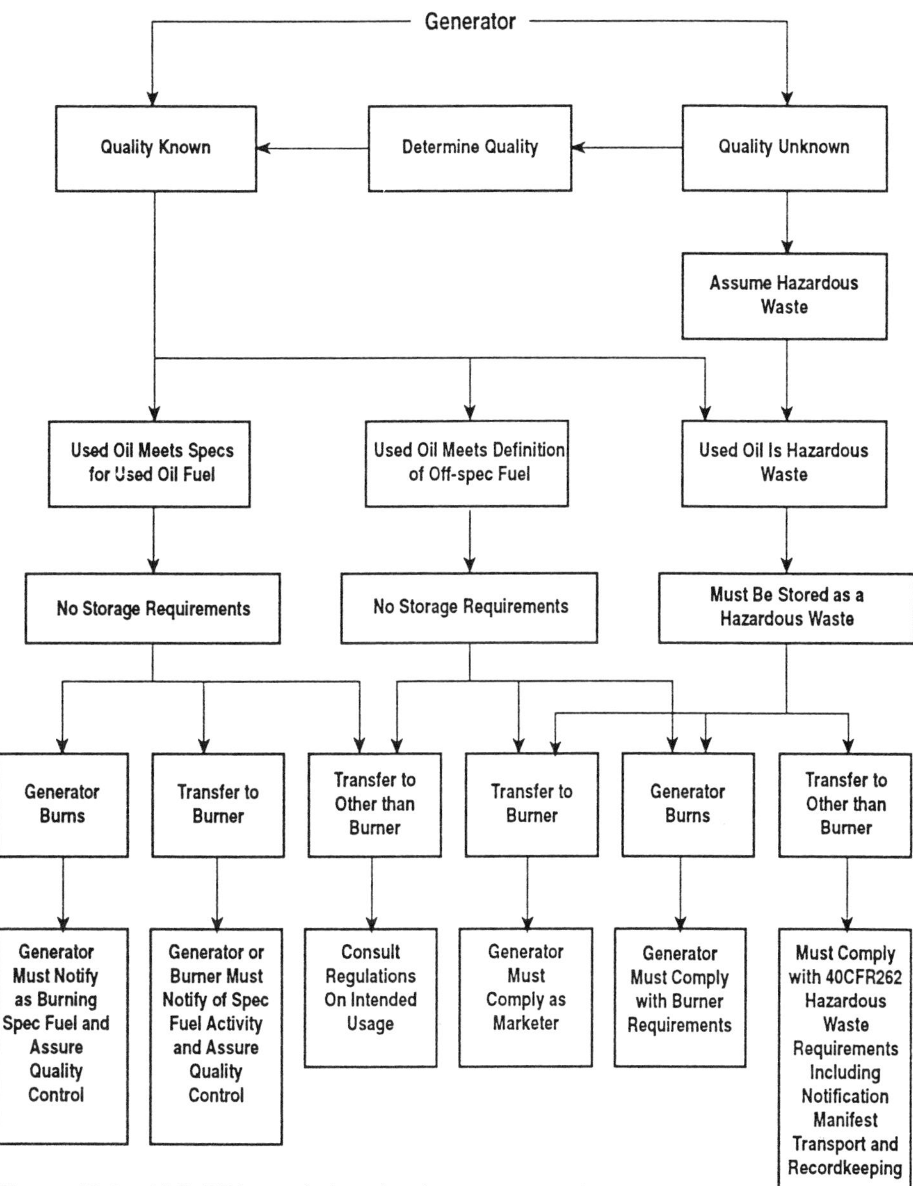

Figure 15.2. U.S. EPA regulations for the generator of used oil.

Onsite Burning in Space Heaters

Specially designed space heaters are available that can burn used oil and similar waste streams, such as automatic transmission fluid, with little or no pretreatment. The heat generated from burning the waste is used for space heating. This practice not only saves on disposal costs, but also lowers heating fuel costs by reducing or eliminating the need for other fuel for heating.

A typical space heater that burns used oil can generate between 150,000 and 500,000 Btus per hour, which would be capable of adequately heating a 60,000 to 200,000 ft^3 space. A heater of this size would burn between 1 and 3.6 gal of oil per

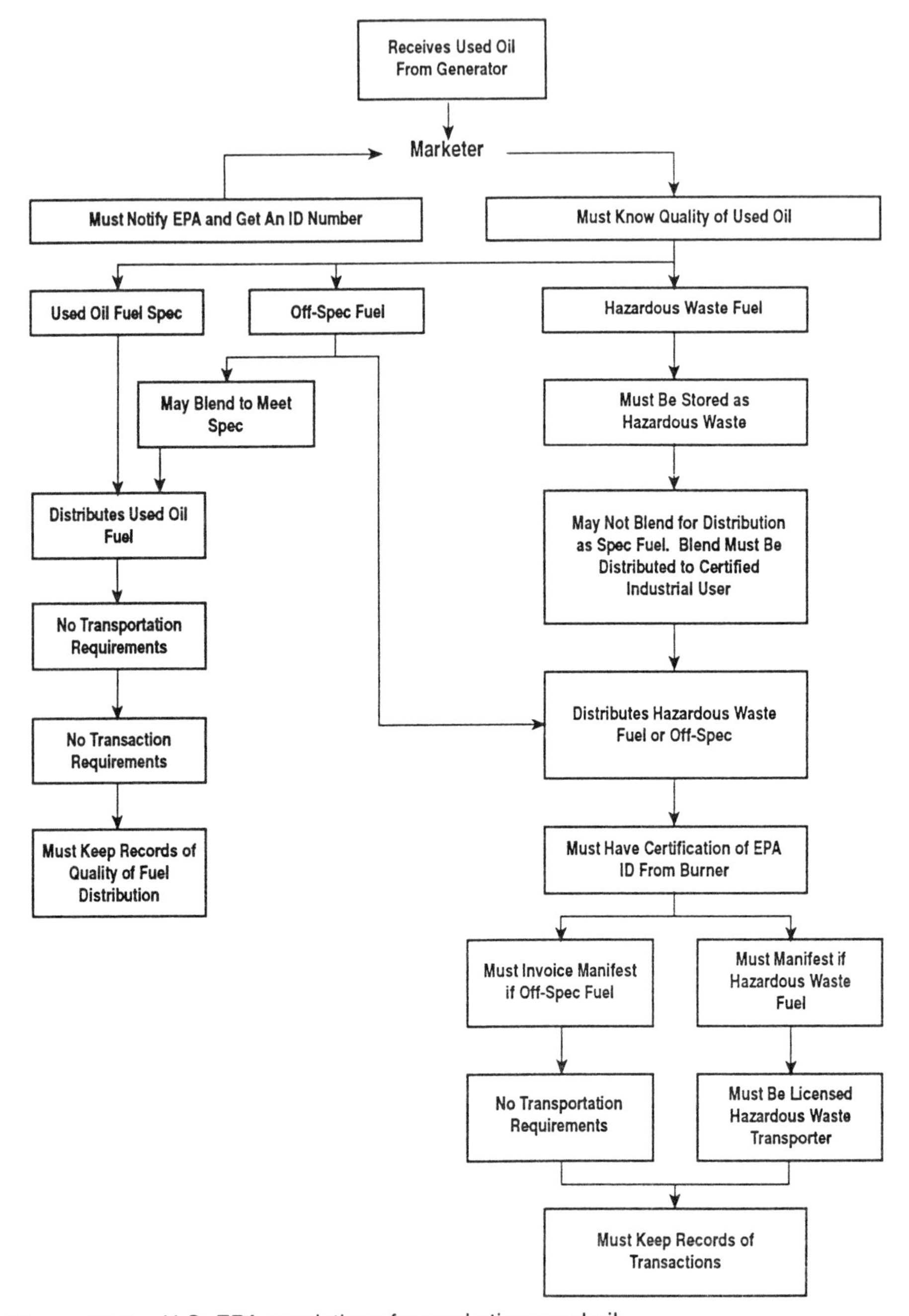

Figure 15.3. U.S. EPA regulations for marketing used oil.

hour. One space heater operating at the lower end of this range for 40 hr per week and 6 months of the year would burn approximately 1,000 gal of oil annually.

Vendors of space heaters that burn used oil estimate that 1 gal of used oil contains approximately 140,000 Btu, the same heat value as $0.50 worth of natural gas. This estimate would, of course, vary depending on the local price of natural gas. Savings when compared with propane or fuel oil can be expected to be even greater.

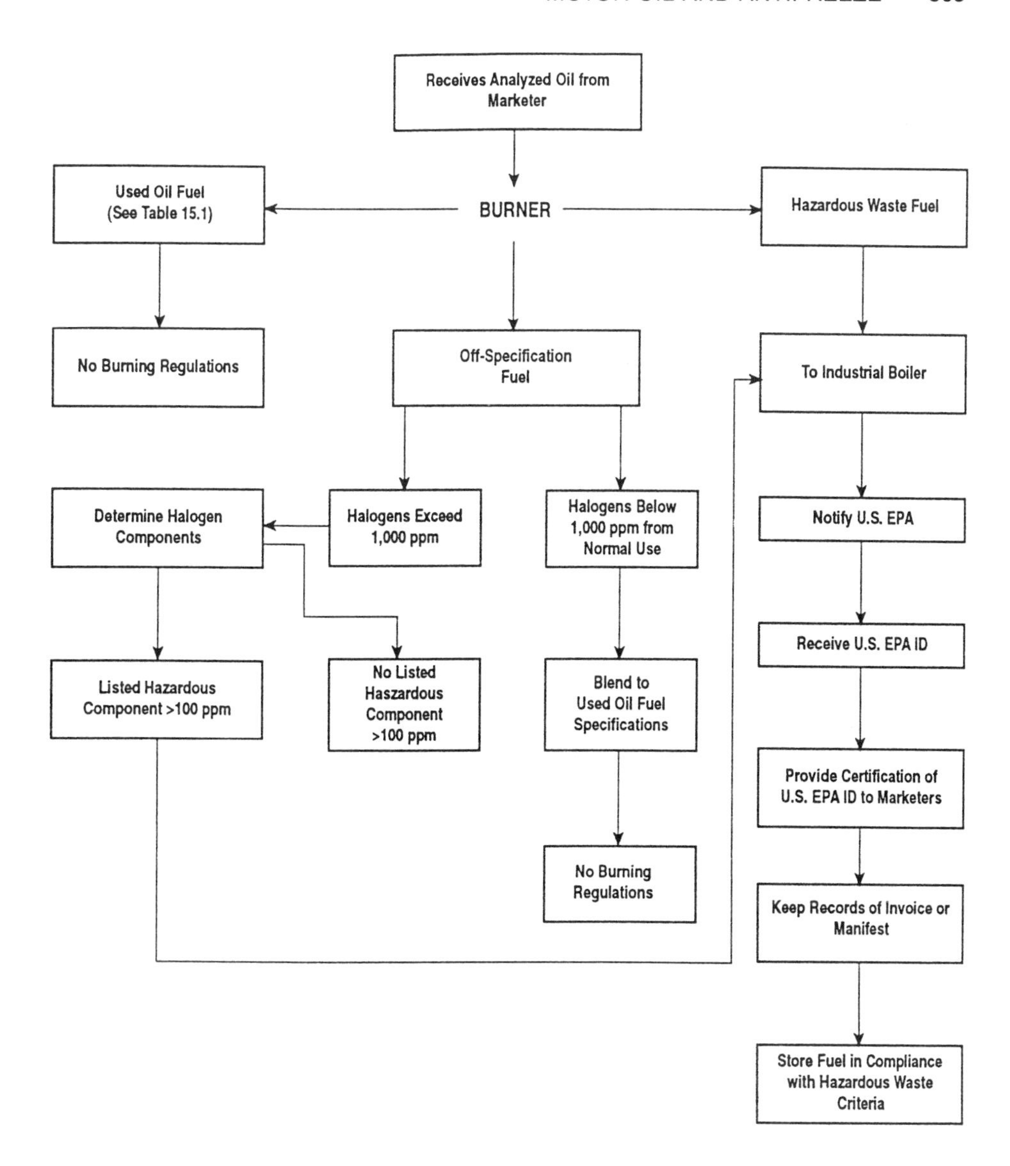

Figure 15.4. U.S. EPA regulations for burning used oil.

These heaters can be ideal for heating industrial work places such as automobile maintenance garages, service stations, parking garages, airplane hangars, and other places where used oil is generated.

The units can be relatively small in size. A 200,000 Btu/hr, Black Gold heater, manufactured by the Robert Sun Company, is 26 in. by 26 in. by 80 in. and weighs approximately 500 lb. A larger unit manufactured by Clean Burn, Inc., is 38 in. by 75 in. by 93 in., weighs 1,300 lb, and can generate 500,000 Btu/hr. These units can be installed directly in the area to be heated and vented to the outside, or they can be installed remotely and ducted to the area to be heated. Waste lubricating oil conceivably can be drained directly from an automobile engine to the heater's storage tank.

Accessories for units such as the Black Gold heater include an oil storage tank that also can be used as a work bench (see Figure 15.5) and a module that can be used to heat water. The 1991 price of a Black Gold heater ranged from $4,200 to $7,500 plus installation, depending on the accessories desired.

Federal regulations for burning used oil in space heaters are addressed in 40 Code of Federal Regulations (CFR) 279 Subpart C. On-specification oil generated on the site may be burned in these units without restriction. On-specification oil generated

Figure 15.5. Used-oil fueled space heater and fuel storage tank/work bench (Photo courtesy of Robert Sun Company, Nashville, Tennessee).

off the site also may be burned in these units; however, the owner would then be required to notify the U.S. EPA, analyze the oil to show it is on-specification, and maintain records for 3 years.

Off-specification used oil may be burned in space heaters only if the oil is generated on the site. The heater can have a maximum capacity of 0.5 million Btu/hr, and combustion gases must be vented to the ambient air. Hazardous used oil cannot be burned in space heaters. New Jersey and California prohibit burning of any used oil in space heaters.

Burning in Onsite Boilers and Furnaces

Under certain conditions, used motor oil also may be burned as fuel in onsite boilers and furnaces that were initially designed to burn natural gas, propane, or virgin heating oil. As with other burning-as-fuel technologies, savings would be realized by reducing the demand for virgin fuel and eliminating the disposal costs for used oil.

This application usually requires some type of pretreatment to remove water, sediment, and other contaminants that can foul furnaces and increase maintenance costs. Pretreatment can vary from simply removing water and sediment through settling or filtration to removing contaminants and neutralizing acidic compounds through comprehensive treatment such as flash distillation and chemical additives. After treatment, the used oil is blended with virgin fuel oil at concentrations of 10% or less before it is fed to the burner. Blending is necessary because most boilers and furnaces are not equipped to burn straight motor oil without causing serious maintenance problems.

Burning recycled fuels in furnaces and boilers requires regular analysis of the used oil. Burning of used oil that does not meet the specifications listed in Table 15.1 is regulated by 40 CFR 279 and is limited to facilities that meet the definition of industrial furnaces or industrial or utility boilers (and space heaters as discussed previously). On-specification used oil may be burned in any type of facility.

For hazardous waste to be burned in industrial boilers, the entire unit, including the waste storage area, has to be permitted under RCRA as a hazardous waste boiler. This is required by the February 21, 1991, Boiler and Industrial Furnace Regulations. These regulations limit emissions of carbon monoxide, hydrocarbons, toxic metals, particulates, hydrogen chloride, and chlorine gas. Under some circumstances, they require a test burn and destruction and removal efficiencies of 99.99%.

A small-quantity burner exemption allows burning of a small quantity of hazardous waste (up to 1% of the total fuel burned in a given boiler) without having to obtain a permit. Hazardous wastes burned under this exemption are limited to those generated on the site.

The feasibility of burning used oil depends on the degree of processing (and the associated cost) required to produce a used-oil fuel that does not cause excessive fouling, flameouts, and subsequent maintenance problems.

The National Institute for Petroleum and Energy Research (NIPER) attempted to document the effect of burning used motor oil on the maintenance costs of boilers. Although NIPER found little quantitative data, they concluded that burning used oil on a long-term basis results in clogging of feed pipes and nozzles, corrosion of pipes and tanks, and reduced heat-transfer efficiency.[1] Figure 15.6 shows the estimated effect of used-oil fuel on boiler maintenance costs. The graph indicates that burning blends of less than 10% used oil will have a negligible effect on boiler

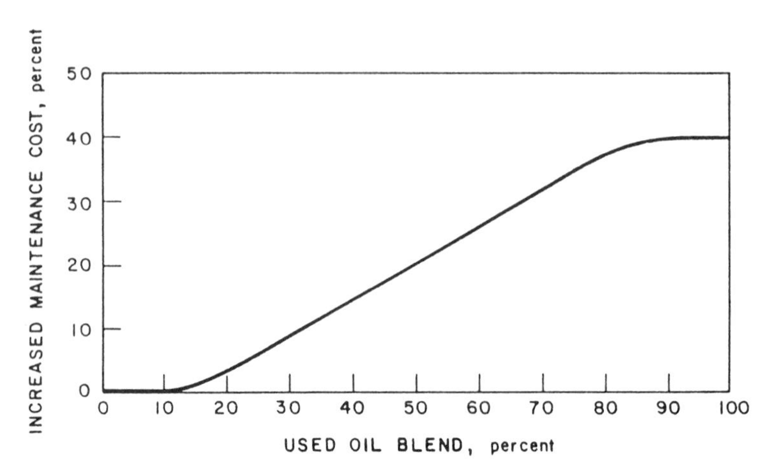

Figure 15.6. Increased boiler maintenance resulting from used lubricating oil.

performance and maintenance costs, provided the used oil receives at least minimal pretreatment to remove water and sediment. It also should be realized that the cost advantages of reusing motor oil are offset somewhat by the cost of disposing of waste such as separated water, sediment, and sludge.

Case Study 15.2: Burning Used Oil in an Onsite Natural Gas Boiler

In 1981, Fort Benning in Columbus, Georgia, began burning used oil mixed with No. 2 and No. 6 oil in three water-tube boilers. The boilers operate on natural gas 90% to 95% of the time and on a mixture of used oil and No. 2 or No. 6 fuel oil 5% to 10% of the time. Approximately 50,000 gal of blended oil fuel is burned per year. The used oil fuel, which is mostly motor oil, is filtered through two wire-screen strainers, dewatered in an oil-water separator, and stored in a 500,000-gal tank.

Before 1981, virgin oil was used as backup, which tended to increase maintenance costs over using natural gas only. Since switching to used oil, operating and maintenance costs have risen 50%. The plant had to be shut down twice a year rather than once a year as was the case previously, and problems with sticking valves, pump failures, and increased housecleaning occurred more frequently. By 1984, a buildup of calcium and sulfate scale on boiler tubes had drastically decreased boiler performance.

A subsequent inspection found that the fuel oil had a sediment and water content of 30% and a free water content of 13.5%. It was further discovered that the oil was pumped directly from the bottom of the 500,000-gal holding tank, which had accumulated a 2-foot layer of water and sediment.

Inventory records also revealed that virgin fuel oil had not been added to the tank since 1981. As a result, the feed tank contained only waste oil with a layer of water and sediment at the bottom, and the boilers were at times being fed water. Inspectors made several recommendations, including providing better oil/water separation and installing a riser on the tank feed draw so that fuel is pumped from a level several feet above the bottom of the tank.[1]

Burning in Offsite Boilers or Furnaces

Offsite burning of used oil is typically handled by selling the oil to, or paying to have the oil removed by, a burner or a marketer. The burner or marketer treats and blends the oil as necessary before burning it or reselling it as a fuel.

This method of reusing used oil requires minimal effort by the generator. It is usually handled by a contract between the generator and the burner or marketer. These contracts typically require that the generator provide testing results for regulated compounds and characteristics such as arsenic, cadmium, chromium, lead, total halogens, PCBs, flash point, and pH to show that the oil meets certain specifications and is not hazardous. Indemnification agreements or monetary assurances also are common for handling the occasions when a "bad" load is picked up.

The price a burner or marketer will pay (or charge) for used oil depends on the condition of the oil, the current price of oil, its location, and current regulations. The major characteristic that affects the value of the oil, aside from its regulatory classification of on- or off-specification or hazardous, is the bottom sediment and water (BS&W) content. The 1991 price per gallon paid for used motor oil by a marketer in Baltimore, Maryland, was based on BS&W content, as shown in Table 15.3.

As of 1992, oil marketers and burners in most states were paying generators for their used on-specification oil. It is important to note, however, that marketers in Alaska charged an average of $0.60 per gal to accept used on-specification oil. The direct correlation between the condition of used oil and its market price or acceptability emphasizes the value of implementing effective segregation practices and providing proper storage facilities.

Re-Refining Used Motor Oil

Re-refining involves processing used lubricating oil to return it to virgin oil specifications so it can be reused as a motor oil. Re-refining also involves relatively elaborate and capital-intensive processes that would not be considered feasible as an onsite technology for most used-oil generators. Realistically, re-refining should be considered only as an offsite disposal option, handled either through a broker or directly with a re-refining facility, an arrangement similar to that used with used-oil fuel processors.

From an environmental, conservation, or regulatory viewpoint, re-refining is preferred over burning used oil. However, because re-refining costs are higher and

Table 15.3. Fuel Prices Based on Content

Percent Bottom Solids and Water Content	Price Paid to Generator[a]
0–7	13¢/gal
7–15	13¢/gal on net[b]
15–20	6–1/2¢/gal on net[b]
20–25	4¢/gal on net[b]
>25	Generator pays 25¢/gal

[a]Marketer in Baltimore, Maryland, in 1991.
[b]Total volume of oil minus the BS&W.

the range of used oil found acceptable is more restricted, re-refiners often find it difficult to compete economically with fuel blenders. As a result, there are many more fuel blenders than active re-refiners in the United States. Proposed federal regulations are being considered that recognize the environmental benefits of re-refining over burning. These regulations would favor re-refining by placing further restrictions on burning of used oil as a fuel. If these regulations were enacted, the demand for re-refining facilities would likely increase.

Re-refining technologies include older methods, such as acid/clay treatment; contemporary methods, such as vacuum evaporation; and newly emerging processes, such as direct-contact hydrogenation. Acid/clay treatment is capable of removing organic and inorganic contaminants from oil but creates large quantities of hazardous waste sludge and clay. This process was all but abandoned in the 1970s when RCRA was enacted. Vacuum evaporation involves a complex series of unit operations including filtration, heating, settling, flash dehydration, vacuum stripping, and vacuum distillation. The complexity, high capital costs, and operating and maintenance requirements of this process have resulted in a decrease in the attractiveness of this alternative compared to competing technologies such as fuel blending. Direct-contact hydrogenation uses heated hydrogen gas in direct contact with the oil to destroy organic contaminants, and a proprietary separation process to remove metals and other solids.

The effectiveness of today's re-refining technologies for lubricating oils has been technically certified on many occasions. Offsite re-refining is currently practiced by a number of U.S. government agencies, including the military and postal services, with very good technical results. However, because of the economic factors, only 5% to 6% of the 1.2 billion gal of used oil generated annually in the U.S. is re-refined.[2]

Example. Evergreen Oil Co. of Newark, California, operates a state-of-the-art re-refinery that processes 30,000 gal of used oil daily. The plant produces 8 million gal of high-quality base oil each year that is used in commercial products such as motor oil, automatic transmission fluid, and industrial machining and hydraulic oils.[6]

OIL FILTER RECYCLING

The need to dispose of spent oil filters usually goes hand in hand with the need to dispose of used oil. Federal regulations governing the disposal of used oil filters are contained in 40 CFR 261. The regulations require that used oil filters be tested for toxicity characteristics and disposed of accordingly. However, an exemption for testing is provided for non-terne-(tin and lead alloy)plated used oil filters that are drained in any of the following ways:[7]

- puncturing the filter and hot-draining for 12 hours
- hot-draining for 12 hours and crushing
- dismantling and hot-draining for 12 hours
- any equivalent hot-draining method that removes used oil

A common practice for minimizing the volume of waste fuel filters to be disposed of is to crush the filters after draining. This can be done using specially designed hydraulic filter crushers (Figure 15.7). This practice maximizes the number of filters that can be placed in a disposal drum or other receptacle, which can drastically reduce disposal costs if costs are based on volume.

The need to dispose of filters can be eliminated altogether by recycling. Oil filter recycling vendors operate in many parts of the country. They process spent filters by shredding them and then separating the paper filter element from the metal casing. The metal casing is recycled as scrap metal, and the paper is disposed of or burned as fuel. Recycling facilities typically charge generators for accepting used filters. A Chicago recycler charges $0.42 per lb for filters. Most recyclers require that filters be drained for 12 hours and that they not be crushed. Many operations furnish customers with dedicated collection bins.

Another option for generators of used oil filters is to separate the filter elements onsite. This can be done by shredding or cutting the filters open and removing the paper element from the casing. The metal casing can then be drained and sold as scrap metal. CTEC, Inc., of Salt Lake City, Utah, manufacturers an oil filter cutter that, like a large can opener, cuts the bottom off the filter casing so the element can be removed. The cutter is operated manually and adjusts to most filters. Operators at a drive-in oil-change facility that has used the CTEC device stated that the cutter is easy to use and that filters can be opened at a rate of 60 per hour.

Figure 15.7. Oil filter crusher at fleet operations, Hanford, WA. As the filter is crushed, the oil is collected and recycled. The crushed filter is recycled into rebar. (Photo courtesy of Westinghouse Hanford Company/BCSR, Richland, WA).

TECHNIQUES FOR REDUCING AND REUSING USED ANTIFREEZE

Many of the basic waste reduction techniques discussed for used motor oil also can be applied, with slight modifications, to used antifreeze. Reuse applications differ significantly, however, because antifreeze has little if any heating value and could not be burned as fuel. Reuse technologies for antifreeze focus instead on recycling and further use as a coolant. The discussion that follows focuses on practices that are applicable to a typical engine repair shop. Types of practices addressed are:

- improving housekeeping practices
- extending product life
- segregating waste streams
- recycling used product

Examples of vendors that furnish equipment and technologies for implementing some of these techniques are listed in Table 15.4.

Improving Antifreeze Handling and Housekeeping Practices

Purchasing, inventory, handling, and storage practices such as those discussed for motor oil also can be implemented to reduce the disposal of new or unused antifreeze and to minimize antifreeze-related waste generated as a result of spills or leaks of product.

Shelf life is not a particular concern with antifreeze because ethylene glycol tends to be a very stable compound. Additives may settle somewhat if virgin products sit for long periods (several years). Shaking or stirring is effective for resuspending or redissolving additives.

As with motor oil, antifreeze should be stored in sealed containers and kept free of contaminants such as dirt and oil. Storage temperature is not a concern because ethylene glycol and the additives in antifreeze remain stable within the range of temperatures that may be experienced under ambient conditions.

Table 15.4. Technology Vendors: Used Antifreeze

Company	Address	Telephone
Antifreeze Ultrafiltration		
Kleer-Flo	15151 Technology Drive Eden Prairie, Minnesota 55344	612/934–2555
Finish Thompson, Inc.	921 Greengarden Road Erie, Pennsylvania 16501	814/455–4478
C.B. Mills	1225 Busch Parkway Buffalo Grove, Illinois 60089	708/459–0007
Antifreeze Recyclers		
C&R Industries, Inc.	5555 Branchville Road College Park, Maryland 20740	301/441–4824

Extending Product Life

The effective life of antifreeze is the length of time that the antifreeze retains the properties that make it an effective engine protectant. Usually, a time or mileage standard for replacement is established by the engine manufacturers on the basis of test data and estimated average operating conditions.

In some circumstances, it can be cost-effective to implement measures to extend the useful life of antifreeze. Maintenance managers of large fleets of vehicles have often found this to be the case. Two approaches that work are antifreeze testing and onboard recycling.

Antifreeze Testing

Antifreeze properties that typically would be tested include specific gravity, freezing point, boiling point, pH, general corrosivity, aluminum corrosivity, and foaming. The American Society for Testing and Materials (ASTM) has established standard specifications for properties of antifreeze for automobile and light-duty service engines, as well as for heavy-duty engines. These standards and the ASTM test methods used to determine them are shown in Table 15.5.

Antifreeze testing can be applied through several types of programs. For large engines, such as power generators, ships, or trains, when the appropriate laboratory facilities are available, it may be cost-effective to test antifreeze on an engine-by-engine basis. At regular intervals, samples would be collected and tested. Antifreeze

Table 15.5. ASTM Antifreeze Specifications

Parameter	Range	ASTM Test Method
Freezing point °F (°C) (50% volume in distilled water)	–34 (–37)	D 1177
Boiling point °F (°C) (50% volume in distilled water)	226 (107)	D 1120
pH (50% volume in distilled water)	7.5 to 11.0	D 1287
Chloride (ppm)	25 maximum	D 3634
Corrosion in glassware (weight loss, mg/specimen)		D 1384
Copper	10 maximum	
Solder	30 maximum	
Brass	10 maximum	
Steel	10 maximum	
Cast Iron	10 maximum	
Aluminum	30 maximum	
Corrosion of cast aluminum ($mg/cm^2/week$)	1.0 maximum	D 4340
Foaming tendency (volume, ml)	150 maximum	D 1881

Source: Reference 8

not meeting specifications would be replaced, and the used antifreeze would be recycled or disposed of.

A less extensive program would involve testing antifreeze samples from representative vehicles in a motor pool to establish an average antifreeze life for each vehicle class. On the basis of test results, a schedule for changing the antifreeze of all vehicles in that class would then be implemented.

A third type of program would involve changing antifreeze on a regular basis and consolidating and testing the used antifreeze from a number of vehicles. Test results would govern whether the antifreeze would be reused, recycled, or disposed of.

On-Board Recycling

On-board recycling involves the use of filter systems that are installed on the engine. Antifreeze traveling through the engine cooling system is filtered to remove contaminants that adversely affect its properties. On-board filtration systems can be either full-flow or bypass, as discussed for motor oil. Supplemental coolant additives can be added on a regular basis to replace corrosion and foam inhibitors.

On-board filtration is particularly effective and often necessary for long-running heavy-duty gasoline and diesel engines where antifreeze must be maintained in prime condition to protect expensive components and reduce maintenance costs and vehicle downtime. This technology has been used in this application for more than 25 years. Most engines are not factory equipped with onboard filters; typically, the filters are installed by the owner.

Recommended coolant life can be as long as 240,000 engine miles when on-board recycling is used, compared to 36,000 to 40,000 miles when it is not. On-board filtration has not been used much on automobiles and light trucks because it is not cost-effective. For small engines, off-board recycling technologies would appear to be more useful. These technologies are discussed in a later section.

Segregation of Waste Streams

Antifreeze that no longer meets specifications because of normal use in an engine is not usually a hazardous waste, and because it is readily biodegradable, can often be treated inexpensively in a biological wastewater treatment plant. However, if mixed with used motor oil, gasoline, solvents, transmission fluid, or other waste streams common to the automotive maintenance industry, antifreeze can become nonrecyclable, costly to treat, or even hazardous. To prevent this, radiator fluid should be drained and stored in dedicated and clearly marked containers. Handling procedures should be implemented that minimize spills and mixing of wastes.

Antifreeze Recycling

Off-specification antifreeze often can be restored through simple physical processes that remove contaminants, and through replacement of chemical additives (Figure 15.8). Specific recycling methods that are seeing relatively widespread use include:

- standard particle filtration
- ultrafiltration
- distillation

Figure 15.8. A Glyclean Recycling Process Unit annually recycles 2,000 gallons of ethylene glycol at fleet operations, Hanford, WA (Photo courtesy of Westinghouse Hanford Company/BCSR, Richland, WA).

Standard Particle Filtration

In standard particle filtration, multistage filters in the 5-to 25-micron range are used to remove solids such as dirt, rust, and suspended metals that can act as abrasives and cause engine wear. This can be followed up with ion exchange, which removes dissolved metals that cause corrosion, or with aeration/filtration, which removes oils that can affect the freezing and boiling point or increase foaming. Additives usually must be added to precipitate out metals, reduce foam, and restore color. Virgin antifreeze can be added to lower the freezing point.

Several types of standard filtration systems are available in the automotive repair industry. Large, stationary, fleet-sized treatment units operate in 50-to 100-gal batches. These units can be set up to draw used antifreeze from a feed drum or small tank, treat it, and discharge the recycled product to a second container. Portable units are available that can be hooked up directly to a vehicle's radiator. With all of these filtration units, filter elements must be replaced regularly and often have to be disposed of as a hazardous waste.

Another type of filtration system for antifreeze has recently been developed by Gulf Coast Filters of Gulfport, Mississippi, at the request of the U.S. Air Force. This relatively simple system uses rolls of household paper towels as elements and can remove particles smaller than 1 micron. The filters can be mounted on a portable skid or trailer and used throughout a maintenance facility to recycle antifreeze that has been collected in containers, or they can be attached directly to the engine to provide continuous filtration during engine use. No data are available on the ability of these filters to return antifreeze to ASTM specifications. Gulf Coast paper towel filters were initially developed as motor oil and fuel filters and are discussed in more detail in the section on motor oil filtration.

Ultrafiltration

Ultrafiltration uses a multistage filtration process where the initial filter is typically in the 5-micron range and the final filter is in the 0.001-micron range. Ultrafiltration is designed to remove molecular-size contaminants, such as sulfates and chlorides, which are the primary causes of corrosion.

Kleer-Flo Company of Eden Prairie, Minnesota, markets an ultrafiltration unit that uses 5- and 0.0025-micron filters. This device is reportedly capable of restoring antifreeze to meet all applicable ASTM standards, including corrosion. As with other antifreeze recycling technologies that use filtration, additives and new antifreeze must be added to restore some properties.

Distillation

Antifreeze distillation is a two-step process. In the first step, water is distilled under atmospheric pressure. In the second step, ethylene glycol is distilled under a vacuum. The two streams are condensed separately and collected in drums as processed ethylene glycol and distilled water. Dissolved and suspended solids and other contaminants remain in the process vessel and are disposed of. The recycled ethylene glycol can then be mixed with the proper amount of distilled water and additives, and can be reused.

A batch-operated still, applicable for use in a maintenance garage, will operate in 15- to 20-gal batches. Each batch takes 10 to 15 hours to treat and generates 0.5 to 1 gal of still bottoms requiring disposal.

Case Study 15.3: Comparison of Antifreeze Recycling Methods

The U.S. EPA Risk Reduction Engineering Laboratory recently conducted a study to evaluate the performance of several standard filtration units and a batch distillation unit currently being marketed for recycling antifreeze. The tests were conducted on spent antifreeze generated from a range of vehicles owned by the New Jersey Department of Transportation. Two filtration units were studied: a fleet-size unit that used 25- and 5-micron filters followed by aeration and refiltration, and a portable single-vehicle unit that used 25- and 5-micron filters followed by ion exchange. Both treatment processes replaced additives specified by the manufacturers. The distillation unit was a 15-gal batch unit that also required final replacement of additives.

The antifreeze was analyzed both before and after treatment using selected tests recommended in ASTM D3306 and ASTM D4984. The results were compared to the ASTM standards for virgin antifreeze. Both filtration treatment processes restored the boiling point, freezing point, pH, and foaming tendency to specified levels. However, neither filtration unit was able to produce recycled antifreeze that passed the general corrosivity or aluminum corrosion test. Subsequent chemical tests showed that corrosive salts such as sulfates and chlorides were not removed by filtration. Recycled coolant from the distillation unit passed both the general corrosivity and aluminum corrosion test, and a chemical characterization showed that corrosion-causing contaminants such as sulfates, chlorides, and dissolved solids had been removed or reduced.

REFERENCES

1. "Final Report: Management of Used Lubricating Oil at Department of Defense Installations." National Institute for Petroleum and Energy Research, June 1986.
2. "Study Shows Most Used Oil Goes Up in Smoke; Coalition Calls for Action to Promote Recycling." *Hazardous & Solid Waste Minimization & Recycling Report*, February 1992.
3. Hazardous Waste Management System; Identification and Listing of Hazardous Waste; Recycled Used Oil Management Standards; Final Rule. *Federal Register* 57(176): (September 10, 1992).
4. "Waste Minimization Plan for the Department of the Air Force, Elmendorf Air Force Base, Anchorage, Alaska." Elmendorf Air Force Base, December 1991.
5. Jowers, L., Dairy Fresh Corporation, Letter to C. Sims, Gulf Coast Filters, Inc., 1980.
6. "California Oil Rerefinery Aims to Prevent Environmental Pollution by Recycling Oil." *Hazardous & Solid Waste Minimization & Recycling Report*, December 1990.
7. Hazardous Waste Management System: General; Identification and Listing of Hazardous Waste; Used Oil. *Federal Register* 57(98):(May 20, 1992).
8. *1991 Annual Book of ASTM Standards*, American Society for Testing and Materials, 1991.

GENERAL REFERENCES

Code of Federal Regulations (CFR) Section 40, Chapters 261–270; RCRA Regulations.

Gavaskar, A.R., R.F. Olfenbuttel, J.A. Jones and P.M. Randall. "Automotive Engine Coolant Recycling," presented at the Utah State University, Waste Management Conference 1992.

Kalnes, T.N., K.J. Youtsey, R.B. James and D.R. Hedden "Recycling Waste Lube Oils for Profit (UOP Direct Contact Hydrogenation Process)" *Hazardous Waste & Hazardous Materials* 6(1):(1989).

"Recycled Coolant Meets Auto-Maker Standards." *Hazardous & Solid Waste Minimization & Recycling Report*, April 1992.

"Technical Bulletin: Myths and Realities of Recycling Anti-Freeze," Kleen-Flo Company, July 1990.

CHAPTER 16

Aluminum Industry

*Jim Mavis**

ALUMINUM INDUSTRY AND WASTE SOURCES

The aluminum industry is a mature industry composed of three principal segments plus a growing recycling segment. The three main segments are bauxite processing (sometimes associated with an adjacent mine), reduction, and forming. Each segment is described briefly, along with origins of production wastes.

Bauxite Processing

Bauxite is processed using the 100-year-old Bayer process which produces alumina (Al_2O_3) that is in turn used in reduction plants. The main steps in the Bayer process are digestion with caustic soda (NaOH) at high temperature (145 to 250°C); clarification to remove iron-rich "red mud," producing green liquor; cooling of the liquor to precipitate aluminum hydrate; and calcination to convert the hydrate to alumina.

Red mud, the highly alkaline, insoluble residue from bauxite digestion, constitutes the principal waste product from Bayer process plants. Because opportunities for large-scale waste reduction of this by-product are unlikely, and there is no bauxite processing industry in the United States, this segment of the aluminum industry is not discussed further.

Reduction

Aluminum metal is produced by electrolyzing alumina that is dissolved in a bath of molten cryolite (Na_3AlF_6). Electrolysis is carried out in long rows or lines of individual cells or pots (potlines).

Each pot is a shallow, rectangular, covered steel vessel that is lined internally with a thermal and electrical insulation layer (refractory), inside of which is placed a layer of carbon blocks. The carbon blocks act as the cell cathode (negative electrode). The combined insulator/carbon block interior is called the potliner. The cross section of a typical pot is shown schematically in Figure 16.1.

---

*Mr. Mavis is a senior consultant in CH2M HILL's Seattle, Washington, office and may be reached at (206) 453-5000.

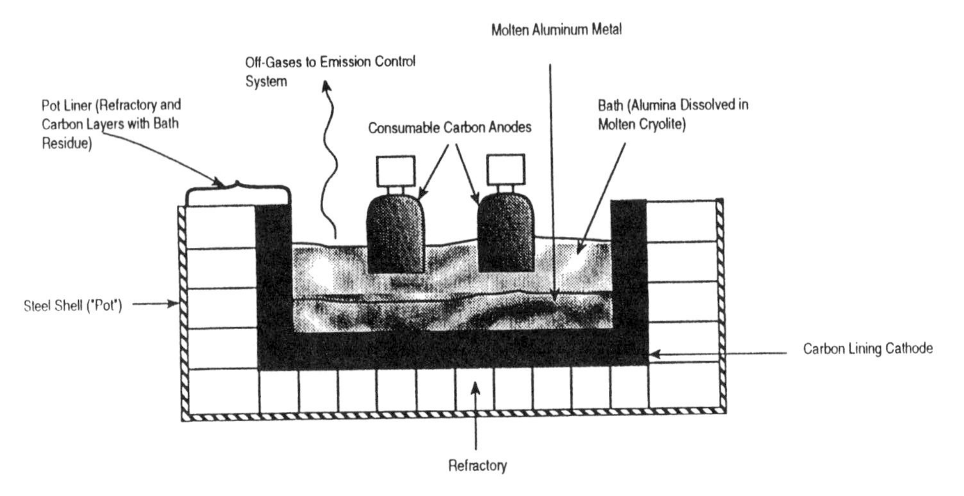

Figure 16.1. Schematic cross section of a typical pot for aluminum production.

Spent Potliner

After operating for several years, the potliner typically fails and the pot must be removed from service for relining. The solid waste from an old pot is called "spent potliner," a listed hazardous waste (K088). Spent potliner is the most troublesome waste produced in aluminum reduction because of the large volumes that have historically been produced and because of its characteristics. It is estimated that the United States reduction industry produces more than 200,000 tons per year of spent potliner.[1] The rate of production of spent potliner has been estimated as ranging from about 65 to 100 pounds per ton of aluminum produced.[2] Spent potliner is composed of roughly equal parts of carbon (cathode material) and thermal insulator (refractory), with solidified bath electrolyte. The composition of this waste stream is somewhat variable, with concentrations of major and minor constituents falling into the ranges shown in Table 16.1.

Table 16.1. Typical Characteristics of Spent Potliner

Characteristics	Percent
Carbon	18 to 59
Fluoride	11 to 16
Aluminum	3 to 15
Sodium	3 to 14
Calcium	1 to 2
Silicon	1 to 2.5
Cyanide	0.08 to 0.3
Sulfur	0.1 to 0.25
Iron	0.4 to 1
pH (aqueous slurry)	10.5 to 12
Lithium, Magnesium, misc.	—
Physical State	Powder through hard chunks >1 foot

Spent potliner is listed as a hazardous waste because it contains cyanide. Cyanide is not used in the aluminum reduction process itself—rather it is formed in or on the carbonaceous cathode by mechanisms that are not understood, but may involve starved-air oxidation of carbon in confined spaces. When spent potliners are exposed to atmospheric precipitation, groundwater, or high humidity, the cyanide may dissolve and undergo a number of chemical reactions. The most significant reaction is the combination of cyanide ion with iron to form ferrocyanide, $Fe(CN)_6^{4-}$. Iron occurs both as an impurity in spent potliner and in most subsurface soils beneath disposal sites. Whereas simple cyanide such as CN^- is widely recognized as toxic, iron-complexed cyanide is considerably less toxic. In fact, it has been used medically to test kidney function.[3]

Nevertheless, the presence of cyanide is the basis for its listing as a hazardous waste (K088) by the U.S. EPA. Speciation of cyanide forms in leachate from a typical spent potliner is shown in Table 16.2.[4]

Wet-Scrubber Blowdown

A bath of molten cryolite is maintained inside the lined pot as the nonconsumable electrolyte. Alumina from a Bayer plant is dissolved in the cryolite and is the consumable material that electrolytically is reduced to form aluminum metal. The positive-cell electrode is a consumable carbon anode that is lowered into the molten-salt bath from above. Oxygen from the decomposing alumina combines with carbon from the anode and is released as carbon dioxide and carbon monoxide. In addition, small amounts of hydrofluoric acid, sulfur dioxide, and heavy organic compounds are emitted as gases and captured either in wet scrubbers or in dry beds of granular aluminum oxide. Blowdown from wet scrubbers must be treated before being discharged to the environment. Alumina from dry scrubbers is fed to the pots as part of the alumina feedstock. The approximate composition of blowdown from potline wet scrubbers is shown in Table 16.3.

The carbon anodes may be precured carbon blocks or a mixture of coke and pitch that cures when exposed to the high-temperature bath (Soderberg process) by baking-off semivolatile organic matter from the pitch binder. In prebake plants, carbon and pitch are pressed in molds and heated to drive off volatile organics before the anodes are installed in the pots.

As electrolysis of the alumina proceeds, molten aluminum accumulates in the bottom of the pots and is periodically withdrawn into crucibles and transported to the cast house. Casting of molten aluminum is a part of both the reduction and forming segments of the industry.

Table 16.2. Speciation of Cyanide in Spent Potliner Leachate

Species	Concentration (mg/L)	Percentage of Total Cyanides
Free CN^-	20	0.4
$Fe(CN)_6^{4-}$	4,280	88.8
$Fe(CN)_6^{3-}$	<10	<0.2
SCN^- [a]	510	10.6

[a]Strictly speaking, thiocyanate (SCN^-) is only a cyanide derivative. It does not behave as a cyanide or exhibit cyanide's toxicity.

Table 16.3. Approximate Composition of Potliner Wet-Scrubber Blowdown

Constituent	Concentration
Sodium, mg/L	3,500
Potassium, mg/L	400
Calcium, mg/L	19
Magnesium, mg/L	3.6
Ammonia (total), mg/L	3.6
Chloride, mg/L	2,900
Sulfate, mg/L	3,650
Fluoride, mg/L	180–440
Phosphate, mg/L	0.4
Cyanide (total), mg/L	1.7–38
Cyanide (amenable to chlorination), mg/L	0.6
Alkalinity (total), mg/L as $CaCO_3$	150
Alkalinity (phenolphthalein), mg/L as $CaCO_3$	110
Silica (SiO_2), mg/L	15
Aluminum, mg/L	9–51
Antimony, mg/L	0.011–0.036
Arsenic, mg/L	0.47
Chromium, mg/L	0[a]
Copper, mg/L	0[a]
Iron, mg/L	1
Nickel, mg/L	0.25–0.51
Zinc, mg/L	0.1
Benzo(a)pyrene, μg/L	0[a]–0.047
Fluoranthene, μg/L	3.1
Phenol, μg/L	9
Pyrene, μg/L	1.7
Oil and Grease, mg/L	2; <5
Total Suspended Solids, mg/L	100–257
Electrical Conductivity, μmhos/cm	32,400–36,000
pH	8.6–9.0

[a]Below unspecified detection limit.

Many plants contain both reduction and forming facilities, making certain operations such as casting difficult to assign solely to one industry segment. Because waste products from casting within either segment are similar, casting is described in this chapter as part of reduction, rather than introducing unnecessary complexity into the discussion.

Aluminum may be cast into stationary molds (pigs and sows), drop cast into logs or tees that are cooled by direct-contact cooling water, or cast into other shapes such as continuous-cast rod, usually with the application of direct-contact cooling water. Before actual casting, the molten metal may be purified, filtered, or alloyed and held in a melt furnace. Contact cooling water is usually contaminated with oil and particulate matter. Molten aluminum held in casting furnaces is usually blanketed with salt to prevent oxidation. The discarded salt layer (dross) contains aluminum oxide and some metallic aluminum.

Forming

Aluminum forming encompasses a group of operations that convert aluminum and its alloys into intermediate or semifinished products. The U.S. EPA recognizes five forming subcategories and an additional group of ancillary operations. The subcategories are:

- rolling with neat oils
- rolling with emulsions
- extrusion
- forging
- drawing with emulsions or soaps

Although the U.S. EPA also identifies a set of ancillary operations likely to be associated with each forming subcategory, it is more useful to consider forming in terms of its waste products: oil, metals, chromate, and caustic etch bath.

Oil

A number of forming operations use oils or emulsions as lubricants and may contaminate process water with oil, grease, or emulsions. Rolling and draining operations normally use oil-based lubricants which enter the plant wastewater stream. In addition, oil frequently leaks into wastewater from extrusion presses.

Metals

Both forming and subsequent finishing operations can introduce metals into plant wastewater. Fine particulate matter from these operations contributes metal to some plant waste streams, whereas water rinses, following operations such as sealing with nickel acetate, introduce soluble metal to the wastewater. A less obvious source of metal in wastewater may be the raw-water supply serving the plant. For example, zinc was found in the incoming plant process and service water supply to one plant that drew its water from a nearby river.

Chromate

Although not strictly part of forming, chromate conversion coating of formed aluminum products introduces both trivalent and hexavalent chromium into plant wastewater. The chromium in chromate is a heavy metal, yet it is considered separately from other heavy metals such as nickel and zinc because they are positively charged in aqueous solution whereas chromate is negatively charged. Treatment of chromate-containing wastewater requires different techniques than those used for cationic (positively charged) metals.

Caustic Etch Bath

Formed aluminum products are sometimes etched as part of their surface conditioning for subsequent application of coatings or finishes. After a period of operation, the caustic etch bath becomes contaminated with aluminum removed during etching and with other impurities. Unless the impurities are removed, the etch bath must be periodically discarded and replaced with fresh bath. Spent etch bath may be bled into the wastewater stream and neutralized before being discharged.

The foregoing waste types and constituents from aluminum-forming operations do not represent a comprehensive treatment of the subject. Other types of waste are produced at various facilities, but a comprehensive discussion is beyond the scope of this chapter.

WASTE MANAGEMENT AT REDUCTION PLANTS

Spent Potliner

The aluminum industry has tried long and hard to find an acceptable means of managing spent potliner. These efforts fall under three strategies: waste reduction, waste reuse, and waste treatment for disposal. Each strategy is discussed below.

Waste Reduction

The U.S. EPA's listing of spent potliner waste as "hazardous" (K088) has added urgency to efforts already under way among primary aluminum producers to increase the service life of potliners.[5] Potliner design improvements and increases in life expectancy vary with individual plants because of differences in the ages and configurations of the installations.

Waste Reuse

Before spent potliner was listed as a hazardous waste, most primary aluminum producers had already exhausted in-house reuse options and were exploring offsite reuse alternatives in nonaluminum-producing industries. The listing of spent potliner as a hazardous waste has effectively blocked offsite reuse by other industries because of the onerous permitting, reporting, and special handling requirements and the stigma attached to products made from "hazardous waste." A few of the reuse options that were used or considered before it was listed are summarized below.

Spent potliner has been processed on the site to recover materials for reuse in aluminum reduction plants either on or off the site. To a limited extent, this is still being practiced.

Example. One facility used a wet process for recovering cryolite (Na_3AlF_6) for use as the initial charge of electrolyte (molten salt) when a new reduction plant (or a new pot) is placed into service. The cryolite recovery plant has been decommissioned because its production capacity greatly exceeded current and expected future demands. A description of wet cryolite processing can be found in a report by K.E. Aunsholt.[6]

Example. At a second facility, a small fraction of the spent potliner is calcined to remove the carbonaceous fraction, and the recovered residue containing cryolite and alumina is returned to the pots. The amount of calcined product that may be returned to the pots is limited by constraints on the levels of impurities that can be tolerated in the aluminum product.

Investigations into reusing the carbonaceous fraction of spent potliners as a supplement in Soderberg anodes also revealed that product quality was affected, and the quantity recycled was severely limited.[7]

One of the most promising offsite uses of spent potliner is as a cement kiln feed supplement. The high carbon content provides fuel value and the process destroys cyanide, the contaminant that prompted the U.S. EPA to list the waste. Aluminum industry pursuit of this potential solution was stopped by nontechnical constraints of the current federal hazardous waste regulations, particularly RCRA recycling rules and Boiler and Industrial Furnace (BIF) rules.

Spent potliner was also evaluated for use as a flux for steel-making in open-hearth furnaces.[8] Although early testing showed some potential use as a substitute for fluorspar, the market disappeared and technical development was discontinued as a result of changes in hazardous waste regulations.

Other options for use of spent potliner are in mineral wool manufacturer or as a fluoride source. For a combination of technical, market, economic, regulatory, and other reasons, these options have not proven feasible. Reuse options for spent potliner are reportedly still under investigation, but none is believed to be close to commercialization.

Waste Treatment and Disposal

Large amounts of spent potliner have accumulated over the decades, and waste production continues, though probably at a decreasing rate. The previous method for managing spent potliner was by placement in a properly constructed, licensed, and operated landfill. After May 1993, federal land disposal restrictions prohibit placement of spent potliner if it contains toxic forms of cyanide at a concentration above 500 mg/kg.

One aluminum company has received U.S. EPA approval to operate an incinerator for treating spent potliner at a site near Gum Springs, Arkansas. This facility will be used to treat residues from other primary aluminum producers, and appears to be capable of meeting the 500 mg/kg criterion. Several commercial hazardous waste disposal companies have landfills suitable for impoundment of spent potliner, and at least one reduction facility has a dedicated storage facility on the site.

Gaseous Emissions

Aluminum Reduction

A small fraction of the fluoride in aluminum reduction plant electrolytic cells is released as hydrogen fluoride when moisture in the alumina feed reacts with components of the molten salt bath. Additional fluoride is lost as airborne particulate matter. This lost fluoride causes environmental damage and depletes the fluoride content of the electrolytic cells, necessitating addition of supplemental fluoride.

The older practice of controlling fluoride emissions with wet scrubbers has been largely superseded by newer dry-scrubber technology, which captures the fluoride with in-duct injection, or passes the gas through a fluidized bed of alumina, as shown schematically in Figures 16.2 and 16.3. The fluoride-laden alumina is then fed into the reduction cells. In this way, up to 99% of the released fluoride is recovered and returned to the process with the alumina feedstock. Dry capture of fluoride from electrolytic cells eliminates a significant source of wastewater from primary aluminum manufacture, but a small quantity of alumina dust usually escapes, normally depositing within a short distance of the dry scrubber, and contributing suspended solids to stormwater runoff.

Prebake Anode Production

Prebake anodes for aluminum reduction plants are produced by blending petroleum coke and pitch at an elevated temperature and casting the mixture into molds (green anodes), then curing the castings by baking them at a high temperature. Anode butts that have been removed from service are cleaned, crushed, and recycled to the green anode production step. Hydrocarbon vapors and coke dust are generated during green anode preparation, and hydrogen fluoride and hydrocarbon vapors are emitted when the anodes are baked.

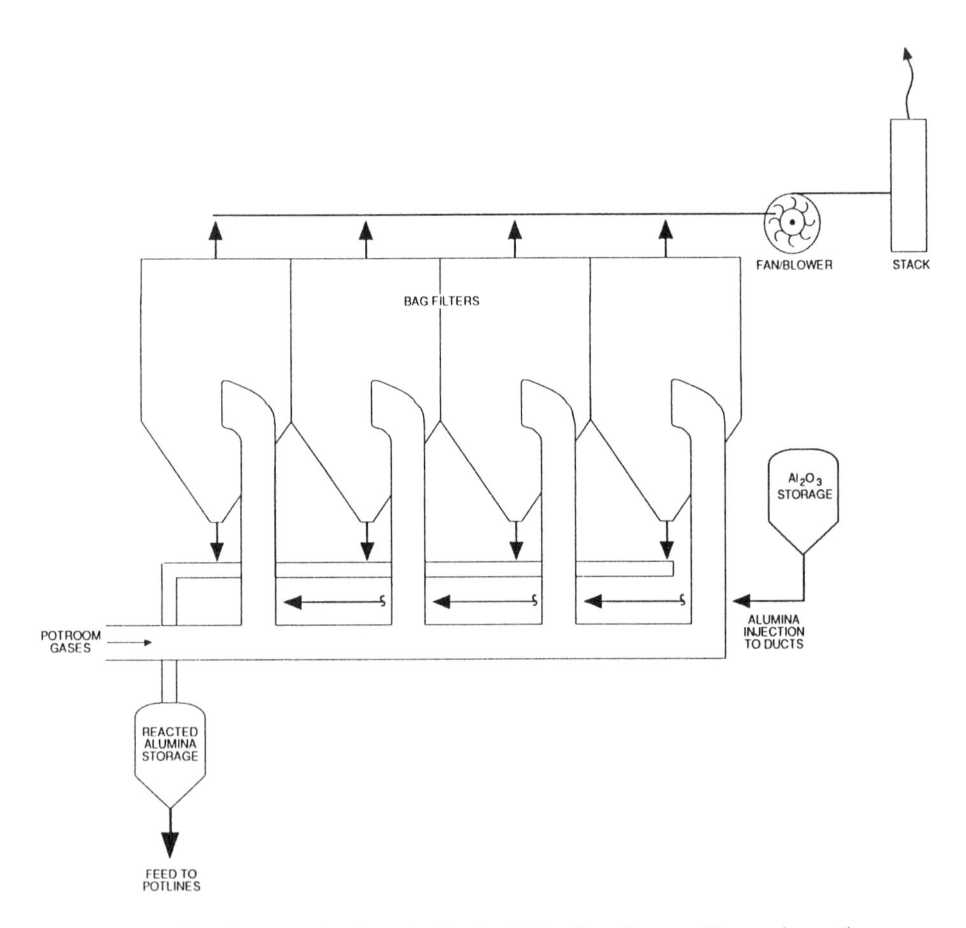

Figure 16.2. Aluminum reduction plant—duct injection dry-scrubber schematic.

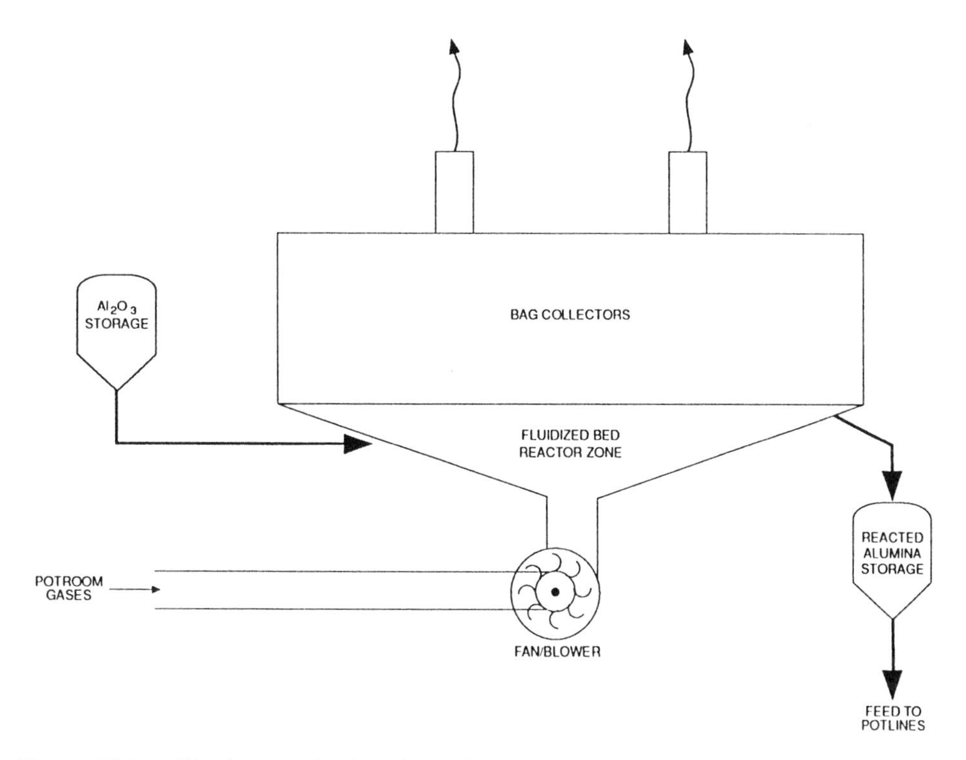

Figure 16.3. Aluminum reduction plant—fluidized-bed dry-scrubber schematic.

Coke dust and hydrocarbon vapors emitted from green anode production may be captured in a bag filter that has been precoated with coke fines. The recovered coke dust and hydrocarbon tars, along with the precoating material, may be used in the preparation of green anodes.

In the past, wet scrubbers and wet electrostatic precipitators have been used to capture emissions from anode baking, but dry-scrubber technology is widely used today. Dry scrubbing of gases emitted from baking furnaces (offgas) requires that the gas stream first be cooled to condense the hydrocarbon into particles that can be captured on activated alumina that is fed into either a duct-injection scrubber or a fluidized-bed system. The reacted alumina is used to supplement the feed to the aluminum reduction cells, either as is or after the alumina is heated to destroy the accumulated hydrocarbons. Baking furnaces, which can be operated with low air and high tar, permit hydrocarbons to be recovered for reuse in green anode preparation. Hydrocarbons are recovered in a system upstream of the alumina-based dry fluoride recovery step. Hydrocarbon tar is condensed in an indirectly cooled tube-and-shell surface condenser and collected in a vessel that is heated to keep the product liquid. Fine hydrocarbon tar aerosols not captured in the surface condenser are removed in an electrostatic precipitator, and the cooled gas stream is then sent to a dry alumina scrubber for fluoride recovery.

Wastewater

Federal effluent guidelines for the primary aluminum (reduction) industry restrict the discharge of contaminants on the basis of mass load rather than concentration of contaminants in the effluent stream. Reduction facilities have frequently found it necessary to reduce effluent flows so that the contaminant concentrations are high enough to be efficiently removed with conventional treatment technology. Thus, waste reduction as applied to wastewater from a primary aluminum plant is nearly always the result of a need to reduce the contaminant mass loading in the effluent stream.

Strategies for reducing the amount of industrial wastewater have been developed extensively during the past decade and can be summarized briefly. The first step is to identify the sources of the regulated contaminants and to segregate contaminated from noncontaminated streams. The second step is reducing the flow of the contaminated streams. This may be accomplished in many ways, such as recirculating once-through cooling water through a cooling tower for reuse, or using the effluent from one process as the water supply for another process (cascade reuse). Finally, the concentrated waste streams are treated to remove regulated constituents before being discharged.

Various methods for reducing wastewater flow from industrial facilities are in widespread use. Because they have been reported in technical literature for many years, they will not be further discussed here.

WASTE MANAGEMENT AT FORMING PLANTS

Dross

Dross, the salt flux that provides a floating protective layer over molten aluminum in a casting furnace, contains aluminum oxide, along with fine metallic aluminum droplets and other impurities. The aluminum metal content of dross represents an economic loss and poses a safety hazard if the aluminum particles ignite and start a thermiting reaction.

Safety issues and more efficient metal recovery from dross have been addressed in a variety of ways in the aluminum industry, most commonly by speeding the rate of dross cooling. At least one system for cooling dross and improving metal recovery and safety is commercially available. The heart of the system is an external water-cooled rotary cooler. On the basis of a survey of four users, the manufacturer claims improvements in metallic aluminum recovery of approximately 10% to 30% over prior practices.[9]

Wastewater

Flow Reduction

Waste minimization opportunities associated with water use in the aluminum-forming industry are diverse and vary with each facility. Casting is a common operation using cooling water, and wastewater usage can be reduced by use of a recirculating system with an evaporative cooling tower. Cooling water used in recirculating systems must usually be conditioned to prevent scale formation, corrosion, fouling, and other problems.[10]

Another common method of reducing the wastewater volume from some aluminum-forming (and other metal-finishing) operations is to install countercurrent rinsing. This rinsing configuration and its benefits are discussed in the chapter on metal plating and surface finishing and will not be repeated here.

Treatment

Plants that use multistage countercurrent rinsing with caustic etching may have an opportunity to reduce waste by installing a caustic recovery system. Over the life of a typical caustic etch bath, the aluminum content increases until the bath must be discarded and a fresh bath prepared. To extend the life of the bath by "caustic recovery," the caustic etch is diluted and chilled to form crystals of hydrated aluminum oxide, which are filtered out and sold, leaving the caustic portion of the bath in solution. This reconstituted solution of caustic soda can be returned to the etch tank.

REFERENCES

1. Rickman, W.S., "Circulating Bed Combustion of Spent Potliners." *Light Metals 1988.* Proceedings of the Technical Sessions by the TMS Light Metals Committee, 117th TMS Annual Meeting. Phoenix, Arizona, January 25–28, 1988.
2. Filho, A.C. Braut, et al., "Use of Spent Potlining in the Red Brick Ceramic Industry." *Light Metals 1988.* Proceedings of the Technical Sessions by the TMS Light Metals Committee, 117th TMS Annual Meeting. Phoenix, Arizona, January 25–28, 1988.
3. Stieglitz, E.V., and A.A. Knight, *J.A.M.A.*, 103:1760–1764 (1934); Plotz, M., M. Rothenberger, *J. Lab. Clin. Med.*, 24:884–887 (1939); Kleeman, C.R., and F.H. Epstein. *Proc. Soc. Exp. Biol. and Med.*, 93:228–233 (1956).
4. Kimmerle, F.M., et al., "Cyanide Destruction in Spent Potlining." *Light Metals 1989.* Proceedings of the Technical Sessions by the TMS Light Metals Committee, 118th TMS Annual Meeting. Las Vegas, Nevada, February 27-March 3, 1989.
5. Peyeau, J.M., "Design of Highly Reliable Pot Linings," *Light Metals 1989.* Proceedings of the Technical Sessions by the TMS Light Metals Committee, 118th TMS Annual Meeting, Las Vegas, Nevada, February 27-March 3, 1989.
6. Aunsholt, K.E., "Plant for Regeneration of Cryolite from Spent Potlining," *Light Metals 1988*, Proceedings of the Technical Sessions by the TMS Light Metals Committee, 117th TMS Annual Meeting, Phoenix, Arizona, January 25–28, 1988.
7. Cutshall, E.R., et al., "Recycle of Spent Potliner through Soderberg Anodes," *Light Metals 1989*, Proceedings of the Technical Sessions by the TMS Light Metals Committee, 118th TMS Annual Meeting, Las Vegas, Nevada, February 27-March 3, 1989.
8. Augood, D.K., and J.R. Keiser, "The Use of Spent Potlining as Flux in Making Steel," *Light Metals 1989*, Proceedings of the TMS Light Metals Committee, 118th TMS Annual Meeting, Las Vegas, Nevada, February 27-March 3, 1989.
9. McMahon, J.P., and B.J. Davies, "A Safe, Efficient and Economic Method of Cooling Aluminum Dross," *Light Metals 1989*, Proceedings of the Technical Sessions by the TMS Light Metals Committee, 118th TMS Annual Meeting, Las Vegas, Nevada, February 27 to March 3, 1989.
10. Smyrniotis, C.R., "Water Treatment Concerns in Aluminum Casting," *Light Metals 1988*, Proceedings of the Technical Sessions by the TMS Light Metals Committee, 117th TMS Annual Meeting, Phoenix, Arizona, January 25 to 28, 1988.

CHAPTER 17

Construction and Demolition

*Paula Magdich**

THE CONSTRUCTION AND DEMOLITION INDUSTRY

The construction and demolition (C&D) industry generates a significant quantity of waste, most of which ends up in landfills. Figures vary, but estimates indicate that C&D waste accounts for up to 25% of all waste delivered to some landfills.[1] Although C&D waste is largely inert, it is of concern because the volumes generated consume a large portion of our limited landfill space.

C&D activities include demolition; commercial leasehold improvements (CLI); and construction of new residential, commercial, and industrial facilities. Demolition and CLI activities generate up to 10 times as much waste as new construction, and represent the greatest potential for waste reduction.

It is nearly impossible to define a typical C&D site; sites vary depending on the nature of the project and the activities involved. C&D companies handle a wide range of contracts including building roads, bridges, pipelines, hospitals, offices, shopping centers, and multi-family and single-family residences. All these projects may include demolition, renovation, and new construction. A C&D project can be divided into preconstruction, rough construction, and finishing, with various subcontractors being involved at each phase, as illustrated in Table 17.1.[2]

C&D WASTE

The C&D industry uses a wide variety of materials, including lumber and wood, concrete and other masonry products, gypsum drywall, and metal products. In addition, hazardous materials such as paints, solvents, and adhesives are used. Many of these materials eventually become waste; typically up to 10% of the materials delivered to a construction site become waste.[3]

The components of C&D waste differ depending on the nature of the project. This is illustrated in Table 17.2, which lists components of new construction waste and demolition waste.[4] This difference in waste stream composition is shown graph-

---

*Ms. Magdich is an environmental engineer in CH2M HILL's Calgary, Alberta, Canada, office and may be reached at (403) 237-9300.

Table 17.1. Stages of a C&D Project

Preconstruction
- Site selection
- Demolition
- Clearing
- Staking out
- Excavation

Rough Construction
- Footings
- Foundation walls/slab
- Posts and girders
- Floor joist framing
- Floor openings framed
- Subflooring
- Wall framing
- Wall sheathing
- Windows
- Exterior doors
- Siding
- Ceiling framing
- Roof framing
- Plumbing
- Electrical

Finishing
- Exterior trim
- Exterior painting/finishing
- Exterior fixtures
- Roofing
- Interior doors
- Cabinets
- Interior trim
- Interior wall painting/finishing
- Interior fixtures
- Flooring/carpeting
- Touch up
- Final cleanup

Source: Adapted from "Waste Audit Study—Building Construction Industry," California Department of Health Services (1990).

ically in Figure 17.1.[5] Geographical location may also influence the composition of the waste stream. For example, brick construction is more common in some areas of North America, whereas wood construction is more common in other areas.

Waste from new construction is composed primarily of a mixture of unused or damaged raw materials, as well as offcuts (discarded cut material) and packaging. The largest component of waste from new construction is lumber and wood, followed by gypsum wallboard, cardboard, plastic packaging, and metal. Demolition waste includes actual building components, such as full-length studs and concrete slabs. The largest component of demolition waste is concrete, followed by wood, brick and clay, and metal.

Waste materials from new construction are usually clean and relatively uncontam-

Table 17.2. Major Components of C&D Waste

New Construction	Demolition and CLI
Dimension lumber	Dimension lumber
Plywood	Plywood
Concrete/masonry	Concrete/masonry
Metals	Asphalt
Drywall plastics	Reusable fixtures
Carpet	Metals
Cardboard	White goods (appliances)
Foam insulation	Plastics
Fiberglass	Drywall
Soil and land-clearing waste	Carpet
Food/organic waste	Other
Hazardous waste (solvents/oils)	
Other	

Source: Adapted from "Construction and Demolition Industry Waste Audit Study," CH2M HILL, Ltd. (March 1992)

inated, whereas demolition waste materials are often dirty or contaminated and are mixed with other materials. These differences between C&D wastes create specific opportunities and challenges for waste reduction. C&D waste-stream compositions also vary with changing trends in building materials, such as the trend toward using more plastic and composite materials.

The generation rate of specific C&D wastes varies through the stages of a project. During initial stages of new construction, most of the waste is wood from rough carpentry, whereas finishing stages generate more drywall, cardboard, and plastic packaging. During demolition, large quantities of waste are generated in a relatively short time period.

Although C&D waste is generally nonhazardous and inert, a number of hazardous or potentially hazardous materials are used. These are listed in Table 17.3.[2] Many of the materials on this list are transformed during the construction process into inert materials as they dry (e.g., paints, adhesives, caulking); however, residual material in partially used containers requires proper management and handling. A large percentage of C&D hazardous wastes are generated during demolition and renovation when asbestos insulation, PCB-containing light ballasts and transformers, and lead-based paint are removed. Hazardous wastes should be managed in accordance with all applicable regulations.

For decades, C&D waste was landfilled without raising concerns, given that the materials are relatively inert. Special "dry waste" or "C&D waste" landfills emerged to handle the large volumes of C&D waste, and tipping fees were relatively low. However, landfills are filling to capacity and tipping fees are rising—to the point where alternatives to land disposal are being sought.

OPPORTUNITIES FOR REDUCING C&D WASTE

A number of opportunities for reducing waste are available to the C&D industry. One of the best ways to identify these opportunities is to conduct a waste audit, which involves tracking the wastes generated and the disposal costs. Most waste

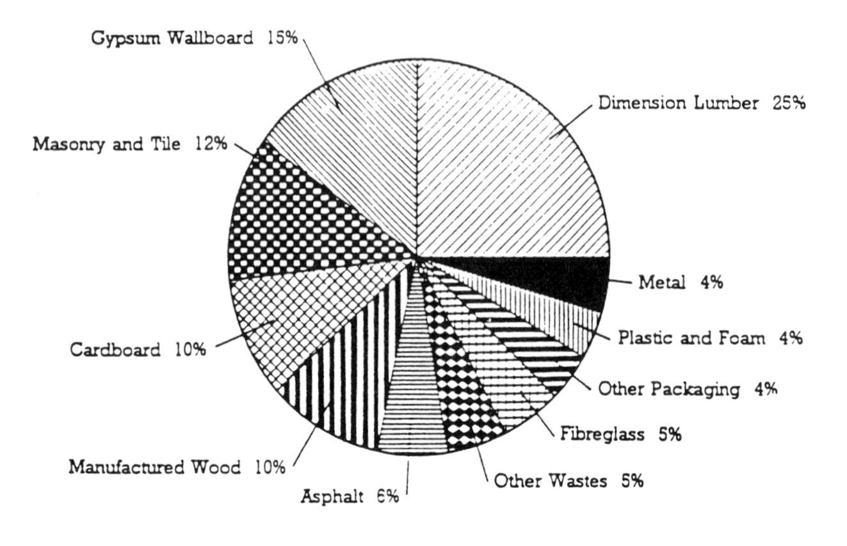

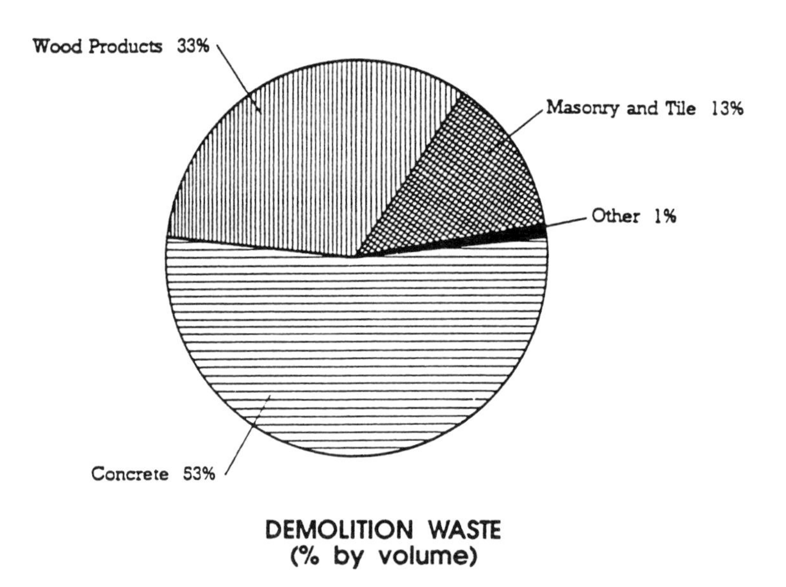

Figure 17.1. Estimated composition of new construction and demolition wastes.

reduction measures are relatively low technology and low cost; however, they often require operational or procedural changes. Organization and effort are required for implementing measures for reducing waste throughout the phases of a C&D project.

Design Phase

C&D waste reduction begins at the design phase. Careful consideration should be given to design and material specifications because they affect the quantity of waste

Table 17.3. Hazardous and Potentially Hazardous C&D Materials

Acetone	Glues
Acetylene gas	Greases
Adhesives	Helium (in cylinders)
Ammonia	Hydraulic brake fluid
Antifreeze	Hydrochloric acid
Asphalt	Insulations
Benzene	Kerosene
Bleaching agents	Lime
Carbon black	Lubricating oils
Carbon dioxide (in cylinders)	Lye
Caulking, sealant agents	Metals
Caustic soda (sodium hydroxide)	Methyl ethyl ketone
Chromate salts	Motor oil additives
Chromium	Paint remover
Cleaning agents	Paint stripper
Coal tar pitch	Paint/lacquers
Coatings	Pentachlorophenol
Cobalt	Polishes for metal floors
Concrete curing compounds	Putty
Creosol	Resins, epoxies
Cutting oil	Sealers
De-emulsifier for oil	Shellac
Diesel fuel oil	Solder flux
Diesel lube oil	Solder, soft (lead)
Etching agents	Solder, other
Ethyl alcohol	Solvents
Fiberglass, mineral wool	Sulphuric acid
Foam insulation	Transit pipe
Freon	Varnishes
Gasoline	Waterproofing agents
	Wood preservatives

Source: Adapted from "Waste Audit Study—Building Construction Industry," California Department of Health Services (1990).

generated and the hazard potential. Communication between the designer and builder is important so that each realizes how design decisions affect onsite operation and waste generation. Some opportunities for waste reduction during the design phase include:

- Specify durable and maintainable materials so that frequent replacement is avoided.
- Specify reused or recycled materials to prevent those materials from entering landfills.
- Specify materials that can be easily reused or recycled.
- Specify locally made products to reduce transportation costs and fuel consumption.
- Favor building renovation over demolition where appropriate to reduce demolition waste.

- Design using a standard module that corresponds to material dimensions to minimize cutting waste (8′ or 10′ grid).
- Specify water-based adhesives and paints where practical.
- Discourage labor-only contracts with subcontractors. Make the subcontractors responsible for materials and waste management.

Purchasing and Inventory

Purchasing and inventory should be carefully controlled to prevent waste of raw materials. Consideration also should be given to the types of materials purchased. Waste reduction at this stage of a project translates directly into cost savings. Some of these measures include the following:

- Institute just-in-time ordering so that materials arrive when they are needed. This minimizes damage of materials that are stored and moved around the site.
- Order appropriate material sizes to reduce the amount of cutting waste.
- Purchase materials in bulk where possible to minimize the amount of packaging waste (e.g., buying larger-size boxes of screws).
- Favor suppliers who are willing to take back excess packaging, such as pallets and cartons from appliances and fixtures.
- Favor suppliers who will take back excess material (e.g., unused bricks).
- Avoid purchase of toxic materials; if they are necessary, buy in small quantities to avoid disposing of the excess.
- Check materials thoroughly on arrival and return damaged goods so they do not become onsite waste.
- Ensure that materials susceptible to weather damage are stored properly (indoors or under polyethylene); train employees to properly handle and store materials to prevent damage.
- Provide secure fencing around the site to prevent vandalism.

Onsite Activities

Onsite C&D activities include demolition and site preparation, as well as construction activities. Waste reduction measures can be implemented during each of these activities as discussed below.

Demolition

Waste reduction during demolition is commonly referred to as materials salvaging (Figure 17.2). C&D waste may be significantly reduced at this stage because demolition and CLI renovations create about 10 times the waste of new construction. To be profitable, careful planning and removal of reusable or recyclable construction materials are required. The demolition process may take longer because of the

Figure 17.2. Demolition of a pump house resulted in 900 tons of recycled steel and 13,000 tons of crushed concrete and block because of this universal processor which includes a hydraulic sheer, concrete breaker, and concrete pulverizer to aid in demolition and material recovery. (Photo courtesy of Westinghouse Hanford Company/BCSR, Richland, WA).

salvaging of materials, but the economic and environmental benefits are significant. Some waste reduction measures include the following:

- Identify reusable and resalable materials before demolition (e.g., plumbing and lighting fixtures, appliances, doors, windows, large-dimension timber, molding, and masonry); consider a separate contract for predemolition salvaging.
- Inform workers of the materials to be salvaged and the appropriate strategies for removing items intact.
- Identify uses or onsite or offsite recycle markets for salvage materials (e.g., building material exchanges or charities).
- Be aware of hazardous components of old buildings (such as light ballasts containing PCBs, lead-based paints, and asbestos insulation) and obtain technical advice on handling these materials. Some hazardous waste materials may be reusable.

Case Study 17.1: The Architectural Clearinghouse

With the help of a computerized inventory database and a recycling depot, Envirocycle Expediting operates the Architectural Clearinghouse in Edmonton, Alberta, where a broad range of building materials is collected and redistributed. A system of predemolition marketing has been developed that is more extensive than most regular salvage operations. An onsite audit is undertaken and a contract is negotiated with the property owner or manager to remove recyclable materials before demolition. The end user of the materials supplies the labor, equipment, and transportation required to remove the materials. The Architectural Clearinghouse has demonstrated that more than 100 metric tons of material destined for landfills per month can be diverted for reuse by contractors, renovators, and individuals. It estimates that more than 3,000 metric tons of building materials have been diverted from landfills, representing a savings of more than $1.5 million Canadian dollars in materials based on current retail replacement costs.[6]

Site Preparation

During site preparation, an effort should be made to minimize the amount of excavation and vegetation disturbance. This will reduce the amount of excavated soil and vegetation that must be disposed of in landfills as well as reduce the amount of clean fill and new vegetation that must be brought in. Some waste reduction measures include the following:

- Use existing landscaping and grades to avoid over-excavation.
- Maintain existing vegetation to minimize soil erosion.
- Design roadways that avoid damage to vegetation.
- Use equipment with special tires to avoid soil compaction and erosion.
- Employ an appropriately sized bucket to avoid digging a bigger trench than necessary.
- Shore trenches to enhance safety and to minimize excavation.
- Investigate the use of recycled aggregates as backfill.
- Maximize the use of excavated soil from the site for landscaping and regrading.
- Chip and shred vegetative material from clearing land to use as mulch in landscaped areas.

Concrete and Asphalt

Concrete and asphalt wastes include concrete forms, excess concrete or asphalt, release oils, concrete sealants, and slabs or blocks from demolition activities. Some opportunities for reducing the volume of concrete and asphalt wastes include the following:

- Reuse concrete forms where possible.
- Cut and measure forms carefully to ensure minimum cutting and maximum reuse.
- Use excess concrete on the site where possible (e.g., for parking curbs and planters).

- Handle release oils from concrete forms carefully to avoid contamination of soil.
- Prevent concrete washings from entering storm sewers and surface water because the high pH is toxic to fish.
- Use water-based concrete sealant instead of solvent epoxies, which are hazardous and generate more noxious fumes.
- Use recycled aggregate for concrete and asphalt where possible.
- Recycle concrete and asphalt.

Case Study 17.2: Concrete and Asphalt Recycling

Recycling of waste concrete and asphalt is a well-established practice in North America. Waste materials are crushed in stages, and metal such as steel rebar is removed using magnetic separation. The aggregate is produced to meet specifications for road base or fill material. For example, concrete and asphalt recyclers in the Seattle area have supplied recycled concrete aggregate for major projects at the Seattle-Tacoma airport and local municipalities. The Asphalt Recycling and Reclaiming Association supports and promotes asphalt recycling programs. The Los Angeles Bureau of Street Maintenance has an asphalt-recycling program, one of the largest in the world. Concrete and asphalt recycling makes good economic sense, because the recyclers generally charge significantly lower tipping fees than landfills, and the recycled aggregate is less expensive than virgin aggregate.

Masonry

Construction masonry, a significant portion of new construction waste, includes brick, stone, clay, and tile. Reduction measures include the following:

- Store pallets of bricks on firm level ground with metal banding or shrink-wrap plastic covers to prevent breakage.
- Store mortar bags under cover to prevent weather damage.
- Mix only as much mortar as can be used that day.
- Reuse bricks and other masonry products.
- Recycle unused and damaged masonry products (most concrete recyclers accept other masonry materials).

Wood Products

Wood waste is one of the largest components of waste leaving a construction site. It is composed of offcuts, discarded pallets, concrete forms, and damaged or warped lumber. Some ways to reduce wood waste include the following:

- Designate a central cutting area so reusable pieces can be stacked and easily located for reuse (e.g., short pieces of wood can be used for bridging, backing, or stakes).

- Identify scrap wood that is available for reuse by employees (e.g., post a sign such as "free wood").
- Carefully stack and protect wood to prevent damage and reduce warpage.
- Favor water-based adhesives where appropriate.
- Recycle wood waste where possible.

Onsite recycling may involve grinding wood into chips or dust for mulch. Offsite recyclers may also be contacted, such as recyclers who manufacture roofing felt and fiberboard. In addition, some municipalities are experimenting with mixing wood waste and sewage sludge and using it as a compost material. Bulk burning of wood waste is not recommended and most municipalities prohibit this practice.

Metals

Recycling of nonferrous and ferrous metals is a well-established practice throughout North America. Listings of metal recyclers are readily available in the yellow pages. Metal C&D waste that may be recyclable include steel rebar, aluminum siding, plumbing fixtures, aluminum and copper wire, piping, and metal banding. In addition, some metal appliances can be sold to scrap dealers. To reduce the amount of metal waste generated, an onsite storage area should be designated so that reusable lengths can easily be found.

Plastics and Vinyl

The use of plastic materials and packaging is increasingly common in the industry. Plastic recycling is available, but it is complicated somewhat by the need to segregate plastic by resin type and the fluctuating markets for recycled plastic products. Recyclers may accept only specific types of plastic, depending on their recycling process, and they may require the plastic waste to be clean and uncontaminated. As with all recycling, proper waste segregation is key. Plastic wastes that can be recycled include plastic shrink wrap, styrofoam, bubble wrap, and vinyl siding. These waste plastics are reprocessed into a variety of low-grade plastic products such as garbage bags, plastic piping, and wood substitutes (Superwood). Some of these recycled products also can be used on C&D sites.

Paper and Cardboard

Cardboard and paper, which constitute approximately 5% to 10% of construction waste by volume, are readily recyclable. A growing number of jurisdictions have encouraged cardboard recycling by placing surcharges or bans on disposal of cardboard in landfills. Cardboard and paper are reprocessed into new paper products.

Thermal and Moisture Protection

Thermal and moisture-protection wastes include roofing, siding, and insulation materials. Some waste reduction options include the following:

- Favor a membrane roofing system over a hot-tar roofing system. The membrane system uses fewer potentially toxic materials and can reduce the quantities of solvents used to clean tools and equipment.
- Collect excess insulation to reuse or to give away.
- Recycle insulation and roofing shingles where possible. Rigid insulation can be reprocessed into blown insulation and packing materials; roofing shingles can be incorporated into cold-mix asphalt recycling processes.
- Favor insulation made from recycled products such as blown cellulose insulation from recycled paper or mineral fiber insulation from mining waste.

Asbestos may be encountered during renovation or demolition projects. All health and safety precautions should be followed to ensure that asbestos is handled and disposed of in accordance with applicable regulations.

Doors and Windows

Doors and windows are prime candidates for reuse. Building material clearinghouses or exchanges keep inventories of the supply and demand for used building materials. Although used windows may not be suitable for external use because of energy efficiency considerations, they may be useful for interior and decorating applications. Also, during construction cleanup, film-free detergents are available for cleaning exterior windows before applying sealants, thus avoiding the use of petroleum-based solvents.

Finishes

Finishing includes installing drywall, carpeting, and painting. Some ways of reducing waste from these activities include the following:

- Handle and store drywall to prevent damage.
- Order drywall to correspond to the most appropriate dimensions (custom lengths).
- Store drywall cutoffs for reuse and patching.
- Use water-based paints where appropriate.
- Avoid purchase of excess paint and use excess paint as an undercoat in future jobs where appropriate.
- Reuse paint solvents where possible.
- Use carpet tiles to reduce cutting waste. The tiles are easily lifted and reused.
- Recycle waste drywall, paints, and solvents where possible. Drywall can be recycled by removing the paper coating and reprocessing the stripped board into new wallboard. Some companies also are investigating the use of chipped waste gypsum wallboard as a soil amendment to improve acidic or saline conditions. Spent solvents and paints are readily recycled as discussed in other chapters of this book.

Mechanical

These activities include installing heating, ventilation, and air conditioning systems and plumbing. Some waste reduction opportunities include the following:

- Measure and cut plumbing pipe carefully to avoid waste.
- Reuse plumbing hardware where possible.
- Recover and reclaim Freon (CFCs) when installing or servicing air conditioning systems and refrigeration equipment.

Electrical

The main wastes generated by the electrical trades are used fixtures and metal wires. Waste reduction opportunities include the following:

- Reuse fixtures where possible.
- Save reusable lengths of cable.
- Ensure that light ballasts and transformers containing PCBs are properly disposed of (pre-1978 ballasts may contain PCBs).

REFERENCES

1. "C&D Debris: A Crisis is Building," *Waste Age* (January 1992): pp. 26–32.
2. "Waste Audit Study-Building Construction Industry," California Department of Health Services, 1990.
3. Skoyles, E.R., and J.R. Skoyles, "Waste Prevention on Site," (London: Mitchell, 1987), p. 28.
4. "Construction and Demolition Industry Waste Audit Study," prepared for Alberta Environment by CH2M HILL Ltd., March 1992.
5. *Construction Waste Management Report*, Science Council of British Columbia, 1991.
6. Gerrand, S.B., "Building Toward Change: Recycled Architectural Building Materials," Envirocycle Expediting, May, 1992.

GENERAL REFERENCES

"Construction and Demolition Waste Minimization Guidance Manual," Alberta Environment, Action on Waste, May 1992.

"Making a Molehill Out of a Mountain," Canada Mortgage and Housing Corporation, 1991.

Chattergee, S. et al., "Predictive Criteria for Construction/Demolition Solid Waste Management," U.S. Army Report N-14, Construction Engineering Research Laboratory, Columbus, OH, 1977.

"Managing Construction and Demolition Debris: Trends, Problems, and Answers," General Building Contractors of New York, March 1990.

Recycling Construction and Demolition Waste in Vermont, Final Report, Vermont Agency of Natural Resources, December 1990.

CHAPTER 18

Electric Utilities

*Courtney H. Brown and Linda Kunitsky**

This chapter describes wastes generated by electric utilities, outlines common pollution prevention strategies, and presents several case studies of waste reduction projects that have been implemented by electric utilities. Coal- and oil-fired, hydroelectric, and small gas or diesel electric power generating stations are addressed in this chapter. Reduction of wastes produced in maintenance of utility rights-of-way also is presented.

DESCRIPTION OF WASTE GENERATED BY ELECTRIC UTILITIES

Wastes produced in generating electricity vary, depending on the source of energy. Wastes generated by coal- and oil-fired steam electric generating stations, hydroelectric stations, small-capacity gas-turbine and diesel generating stations, and hydroelectric stations are described below.

Coal- and oil-fired stations generate various wastewaters, solid fuel by-products, and air emissions that often are challenging to manage.

Hydroelectric power is produced by capturing the energy of flowing water without consuming or altering the composition of the water source, so it is a very clean method of generating power. Typically, in hydroelectric generating stations as well as in small-capacity fossil-fuel stations, the largest volume of waste results from maintenance and repair of equipment used for generating and distributing electricity.

Gas-turbine or diesel engines often are used for generating electricity in remote areas having insufficient demand to justify larger, more efficient generators or to meet peak demands in urban areas. Diesel generators also are used as backup equipment when regular power supplies are interrupted. Another fuel used to provide the required energy source is natural gas in areas that have sufficient local supplies. Wastes generated by these facilities include air emissions and waste from

*Ms. Brown is an environmental engineer in CH2M HILL's Reston, Virginia, office and may be reached at (703) 471-1441. Ms. Kunitsky is an environmental engineer at Delmarva Power in Newark, New Jersey, and may be reached at (302) 452-6036. Evan Jones wrote the case study on B.C. Hydro for this chapter.

the maintenance and repair of equipment used for generating and distributing electricity.

Wastewater

Coal- and oil-fired steam electric power stations produce wastewaters that typically are treated and discharged to a surface water body. The wastewaters include low-volume wastewaters, cooling tower blowdown, once-through cooling water, nonchemical cleaning wastes, chemical metal-cleaning waste, and bottom- and fly-ash transport water. Small-capacity fossil-fuel-fired stations and hydroelectric stations generate less significant quantities of wastewater.

Low-Volume Wastewaters

Low-volume wastewaters generated at coal- and oil-fired steam electric plants include demineralizer regeneration waste, fly-ash unloading wash water, coal-pile runoff, boiler blowdown, floor-drain waste, cooling-tower basin cleaning wastes, and miscellaneous wastes generated from water treatment. The most common problem wastes are described below.

- **Demineralizer Regeneration Waste.** Ion exchange is used to produce demineralized boiler feed water. Periodically, the ion exchange resins are regenerated using acid and caustic. The regeneration waste is characterized by high acidity or alkalinity and high concentrations of dissolved solids. The waste also includes rinse waters that are relatively clean. Rinse waters can be recycled for use as cooling tower makeup or treated and used as boiler feed makeup.

- **Fly-Ash Unloading Wash Water.** Water is sprayed in fly-ash unloading areas for dust suppression. In addition, trucks usually are washed after filling and the unloading area is washed to remove accumulated ash. This wastewater typically contains very high concentrations of suspended solids and some metals from the fly ash.

- **Coal-Pile Runoff.** Stormwater runoff from coal piles can contain high concentrations of metals, particularly iron, and acidity proportional to the sulfur content of the coal. The characteristics of the runoff depend on the type of coal and the configuration and size of the coal pile. The longer that stormwater is in contact with the coal, the more concentrated the runoff will be. The acid in coal-pile runoff results from bacterial oxidation of the sulfur in coal to sulfuric acid.

- **Floor-Drain Waste.** Floor-drain waste includes leakage from pipes and equipment, waste from washing floors, and spills. Typically, the waste has high concentrations of suspended solids and often has significant concentrations of oil and grease.

Cooling-Tower Blowdown

Cooling-tower blowdown is a high-volume waste stream that usually is managed separately from other wastewaters. It contains suspended and dissolved solids and can contain metals, such as copper, produced by corrosion of condenser tubes or concentrated from the raw water throughout evaporation. Chlorine and other chemicals added to control biological growth and inhibit scaling or corrosion in condensers and cooling towers also are in the blowdown.

Once-Through Cooling Water

Once-through cooling water is a high-volume waste in comparison to other wastes. It can contain residual chlorine and added metals, primarily copper and zinc, from condenser-tube corrosion.

Nonchemical Cleaning Wastes

These wastes are produced by using high-pressure water for cleaning air heaters, boiler firesides, coolers, and condensers. The wash waters are generated intermittently, are usually acidic, and contain high concentrations of suspended solids and heavy metals.

Chemical Metal-Cleaning Waste

Boilers, feedwater heaters, condensers, and other process equipment is periodically cleaned with acid to remove scale deposits. The chemical cleaners can contain a variety of acids and chelating agents. The waste usually has high acidity and very high concentrations of heavy metals. For example, iron concentrations are typically on the order of 5,000 ppm. Boiler cleaning waste can be very difficult to treat if chelating agents are used for cleaning.

Bottom-Ash and Fly-Ash Transport Water

Waters used to sluice bottom ash from the boiler and to suspend fly ash to make it pumpable typically are large-volume waste streams that are difficult to manage. Fly-ash transport wastewater often is acidic and has significant concentrations of heavy metals. Bottom-ash transport wastewater typically does not have an extreme pH and contains much lower concentrations of heavy metals.

Solid Wastes Produced at Generating Stations

Asbestos-Containing Material

This category consists of insulation and boards that contain the mineral asbestos. Other miscellaneous asbestos-containing wastes include containment materials and discarded protective suits used for handling asbestos. These wastes are generated during renovation, demolition, or maintenance activities, such as pipe repair. Workers are required to segregate these wastes and place them into yellow or other specially colored bags that are clearly labeled. The wastes are stored in a separate dumpster and are removed to a landfill that is approved for disposal of asbestos materials.

Ash

Fly ash and bottom ash are produced at both coal- and oil-fired steam electric power plants, but the volume is much higher at coal-fired plants. Fly ash, collected from flue gas, has small particles. Bottom ash consists of larger particles that collect in the bottom of the boiler. Both fly ash and bottom ash contain heavy metals. The leachability of the metals depends on the disposal conditions and the original sulfur content of the coal. High-sulfur ash will tend to produce highly acidic leachate or runoff that can solubilize heavy metals. The volume of ash produced from coal burning is about 10% of the feed coal, depending on the ash content of the coal.[1] The primary components of ash are oxides of silicon, aluminum, iron, and calcium.

Blasting Media

Blasting sand and other materials are generated during maintenance (in surface preparation before painting) to remove rust and other deposits from metal equipment and parts, such as turbines. The materials used for blasting are granular and are silica-based (glass or sand), metal-based (steel or aluminum), or slag-based (Black Beauty). The waste materials usually are classified as residual but occasionally have tested as hazardous because of elevated levels of chromium, lead, or cadmium. Usually, municipal or residual waste landfills require testing of each load before accepting the load as nonhazardous.

Boiler Slag

Boiler slag is melted bottom ash that is reformed into a solid and recovered from the boiler. Typically, slag resembles large chunks of rock.

Chemicals

This waste stream comprises hazardous and residual wastes that are generated either in onsite laboratories, as a result of cleaning out expired chemicals, or from the disposal of unusable metering or other equipment that contains mercury. Small expended laboratory chemical containers are placed in a special drum (called a lab pack) that specifies certain compatible materials that can be combined for disposal of hazardous waste off the site. Other laboratory wastes fall under this category and may not be hazardous. Depending on the materials that are expended, these wastes are removed in drums by a licensed hauler to an approved disposal facility for hazardous or other chemical wastes.

Coal Mill Rejects

Coal-fired power stations that use roller mills generate coal mill rejects, but stations that have ball mills do not. During pulverization of the coal, the parts of the material that are high in mineral content are separated and rejected (by removal of the heavier parts). These residual wastes also are referred to as "pyrites" because of their high percentage of iron sulfides by weight. These mill rejects fall out to the bottom of the mills and either are sluiced to a storage tank or remain dry and are stored in (or conveyed to) a stockpile until removal to an onsite basin or landfill or to an offsite landfill. Plants burning bituminous coal of medium sulfur content generate pyrites ranging from 750 to 1200 tons per year per 100 megawatt capacity.

Cooling Tower Sediment

This waste stream is the result of a continuous blowdown wastewater stream that flows through cooling towers, which contains dirt and other materials in the water source. An estimate of the generation of cooling tower sediment is 0.02 cubic yards per megawatt-year.[2] Current management methods include removal during sump cleanout or outages, if needed. Disposal usually is by dewatering and disposal in an onsite or offsite impoundment or landfill.

Flue Gas Desulfurization (FGD) Sludge

FGD scrubbers are used to remove sulfur dioxide from flue gas. Lime or sodium hydroxide is used to absorb sulfur dioxide from the flue gas. The sludge produced from the scrubbers is composed primarily of gypsum (calcium sulfate) or of sodium sulfate with lower concentrations of ash.

Paint Waste

This waste stream includes paint and related coating materials resulting from painting of equipment or from cleaning out old paint stock. Paints can be latex or oil-based. Paint wastes can be hazardous because of ignitability or toxicity. If they are not ignitable or toxic, they are residual wastes, possibly small quantities. Paint cans sometimes are placed in drums for disposal as hazardous waste. Non-hazardous empty paint cans sometimes are crushed and placed in the incidental maintenance waste dumpster. Some companies have arranged for vendors to remove half-empty paint cans.

PCB-Containing Wastes

This waste stream consists of polychlorinated biphenyl (PCB)-containing liquids and solids, defined as wastes having concentrations of less than 50 ppm but more than 2 ppm of PCBs. PCB-containing liquids are a subset of this waste stream and include oils and other liquids from draining transformers or other electrical equipment. These liquids are placed in drums and removed for approved disposal. PCB-containing solids include empty drums or other containers that held PCB-containing liquids, or oily debris from cleanup of PCB-containing liquids. These wastes also are put in drums, similar to the oily-debris waste stream, for approved offsite disposal.

Petroleum-Based Lubricants and Fluids

These types of waste account for the largest volume of waste liquids generated at hydroelectric power facilities. Hydraulic oil, insulating oil, and breaker oil or switch oil wastes result from maintenance of electrical equipment. Waste lubricants and greases are generated from maintenance of generation equipment. In most cases, these wastes are typical of waste petroleum products generated by mechanical equipment, and they may contain pollutants, such as heavy metals. A small proportion of waste petroleum products may be contaminated with solvents or other materials. Similar to petroleum-based wastes are used oil filters, which are replaced during routine equipment maintenance.

Photographic Processing and Reprographic Wastes

This waste stream includes containers of toner, antistatic fluid, developer, lubricant, and other wastes generated from photographic development and microfilm processing and from maintenance of photocopiers. These wastes could be hazardous depending on the equipment and chemicals used.

Metals

Scrap copper wiring, both insulated and bare, is a large component of the metal waste stream generated at hydroelectric facilities. Other metal wastes include lead, motors, mixed ferrous metals, and equipment hulks. To a lesser extent, metal wastes, such as scrap aluminum and metal containers, also are produced. These metals are discarded during the repair and replacement of electrical equipment. Before disposal, the equipment often is disassembled, and fluids they contain are drained and collected. Historically, some of this electrical equipment contained PCB fluids, which may result in the emptied metal hulk being considered a PCB waste material.

Miscellaneous Solid Wastes

Most of the other waste produced at hydroelectric facilities and small-capacity fossil-fuel stations is solid waste from material packaging and shipping, office operations, construction and demolition activities, and routine maintenance. Wood pallets from shipments to hydroelectric sites accumulate at the facilities, as do other packaging wastes, such as cardboard and plastic wrappings. Rags used in cleaning and maintaining equipment are generated in large volumes. Paper products from office activities, including fine paper, computer paper, and mixed waste paper, are a smaller portion of the total solid waste stream than in typical municipal or industrial waste streams, but their volume is nevertheless significant. Construction, demolition, and maintenance work, such as equipment repair, generate mixed building and disposable work clothing debris. In addition, because many hydroelectric projects incorporate recreational activities involving the created water body, a significant amount of organic debris, such as lawn trimmings, brush cuttings, and driftwood, is generated at project sites.

Solid Wastes from Operation and Maintenance of Distribution Systems

The waste-generating activities related to power distribution include right-of-way maintenance and equipment maintenance and repair in linerooms and substations. Support activities, such as office duties, construction, and shipping/receiving, also generate wastes. Because of the number of intermediate facilities required and the length of typical distribution systems, a large network can result in large geographically dispersed volumes of waste.

Light Bulbs

Major change outs of street lamps are the source of this waste stream. In general, the bulbs are high-pressure sodium, metal halide, and perhaps mercury vapor (from older installations) lamps that are replaced as needed or on a regular schedule. The bulbs usually are kept separate from other waste streams and are removed for offsite disposal. The bulbs may test hazardous for lead.

Organic Debris

Typically, the largest volume of waste generated at distribution systems results from the need to maintain access to the distribution grid and from maintaining clearance between power lines and vegetation. Ground clearing and tree trimming or removal can result in a large volume of waste that, in most urban areas and some rural areas, must be disposed of in designated landfills.

Pesticides and Containers

Pesticides often are used to control vegetation in rights-of-way. This practice results in waste, including emptied pesticide containers, other materials contaminated during application, and expired materials that must be disposed of. Although the volume of material is not large, the nature of the waste requires special handling and disposal.

Petroleum-Based Lubricants and Fluids

The largest volume of liquid wastes generated by linerooms and substations typically are lubricating and insulating oils. Breaker oil and switch oil also are generated in large amounts. There is little difference in the nature of the oils generated at production facilities and those generated at distribution facilities, but the relative proportions of the types of oils will vary. Insulating oils are most common at distribution facilities.

Metals

Maintenance and repair of electrical equipment generates large amounts of metal wastes, including scrap aluminum and copper, metal equipment hulks, and miscellaneous ferrous and nonferrous metals. These wastes typically are drained and sorted by metal.

Batteries

Banks of batteries are used in substations to provide power in a power failure. Periodic replacement of dead batteries results in a waste stream that may be composed of a wide variety of batteries, including lead-acid, lead-antimony, lead-calcium, potash, or nickel-cadmium. A significant number of dry-cell batteries also are discarded.

Miscellaneous Solid Wastes

Distribution systems generate the same assortment of mixed solid wastes as production systems do, including construction and demolition debris, disposable work clothes and protective equipment, pallets and packaging wastes, and office wastes. Other solid wastes unique to distribution systems include ceramic insulators, treated lumber and poles, street lamp bulbs, and various building materials and paints.

Air Emissions

Combustion of fossil fuels for producing electricity naturally results in the generation of air emissions. The emissions are typical of any combustion process, but the choice of fuels has a significant effect on the quality of the emission. Air pollution control equipment, such as bag houses, precipitators, and scrubbers, is used at large coal-fired stations. Air pollution control equipment typically is not added to small-capacity generating facilities because of the relatively small volume of emissions.

TARGET WASTE TYPES AND REDUCTION STRATEGIES

Wastewater

Wastewater reduction can have several benefits: (1) Reducing the volume of wastewater that requires treatment lowers capital costs for new wastewater treatment systems or upgrades of existing systems and lowers operation and maintenance costs. (2) Reducing the volume of wastewater decreases the mass of pollutants discharged to surface water bodies. (3) Reducing the level of treatment required minimizes the complexity of a treatment system. Target strategies for waste reduction for industrial wastewater are listed below.

Housekeeping Measures

- Prevent oil from entering wastewater collection systems by installing curbs or containment structures around equipment and tanks that may leak oil.
- Eliminate or plug floor drains. Floor drains collect waste flows or streams from washing operations. At many locations, floor drains are not necessary. Other dry or wet systems for floor cleaning are available, such as floor-sweeping machines.
- Eliminate accessible hose bibs and fire hoses for washing floors. Avoid using fire hoses for washdown. Smaller hoses can be effective, and they generate much less wastewater.
- Limit the quantity of contaminated process-area stormwater by constructing curbs or berms around dirty areas. The contaminated stormwater then can be discharged to a treatment system with other wastes, treated separately from other wastes, or pretreated for removal of large quantities of ash or oil,

then combined with other waste streams for treatment. Clean stormwater from nonprocess areas can be discharged without treatment.

- Fly-ash unloading areas typically produce a significant volume of wastewater with a very high solids content. The volume of wastewater and the quantity of solids discharged to a wastewater treatment system can be reduced in several ways. Wash the area as little as possible. Remove accumulated ash using other means, such as a shovel or a mechanical loader. Regrade fly-ash unloading areas to facilitate pickup of solids by a loader or other equipment.

Equipment and Process Modifications

- Replace water pump seals with mechanical seals where possible to eliminate or minimize waste seal water.
- Switch from wet sluicing of bottom and fly ash to dry-ash transport systems.
- Switch from a once-through cooling system to a closed-loop system for small service-water-type cooling needs. This can eliminate potential contamination of water by oil and leaks from pump seals.
- Minimize air-heater wash. This can be accomplished by using a closed-loop system that treats air-heater wash as it is being generated. Dirty wash water is neutralized, solids and metals are removed, and the air-heater wash is recycled directly back to the washwater headers. This practice has been adopted by several power plants. Alternatively, treated effluent can be used as air-heater washwater and for other washing operations.
- Recycle as much process water as possible. Boiler blowdown and demineralizer regeneration rinse waters (not including the acidic or caustic first rinses) are good candidates for reuse because they are relatively clean wastewaters. Effluent from wastewater treatment plants also can be recycled. This reduces the amount of raw water required for power-plant operations and reduces the quantity of water and pollutants discharged to surface waters.
- Minimize the area used for coal storage. Provide good drainage of coal piles to minimize contact of rain water with the coal and to minimize the formation of sulfuric acid.
- Increase cycles of concentration on cooling towers, where possible, to limit the quantity of cooling-tower blowdown.
- Convert from an open bottom-ash sluicing system to a closed system involving minimal water discharge.
- Use reverse osmosis before ion exchange for demineralization to reduce the frequency of regeneration and thus reduce the volume of acid and caustic used and the volume of regenerant produced.

Solid Wastes

Opportunities for reducing solid wastes include numerous recycling options. Reducing the volume of solid waste streams generally is not feasible without changing fuels; e.g., a coal-fired plant converting to oil, which produces far less ash.

Fly Ash

Fly ash often is disposed of in landfills, but if it is of high quality, it can be reused. The most common use for fly ash is as a cement substitute or an aggregate in concrete. According to one estimate, 50 million tons of fly ash are reused annually.[3] Approximately 30% of coal ashes are reused.[4] Of the quantity reused, about 40% is used in concrete and 4% is used as highway fill.[3] The primary concern with reusing coal ashes is the potential for leaching of heavy metals into the environment. Fly-ash uses are summarized below.

- concrete and cement products
- road base, soil stabilizer, and clean fill
- filler in asphalt
- metal recovery—silicon, aluminum, iron, calcium, magnesium, and titanium (largely experimental)
- use as mineral filler, after treatment by direct-acid leaching, in place of calcium carbonate or gypsum in the following applications: asphalt roofing shingles, wall board, joint compounds, carpet backing, vinyl flooring, plastics, industrial coatings, aluminum alloys

Fly ash usually must be handled using dry conveyance systems if it is to be reused. Use of fly ash as a cement substitute is the most common reuse of the material. The resulting concrete is equal or superior in strength to concrete made exclusively with portland cement.

Research of processes for reclaiming metals from fly ash has been conducted for more than a decade. The metals are recovered using extraction processes. Iron is recovered from bituminous fly ash by magnetic extraction. Titanium is recovered by high-temperature chlorination. Processes for extracting other metals, including silicon, aluminum, calcium, and magnesium, also are being developed.[5]

Bottom Ash

It is estimated that one third of bottom ash produced at coal-fired plants is reused.[1] Bottom ash is used as:

- construction and road fill material
- abrasive blasting grit
- granules on asphalt roofing shingles
- aggregate for concrete and masonry blocks

- substitute for sand for traction on icy roadways
- soil amendment to increase permeability

For additional information on the reuse of fly ash and bottom ash, contact the American Coal Ash Association in Washington, D.C.

American Coal Ash Association, Inc.
1913 I Street, N.W., Sixth Floor
Washington, D.C. 20006
(202)659-2303

Oil Ash

Ash from oil-fired power plants can contain high concentrations of vanadium. Vanadium has been recovered and used as a hardening agent in the production of steel alloys. Oil ash can be marketed to vanadium reclaimers. However, the market for vanadium is not as big as it once was, and reclaiming vanadium from oil ash has become less popular.[3]

Flue-Gas-Desulfurization Sludge

There has been little reuse of scrubber sludge in the United States because as yet there are few operating FGD facilities. Less than 1% of FGD sludges are reused.[1] FGD sludge is reused extensively in Japan and Germany. Gypsum, the main component of the sludge, is used in wall board, as cement, as structural fill, as an agronomic soil amendment, and as a source of sulfur in manufacturing. The high cost of preparing the sludge for reuse often makes reuse uneconomical. In addition, the market for gypsum probably will become saturated very quickly because the quantity to be produced by planned FGD facilities is large in comparison to the demand.

Organic Debris

Implementing alternative practices for disposing of tree trimmings, right-of-way maintenance wastes, and other brush cuttings has the potential for diverting a significant amount of solid waste from landfills. Utilities typically pay contractors to perform this work, and often the contractor must arrange for disposal. The utility can control this waste stream by specifying appropriate disposal practices in the contract for services. Some alternative practices include:

- Chip the trimmings on the site and use them as a ground cover (on the right-of-way).
- Chip the trimmings and deliver them to a central processing area for distribution to public or charitable organizations for use or resale as mulch.
- Participate in, or develop, regional composting programs with municipal or community organizations.
- Require contractors to arrange for composting as part of the disposal contract.

In addition, less trimming may be needed if public incentive programs and vegetation-control strategies are implemented. Some of these alternatives include:

- Species selection may be used to discourage the growth of problem trees. By planting low, dense groundcover in right-of-way areas, the reintroduction of tall-growth vegetation by seed is hindered, reducing the growth of vegetation requiring trimming.
- Herbicides may be used in specific areas where environmentally appropriate. These chemical controls typically are combined with physical controls (e.g., mechanical trimming) to retard regrowth and reduce the frequency of maintenance programs. This approach is particularly applicable in areas of deciduous growth.
- Secondary uses of right-of-way areas can replace or control vegetation while providing space for other programs. Typical secondary uses include playing fields, golf courses, nursery projects (including Christmas tree farms), and agricultural production.
- In urban areas, information can be given to the public about appropriate species of ornamental trees that can be planted near electric lines to reduce future problems. This approach may include giving rebates to customers who purchase appropriate trees for replacing problem trees.

Waste Petroleum-Based Lubricants and Fluids

Reduction of waste petroleum-based lubricants and other fluids is achieved primarily through recycling programs. Recycling waste oil has become an established practice in most areas, and many of the petroleum-based lubricants generated by electric utilities are suitable for inclusion in these programs. In general, electric utilities do not generate sufficient quantities of waste lubricating oils to justify in-house recycling.

Other petroleum-based fluids, such as breaker or switch oils and insulating oils, are used in large quantities by utilities but are not used extensively outside of the electric industry. Recycling programs for these materials often will have to be developed internally, and the cost of recycling needs to be measured against the cost of disposal and replacement of the oils. Recycling may be as simple as filtration followed by blending with new oils. When the insulating oil is drained from equipment that has undergone a failure involving substantial heat generation, however, the oil can require such substantial reprocessing that disposal becomes a more cost-effective alternative.

Another alternative for electric utilities is the establishment of material-exchange programs with suppliers. In these programs, the utility essentially "rents" the material from a supplier and returns the used material for recycling. The availability of these programs is limited, but increasing demand is expected to result in increased availability of such services.

Metals

Electric utilities generate large amounts of waste metals, including copper, aluminum, lead, and ferrous metals. There are two primary methods by which this waste stream may be reduced: internal material exchanges and recycling. The greatest potential for volume reduction comes from recycling, although the strength of the markets for different metals varies considerably. Aluminum and copper are readily marketable, but ferrous metals command lower prices. The typical electric utility may generate metal waste in areas far removed from the secondary materials markets. In addition, recycling of electric wiring and equipment can be complicated because of factors such as insulation coverings, oily contamination, and mixed high- and low-grade metals. The marketing of these materials to recyclers may be enhanced through any of the following measures:

- "leveling" of material values by including a requirement for removing and recycling low-grade metal wastes in the sale of the more desirable high-grade metals
- "leveling" of unit values in geographic areas by requiring the collection and recycling of desirable metal wastes in remote areas at the same unit price as the recycler pays in central areas
- internal metals processing, including sorting and cleaning of waste metals where the value of the recovered metals justifies the effort

Other Solid Wastes

The volume of other wastes generated by utilities also may be reduced through source reduction, reuse, or recycling programs. Purchasing programs for designated materials may be established that encourage the use of recycled or recyclable materials, which in turn increase the ability of the utility to divert wastes from landfills. Some of these alternatives are described below.

Rags. Often, rags used for cleaning can be cleaned and reused (Figure 18.1). Constraints on this practice can arise either from purity requirements, in the case of cleaning delicate equipment, or if the rag has been used with a material that is not easily removed (such as heavy oil, sludge, and paint). Rags can be cleaned and reused through supplier return programs (the supplier cleans the rag and replaces it) or by purchasing and operating internal cleaning equipment (Figure 18.2).

Pallets. Electric facilities often receive shipments of equipment and supplies on pallets but rarely need the pallets to ship materials out. Pallets tend to accumulate and often are disposed of in landfills or are burned. Internal reuse may be enhanced through standardization of pallet sizes. If the standardized sizes are also the sizes that are popular with external users, a market often may be found for pallets in good condition. Other options include using pallets made from recycled plastics and pallets designed for denser stacking, which lower shipping costs.

Oil Filters. Oil-filter recycling is available in many areas. Oil filters are crushed, the oil is recovered and recycled, and the metal is recycled. The paper residue usually is incinerated. Oil-filter recycling units may be inexpensive enough to operate in-house. However, because some of the oil filters used by generating equipment are not self-enclosed (that is, the paper component is removable but the metal casing is reused), this alternative is not suitable for all applications.

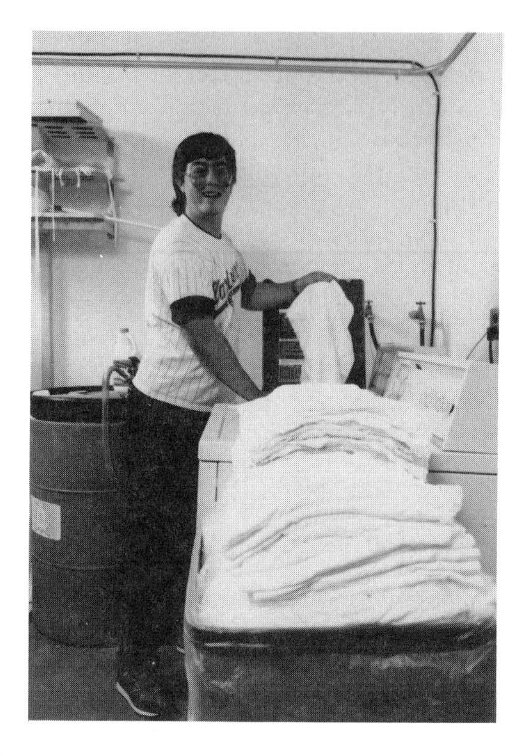

Figure 18.1. Cloth diapers purchased by custodial services to replace paper cleaning towels. The cloths are used for basic office cleaning and washed in washing machines by janitors, saving $84,000 in cost of paper towels each year. (Photo courtesy of Westinghouse Hanford Company/BCSR, Richland, WA).

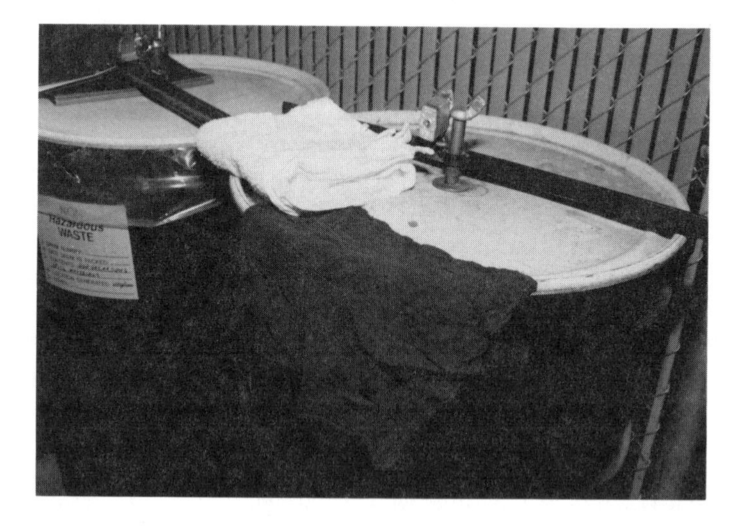

Figure 18.2. Use of color-coded rags. Red rags are used with regulated chemicals and white rags are used with nonregulated, thus preventing mixtures of the two wastes. (Photo courtesy of Westinghouse Hanford Company/BCSR, Richland, WA).

Case Study 18.1: B.C. Hydro Comprehensive Waste Management Strategy

British Columbia Hydro and Power Authority (B.C. Hydro) supplies power to the majority of the residents of the province of British Columbia (B.C.), Canada. B.C. Hydro's customers are located in communities varying from the Greater Vancouver area, population 1,600,000, near the Canada-Washington State border, to the tiny community of Atlin, population less than 500, near the B.C.-Alaska border. Because of the mountainous terrain of B.C., the majority of the electric power in the province comes from hydroelectric generation. Remote and smaller communities must receive their power from diesel and gas-fueled generators. Wastes generated by B.C. Hydro are primarily the result of power distribution rather than generation.

In 1990, B.C. Hydro established the goal of reducing waste disposal 50% by the year 2000. A comprehensive waste management strategy (CWMS) was developed to achieve the goal. The first phase in developing the strategy was inventorying all liquid and solid wastes generated throughout B.C. Hydro operations. Using this inventory, target waste types were identified that, if included in waste reduction programs, would reduce waste significantly. Programs for reducing the waste were then developed, considering the geographical distribution of B.C. Hydro's facilities. An estimate was made of the extent of waste diversion from landfills that the CWMS would achieve, and a cost-benefit analysis of the program was performed. The major results of the research and analysis include the following:

- B.C. Hydro could achieve more than a 50% reduction of waste requiring disposal through the CWMS by targeting the following waste types generated in large volumes: recyclable paper products, pallets, scrap metal, tree trimmings and brush, and waste oil.

- Additional programs could be included in the CWMS that address areas of concern without significant additional effort, including programs for the following waste types: glass containers, PET/HDPE bottles, lead-acid batteries, rags, antifreeze, and paints. These programs would contribute a minor amount to the overall percentage of diverted waste.

- Recycling of metals alone would provide revenues estimated to be much higher than the cost of collection and transport. Other materials that could be recycled profitably include fine paper and cardboard.

- The cost-benefit analysis of the CWMS included an assessment of external benefits, including societal benefits and public-image benefits. Overall, the cost-benefit analysis assisted B.C. Hydro in identifying areas of primary importance in the CWMS programs.

- Many local offices within B.C. Hydro have begun small efforts at waste reduction that have achieved promising results. Rags are being washed in equipment purchased at one generating station, with a projected net saving over replacement. Other facilities have developed waste reduction programs for pallets, batteries, insulating oils, and lubricating fluids.

B.C. Hydro realizes that it holds substantial purchasing power. It has used that power in the past in encouraging suppliers to provide environmentally friendly alternative products. Part of the CWMS was a recommendation that B.C. Hydro include provisions for encouraging the development of new materials and secondary material markets. B.C. Hydro has been successful in developing new product standards for recycled paints and recycled paper products, among other materials.

Case Study 18.2: Wastewater Reduction at Coal-Fired Power Stations

A mid-Atlantic electric utility conducted wastewater management evaluations at three coal-fired power stations. As part of the evaluation, wastewater treatment upgrades were proposed and wastewater reduction measures were implemented. The waste reduction measures resulted in significant decreases in wastewater flow and savings in construction and operation costs for the upgraded treatment plants. The three target areas for waste reduction were eliminating clean water from the wastewater treatment systems, converting the bottom-ash system to a closed-loop system, and increasing the reuse of effluent from the industrial wastewater treatment plant.

Eliminate Clean Water

Often, water that did not require treatment was discharged to a treatment plant because of operational procedures or existing piping or other conditions. Eliminating these clean streams from wastewater treatment can be an easy method of waste reduction.

At one power plant, part of the noncontact cooling water was discharged to the wastewater treatment plant. Bottlenecks in the once-through cooling system caused problems with cooling in summer. To increase the flow of cooling water, a portion of the water was diverted to the industrial wastewater treatment system through the floor-drain system. During the summer, some noncontact cooling water was discharged to floor drains to increase the cooling water flow, thereby increasing cooling capacity. Much of the time that the water was diverted, cooling capacity was not a problem, but as a result of operational procedures the water was diverted continuously anyway. By revising their operational procedures, the utility significantly reduced wastewater flow to the treatment plant. At one time, the noncontact cooling water represented 25% to 50% of wastewater flow.

At another power plant, instrumentation and control were modified to minimize wastewater production. Several tanks routinely overflowed to the wastewater treatment plant unnecessarily because of poor level control. The controls on the tanks were upgraded to minimize the incidence of overflows and reduce the volume of wastewater treated.

Convert to Closed-Loop System

At a power plant with two coal-fired units rated at 140 and 300 megawatts, the bottom-ash system was converted to a closed-loop system. In the new closed-loop system, bottom-ash sluice water is treated in a clarifier to remove solids and is reused for sluicing. The wastewater generated was reduced by an estimated 170 gpm, which minimized the size of unit processes for an upgraded wastewater treatment plant that was constructed recently. Figure 18.3 is a process flow diagram for the treatment system. Table 18.1 presents an equipment list and design criteria for the

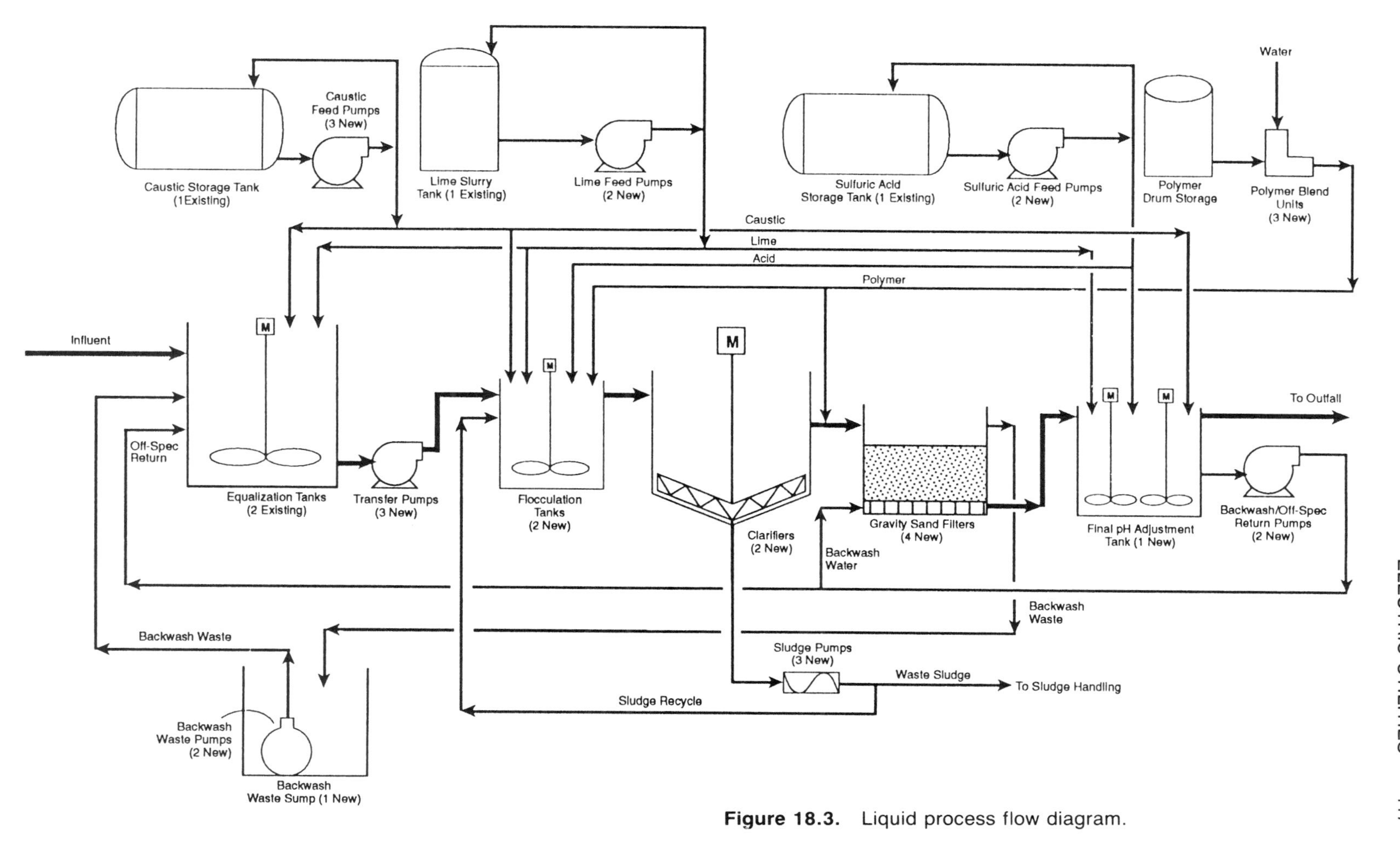

Figure 18.3. Liquid process flow diagram.

Table 18.1. Equipment List and Design Criteria

Equipment Item/Criteria	Alternative 1 Closed Bottom-Ash System	Alternative 2 Open Bottom-Ash System
Design Flow	1.5 mgd	2.83 mgd
Equalization		
No. of tanks	2 (existing)	3 (2 existing)
Tank volume, each	750,000 gals	750,000 gals
Detention time	24 hours	19 hours
Mixer horsepower	75 hp	75 hp
Transfer Pumping		
No. of pumps	3	5
Capacity/head	600 gpm @ 30 ft	1,200 gpm @ 30 ft
Horsepower	75 hp	75 hp
Flocculation		
No. of tanks	2	2
Tank volume, each	7,900 gals	15,000 gals
Detention time	15.1 min	15.3 min
No. of mixers, per tank	1	1
Mixer horsepower	2 hp	3 hp
Clarification		
No. of clarifiers	2	2
Diameter	60 ft	80 ft
Effective settling area	2,545 ft^2	4,524 ft^2
Overflow rate	295 gpd/ft^2	313 gpd/ft^2
Mechanism horsepower	0.75 hp	1 hp
Sludge Pumping		
No. of pumps	3	3
Capacity/head	60 gpm @ 30 ft	120 gpm @ 30 ft
Horsepower	3 hp	5 hp
Sludge Thickening		
No. of thickeners	2	2
Volume, each	51,400 gals	51,400 gals
Diameter	27 ft	27 ft
Detention time (@ 10% solids)	19 days	4 days
Mechanism horsepower	0.5 hp	0.5 hp
Sludge Dewatering		
No. of vacuum filters (existing)	2	
Filtration area	140 ft^2	
Loading rate	5 $lbs/hr/ft^2$	
Capacity	700 lbs/hr	
Filtration		
No. of cells	4	4
Filtration area, per cell	91 ft^2	160 ft^2
Loading rate	2.9 gpm/ft^2	3.1 gpm/ft^2
Backwash rate, normal	8 gpm/ft^2	8 gpm/ft^2
Backwash rate, maximum	18 gpm/ft^2	18 gpm/ft^2
No. of vacuum pumps	2	2
Vacuum pump horsepower	1 hp	1 hp
No. of blowers	2	2
Blower horsepower	15 hp	25 hp

Table 18.1. Equipment List and Design Criteria (Continued)

Equipment Item/Criteria	Alternative 1 Closed Bottom-Ash System	Alternative 2 Open Bottom-Ash System
Backwash Waste Pumping		
Volume of sump	7,850 gals	13,100 gals
No. of pumps	2	2
Capacity/head	800 gpm @ 60 ft	1,500 gpm @ 60 ft
Horsepower	20 hp	30 hp
Final pH Neutralization		
No. of tanks	1	1
Tank volume	15,700 gals	30,200 gals
Detention time	15.1 min	15.4 min
No. of mixers	2	2
Mixer horsepower	2 hp	3 hp
Backwash Pumping		
No. of pumps	2	2
Capacity/head	1,100 gpm @ 60 ft	2,000 gpm @ 60 ft
Horsepower	20 hp	40 hp
Lime Slurry Pumping		
No. of pumps	2	2
Horsepower	5 hp	5 hp
Caustic Pumping		
No. of pumps	2	2
Horsepower	5 hp	5 hp
Sulfuric Acid Pumping		
No. of pumps	2	2
Horsepower	5 hp	5 hp
Polymer Blending		
No. of units	3	3
Horsepower	0.5 hp	0.5 hp

alternatives of closing the bottom-ash system or leaving it open. Tables 18.2 and 18.3 present construction costs and O&M costs, respectively, of the two alternatives. Closing the bottom-ash system reduced the capital cost from $7.2 million to $4.7 million and reduced annual O&M cost from $1.1 million to $0.7 million.

Reuse Treatment Plant Effluent

Treatment plant effluent is reused at all three power stations. Uses include make-up water to the bottom-ash surge-tank system, seal water for pumps, clean water supply for backwashing wastewater treatment filters, and miscellaneous in-plant uses. Reuse of wastewater has reduced the quantity of city water used in the plants and has reduced the effluent discharge significantly.

Table 18.2. Construction Cost Estimate

Item	Alternative 1—Closed Bottom Ash System	Alternative 2—Open Bottom Ash System
Equalization tanks	$ 230,000	$ 565,000
Transfer pumps	15,000	50,000
Flocculation tanks	50,000	75,000
Clarifiers	500,000	770,000
Filters/post-pH adjustment tank	385,000	500,000
Recycle/backwash water pumps	15,000	25,000
Backwash waste pumps	20,000	25,000
Chemical feed systems	40,000	40,000
Sludge pumps	15,000	15,000
Sludge thickeners	175,000	175,000
Concrete-lined ash pits	0	140,000
Seal trough pump station	0	50,000
Sump modifications	20,000	20,000
Sump pumps	130,000	130,000
Sump overflow valves	120,000	120,000
Condenser pit modifications	10,000	10,000
Demolition of existing equipment	20,000	20,000
General conditions (5%)	150,000	235,000
Finishes (2%)	61,000	94,000
Yard piping and mechanical (20%)	602,000	942,000
Electrical (10%)	301,000	471,000
Instrumentation and control (5%)	150,000	235,000
Subtotal	**$3,009,000**	**$4,707,000**
Mobilization/bonding/insurance (5%)	150,000	235,000
Overhead and profit (20%)	602,000	942,000
Contingency (25%)	752,000	1,177,000
Total Capital Costs	**$4,513,000**	**$7,061,000**
Repainting equalization tanks	160,000	160,000
Total Construction Costs	**$4,673,000**	**$7,221,000**
Use	**$4,700,000**	**$7,200,000**

Table 18.3. Annual O&M Cost Estimate

	Alternative 1	Alternative 2
Item	Closed Bottom-Ash System	Open Bottom-Ash System
Power	$ 39,000	$ 54,000
Labor—Operations	140,000	120,000
Labor—Maintenance	84,000	67,000
Lime	81,000	153,000
Caustic	96,000	182,000
Sulfuric acid	26,000	48,000
Polymer	22,000	42,000
Sludge disposal	185,000	450,000
Total	**$673,000**	**$1,116,000**

Case Study 18.3: Source Reduction at Power-Generating Facilities in Pennsylvania

Introduction

This case study involves evaluating opportunities for reducing residual and hazardous wastes that are commonly produced at facilities that generate electric power from fossil fuel and nuclear power. CH2M HILL conducted the evaluations for an association of electric utilities in Pennsylvania. The state passed legislation that requires industrial facilities to prepare strategy reports for reducing residual and hazardous waste streams at the source.

The case study presents an overview of waste reduction and examples of waste-stream source reduction evaluations, including waste description, source reduction alternatives, and evaluation results. The quantitative information in this case study reflects projections resulting from the evaluations. The associated electric utilities are in the process of quantifying their individual wastes and implementing recommendations for source reduction.

Overview of Requirements for Source Reduction

Pennsylvania industries are required to comply with recent state legislation addressing management of residual and hazardous waste. The U.S. EPA defines waste reduction as "the elimination or reduction, to the extent feasible, of hazardous waste that is generated and would otherwise be subsequently treated, stored, or disposed of."[6] Pennsylvania has expanded the definition to include the regulation of residual waste.

Waste reduction alternatives are classified as either source reduction or recycling opportunities. Source reduction is any activity that reduces or eliminates the generation of hazardous wastes at the source, usually within a process. Recycling is the next level down in the hierarchy of waste reduction alternatives established by the U.S. EPA and includes activities in which material is reused or reclaimed. Pennsylvania regulations focus only on source reduction efforts that reduce the weight or toxicity of residual wastes. Thus, recycling, unless directly into an ongoing onsite manufacturing process, and other beneficial-use activities are not considered source reduction for reporting purposes.

In general, most of the common residual and hazardous waste streams have implementable source reduction strategies. In many instances, low-cost, easily implementable options, such as improved maintenance procedures and material-handling training programs, can achieve reductions in waste generation. In addition, reductions in waste stream weight or toxicity would probably result if material purchasing was limited to fewer products.

Waste streams constituting the bulk of the total weight of wastes generated at utilities (bottom ash, fly ash, boiler cleanings) present limited opportunities for source reduction. In those cases, other waste minimization techniques (beneficial use and recycling) may be preferred alternatives, even though credit cannot be claimed for source reduction in Pennsylvania. Assessments of alternatives for source reduction of other wastes follow.

Assessments of Alternatives for Source Reduction

This section presents assessments of specific alternatives for source reduction of selected waste streams. The assessments are based on the following:

- nonquantitative evaluation using evaluation criteria
- cost-benefit analysis using cost worksheets

Electric utilities can establish qualitative evaluation criteria for selecting alternatives for source reduction. Alternatives should be evaluated on the basis of the following criteria:

Technical

- proven technology
- minimal site constraints
- minimal space requirements
- minimal effects on operations
- minimal manpower requirements
- potential reduction in waste stream (%)
- potential reduction in total wastes generated (%)
- compatibility with corporate waste reduction goals
- feasible implementation schedule

Environmental, Health and Safety

- reduction in toxicity
- minimal permit requirements
- compatibility with other environmental regulations and programs
- acceptance by the community
- minimal effect on employee health and safety

Economic

- capital costs
- direct and indirect operation and maintenance costs
- payback period

The nonquantitative and non-site-specific criteria from this list were used in the section that follows to compare alternatives for source reduction and waste reduction. Each alternative was evaluated according to nonquantitative and non-site-specific criteria.

Cost-Benefit Analysis of Alternatives for Source Reduction and Waste Reduction

For waste streams where capital and other cost information was available, a cost-benefit analysis was completed.

In most cases, the cost analysis will be site-specific and will depend on the effectiveness and cost of present waste generation, management, and disposal functions, the facility type and age, and the existing infrastructure.

Cost-benefit analysis has been conducted as part of the source reduction analysis. The worksheets presented for each alternative serve to illustrate one approach to preparing cost analyses. The approach is appropriate for determining whether a particular option will be cost neutral or will entail significant savings or additional costs in comparison to existing practices. Approximate cost factors have been estimated from vendor information, literature, previous research, and professional communication.[7] The cost assumptions used for waste disposal throughout this chapter are listed in Table 18.4.

The next section presents detailed source reduction analyses, including cost-benefit analysis, for the following waste streams:

- asbestos-containing material
- blasting media
- demineralizer regenerant
- firebrick and refractory
- light bulbs
- paint wastes

Example 1: Asbestos-Containing Material

Description of Waste

By definition, asbestos-containing wastes are classified as residual in Pennsylvania. Asbestos-containing wastes include insulation, boards, flooring, containment materials, and discarded protective suits used in handling asbestos. The wastes are generated during renovation, demolition, and maintenance activities, such as pipe repair. Workers are required to segregate the wastes and place them in yellow or other specially colored bags that are clearly labeled. The wastes are stored separately and are disposed of in an approved landfill.

Characterization of Typical Waste

This waste stream is composed primarily of asbestos. Additional characterization data are scarce. One utility reported that the waste contains small concentrations of arsenic, barium, chromium, copper, nickel, selenium, and zinc. Several toxicity characteristic leaching procedure (TCLP) tests indicated that except for phenol (0.4 mg/L) none of the tested parameters exceeded detectable limits.

Table 18.4. Assumptions Regarding Disposal Costs

Waste Characteristics	Treatment or Disposal	Cost
Residuals	Disposal in municipal landfill	$65/ton
Hazardous[a]	Removal and disposal	$250/drum ($900/ton[b])
Asbestos-containing material	Disposal in approved landfill	$75/ton
Hazardous liquid	Removal in tank truck	$0.27/gal
Liquid	Treatment in onsite impoundment	$0.02/gal
Paint waste	Shipping and disposal in drums	$12/drum (2,400/yd^3)
Hazardous solvent	Shipping and disposal	$12/gal
Nonhazardous solvent	Shipping and disposal	$4/gal
Antifreeze	Removal for recycling	$1.50/gal

[a]For example, blasting media
[b]Assuming 75 lb/yd^3

Alternatives for Source Reduction

The following waste reduction alternatives are already being implemented at utilities.

Use a Substitute Product. Asbestos-containing materials have been replaced by products containing calcium silicate or fiberglass insulation, which present their own health and waste-handling problems. For cooling-tower lattice, asbestos boards have been replaced with polyvinylchloride (PVC) board.

Product substitution may occur in two forms: planned aggressive removal of asbestos-containing materials and removal as needed (e.g., when replacing a pipe or valve, asbestos insulation would be removed and replaced with nonasbestos insulation.) Aggressive removal has the disadvantage of potentially increasing worker exposure.

Encapsulate the Waste. Corporate policies will dictate whether encapsulation, which minimizes future generation of asbestos waste, is desirable. The purpose of encapsulation is to enclose the asbestos materials indefinitely.

Other Methods for Managing Waste

Solidification. Combining asbestos with pozzolanic materials (such as fly ash or portland cement) stabilizes the materials, forming a concrete block. Although the waste is nonfriable and can be disposed of in landfills, it is classified as asbestos-containing material, which does not remove perpetual liability. The method has not proven to be economically feasible for mixtures of asbestos at 25 to 50%, which may not be sufficiently stable for the long term.[7]

Chemical Treatment. Asbestos fibers can be destroyed chemically. Only one commercial chemical treatment system is available: ABCOV. The primary advantage of the system is that it destroys the fibers and the material is no longer hazardous, although the process would still be regulated under national environmental safety and health regulations covering processes that convert asbestos to nonasbestos materials. The costs of this process have been found to be greater than bagging and landfilling.[8]

Vitrification. Vitrification also destroys asbestos fibers and can produce a usable glass-like product by heating the asbestos materials to between 400 and 1,100°F. Several commercial applications for vitrification of asbestos may be available soon.[8]

Assessment of Alternatives

Table 18.5 provides a nonquantitative evaluation of the two source reduction alternatives for asbestos waste. This charge was rated as either having a beneficial effect (+), no effect (0), or a negative effect (-), on each of 12 evaluation criteria. Continuing removal of asbestos-containing materials through regular maintenance, as needed, minimizes labor and employee exposure, assuming that leaving existing materials in place does not jeopardize employee health.

If in-place materials are perceived as a health risk, encapsulation offers an alternative to mass removal. Encapsulation stabilizes existing asbestos-containing materials, allowing them to be removed during regular facility maintenance. This approach will steadily deplete asbestos-containing material and prevent generation of large amounts of waste in the short term. As shown in Table 18.6, encapsulation makes possible a net saving if the assumption is that the labor needed for encapsulation equals the labor that otherwise would have been needed to replace the material.

Table 18.5. Qualitative Assessment of Alternatives for Source Reduction for Asbestos-Containing Material

	Source Reduction Alternative	
Evaluation Criteria	**Use Substitute Product**	**Encapsulate**
Waste Reduction Potential (% of waste by weight)	+	+
Waste Reduction Potential (% of total waste by weight)	0	0
Space Requirements	0	0
Modifications of Plant Processes	0	0
Impact on Operations	0	0
Manpower Requirements	0	–
Permit Requirements	0	0
Proven Techniques	+	0
Reduction in Toxicity	+	0
Employee Health and Safety	+	0
Community Acceptance	+	0
Consistent with Other Environmental Programs	+	0

Table 18.6. Cost/Benefit Analysis for Asbestos-Containing Material

	Option 1 Product Substitution	Option 2 Encapsulation
Quantity of Waste Generated Annually, ___ units/yr Q = unit = tons	30 units/year	30 units/year
Annualized Cost		
1. Cost of Process Modification and/or Capital Investment $___ × CR = $___/yr	$0/yr	$0/yr
Payback Period (check one) — CR[a]		
5 ___ — .24		
10 ___ — .14		
15 ___ — .10		
20 ___ — .087		
2. Additional Variable O&M Costs		
Material Cost = $Q \times [(1 - R/100) \times F_p - I_p]$ = ___/yr	1,500/yr	NA
where:		
I_p = initial price of material per unit ___	50	NA
F_p = future price of material per unit ___	100	
R^b = estimated reduction in material requirement per unit ___	0%	0%
Other Additional O&M Cost $___/unit × Q = ___/yr	0/yr	
3. Additional Fixed O&M Costs (i.e. labor and administration) ___/yr Assume labor needed to encapsulate equals maintenance of otherwise removed materials.	0/yr	0/yr
4. Additional Contract Disposal Costs	–300/yr	0/yr
$Q \times [(1 - R/100) \times F_e) - I_e]$ = ___/yr	75 65 0%	75 0 100%
I_c = current disposal cost per unit ___		
F_c = disposal cost per unit after SRS ___		
R^b = estimated reduction in waste ___		
Total Additional Cost $___/yr	**$1,200/yr**	**$2,250/yr**

[a]Capital Replacement factor based on 6% rate-of-return

[b]Where reduction is derived from increase in product life-expectancy;
R = 1 – [(average previous life exp./average future life exp.) × 100]

Example 2: Blasting Media

Description of Waste

Various blasting media are used to remove rust and other deposits from metal equipment and parts, such as turbines, to prepare the surfaces for painting. The materials used for blasting are granular and are silica-based (glass or sand), metal-based (steel or aluminum), or slag-based (Black Beauty). Waste materials from blasting usually are classified as residual but occasionally have tested as hazardous because of elevated levels of chromium, lead, or cadmium. Usually, municipal or residual waste landfills require each load to be tested before accepting it as non-hazardous.

Characterization of Typical Waste

Blasting media tested on an as-received basis varied widely, depending on the source of the material. Expected metals, such as zinc, nickel, lead, copper, barium, and cadmium, were detected. Toxicity characteristic leaching procedure (TCLP) tests showed low concentrations of most of these metals, but test results for lead, zinc, and cadmium were above 5 mg/L for some materials.

Alternatives for Source Reduction

Use a Substitute Product. Slag-based media are the least expensive materials for blasting. However, the material can be high in metals and can leach toxics at concentrations high enough for the waste to be classified as hazardous. Alternatives to slag-based media include sand, steel shot, glass beads, plastic media, and walnut shells. The specific application dictates the media that can be used successfully. An additional alternative is garnet. Mined in Australia, garnet is crystalline silica, free of toxic metals. It is readily recycled and has the advantage of creating minimal dust. Selecting replacement blasting media may reduce waste toxicity, but it will not necessarily lessen the volume of waste generated.

Change Operations. Mechanical cleaning techniques may replace blasting in some applications, although they may be operationally intensive. These methods also produce wastes (e.g., paint flakes, metal fragments). Examples of several technologies are described below.

- **High-pressure water.** In this method, pressurized water (up to 35,000 psi) is sprayed against a close surface to remove paint or scale. Process wastewater usually is collected for reuse. This method has advantages over conventional hydroblasting, producing as much as 90% less wastewater, and additives may not be needed.
- **Cryogenic blasting.** The use of pelletized dry ice as a blasting medium is a unique alternative to conventional blasting. The media sublimes without contributing to the waste stream. However, capital costs, as estimated in Table 18.7, may be difficult to justify.
- **Power-tool cleaning.** Scarifying or pneumatic-impact tools are effective in

Table 18.7. Cost/Benefit Analysis for Blasting Media

		Option 1 Change process (CO_2 system)	Option 2 Steel shot recycling
Quantity of Waste Generated Annually, Q = unit = tons	___ units/yr	100 units/year	100 units/year
Annualized Cost			
1. Cost of Process Modification and/or Capital Investment $___ × CR = $___/yr		158,000[a] .10 $15,800/yr	50,000[a].1 $5,000/yr
Payback Period (check one)	CR[a]	one pellet maker and two mobile stations	two new blasting systems with reuse capability
5 ____	.24		
10 ____	.14		
15 ____	.10		
20 ____	.087		
2. Additional Variable O&M Costs			
Material Cost = $Q \times [(1 - R/100) \times F_p - I_p]$ =	___/yr	–500/yr	–1,620/yr
where:			
I_p = initial price of material per unit	____	40 (sand)	40 (sand)
F_p = future price of material per unit	____	70 (liq. CO_2)	70 (shot)
R^b = estimated reduction in material requirement per unit	____	50%	66%
Other Additional O&M Cost $___/unit × Q =	___/yr	0/yr	
3. Additional Fixed O&M Costs (i.e. labor and administration) Assume labor needed to encapsulate equals maintenance of otherwise removed materials.	___/yr	1,800/yr	0/yr (vendor maintenance)
4. Additional Contract Disposal Costs		2,500/yr	29,500/yr
$Q \times [(1 - R/100) \times F_e) - I_e]$ =	___/yr	65 900 90%	65 900 60%
I_c = current disposal cost per unit	____		
F_c = disposal cost per unit after SRS	____		
R^b = estimated reduction in waste	____		
Total Additional Cost $___/yr		**$19,600/yr**	**$32,880/yr**

[a]Capital Replacement factor based on 6% rate-of-return
[b]Where reduction is derived from increase in product life-expectancy;
R = 1 – [(average previous life exp./average future life exp.) × 100]

removing paint from flat surfaces. With vacuum attachments, enclosure or containment typically is not necessary.

Reuse or Recycle Products. Some products can be reused because of their durability. They include steel grit and aluminum oxide, which are collected, separated from the contaminants, and reused.

Improve Documentation of Painting Activities. By stenciling areas that are painted with lead-free paint, future testing of paint waste for disposal will be minimized.

Data are available for comparing the costs of alternative blasting systems, but application of costs is based on project scale.[8] The report indicates some cost advantages of less traditional methods (such as CO_2 blasting and high-pressure washing) over sand blasting and bead blasting.

Of importance to note is that using a substitute product and recycling rely on the implementation of stringent purchasing specifications for blasting media.

Other Methods for Managing Waste

Blasting operations inside boilers may pose special waste-handling problems. Removal of blasting wastes inside the boilers is time-consuming. Therefore, using a material that can be subsequently sluiced with bottom ash under permitted applications minimizes wastes reported as blasting wastes.

Adequate mixing of samples of blasting media waste could ensure that the waste is nonhazardous, whereas grab samples that are not composited may contain high metal concentrations.

Additives can be mixed with blasting materials to make the waste nonhazardous. Steel shot, iron filings, or specific chemicals designed for this purpose help lower the levels of leachable lead. Although this is not source reduction, it may allow management of the waste as residual instead of hazardous.

A final waste management consideration is using blasting media as an ingredient in asphalt and concrete, which does not constitute disposal under federal regulations.

Assessment of Alternatives

The first three source reduction alternatives were evaluated using nonquantitative criteria (see Table 18.8). Product replacement to reduce toxicity is the most easily implementable option and has no negative effects. Depending on the replacement media selected, however, the waste volume may not be reduced. The other two source reduction strategies have related benefits but with some important disadvantages.

By specifying media that are durable as well as nontoxic, waste can be reduced significantly (as much as 66%) by recycling. Special equipment will be required to recycle the media. The disadvantages of this alternative are the capital investment for recycling equipment and the cost of training operators.

Replacing sand blasting with a new blasting process, such as a dry-ice system, can reduce the quantity of waste by 90% or more. The disadvantages of such a system include the training of operators and the capital and operational costs of implementation, as shown in Table 18.8. Versatile dry-ice systems, in particular, are expensive

Table 18.8. Qualitative Assessment of Alternatives for Source Reduction for Blasting Media

	Source Reduction Alternative		
Evaluation Criteria	**Use Substitute Product (Toxicity)**	**Change Operations (e.g., CO_2)**	**Reuse or Recycle Products (steel shot)**
Waste Reduction Potential (% of waste by weight)	0	+	+
Waste Reduction Potential (% of total waste by weight)	0	0	0
Space Requirements	0	0	0
Modifications of Plant Processes	0	–	–
Impact on Operations	0	0	0
Manpower Requirements	0	–	–
Permit Requirements	0	0	0
Proven Techniques	+	+	+
Reduction in Toxicity	+	–	0
Employee Health and Safety	+	0	0
Community Acceptance	+	+	+
Consistent with Other Environmental Programs	+	0	+

to implement in comparison to other options. Another disadvantage is that the blasting waste will consist entirely of flaked metal, paint, and scale and thus is more likely to be a hazardous waste.

Example 3: Demineralizer Regenerant

Description of Waste

Regeneration of the demineralizer's cation, anion, and mixed-bed reactors occurs regularly, resulting in a large volume of wastewater. Regeneration waste usually is handled in onsite basins or other permitted wastewater treatment facilities. Because the pH of this waste stream is intermittently below 2.0 or above 12.5, the waste is sometimes classified as hazardous. The flow rate for this waste stream is about 80 to 110 gallons per megawatt-day.[2]

Characterization of Typical Waste

The results of testing from two sampling events for one utility company showed these maximum concentrations: aluminum (0.613 mg/L), tin (0.25 mg/L), antimony (0.030 mg/L), hexavalent chromium (0.020 mg/L), copper (0.0214 mg/L), nickel (0.020 mg/L), selenium (0.025 mg/L), and zinc (0.0391 mg/L). Arsenic, cadmium, copper, lead, silver, thallium, and beryllium were less than or equal to 5 parts per billion (ppb).

Alternatives for Source Reduction

Change Operations. This waste stream can be reduced if demineralization is based on the need for regeneration rather than on a planned schedule. Implementing this alternative would require using a conductivity meter or another total dissolved solids (TDS) meter to determine when regeneration is required.

Pretreat. To reduce the volume of waste from demineralization regeneration, reverse osmosis can be used to remove up to 90% of the TDS before demineralization.

Eliminate Process. Instead of performing demineralization, utilities may use a water vendor, such as Arrowhead or Ecolochem, to provide demineralized water through onsite mobile units. The water supplier would be responsible for disposing of regeneration waste (as well as replacing and disposing of resin). Source reduction in this case would be realized by only the utility company, which would be relieved of generator status and the accompanying responsibility. Even though the utility would no longer be the responsible generator, the waste stream would continue to be generated.

Assessment of Alternatives

A qualitative evaluation of the three alternatives for source reduction is presented in Table 18.9. Although the alternatives reduce waste, the costs of implementation may be excessive. Table 18.10 shows that the costs of implementing the vendor-

Table 18.9. Qualitative Assessment of Alternatives for Source Reduction for Demineralization Regenerant

	Source Reduction Alternative		
Evaluation Criteria	Change Operations	Pretreat	Eliminate Process
Waste Reduction Potential (% of waste by weight)	+	+	+
Waste Reduction Potential (% of total waste by weight)	+	+	+
Space Requirements	0	–	–
Modifications of Plant Processes	0	–	0
Impact on Operations	0	–	+
Manpower Requirements	0	–	+
Permit Requirements	0	0	0
Proven Techniques	+	+	+
Reduction in Toxicity	0	0	0
Employee Health and Safety	0	0	+
Community Acceptance	+	+	+
Consistent with Other Environmental Programs	0	0	0

Table 18.10. Cost/Benefit Analysis for Demineralizer Regenerant

		Option 1
		Vendor to supply water
Quantity of Waste Generated Annually, Q = unit = 1,000 gal (assuming 20,000 gpd)	___ units/yr	7,300 units/year
Annualized Cost		
1. Cost of Process Modification and/or Capital Investment	$___ × CR = $___/yr	$0/yr

Payback Period (check one)	CR[a]
5 _____	.24
10 _____	.14
15 _____	.10
20 _____	.087

2. Additional Variable O&M Costs		
Material Cost = $Q \times [(1 - R/100) \times F_p - I_p]$ =	___/yr	0/yr
where:		
I_p = initial price of material per unit _____		NA
F_p = future price of material per unit _____		$
R^b = estimated reduction in material requirement per unit _____		NA
Other Additional O&M Cost $___/unit × Q =	___/yr	146,000/yr (@$20/1,000 gal)
3. Additional Fixed O&M Costs (i.e. labor and administration) Assume labor needed to encapsulate equals maintenance of otherwise removed materials.	___/yr	0/yr
4. Additional Contract Disposal Costs		
$Q \times [(1 - R/100) \times F_e) - I_e]$ =	___/yr	−146,000/yr
I_c = current disposal cost per unit _____		0
F_c = disposal cost per unit after SRS _____		0.02 (onsite basin)
R^b = estimated reduction in waste _____		100%
Total Additional Cost	**$___/yr**	**$0/yr**

[a]Capital Replacement factor based on 6% rate-of-return

[b]Where reduction is derived from increase in product life-expectancy;
R = 1 − [(average previous life exp./average future life exp.) × 100]

supplied water option is paid back if 20,000 gallons per day (gpd) of regenerant is generated. Using a vendor has several benefits:

- The utility is no longer generating hazardous waste, resin, or other system-related wastes (expired chemicals).
- The risk to employees is decreased because employees no longer operate a demineralization system.
- Operating and permitting costs will decrease for onsite treatment facilities.

Example 4: Firebrick and Refractory

Description of Waste

Refractory and other protective materials are used to line the bottoms and doors of boilers. Other names for these materials are "firebrick," "Super 3000," "Blue Ram," and "Hot Stop." These materials become wastes when they are removed from the boilers because of damage from heating and cooling or from large falls of slag. Unused expired refractory also may be part of this waste stream. This material usually tests as nonhazardous but may test as hazardous because some refractories, especially older ones, contain chromium. Firebrick and refractory wastes are stored on the site and usually are disposed of in onsite basins or landfills or in offsite landfills.

Characterization of Typical Waste

On the basis of elemental analysis, refractory wastes contain mainly silicon, aluminum, calcium, and iron and less than 1% each of chromium, lead, zirconium, and other metals. TCLP test results have shown about 100 mg/L of iron, 1,000 mg/L of sodium and calcium, and 8.6 mg/L of zinc, but less than 2 mg/L each of heavy metals, including chromium.

Alternatives for Source Reduction

Use Durable Materials. More-durable refractory will need to be removed and replaced less frequently, resulting in less waste.

Use Less-Toxic Materials. Buying refractory materials that do not contain toxic substances will minimize the chance that waste boiler rubble will test as hazardous. Although most companies still produce at least one type of refractory containing chromium, most refractory is now chromium-free.

Change Material Specifications. Where possible, refractory that can be installed in small sections should be purchased. If damaged, only a section or two will have to be disposed of instead of an entire wall or a large piece.

Table 18.11. Qualitative Assessment of Alternatives for Source Reduction for Firebrick and Refractories

	Source Reduction Alternative			
Evaluation Criteria	**Use Durable Materials**	**Use Less-Toxic Materials**	**Material Specifications**	**Maintenance Procedures**
Waste Reduction Potential (% of waste by weight)	+	0	+	+
Waste Reduction Potential (% of total waste by weight)	+	0	+	+
Space Requirements	0	0	0	0
Modifications of Plant Processes	0	0	0	0
Impact on Operations	+	+	+	+
Manpower Requirements	0	0	0	0
Permit Requirements	0	0	0	0
Proven Techniques	+	+	+	+
Reduction in Toxicity	0	+	0	0
Employee Health and Safety	0	+	0	0
Community Acceptance	+	+	+	+
Consistent with Other Environmental Programs	0	0	0	0

Change Maintenance Procedures. Extending time between outages will result in fewer refractory-removal projects.

Assessment of Alternatives

Using substitute material to increase durability or reduce toxicity is beneficial and, on the basis of the evaluation criteria presented in Table 18.11, presents no disadvantages. Likewise, using sections of materials and increasing the time between outages provide benefits with no discernable disadvantages. However, the feasibility of these alternatives depends on the boiler characteristics and operation. Table 18.12 presents a comparison of the cost of using material with less aluminum silicate (aluminate) vs. material with a higher content of aluminate. The higher content of aluminate makes the refractory more durable but also more expensive. The reduction in materials and wastes generated does not justify the additional cost of using refractory material containing more aluminate.

Example 5: Light Bulbs (Bulk)

Description of Waste

Major changeouts of street lamps are the source of this waste stream. In general, bulbs are high-pressure sodium, metal halide, and perhaps mercury vapor (from

Table 18.12. Cost/Benefit Analysis for Firebrick/Refractories

		Option 1
		Replace lower with higher aluminate content
Quantity of Waste Generated Annually,	___ units/yr	70 units/year
Q = unit = ton, assuming 5–40 CY rolloffd @ 700 lb/CY		
Annualized Cost		
1. Cost of Process Modification and/or Capital Investment	\$___ × CR = \$___/yr	\$0/yr

Payback Period (check one)	CR[a]
5 _____	.24
10 _____	.14
15 _____	.10
20 _____	.087

		Option 1
2. Additional Variable O&M Costs		
Material Cost = $Q \times [(1 - R/100) \times F_p - I_p]$ =	___/yr	36,750/yr
where:		
I_p = initial price of material per unit _____		1,200/ton
F_p = future price of material per unit _____		675/ton
R^b = estimated reduction in material requirement per unit _____		100%
Other Additional O&M Cost \$___/unit × Q =	___/yr	0/yr
3. Additional Fixed O&M Costs (i.e. labor and administration)	___/yr	0/yr
Assume labor needed to encapsulate equals maintenance of otherwise removed materials.		
4. Additional Contract Disposal Costs		−1,360/yr
$Q \times [(1 - R/100) \times F_e) - I_e]$ =	___/yr	65/ton 65/ton 30%
I_c = current disposal cost per unit _____		
F_c = disposal cost per unit after SRS _____		
R^b = estimated reduction in waste _____		
Total Additional Cost \$___/yr		**\$35,380/yr**

[a]Capital Replacement factor based on 6% rate-of-return
[b]Where reduction is derived from increase in product life-expectancy;
R = 1 – [(average previous life exp./average future life exp.) × 100]

older installations) bulbs that are replaced as needed or on a regular schedule. The bulbs usually are kept separate from other waste streams and are removed for offsite disposal. The bulbs may test hazardous for lead.

Typical Waste Characterization

TCLP testing showed mercury-leachate levels of less than 0.25 mg/L but lead leachate levels of around 100 mg/L. The material safety data sheet (MSDS) for light bulbs specifies that they consist of 100% inert materials (glass, wire, aluminum, etc.).

Source Reduction Options

Use Less-Toxic Materials. Many utilities are switching from mercury vapor lamps to less-toxic sodium vapor or metal halide lamps. Sodium vapor and metal halide lamps also are more efficient and more durable than other lamps. Substituting different lamps will require replacing fittings as well, because ballasts are different for each type of lamp.

Avoid Generator Status. Many utilities require that streetlights be owned by the municipality. In that case, the utility continues to service the streetlights but is not responsible for disposal of waste lamps.

Other Alternatives for Waste Reduction

Recycle Mercury. Recovering mercury from mercury vapor lamps could be cost-effective if a market exists. Because the mercury is found in the phosphor powder, the recovery process is difficult. Recent research at a California facility, however, shows successful mercury recovery from the lamps (up to 98%), at an approximate cost of $0.07 to $0.10 per linear foot of lamp.[9]

Assessment of Alternatives

As shown in Table 18.13, both using less-toxic materials and avoiding generator status are beneficial in reducing this hazardous waste stream. A disadvantage of material substitution is the labor and expense of changing the lamps and ballasts. On the basis of the assumptions presented in Table 18.14, the costs of the two options for material substitution do not appear excessive, and the reductions in material and disposal costs produce a small saving after the capital and material costs of using metal halide or sodium vapor lamps are deducted.

Example 6: Paint Wastes

Description of Waste

This waste stream includes waste paint (both latex and oil-based) and coating material. Paint wastes are considered hazardous if they are ignitable or toxic. If they

Table 18.13. Qualitative Assessment of Alternatives for Source Reduction for Light Bulbs

Evaluation Criteria	Source Reduction Alternative		
	Use Less-Toxic Material	Avoid Generator Status	Recycle Mercury
Waste Reduction Potential (% of waste by weight)	+	+	+
Waste Reduction Potential (% of total waste by weight)	0	0	0
Space Requirements	0	0	0
Modifications of Plant Processes	–	+	0
Impact on Operations	–	+	0
Manpower Requirements	0	0	0
Permit Requirements	0	0	0
Proven Techniques	+	+	+
Reduction in Toxicity	+	+	0
Employee Health and Safety	+	+	0
Community Acceptance	+	+	+
Consistent with Other Environmental Programs	+	+	0

are not, they are categorized as residual waste. Paint cans sometimes are placed in drums for disposal as hazardous waste. Nonhazardous empty paint cans sometimes are crushed (Figures 18.4 and 18.5) and disposed of in a maintenance waste dumpster. Some utility companies have successfully arranged for vendors to remove half-empty paint cans. As stipulated in disposal regulations, the cans may be considered small-quantity waste.

Characterization of Typical Waste

TCLP testing has been conducted for paint wastes. Leaching of most parameters has not been detected except for lead (up to 12.3 mg/L), barium (to 5.01 mg/L), chromium (up to 4.1 mg/L), and small amounts of arsenic, cadmium, mercury, and silver (less than 1 mg/L).

Alternatives for Source Reduction

Change Purchasing Practices. Controlling the number, color, and type of paints and related materials available on the site will reduce the number of paints available and thus the number of half-empty containers or expired paints requiring disposal. Minimizing the use of lead-based paint will reduce the amount of toxic material requiring disposal. Purchasing personnel should review the MSDS for each new or changed raw material before accepting it for onsite operations.

Table 18.14. Cost/Benefit Analysis for Light Bulbs

		Option 1 Change from incandescent to sodium lamps	Option 2 Change from incandescent to halide
Quantity of Waste Generated Annually, Q = unit = tons = 1,000 lamps	__ units/yr	1 unit/year	1 unit/year
Annualized Cost			
1. Cost of Process Modification and/or Capital Investment $__ × CR = $__/yr		4,000[a] .14 $560/yr	4,000[a] .14 $560/yr
Payback Period (check one)	CR[a]	4,000 to replace ballasts	4,000 to replace ballasts
5 ____	.24		
10 ____	.14		
15 ____	.10		
20 ____	.087		
2. Additional Variable O&M Costs			
Material Cost = $Q \times [(1 - R/100) \times F_p - I_p]$ =	__/yr	–40/yr	600/yr
where:			
I_p = initial price of material per unit	____	900	900
F_p = future price of material per unit	____	4,300	5000
R^b = estimated reduction in material requirement per unit	____	80% from 3,000 to 15,000 hour life	70% from 3,000 to 10,000 hour life
Other Additional O&M Cost $__/unit × Q =	__/yr	0/yr	0/yr
3. Additional Fixed O&M Costs (i.e. labor and administration) Assume labor needed to encapsulate equals maintenance of otherwise removed materials.	__/yr	0/yr	0/yr
4. Additional Contract Disposal Costs		–720/yr	–630/yr
$Q \times [(1 - R/100) \times F_e) - I_e]$ =	__/yr	900	900
		900	900
I_c = current disposal cost per unit	____	80%	70%
F_c = disposal cost per unit after SRS	____		
R^b = estimated reduction in waste	____		
Total Additional Cost $__/yr		**$-200/yr**	**$530/yr**

[a]Capital Replacement factor based on 6% rate-of-return

[b]Where reduction is derived from increase in product life-expectancy; R = 1 – [(average previous life exp./average future life exp.) × 100]

Figure 18.4. Aerosol can puncture kit punctures can to drain out liquid into drum and collects volatiles in a filter. (Photo courtesy of Westinghouse Hanford Company/BCSR, Richland, WA).

Figure 18.5. Can crusher used to crush punctured aerosol cans and other metal containers, thus reducing the volume of waste. (Photo courtesy of Westinghouse Hanford Company/BCSR, Richland, WA).

Table 18.15. Qualitative Assessment of Alternatives for Source Reduction for Paint Wastes

	Source Reduction Alternative			
Evaluation Criteria	**Change Purchasing Practices**	**Control Inventory**	**Change Contract Provisions**	**Use Can Crushers**
Waste Reduction Potential (% of waste by weight)	+	+	+	+
Waste Reduction Potential (% of total waste by weight)	0	0	0	0
Space Requirements	0	0	0	0
Modifications of Plant Processes	0	0	0	0
Impact on Operations	0	0	+	0
Manpower Requirements	0	0	+	0
Permit Requirements	0	0	0	0
Proven Techniques	+	+	+	+
Reduction in Toxicity	+	+	+	+
Employee Health and Safety	+	+	+	0
Community Acceptance	+	+	+	+
Consistent with Other Environmental Programs	+	+	+	0

Control Inventory. Purchasing or stockroom staff should track and control the paint inventory, monitor expiration dates, and arrange for offsite use of waste material instead of disposal.

Change Contract Provisions. Future contracts for painting jobs could require contractors to use nontoxic paints and dispose of paint wastes and residual paint.

Adopt Alternative Painting Methods. For some applications, electrostatic painting methods may improve quality while reducing waste.[1]

Other Alternatives for Waste Management

Use Can Crushers. Paint-can crushers effectively reduce the volume of this waste stream and allow the waste to be sold to a scrap-metal dealer.

Assessment of Alternatives

According to the evaluation process shown in Table 18.15, source reduction options for paint waste are similar to the chemical waste stream in that there are no significant negative effects. Can crushing, as shown in Table 18.16, can result in savings, even considering the additional labor required, because it reduces the number of drums disposed of off the site as hazardous waste.

Table 18.16. Cost/Benefit Analysis for Paint Wastes

	Option 1
	In-drum paint can compaction
Quantity of Waste Generated Annually, ___ units/yr Q = unit = cubic yard = approx. 4 tons	40 units/year
Annualized Cost	
1. Cost of Process Modification and/or Capital Investment $___ × CR = $___/yr	24,000[a] .24 $5,760/yr

Payback Period (check one)	CR[a]
5 _____	.24
10 _____	.14
15 _____	.10
20 _____	.087

	Option 1
2. Additional Variable O&M Costs	
Material Cost = $Q \times [(1 - R/100) \times F_p - I_p]$ = ___/yr	−2,560/yr
where:	
I_p = initial price of material per unit _____	80
F_p = future price of material per unit _____	80
R^b = estimated reduction in material requirement per unit _____	80% DOT 17H packing drum
Other Additional O&M Cost $___/unit × Q = ___/yr	0/yr
3. Additional Fixed O&M Costs (i.e. labor and administration) ___/yr Assume labor needed to encapsulate equals maintenance of otherwise removed materials.	4,000/yr (4 hrs/wk)
4. Additional Contract Disposal Costs	−76,800/yr
$Q \times [(1 - R/100) \times F_e) - I_e]$ = ___/yr	2,400 2,400 80%
I_c = current disposal cost per unit _____	
F_c = disposal cost per unit after SRS _____	
R^b = estimated reduction in waste _____	
Total Additional Cost $___/yr	**$−69,600/yr**

[a]Capital Replacement factor based on 6% rate-of-return
[b]Where reduction is derived from increase in product life-expectancy;
R = 1 − [(average previous life exp./average future life exp.) × 100]

REFERENCES

1. "Managing Industrial Solid Wastes: From Manufacturing, Mining, Oil and Gas Production, and Utility Coal Combustion," U.S. Congressional Office of Technology Assessment, Washington, D.C., Background Paper, OTA-BP00-82, U.S. Government Printing Office, February 1992.
2. Electric Power Research Institute. *Manual for Management of Low-Volume Wastes from Fossil-Fuel-Fired Plants.* Prepared by Radian Corporation, May 1985.
3. Edison Electric Institute. *Ashes and Scrubber Sludges: Fossil Fuel Combustion By-Products: Origin, Properties, Use and Disposal*, Disposal Committee, Utility Solid Waste Act Group. Prepared by Environmental Management Services, May 1988.
4. Electric Power Research Institute. *Use of Coal Ash in Highway Construction: Michigan Demonstration Project.* Prepared by Consumers Power Co., EPRI GS-7175, Project 2422-7, Interim Report, February 1991.
5. Electric Power Research Institute. *Market Opportunities for Fly Ash Filters in North America.* Prepared by Kline & Co., Inc., EPRI GS-7059, Project 2422-20, Final Report, November 1990.
6. *SARA Subtitle III*, U.S. Environmental Protection Agency, 1986.
7. Electric Power Research Institute. *Options for Handling Noncombustion Waste*, 1990.
8. Electric Power Research Institute. *Noncombustion Waste Seminar Materials*, December 2 to 4, 1992.
9. Hazardous Materials Control Resources Institute (HMCRI), *Proceedings of National Conference on Minimization & Recycling of Industrial & Hazardous Wastes*, 1992.

GENERAL REFERENCES

"Electric Utility Coal Combustion By-Products: A Resource for Recovery and Utilization," Utility Solid Waste Act Group, Resource and Utilization Committee and Envirosphere Co., 1981.

Elliott, Thomas C., Ed. *Standard Handbook of Powerplant Engineering* (New York, NY: McGraw-Hill Publishing Co., 1989).

EPRI. "Characterization of Utility Low-Volume Wastes." Prepared by Radian Corporation, May 1985.

"Hazardous and Industrial Solid Waste Minimization Practices," ASTM, 1989.

"Managing Industrial Solid Wastes from Manufacturing, Mining Oil and Gas Production, and Utility Coal Combustion," U.S. Office of Technology Assessment, Washington, D.C., February 1992.

"Potential Reuse of Petroleum-Contaminated Soil: A Directory of Permitted Recycling Facilities," U.S. Environmental Protection Agency, June 1992.

"The Reduction of Wastes in the Electronics Industry, Waste Reduction Grant Program," Envirosphere Company, June 1988.

Technology Screening Guide for Treatment of CERCLA Soils and Sludges, U.S. Environmental Protection Agency, 1988.

CHAPTER 19

Food Processing

*Bob York and Ellen Bogardus**

INTRODUCTION

During the production of food, vast quantities of waste are generated. For example, it is estimated that, on average, people consume only 20% to 30% of a vegetable plant:[1] the remaining 70% to 80% either is discarded or is used for other purposes.

Fortunately, the waste generated from food processing operations does not create a difficult disposal problem. These wastes generally are nonhazardous and biodegradable. Increasingly, the by-products are beneficially reused or, at a minimum, managed in a way that reduces any detrimental impact on the environment. Beneficial reuse of food processing residuals will continue to become more important as the world's population (and therefore demand for food) grows.

This chapter provides an overview of options for pollution prevention in the food processing industry. Processing of dairy products, fruits and vegetables, meat, poultry, and seafood is included. Several examples are included that illustrate specific programs that have successfully reduced food processing waste through recycling and reuse.

DAIRY PRODUCTS

In the dairy industry, processing plants use milk to produce various products, including fluid milk, evaporated milk, cultured dairy products (such as cheese), and frozen dairy products (ice cream and other confectioneries). Waste in the dairy industry is produced by spills, cleaning and sanitizing operations, products that are off-specification or whose shelf life has expired, and production residuals (e.g., whey from cheese production, distillate from milk evaporation). Figure 19.1 shows a

*Mr. York is a process engineer in CH2M HILL's office in Seattle, Washington, and can be reached at (206) 453-5000. Ms. Bogardus wrote the section on recycling in the food processing industry included in this chapter. She is the director of HDR's Solid Waste Division for Southern California in Irvine, California, and can be reached at (714) 756-6800.

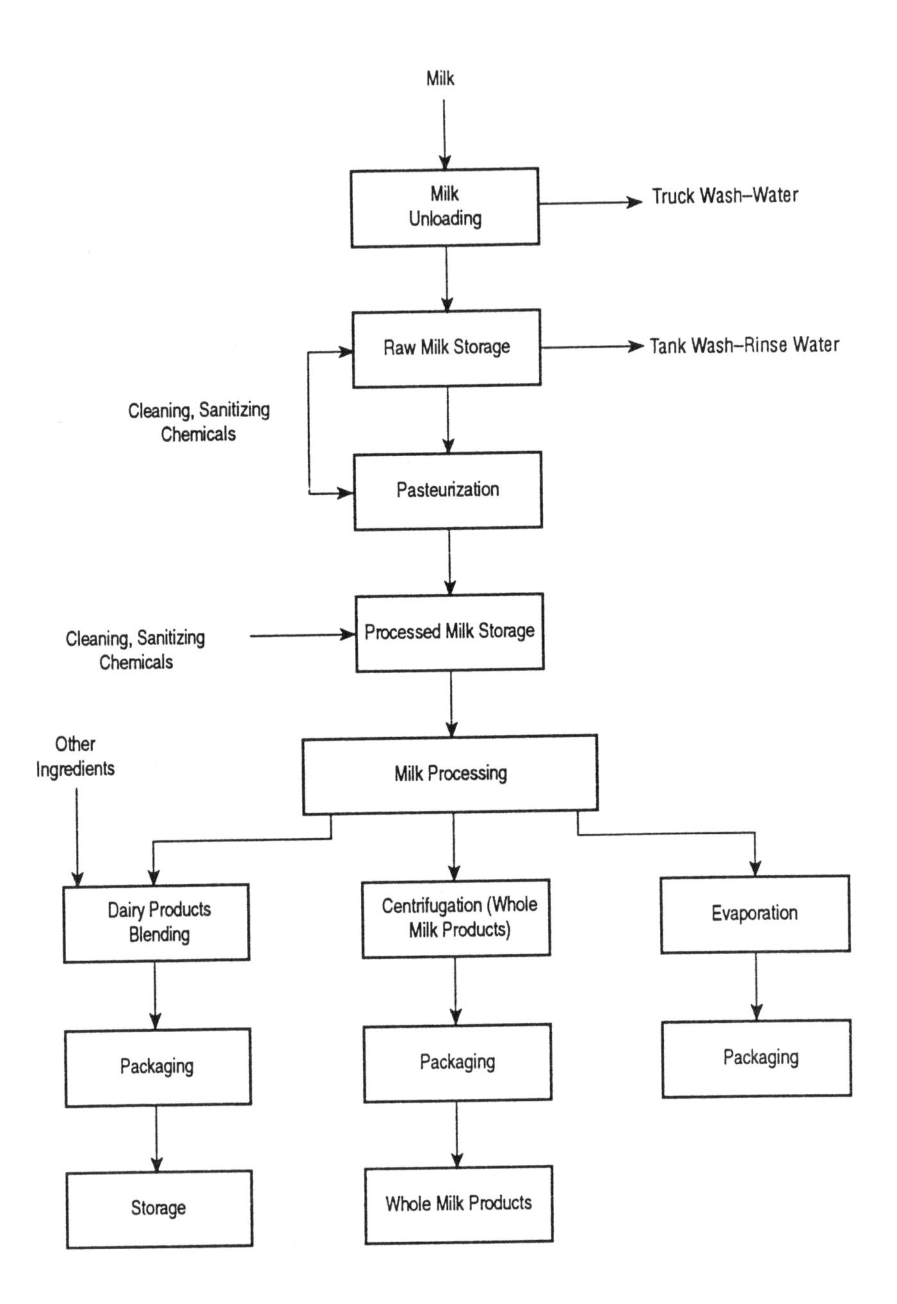

Figure 19.1. Generalized dairy industry processing sequence.

schematic of the process sequence in the dairy industry. Opportunities for minimizing waste include the following:

- minimizing the loss of milk before cleaning and sanitizing operations
- reusing off-specification or unsalable products
- treating and recovering useful commodities from production residuals

Loss Reduction

Loss reduction is a goal common to essentially all milk-processing operations, not only because it maximizes production, but also because of sanitation concerns. Raw milk has a 5-day biochemical oxygen demand (BOD_5) that exceeds 100,000 mg/L, providing a strong incentive to minimize its discharge into sewers and reduce the associated discharge fees. Methods for controlling spills include proper personnel training, equipment and piping maintenance, and placement of containment curbs and pans around equipment and tanks to prevent accidental spills from entering the sewer.

Milk losses can be minimized by designing features into the plant that purge milk piping, vats, and tanks before the clean-in-place (CIP) operation begins. One modern plant does this by air sparging process vessels and pipes. Recovery of the milk reduces sewer losses at the plant to less than 1% (versus the industry average of 2 to 3%), which increases revenue and lowers sewer bills. Chemical use is minimized during the process by reusing the alkaline detergent solution, and water use is minimized by reusing the final alkaline rinse water as the initial rinse for the next CIP operation.

Integrated milk processing plants produce many products, and it is common for the plants to reuse milk products nearing their expiration date in other types of products (such as cultured products). Similarly, milk can be recovered in the final stage of the initial water rinse of a CIP operation and be reused in products where a small quantity of dilution water is acceptable, such as in ice milk or sherbet.

Recovery and Reuse of Milk By-Products

Milk by-products that cannot be reused in products for human consumption can still be reused for animal feed. In these cases, frequent shipments of by-products to feedlots and farms are necessary to minimize spoilage.

One of the most significant sources of organic pollution from dairies is whey, a by-product of cheese making. Approximately 80% of the milk that is used to make cheese is converted to whey, or about 1 gal of whey per pound of cheese. Whey is approximately 6% solids and has a BOD_5 of more than 40,000 mg/L. Few cheese production facilities currently discharge whey into the sewer. Instead, they employ a variety of techniques to make use of the whey.

Case Study 19.1: Whey Reuse at a California Cheese Company

One of the largest and most modern cheese plants in the world is the California Cheese Company facility in Corona, California. This facility converts more than 1 million lb of milk each day into various types of cheese. The whey from cheese-making is first processed to extract the whey butter. The residue is then ultra-filtered to recover a whey protein concentrate. The lactose-rich permeate stream is sent to a fermentation process, where alcohol is produced. Finally, the remaining liquid is sent to an evaporator, which produces a distillate stream that is discharged to the municipal sewer. The salty concentrate from the evaporator is dried and disposed of in a landfill.

Other plants employ similar techniques. New techniques, such as using ion exchange for recovering whey protein, are being researched and may provide more efficient means of extracting usable products and reducing waste.

FRUITS AND VEGETABLES

The fruit and vegetable canning and freezing industry is a diverse industry consisting of a wide variety of processing methods, plant sizes, and locations. Figure 19.2 shows a schematic of the process sequence in the fruit and vegetable industry. In general, processes can be placed in the following categories:

- harvesting, in-field processing, and transporting of produce to a processing plant
- in-plant processing
- handling and packaging of the converted product
- warehousing and distribution

Harvesting, In-Field Processing, and Transporting of Produce

Mechanical harvesting has been practiced with many crops, including green peas, beans, spinach, corn, tomatoes, lettuce, cranberries, cherries, potatoes, beets, carrots, and nuts. During harvesting, plant wastes, such as vines or stalks, are left in the field and used by the farmer as a soil additive or as animal feed. However, much of the inedible component of the crop is not left in the field or orchard but is transported to the processing plant.

By reducing the need for manual harvesting, mechanical harvesting has obvious economic benefits, but it also has several disadvantages in terms of waste management. For example:

- Mechanical harvesting can cause more damage to the usable fruit or vegetable, such as bruising or breaking, or to fruit or nut trees.
- Significant quantities of soil (and soil microbes) are harvested with planted crops, which increases the amount of washing required.
- Yields may be lower and in-plant waste may increase because immature or unripe fruits or vegetables are harvested along with usable crops.

Processing of many crops begins in the field. Raisin grape bunches are cut and then placed on sheets on the ground for natural sun drying. Peas are shelled using mobile viners located at or near the field. The amount of waste that leaves the field can be reduced by using devices that remove stems, sticks, leaves, and soil from crops that have been harvested mechanically. Advantages of in-field processing

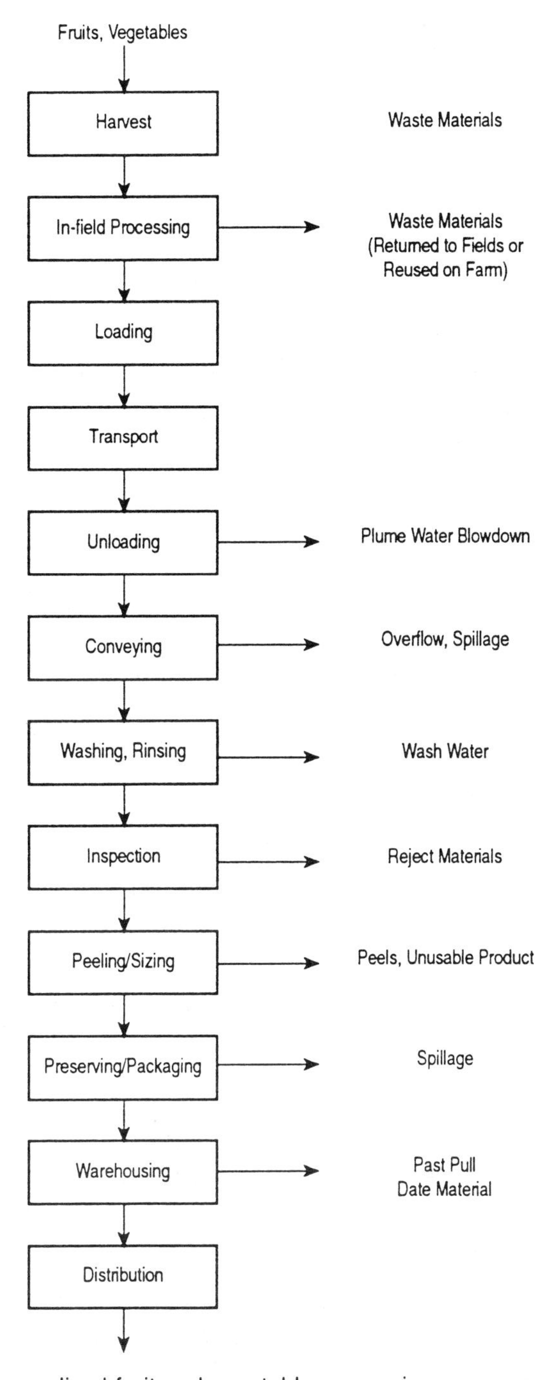

Figure 19.2. Generalized fruit and vegetable processing sequence.

include retention of culls, seeds, peel, and soil for reintroduction into the field soil or for other beneficial reuse at the point of origin. In-field processing thus reduces the amount of waste that is transported to processing facilities.

Each year, more than 30 million tons of fruit and vegetables are transported from the field to processing facilities. A significant recent development has been the direct transfer of crops from the mechanical harvester to dry bulk loading trucks or large bins, eliminating the use of smaller containers such as sacks, baskets, or lug boxes. Integrating mechanical harvesting and transport has decreased the amount of time between the field and the processing plant, thus preserving quality and reducing time-related spoilage.

In-Plant Processing

The processing plant presents many opportunities for reducing waste. The typical steps at the processing plant include unloading and conveying, washing and rinsing, inspection, peeling and sizing, and preserving and packaging. These steps are covered individually in the sections below.

Unloading and Conveying

Unloading the food commodity from the transport vehicle and conveying it into and through the plant are done mechanically, pneumatically, and hydraulically. Mechanical methods include vibrating conveyors, gravity chutes and dumping stations, belt conveyors, and screw conveyors. Pneumatic methods are either pressure or vacuum. Hydraulic methods are fluming and sluicing. The key consideration is to minimize physical damage to the crop.

Water is used extensively in conveying fruits and vegetables at processing facilities because its use is economical, it typically does not damage the crop, and it cools and cleans the product. A significant disadvantage of using water is the leaching of solubles from the product, such as leaching of sugars and organic acids from fruit and leaching of sugars and starch from vegetables. Recent efforts have focused on reducing such losses. Several tomato canneries in California, for example, recently installed treatment systems on their recirculating flume water supply. The systems consist of screening and grit removal using a centrifugal grit-removal device. Reportedly, such systems have cut blow-down quantities in half.

Washing and Rinsing

Washing and rinsing remove residues (soil, dust, pesticides, insects, partial product, and occluded solubles), cool the product, and extract solubles, such as preservative acids or salts. The quantity of water used in washing and rinsing may be as much as 50% of the total water used at the plant.

Using less fresh water for washing and rinsing reduces the volume of wastewater discharged from the plant. Wastewater can be minimized by using fresh water for final washing and rinsing only. The final rinse water can then be reused for preliminary washing and rinsing. Alternatively, a sanitary water source other than fresh water, such as evaporator condensate water in a tomato cannery, can be used for washing and rinsing.

Inspection

Little opportunity exists in the inspection process for additional reduction of waste. Processing plants have systems in place for reusing rejected fruits and vegetables.

Peeling and Sizing

Peeling of fruits and vegetables is necessary for improving product appearance or texture, or for preparing the product for preservation. Peeling can be accomplished by hydraulic pressure, immersion in hot water, contact with caustic solution, steam exposure, mechanical knives, mechanical abrasion, hot air blast, flame exposure, and infrared radiation.

Sizing is defined as processes such as slicing and dicing necessary to shape the products to the desired dimensions. Sizing processes generally use a form of cutting action.

Peeling and sizing are expensive not only as unit processes, but also because of the loss of edible material. Geneticists have had some luck in developing varieties of vegetables and fruit with regular shapes and with thin, smooth skins in order to reduce losses from peeling.

Different peeling technologies produce different kinds of waste. In the potato industry, for example, most processors have shifted from caustic peeling, which produces a high pH, salty wastestream to steam peeling, which produces a less salty stream at a more neutral pH. The savings in chemical use and, therefore, cost, are significant.

Preserving and Packaging

Preserving and packaging result in a ready-to-sell product with an acceptable shelf life. Three basic alternative preserving and packaging methods are fresh pack, canned, and frozen.

In fresh pack, the prepared product by definition requires little or no preservation. Waste can be reduced by using only the minimum packaging required for sanitation, protection, and marketing. In addition, using biodegradable or recyclable packaging materials will reduce the volume of packaging waste contributed to the waste stream.

Canned fruits and vegetables are first placed in the can, after which the can is washed, sealed, and retorted. Waste can be reduced by maintaining and adjusting the filling equipment to minimize or eliminate product losses. Retort water is generally a clean "non-contact" water source, and can be reused either in the retort units or elsewhere in the plant.

Frozen fruits and vegetables are generally blanched before the freezing process. Blanching involves direct contact of the product with hot water, and therefore some quantity of organics is generally leached into the blanching water. Although the volume of water from blanching is small relative to other sources of in-plant wastewater, blanching water typically has high concentrations of dissolved and suspended organic matter.

MEAT, POULTRY, AND SEAFOOD

Processing of meat, poultry, and seafood generates concentrated but biodegradable organic waste with a high percentage of protein and fat. Blood, grease, and byproducts such as viscera, feathers, and inedible parts typify the types of wastes produced. Many of these wastes have reuse value. Other means also are available for reducing the waste produced during processing. Waste reduction alternatives for each food category—meat, poultry, and seafood—are presented below.

Meat Processing

A general flow chart of the processes used in a meat packing plant is shown as Figure 19.3. Primary processes are associated with the production of meat for human consumption. Most meat processing plants also have secondary processes for the by-products of the primary processes, although many do not have edible or inedible rendering on the site.

Blood Recovery and Processing

Because of the high BOD of blood and its effect on water quality, blood recovery is now practiced in nearly all meat processing plants. (In 1950, only 70% of the plants practiced blood recovery.) Blood has an ultimate (20-day) BOD of approximately 405,000 mg/L and a BOD_5 of 200,000 mg/L. Plants that allow blood to escape into the sewer are generally small in size. According to one estimate, blood recovery can reduce the BOD in wastewater by nearly 50%.

Although blood recovery is practiced in nearly every plant, the efficiency of

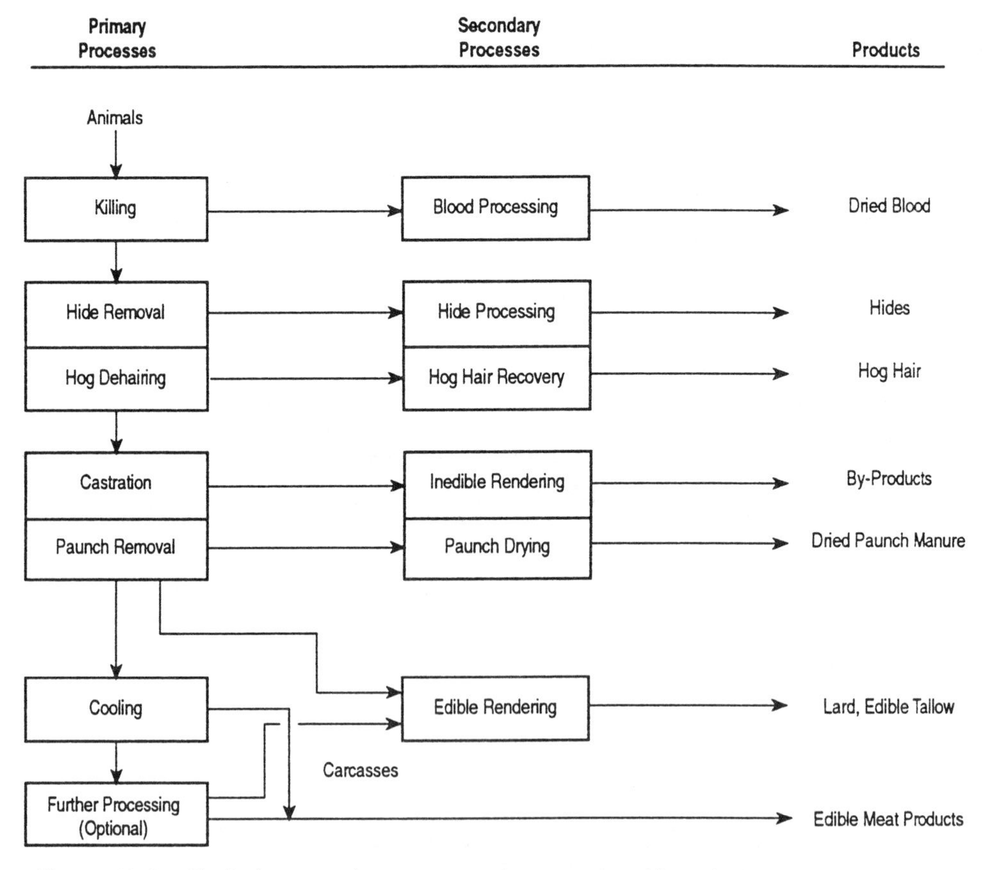

Figure 19.3. Typical processing sequence in a meat packing plant.

recovery varies. The area for collecting blood must be adequate in size to contain the blood before it is conveyed to the next processing step. In recent surveys of several meat processors, sidewalls for containing blood were not always adequate. After the animals have been slaughtered, the blood remaining in the slaughter area should be squeegeed to the blood tank before the area is rinsed with water. Even more blood can be captured by using a low-volume preliminary rinse to clean the area and by channeling the preliminary rinse water through the blood-processing operation. Subsequent rinses would then drain to the sewer.

Processing of blood typically consists of using dryers to extract as much water as possible from the blood. Continuous dryers, using a jacketed vessel with rotating blades to prevent burn-on or using a ring dryer, produce the driest product. Other materials, such as paunch manure or casing slimes, are sometimes dried with the blood. Dried blood can be used as fertilizer.

Hide Removal

Most packing plants remove hides mechanically. Hides are recovered and used to produce leather goods.

Hog Dehairing

Hog hair is loosened by placing the carcass in a scalding tank where the hair is loosened by 140°F water. The hair then is removed by mechanical scrapers. Hog hair has been used in the manufacture of foam rubber.

Evisceration

Viscera are removed and inspected. Some viscera, such as hearts, stomachs, and liver, are edible and are typically handled separately from inedible viscera, which are rendered.

Paunch Manure

Paunch manure is the name given to the undigested or partially digested contents of the "first" stomach of a ruminant, such as cattle and sheep. A typical head of cattle contains 50 to 90 lb of paunch manure. Paunch manure is recovered using either a wet or a dry process. Wet dumping consists of cutting the paunch open in a water flow, discharging to a mechanical screen, and then to the sewer. Paunch "solids" are approximately 75% water, with a BOD_5 that exceeds 100,000 mg/L; therefore, wet handling is not the best practice for reducing waste.

Dry dumping consists of dry discharge to a hopper for ultimate disposal as a waste solid or blending to produce a marketable solid. Paunch material may be used as a soil conditioner or as animal feed.

After the material has been removed from the paunch, the paunch itself can be cleaned and used to produce tripe. If tripe is not processed, the paunch is rendered.

Rendering

There are two types of rendering: edible and inedible. The basic rendering process is similar for both, but the materials rendered are different. Edible rendering converts primarily hog fats to lard and edible tallow. Inedible rendering converts inedible parts of the animal to inedible fats, which are used for soap and animal feed. Animals (or animal parts) that do not pass federal meat inspection also are processed by inedible rendering.

Three commercially used rendering processes are dry rendering, wet rendering, and low-temperature rendering. In dry rendering, tissues are cooked in steam-jacketed tanks under vacuum pressure. Dry rendering requires no contact water; therefore, no wastewater is discharged directly from this process.

In wet rendering, tissues are loaded into a pressure tank, which is then sealed. Steam is introduced into the tank at 40 to 60 psig until the fats are completely freed from the tissues. A three-phase mixture forms in the tank: fat on top, solids on the bottom, and water in the middle. Typically, the fat phase is drawn off, then the water and solids are sent through a press or a centrifuge to be separated. Approximately 75% of the protein content introduced into the wet rendering process is dissolved in the tank water. The water has a high organic load, with a BOD estimated at approximately 30,000 mg/L.

Discharge to the sewer would dramatically increase the BOD of the wastewater, as well as waste a valuable resource. For this reason, most plants with wet rendering processes evaporate the water to a moisture content of about 35%. This material, known as "stickwater," can be used in manufacturing of amino acids or pharmaceuticals, or as animal feed.

Low-temperature rendering is a relatively new process with a maximum temperature of approximately 118°F. Fatty tissues are broken down mechanically with no water added. This process produces less odor and, like dry rendering, has little potential for polluting water.

Poultry Processing

The processing of poultry is shown schematically in Figure 19.4. Live poultry arrives in coops by truck. The birds are stored in their coops for up to several hours, after which they are removed for processing. In general, processing plants schedule deliveries to minimize the time the birds must wait for processing. This practice reduces the amount of manure that must be handled before the birds are processed.

Blood Recovery and Processing

Killing of the birds produces large volumes of blood, which can be a source of high organic loading in processing plant wastewater if the blood is not recovered efficiently. Poultry blood has an approximate BOD_5 of 92,000 mg/L and constitutes approximately 40% of the waste load from a typical processing plant.

Recovered blood can be used by a rendering plant to produce animal feed. However, even with blood recovery, a significant amount of blood may be rinsed into the sewer when the area is cleaned.

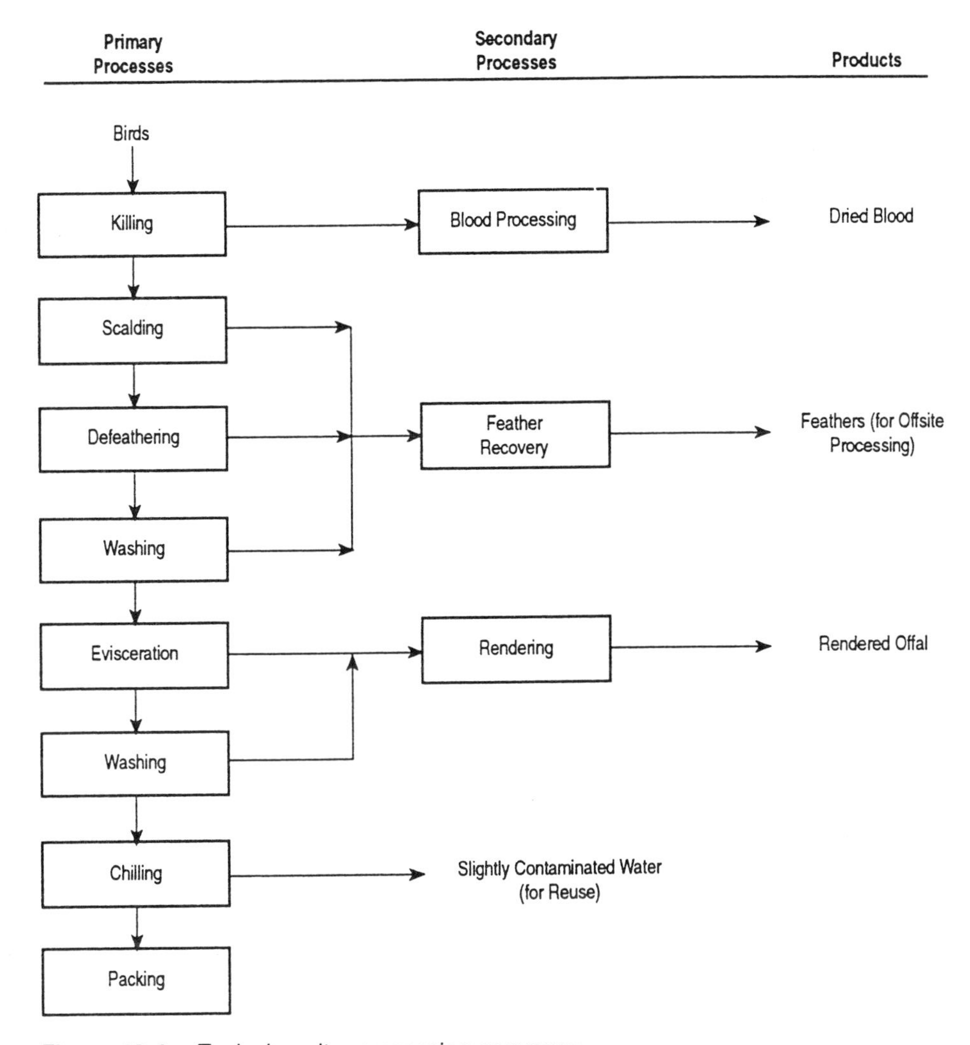

Figure 19.4. Typical poultry processing sequence.

Scalding

After they have been killed, the birds are placed in scalding (128° to 140°F) water to remove blood and dirt and to loosen feathers. The scalding tank is fed water at a rate of 1/4 gallon per bird, as required by federal (U.S. Department of Agriculture [USDA]) regulations. The overflow from the scalding tank is a source of dirt, oil, and organic material: BOD_5, total suspended solids (TSS), and oil and grease in scald overflow have been measured at levels up to approximately 1,200 mg/L, 700 mg/L, and 350 mg/L, respectively.

Make-up water for the scald tank need not be fresh water. To conserve water, relatively clean water from the chilling operation (discussed below) can be used as make-up water.

Defeathering

After scalding, feathers are removed from the birds by a mechanical operation employing rotating rubber drums containing rubber fingers. A continuous stream of water washes the feathers to a flume and then to a screen, where the feathers are separated from the water.

The supply of water for this flume need not be fresh; chiller water, evisceration flume water, and water recycled from the feather flume can be used.

The feathers represent a resource that should be and typically is reused. Feathers may be sent to rendering facilities to be converted to proteinaceous animal feed. Processed feathers also can be used as mulch for agricultural application.

After mechanical defeathering, pin feathers are removed by hand or by wax stripping. The bird is then passed through an arc-gas flame, which singes fine hair and the remaining pin feathers. A continuous spray of water is used to wash the surface of the defeathered bird.

Evisceration

The evisceration area of the plant is separated from the other areas to maintain sanitary conditions. Evisceration involves removal and separate handling of edible and inedible viscera. In general, a water flume is used to carry away evisceration waste because of the convenience and cleanliness associated with this method of transport. After evisceration, the inside of the carcass is washed and inspected. Evisceration flume water and final washwater can be reused in areas of the plant where they will have no contact with edible products.

Chilling

The processed poultry is either chilled or frozen before shipment. Proper chilling is vital for maintaining flavor and increasing shelf life. Chilling and freezing operations are governed by federal regulations. Federal regulations specify that a minimum of $1/2$ gallon of chilling water per frying chicken must be maintained in the first section of the chilling unit. Water may be recirculated from the second and third tank to the first tank. However, regulations specify that all tanks continuously overflow and that the chilling water remain "reasonably clear." Chilling water should be considered for reuse elsewhere in the processing operation, except where USDA regulations require potable water.

Seafood Processing

Unlike the relatively similar processing techniques used in other industries, processing techniques used in the seafood industry are diverse. The scope of waste reduction opportunities for each distinct processing subcategory is beyond the scope of this text. In general, the same themes for waste pollution prevention discussed above for other food processing operations apply to the seafood industry as well. They include good housekeeping for keeping organic materials out of waste stream(s); water conservation and reuse wherever possible; and beneficial reuse of by-products.

OPPORTUNITIES FOR RECYCLING IN THE FOOD PROCESSING INDUSTRY: RESULTS OF A CALIFORNIA STUDY

Results of a study of solid waste generation undertaken by the City of Los Angeles indicate that food processors disposed of 74,209 tons of solid waste in 1990, or 32.5% of the City's disposed industrial waste stream. On the basis of initial surveys, it is estimated that an additional 27,300 tons were diverted. Much of this diversion can be attributed to the reuse of brewer's grains, grain milling by-products, and animal by-products.

Because of its impact on total waste volumes, the food processing industry was targeted by the commercial/industrial recycling program of the City of Los Angeles. The program was undertaken to comply with the Assembly Bill 939 goals of 25% diversion from landfills by 1995 and 50% by 2000. The waste stream from food processors offers potential for source reduction, recycling, and composting. Furthermore, the packaging practices of food processors affect the waste streams of other generators, such as food service operations, grocers, and households.

Recycling Opportunities

Operational and Size Issues

Many food processors interviewed during the study complained that comprehensive hauling service is hard to get for the range of recyclables that they generate. Metals, plastics, wood, and paper require transport to different markets. Coordination and information gathering are time-consuming because available service providers cannot or will not identify markets for some materials generated in large volumes, such as strapping tape.

Although many large producers of food waste already divert much of their waste or have the ability to do so, small generators usually are not as well served because of high collection costs or the lack of a financial incentive to research recycling options. Small generators pursue different, less formal arrangements than large generators. For instance, scrap and excess bread from a large bakery in Oakland was sold to Dext, a national company that processes it into chicken feed. Excess bread from a small bakery in Palo Alto was donated to a charitable group, and coffee grounds were given to gardeners.

Cost Issues

For large food processors, recycling and reuse can reduce disposal costs significantly. In one survey, Kraft USA reported that its recycling and reuse program reduces disposal fees by about $250,000 annually. In addition, revenues from selling recyclables (including corrugated containers, fiber drums, and scrap metal) amount to $50,000 annually. North American Produce (NAP), which packages lettuce and onions for 1,400 McDonald's restaurants, reports annual savings of $100,000 to $125,000 on waste hauling since it began diverting food waste to a pig farmer.

Opportunities for Source Reduction and Reuse

Food processing waste can be reduced through changes in purchasing and packaging practices as well as through recycling. Some items that contribute significantly to the food processing waste stream, such as waxed corrugated containers and food waste, are prime targets for source reduction or reuse. As noted, some companies are testing pallets that are thought to be significantly more durable. Increasingly, food processing companies are seeking alternatives to nonrecyclable waxed cartons.

NAP, in the City of Industry, California, receives all of its produce from one vendor who ships in reusable crates. NAP returns the crates, avoiding use of waxed corrugated containers. This company also participates in a regional pallet exchange program. Miller Brewing in Irwindale, California, reports that changes in the design of rolls of toilet paper and paper towels has significantly reduced waste from these materials. Kraft Foods in Buena Park, California, reused 85,000 corrugated boxes in 1991.

Opportunities for Recycling and Reuse

- The food processing waste stream is highly recyclable or compostable.
- Food processors have the potential to affect the residential and food service industry waste streams.
- Associations representing retailers, food processors, and environmental groups have a heightened awareness of pollution prevention issues and are conducting useful research on recycling and reuse.

Barriers to Recycling and Reuse

- Information is lacking about existing and planned recycling opportunities.
- Composting options are scarce for food waste in urban areas.
- Packaging and product alternatives are unavailable for some items.
- Pests must be handled so they do not become a problem.
- Stable markets for plastics may be difficult to find.
- Collection options are unavailable for some materials.
- Local government franchises may restrict service for some recyclables, particularly food.

Opportunities for Market Development

Real recycling occurs when recovered materials are used to make new products and demand for products containing recycled material stimulates the markets necessary to make collection programs successful. Paper mills and converting operations note that packaging design and content is dictated by their customers. The demand for recycling increases when food manufacturers show a preference for recycled-content packaging, which, surveys show, is increasingly popular with consumers. NAP, whose major customer (McDonald's) requires that its suppliers use recycled-content material, ships its product in corrugated cartons that contain 40 to 50%

recycled material. Food processors, especially large companies, can play an important role in developing a market for recyclables by using their buying power to demand recycled-content packaging from their suppliers.

Opportunities for Pollution Prevention

Some of the best alternatives for pollution prevention associated with food processing are:

- reduce packaging
- donate food
- sell food waste and by-products
- eliminate pallet waste
- use reusable shipping containers

Reduce Packaging

Packaging performs several important functions, including extending product shelf life and protecting the consumer. However, nearly one-third of the solid waste generated in the United States is packaging, and some products are over-packaged. One source estimates that of every dollar spent by consumers in grocery stores, 10 cents are for packaging.

Various manufacturers are making efforts to reduce packaging waste. Sara Lee eliminated paperboard layers in its frozen bagels and removed the aluminum tins from its frozen muffin packages. Shipping cartons were also redesigned to reduce material. Lipton has reduced the weight of many of its outer cartons and canisters. Kraft's microwave entrees no longer have a lid or plastic overwrap.

Donate Food to Charity

Edible but excess or unsalable food can be donated to food banks or charitable organizations. Prolific Oven bakery in Palo Alto donates its day-old bread to a church that provides meals to the needy.

Sell or Donate Food Waste and By-Products for Animal Feed, Alternative Fuel, or Other Uses

The most appropriate use for food waste depends on the product. Some food waste, such as grain from breweries, can be used for livestock feed. This is the practice at the Miller Brewing plant in Irwindale, California. Pits and shells from processed fruits, vegetables, and nuts are often used for alternative fuel. Peach pits have several other uses: Del Monte sells them to peach-growers, and the pits also can be used in fragrance manufacturing. Food waste is often applied as a soil amendment to agricultural and other land. For example, some of Gorton's fish offal is collected by a company that blends it with bakery waste and makes it into chicken feed. These types of practices appear well established in the food industry.

Eliminate Pallet Waste

Products delivered to and shipped from manufacturers are usually delivered or stored on wood pallets. These pallets can vary in durability and value. Different kinds of manufacturers have different relationships with suppliers and distributors and varying capacities to repair their own pallets. Large companies, such as Kraft and Del Monte, have facilities for repairing their own pallets. Smaller companies often rely on local pallet buyers and wholesalers, who may backhaul pallets for repair or reuse. When a mechanism for repairing and reusing pallets does not exist, some small manufacturers have given away their damaged or excess pallets for use as firewood.

Recognizing the value of pallets, a number of manufacturers have developed pallet accounting systems that allow them to reject odd-sized or defective pallets from other sources and/or to charge grocers a fee for pallets not returned. Furthermore, the grocery and pallet industries have been discussing standardizing pallet size and various other options for reducing pallet waste, including creating a national pool of pallets.

Pallets can be made to be highly durable and reusable. Slipsheets and pressed-paper pallets may be recyclable in some areas. Local markets and recyclers should be contacted before a food processor changes pallet suppliers.

Dole Foods wants to encourage the return of pallets to their plant, so they ask for a special agreement by the customer promising to return them. If they are not returned, Dole charges a fee.

Eliminating the use of pallets reduces the weight of shipments and reduces pallet waste. If the customer is agreeable, Dole Foods prefers to use slipsheets instead of pallets for shipping canned goods.

Use Reusable, Readily Repairable Shipping Containers

Dole Foods ships frozen pineapple concentrate in 450-gallon square wooden tote bins, which are disassembled and shipped back for reuse. They are repaired as often as possible before becoming unusable from wear and tear. The sides are all interchangeable, so parts of the container can be repaired and replaced easily without having to rebuild the entire container.

Opportunities for Recycling

Some effective approaches to recycling in the food processing industry are:

- recycle corrugated containers
- recycle stretch wrap
- recycle meat by-products
- use recyclable products
- operate a recycling center
- recycle polycoated paper

Recycle Corrugated Containers

Corrugated containers are a large portion of the waste disposed of by food processing companies. It is important to separate different types of corrugated containers for recycling. Waxed corrugated containers cannot currently be recycled with unwaxed corrugated containers. (The Grocery Industry Committee on Solid Waste recommends composting waxed corrugated containers with food waste.) However, many recyclers will accept kraft paper bags mixed with corrugated containers, as they are made from the same type of fiber. Careful compliance with recyclers' guidelines will maximize the benefits from a recycling program.

Recycle Stretch Wrap

Stretch wrap, a plastic resin known as LLDPE (linear low-density polyethylene), is used to hold multiple boxes together on a pallet and to provide a barrier around food. LLDPE can be recycled in a limited number of locations in the country. Markets for recycled LLDPE in the Los Angeles region have emerged since 1990 and are expected to grow.

Recycle Meat By-Products

Fat and bone together contain tallow, which is used in manufacturing soap. Bone is also used as a protein supplement in agricultural feed. Los Angeles has three major rendering companies that collect fat and bone in special barrels and recycle them as ingredients for new products.

Buy Recyclable and Recycled Products and Packaging

Buying recycled products is the last step in making the entire recycling process work by providing markets for collected materials. For instance, Ben and Jerry's uses a recycled-content outer paper box for its Peace Pop. Del Monte uses recycled-content steel cans and corrugated containers. Reuse and recycling within the food processing industry is limited by U.S. Food and Drug Administration (USFDA) regulations that prohibit the use of certain types of recycled-content packaging that comes in contact with food. The National Food Processors Association is researching this subject and will present its findings to the USFDA for consideration.

Operate an Onsite Buy-Back or Drop-Off Center for Employees or for the Community at Large

Del Monte Foods has instituted drop-off centers at three of its facilities, including the Walnut Creek Research Center in Portage County, Wisconsin, where steel cans are not included in curbside recycling service. The Del Monte plant accepts steel cans from employees, a university, a hospital, and a public school system. This extra 2.5 tons per week of scrap steel is baled and shipped with plant scrap to a steel recycler. At Walnut Creek and in DeKalb, Illinois, employees were given recycling bins to transport their recyclables to the plant.

Recycle Polycoated Paper

Markets for polycoated paper and aseptic packaging are developing, offering opportunities for dairy and other food processing subsectors to divert scrap packaging. Gorton of Massachusetts has found a market that accepts all of the scrap polycoated packaging from its fish processing plant.

Promotion and Education

Including information about waste reduction on packaging and in advertising can inform consumers and enhance the effectiveness of recycling programs. More and more products now indicate that they are recyclable or are made of recycled material. Del Monte recently added the message "Contains Recycled Steel. Recycle Again" to its labels.

Consumers in the United States are showing continued interest in buying products that reduce waste and have recycled or recyclable packaging. Simultaneously, the number of environmental product claims by manufacturers is burgeoning. Unfortunately, a number of these claims are misleading or confusing to the consumer.

Manufacturers can help consumers make well-informed choices by avoiding unsubstantiated claims and working with national certification programs such as Green Cross and Green Seal. These organizations provide a relatively unbiased evaluation of environmental product claims. Green Cross certifies the recycled content of packages and is beginning to use life-cycle analysis, a highly complicated and controversial assessment of all the environmental impacts of a product or package—from the moment the resource is extracted from the natural environment to the moment it is recycled, disposed of, or burned. Green Seal sets standards for specific products, then certifies brands that meet the standards.

GENERAL REFERENCES

Campbell, M.E., and W.M. Glenn, Profit From Pollution Prevention, Pollution Probe Foundation, Toronto, Canada, 1982.

Jones, H.R., Pollution Control in Meat, Poultry, and Seafood Processing, Noyes Data Corporation, 1974.

Jones, H.R., Waste Disposal Control in the Fruit and Vegetable Industry, Noyes Data Corporation, 1973.

CHAPTER 20

Iron and Steel

*Rajeev Krishnan**

The iron and steel industry is one of the oldest and largest manufacturing operations in the United States, with a total production capacity of approximately 117 million tons of steel per year.[1] There are three basic types of steel manufacturers: integrated mills, minimills, and specialty-steel mills.

Integrated steel mills supply about 70% of the U.S. steel market.[2] They use iron ore and coal as the raw materials for manufacturing various steel products. Minimills produce steel from scrap, typically using electric arc furnaces. Specialty-steel mills are similar in operation to minimills; they manufacture stainless, tool, and high-alloy steels. Minimills and specialty-steel mills account for 30% of U.S. steel production.

IRON AND STEEL MAKING PROCESSES

Three process steps—coking, iron making, and steel making—are fundamental to the iron and steel industry.

Coking

Coking is the process of converting coal to pure carbon called coke. Producing coke or coking is typically a batch process carried out at about 1,800 to 2,000°F, and takes approximately 18 hours. Coke provides both energy and carbon for chemical reduction of iron-oxide. Coke is used at a rate of approximately 0.5 ton of coke per ton of finished iron. The coking process yields coke as well as by-products such as coal tar, ammonium sulfate, and light oils. A typical coke plant consists of two major areas: the coke oven batteries and the by-product recovery plant.[3]

*Mr. Krishnan is an environmental process engineer in CH2M HILL's Corvallis, Oregon, office and may be reached at (503) 752-4271.

Iron Making

In iron making, coke is combined with iron ore and limestone and supplied to the top of a blast furnace. Air is heated to about 1,800°F and blown countercurrently from the bottom of the furnace through circumferential openings called tuyeres. In the furnace, coke reacts with iron ore, releasing iron and generating CO and CO_2 gas. Iron sinks to the furnace hearth. Impurities in the charge combine with the limestone, forming slag, which floats to the top of the iron. The combustible exhaust gases from the furnace have considerable heating value and are used to preheat the incoming air and to generate steam. Blast-furnaces produce iron, commonly known as pig iron, with a carbon content of about 4%.[3]

Steel Making

Steel is an alloy of iron, containing less than 1% carbon. Steel is made by three different methods: the basic oxygen furnace (BOF), the open hearth furnace (OHF), and the electric arc furnace (EAF). These furnaces reduce impurities, especially the carbon content by oxidation and add various alloying elements, depending on the desired grade of steel. Because the BOF and EAF are currently the most widely used techniques, only these have been described briefly in the paragraphs that follow.[4]

In the BOF method, hot molten pig iron from the blast furnace, scrap steel, mill scale, and slag conditioning materials (such as limestone, burned lime, dolomite, and fluorspar) are charged to the furnace. Oxygen is injected into the furnace through a lance positioned a few inches over the charge. Oxygen reacts to oxidize the charge which releases large amounts of heat that melts the charge. During the melting process, the fluxing materials, called slag, collect and float on top of the molten metal. The slag is decanted through a spout while the molten steel is either transferred to a teeming station or to a continuous casting station.

EAFs are used for manufacturing common grades of low-carbon steel as well as special alloy and tool steels. A chief advantage of the EAF method is that it is not dependent on hot molten pig iron from blast furnaces and can typically operate, if necessary, with scrap steel as the sole iron source in the initial charge to the furnace. EAFs also need the same or similar flux conditioning materials required by BOFs. Slag and molten steel separation is similar to that in a BOF.[5]

Other Important Iron and Steel Operations

Other common and important operations in an iron and steel plant are sintering, vacuum degassing, casting, rolling, and finishing. For the purposes of this chapter, these operations can be described as follows:

Sintering

Sintering is the cementing of blast furnace fines, mill scale, and flue dust that cannot be directly charged to a blast furnace. In this process, the fines are mixed with fine coal in a traveling gate, which makes them usable as a direct charge to the blast furnace.[4]

Vacuum Degassing

In this process, molten steel is subjected to a vacuum to remove gaseous impurities such as hydrogen, nitrogen, and oxygen.[4]

Casting

At the teeming station, molten steel is poured into molds to form ingots. After the ingots are cooled and the molds are stripped, the ingots are transferred to a primary hot rolling mill where the ingots are reheated in soaking pit furnaces and rolled into billets, blooms, or slabs. The continuous casting operation eliminates the hot rolling operation by casting the molten metal directly into thin slabs or other desired shapes.[4]

Hot Forming

In the secondary hot rolling mill, slabs or blooms are reduced to billets, shapes, or strips.

Cold Rolling

In cold rolling, unheated hot-rolled products are passed through rolls to produce steel with smooth dense surfaces and enhanced mechanical properties. Cold-rolled-steel products are not as thick as those produced by hot rolling.[4]

Surface Preparation

The rolled-steel products are pickled in acid to prepare the surfaces for subsequent application of corrosion-protection coatings. Molten salt baths are used to descale stainless-steel products.[4]

A schematic of the typical operations in an integrated steel mill (excluding the pig iron making operation) is shown in Figure 20.1.

Typical waste sources and applicable pollution prevention practices for the major operations in iron and steel production are discussed below.

COKING OPERATIONS

Air Emissions

The primary environmental problem with coke ovens and by-product plants is air emissions. Air emissions are primarily from oven charging, coke oven door and seal leaks, coke pushing and quenching, and open cooling towers. Emissions include volatile, and semivolatile organic compounds, CO_x, SO_x, NO_x, and particulate matter.[6] Figure 20.2 is a schematic of the various emission points from a coke oven. It is estimated that approximately 72% of the respirable particulate matter (PM10) emissions from the iron and steel industry are generated from coke ovens. The distribution of PM10 emissions from the various coking operations are estimated in Table 20.1.[7]

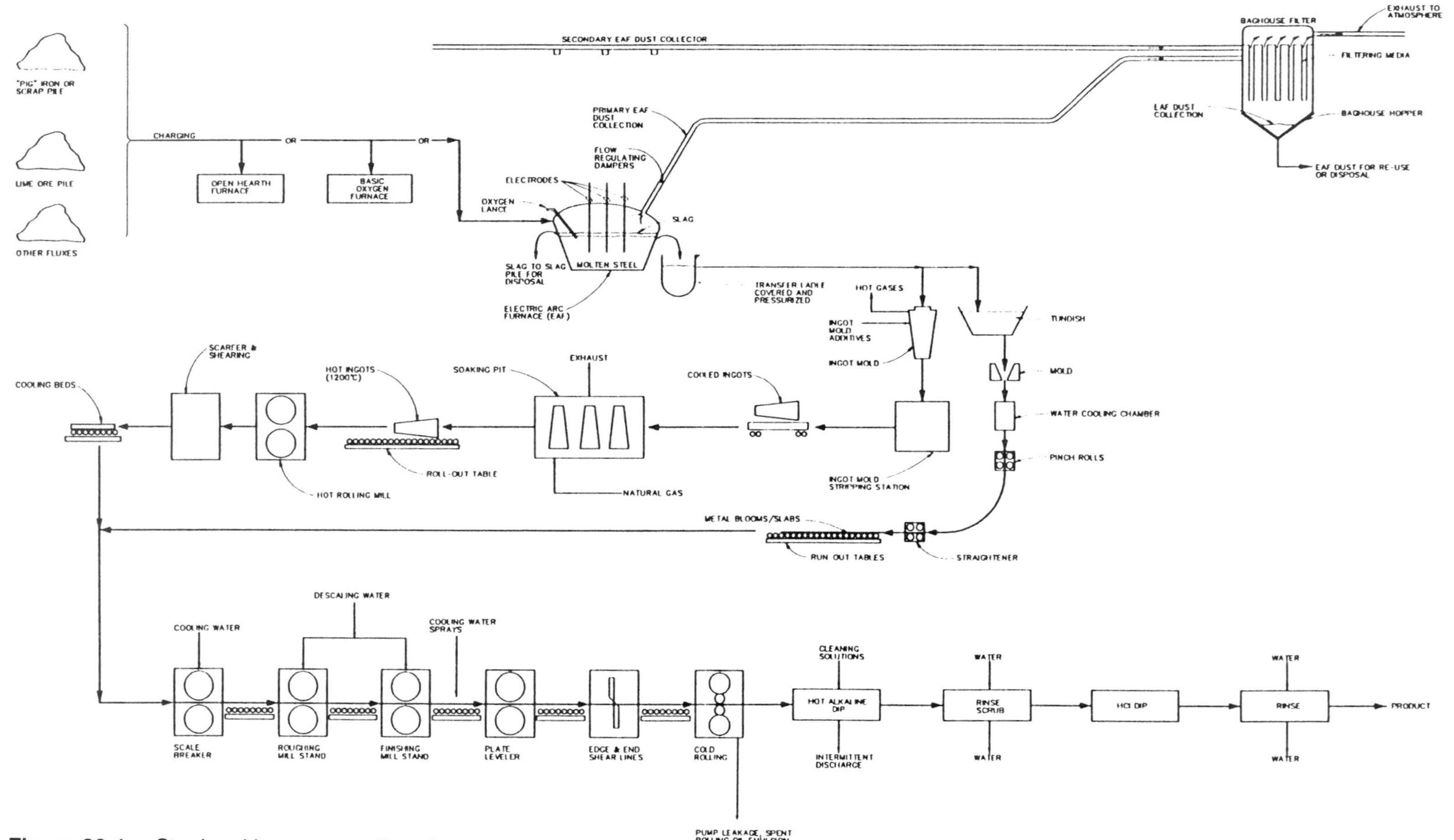

Figure 20.1. Steel-making process flow diagram.

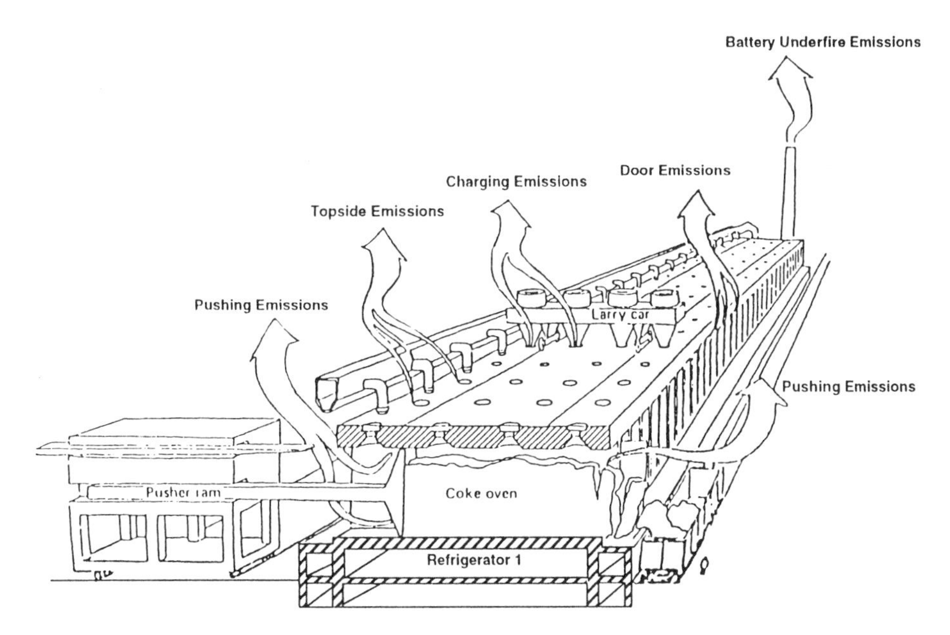

Figure 20.2. Types of air pollution emissions from coke-oven batteries.

Less than 25% of the existing coke plants in the U.S. comply with the provisions of the 1990 Clean Air Act Amendments for various reasons, including the nature of the coking process, obsolescence, heavy capital expenditure required, and extremely competitive market conditions. It is estimated that more than 40% of the coke ovens in North America are nearing the end of their useful life and will have to be refurbished or replaced in the next 10 years.[8] This, coupled with the Clean Air Act Amendments, could force the shutdown of 30% to 40% of U.S. coke-making capacity in the next 15 years unless significant environmental upgrades are implemented.[1]

Potential waste reduction measures for coking operations are summarized below. Some of the measures can be easily implemented whereas others are capital-intensive or are in the research-and-development phase and cannot be readily implemented.

Reducing Emissions from Coking Operations

Oven Leaks

Several technologies and practices can reduce emissions from coking ovens. Use of environmentally safe sodium silicate or other proprietary luting compounds can

Table 20.1. Distribution of PM10 Emissions from Coking Operations

Source	Portion
Leaking doors, lids, and offtakes:	39%
Coke pushing:	11%
Coke quenching:	34%
Underfired combustion:	16%

seal oven doors and lids to significantly reduce emissions. Luting materials and practices should be assessed continuously to keep abreast with the best compounds and systems available. Oven doors can be retrofitted or replaced with state-of-the-art door seals to reduce door leaks.

For instance, Still Otto, makers of "Autoseal" coke-oven doors, claim that their doors can limit emissions to less than 5%. When doors are replaced, innovative designs such as flexible doors (especially for 4- and 6-meter batteries), automatic spotting devices to reduce damage to doors and jambs, and provisions for automatic application of luting compounds can be incorporated to minimize emissions.[7] Inland Steel and Sun Coal Co. are engineering a coking facility for the Indian Harbor works that operates under negative pressure to eliminate door leakage. Similarly, the Enprotech design can minimize leaks. This design incorporates an air gap between the door body and the diaphragm. Because air is a poor conductor, an air gap minimizes heat conduction to the door body and reduces warpage, which reduces emissions.[8]

In addition, cleaning, maintenance, and operating practices for doors, jambs, and lids should be periodically reviewed and updated to reduce emissions.

Desulfurization

Desulfurization of coke-oven gases, with a relatively high sulfur content, can reduce sulfur dioxide emissions. Sulfur-removal technologies such as Sulfiban and Lo-Cat not only reduce sulfur dioxide emissions by over 95% but also generate a marketable sulfur by-product.[8]

Dry Quenching

Dry quenching techniques eliminate the vapor cloud found over typical quench towers. This technique, using inert gases as a medium to transfer heat from red-hot coke to water to make steam for either process or power generation purposes, is being used in Russia, Japan, Europe, and Asia. The heat transfer media (i.e., the gases) is completely enclosed and fully recycled.[8]

Process Alternatives

Four coke-making processes seem to hold promise for the future, namely nonrecovery ovens, formed coke, jumbo coke ovens, and advanced form coke.[9]

Nonrecovery ovens. Jewell Coal and Coke is the only coke plant in the U.S. that uses nonrecovery ovens. This technology eliminates the need for a by-product plant and therefore nearly eliminates hydrocarbon (especially benzene) emissions. These ovens operate under negative pressure and are designed to allow coal gases to burn above the coal. The gases are drawn into flues under the oven, producing coke from two sides. The principal toxic gases are incinerated in the process, leaving relatively small quantities of acid gas and PM10 emissions that require control.

Formed coke. The formed coke process consists of a series of closed reactors that produces a char from noncoking coals. The char is further processed into briquettes (formed coke), a substitute for the traditional coke used in blast furnaces. This process is being developed by Kawasiki Steel, Japan.

Jumbo coke ovens. Jumbo coke ovens, as the name suggests, are larger than typical coke ovens and have less oven surface area to leak per ton of coke produced. They are also equipped with features to reduce emissions, such as coke cassettes for dry cooling and coal preheaters with charging chain conveyors.

Advanced formed coke. This technology is being pursued by the Japanese in a national project. It is based on extracting coal compounds for processing into fuel followed by processing the remaining coal material into coke-type product by briquetting and carbonization.

Eliminate Use of Coke

The use of coke can be eliminated by using technologies such as directly reduced iron (DRI), as discussed under Reducing Emissions from Iron-Making Operations.

Reducing the Quantity and Toxicity of Wastewater

Process waters are the largest fraction of wastewaters from the coking operation and typically account for 60% to 85% of the total flow, which ranges from 120 to 310 gal/ton of coke, depending on recycle. Wastewaters from a coke manufacturing plant can be subgrouped into six main categories:

- tar decanter sludge and wastewater
- waste ammonia liquor (WAL) from primary coolers
- ammonia absorber and crystallizer blowdown
- final cooler wastewater blowdown
- light oil (benzol) plant wastewater
- gas-desulfurizer and cyanide-stripper wastewater[10]

These waste streams contain benzene, phenols, cyanide, ammonia, sulfide, suspended solids, and oil. Examples of waste reduction techniques that have proved useful for some coke plants are listed below.[10-12]

- Directly reuse a high-quality water effluent, without any treatment, for a process or operation requiring low-quality water.
- Recycle excess direct coke-oven gas cooling water (flushing water) after partial treatment. For instance, at Geneva Steel, excess flushing liquors generated in the coke by-products area are biologically treated to remove organics and nitrogen compounds. The treated wastewaters are reused as cooling water in a variety of plant operations.
- Reuse benzol tank cleanouts as fuel in industrial furnaces, provided the fuel has a caloric value of over 5,000 Btu/lb and the plant complies with applicable environmental regulations.
- Recycle tar decanter sludge as a replacement for fuel oil as a bulk density control agent for coal in coke-oven charges. Geneva Steel recycles tar decanter sludge using the method developed by AKJ Industries.
- Quench with coke-plant effluents to maximize the use of water and minimize wastewater generation. However, the quality of water used for quenching

should be such that it does not contribute to toxic or hazardous air emissions.

- Tighten recycling of cooling waters to significantly reduce wastewater generation.
- Use treated municipal sewage water for cooling purposes to substantially reduce the use of raw water.

IRON-MAKING OPERATIONS

Sources of Air Emissions

Typically, the gases generated in the blast furnace are cleaned and used for fuel. Nevertheless, fugitive emissions occur when a blast furnace slip or hot metal transfer occurs in the cast house. A slip is when gas pockets are formed between two charges of feed material. When this gas pocket collapses, pressure increases causing emissions of carbon monoxide and particulates from the relief valve.[6] Slag-handling operations generate H_2S and SO_2 gases.

Reducing Emissions from Iron-Making Operations

Some waste reduction techniques are recommended below for controlling these emissions; however, most are either in the experimental, research and development, implementation, or recently installed stage.

- Use fume-suppression systems, like the USX localized evacuation system, for controlling and reducing casthouse emissions.[8]
- Use the sintering plant for recycling fines and dust accumulated in bag houses and electrostatic precipitators.
- Capture and desulfurize gases generated during slag handling.
- Modify furnaces to use pulverized coal instead of coke in the blast furnace charge. This would reduce emissions related to coking and coke handling. Coal injection is estimated to cut the amount of coal used per ton of hot metal from 900–1,000 lb to 500–600 lb.[1] Armco Steel's Bellefonte furnace at Ashland has been operating using this principle since 1967. An American Iron and Steel Institute (AISI) report indicates that in Japan, 25 out of 33 blast furnaces are equipped with pulverized-coal-injection systems.[9]
- Use DRI technology (a variation of pulverized coal injection) in place of conventional coke ovens and blast furnaces. Typically, iron ore pellets are fed into a fluidized-bed reduction furnace, while coal is injected from the bottom. Partially reduced ore and tar are transferred from this chamber into a smelting chamber. Coal and oxygen are simultaneously injected into the furnace to reduce the ore and produce pig iron. Emissions from this process are estimated to be significantly less than those from separate coking ovens and blast furnaces.[1]

Direct steel making (DSM) is one step beyond DRI in terms of waste reduction and process efficiency. The goal of DSM is to directly make steel (i.e., iron with less than 1% carbon content) from limestone, coal, and iron ore. This technology eliminates the need for coke ovens and blast furnaces. AISI has been testing the process in a 120 ton/day pilot plant located near Pittsburgh.[1] AISI is experimenting with methods to control the smelting process. Once this process is controlled, it will be combined with the refining process in the same vessel.[13]

Reducing the Amount and Toxicity of Wastewater from Iron Making

Water used for noncontact cooling of furnace and stove walls and contact cooling and washing of blast-furnace gases is the chief source of wastewater from iron making. Blast-furnace gas-washing water is similar to coking plant wastewater. In most iron and steel plants, noncontact cooling water (approximately 5,030 gal/ton of iron) is used once and discharged to a receiving body of water.[11] Measures for reducing the quantity and toxicity of process wastewater are discussed below.

Recycling and Reuse

Although not strictly a waste reduction measure, the use of treated municipal sewage for cooling and other purposes can substantially reduce the use of fresh water for industrial purposes. This is practiced particularly in arid areas of the world. For instance, Dunswart Iron and Steel, Ltd., of Benoni, South Africa, produces 18,000 tons per month of rolled steel sections from scrap, pig iron, and ore. It uses about 2,500 cubic meters per day (cmd) of water. Seventy-five percent of this water is treated sewage.[14]

Noncontact cooling water is a relatively high-quality water that can be reused directly in many other operations, such as sintering, slag cooling, coke quenching, and rinsing operations in the surface finishing. A hypothetical reduction in wastewater generation by incorporating this option is shown in Figures 20.3 and 20.4. Blast-furnace gas wash waters can be completely or partially recycled after treatment to remove particulates, ammonia, cyanides, and scale-forming compounds.

Alternatives to Wet Scrubbing

Eliminating or minimizing the need for wet scrubbing can be accomplished by maximizing the use of dry particulate removal systems like cyclones and electrostatic precipitators. Blast-furnace dust and scale can be recycled through the sintering plant.

Use of Slag

Slag generated from blast furnaces is typically disposed of as solid waste. However, in many parts of the world, including Europe and Japan, slag is rapidly cooled and granulated under controlled conditions for use in cements. Experience indicates that water granulation solidifies slag so rapidly that sulfur species are frozen-in to a greater degree than with conventional slow-quenching techniques, with the result that less H_2S is generated.[15]

Research indicates that when blast furnace slag is used as an intergrind in manufacturing portland cement, the resulting product is highly resistant to sulfates (ex-

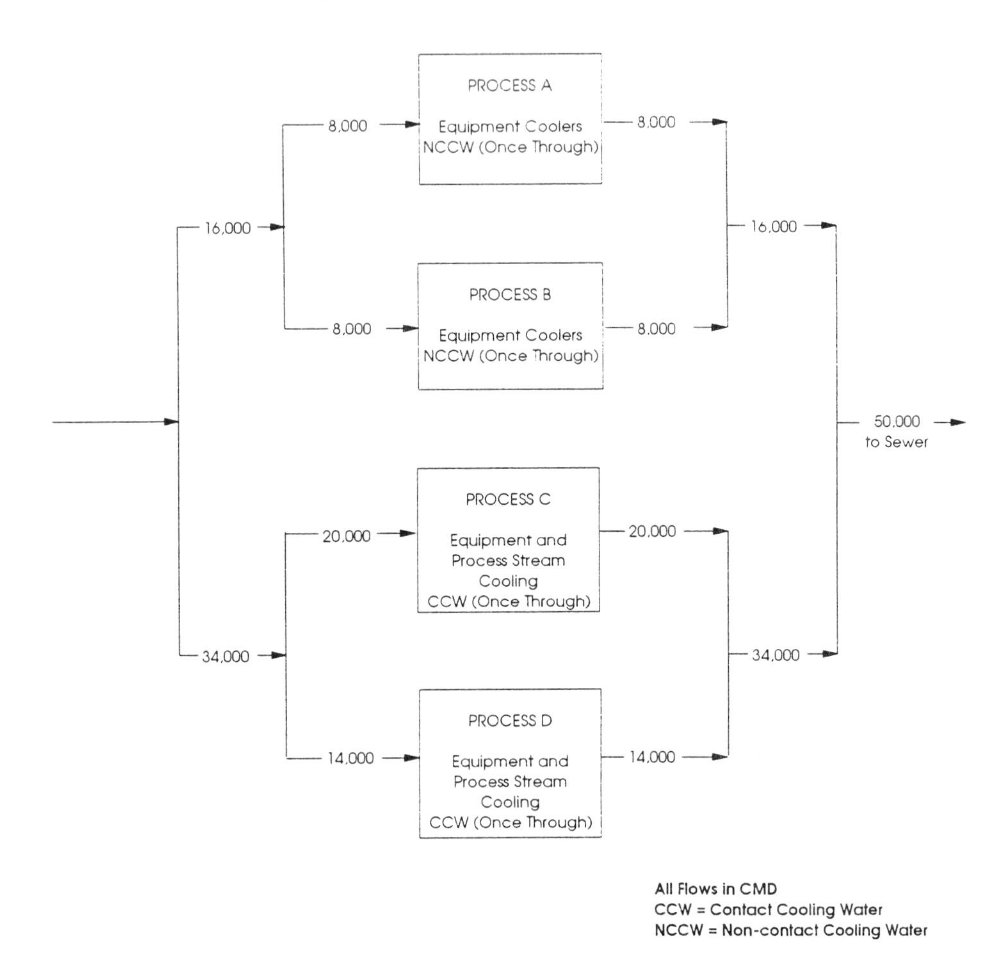

Figure 20.3. Hypothetical example of typical steel plant cooling water use.

cept in sea water), has good weathering properties, and generates little heat of hydration. Conventional portland cement precipitates $Ca(OH)_2$ while forming a CaO-SiO_2-H_2O gel during the hydration process. Slag cements, on the other hand, do not precipitate $Ca(OH)_2$; instead, it is absorbed in the cement matrix in a pozzolanic reaction, forming a stable hydrate that contributes to its favorable properties.[16]

Blast furnace slag also is used in the manufacture of cement clinker, ceramic wares, glazed tiles, roofing tiles, glass, and slag wool. The inclusion of blast furnace slag in the manufacture of the above materials is believed to substantially reduce raw-material costs and fuel consumption. For instance, a mixture of 3% to 4% granulated slag in raw materials has been shown to reduce energy consumption by 6% to 7%, and increase productivity by 10 to 20% in the glass-manufacturing process.[16]

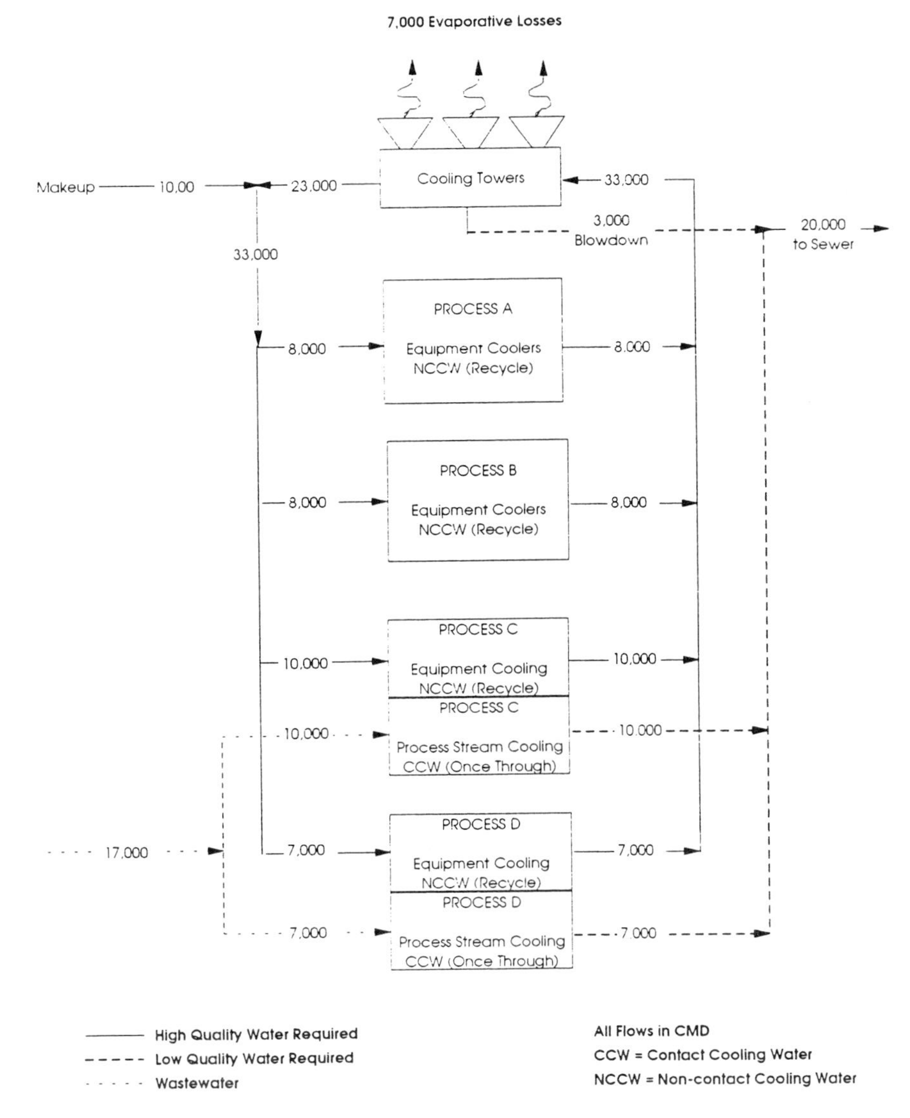

Figure 20.4. Example use of cooling towers to reduce steel plant water throughput.

STEEL MAKING OPERATIONS

Air Emissions

Air emissions from the BOF steel-making process include airborne fluxes, slag, carbon monoxide and dioxide, and submicron iron oxide dust. The source of such emissions are:

- refining
- transferring hot-metal to charging ladle
- charging scrap and hot metal

- dumping slag
- tapping steel[6]

EAF processes generate finer dust particulates containing oxides of iron and zinc. EAF dusts are classified as hazardous wastes because they contain trace quantities of leachable oxides of chromium, lead, cadmium, and arsenic. In addition to particulates, carbon monoxide and carbon dioxide are emitted. Typically, furnaces equipped with direct control evacuation systems (DCES) emit off-gases with an average carbon monoxide concentration of less than 800 ppm. Nevertheless, carbon monoxide peaks of over 7,000 ppm (0.7%) are common. The total air emissions from EAF operations are usually less than 2 pounds per ton of tapped steel.[17] Emissions are generated during the following operations:

- melting and refining
- charging scrap
- dumping slag
- tapping steel[6]

Reducing Wastes from Steel Making

Several potential options for waste reduction are available; however, not all are applicable to all steel plants because they may not be cost-effective or fully developed.

The options are as follows:

- Supplement pig iron or scrap metal with sponge iron or direct reduced iron to reduce fugitive tramp metal emissions and associated impurities.
- Minimize the need for hot metal at integrated steel works by maximizing the use of scrap-based EAFs. This reduces the need for coking and iron making, thus reducing emissions considerably.[5,7]
- Convert EAF dust to greenballs or briquettes and recycle back into the furnace, or inject the dust directly into the metal.[18-20]
- Enrich ZnO content in the dust and recover zinc metal using high-temperature or plasma-arc metal recovery processes.[17-20]
- Use EAF dust as an impervious bottom-liner layer in landfills.[18,19]
- Recycle EAF dust in glass frit manufacture. Glass frit may subsequently be used in the manufacture of tiles, glazes, roofing granules, sand-blasting grit, and cement clinker. In the United States, use of EAF dust in cement clinker may be considered land application of a hazardous waste and, therefore, is restricted.

Various processes are available for manufacturing glass from EAF dust including a system developed at Oregon Steel Mills.[21,22]

Example. A schematic of the glass-manufacturing process using EAF dust as practiced in Oregon Steel Mills, Portland, Oregon, is shown in Figure 20.5. A series of rigorous tests, conducted by Roger B. Ek and Associates, Oregon Steel Mills, and CH2M HILL, has demonstrated that products made from EAF dust are physically and chemically compatible or superior to their counterparts manufactured from natural raw materials.

Oregon Steel Mills has invested over $11 million in research and development of a full-scale glass-manufacturing plant (32,000 tons/year). It should be noted that the cost of the glass-making unit (i.e., the glass furnace) was less than $1,000,000. Depending on site location, accessible markets, and market price for product, a payback period of 2 to 4 years is reasonable for such a plant. The estimated payback period does not account for tax credits that may benefit a steel plant for installing a recycling plant. In addition, the savings realized from eliminating costs associated with disposing of EAF dust were not included in computing the payback.[23]

Recovery of Waste Energy

Approximately 55% of total energy consumption in an integrated iron and steel mill is lost in the form of furnace body radiation; sensible heat in cooling water; and in hot steel products, slag, and waste gases from combustion furnaces.[24] Any changes that reduced energy consumption in the steel-making process would directly decrease emissions of particulate, NO_x, SO_x, and CO_x from combustion of fossil fuels.

For instance, it is estimated that annual CO_2 emissions from worldwide combustion of fossil fuels is approximately equivalent to 6 billion tonnes of carbon. However, it is estimated that if the iron and steel industry were to follow the example of Japan in conserving and recovering its waste energy, the total energy consumption by the world steel industries would be reduced by about 30%. This is about 2% of world CO_2 emissions, or the equivalent of about 100 million tons of carbon. Japan's iron and steel industry presumably has the highest energy-recovery rate per ton of steel produced, approximately 17%.[24]

It is important to realize, however, that energy recovery offers a direct economic incentive to Japanese steel makers because it offsets their dependence on imported raw materials, including coke and iron ores. In the United States, both raw materials are available in relative abundance; therefore, heavy capital investment in energy-saving equipment is less attractive for U.S steel makers.

The coke and iron making processes consume the lion's share of energy in an integrated steel mill. Implementation of coke-dry-quenching (CDQ) equipment, which absorbs the sensible heat of red hot coke, and blast-furnace top-pressure recovery turbine, which recovers the blast furnace gas pressure as electricity, can account for more than 60% of total recovery waste energy.[24] Some important energy-saving equipment and techniques for the iron- and steel-making processes are summarized in Table 20.2.

Other Important Operations

Wastes generated from other operations and the respective options for pollution prevention have been summarized in Table 20.3. Metal surface treatment operations generate considerable quantities of hazardous and solid wastes. A hypothetical case

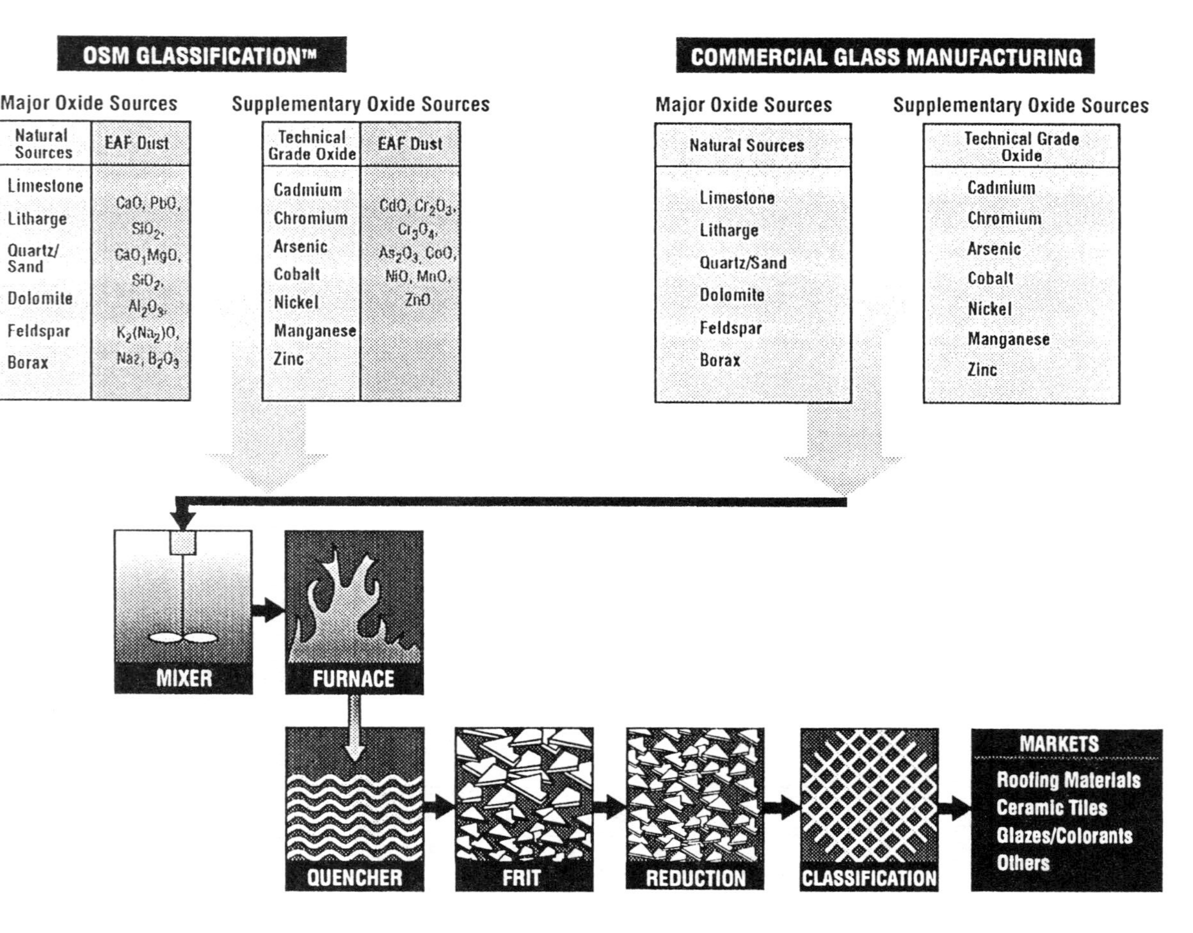

Figure 20.5. Schematic of glass-manufacturing process.

Table 20.2. Summary of Energy-Saving Measures for Iron and Steel Making

Equipment/Technique	Description	Relative Energy Savings/Recovery
DRI continuous hot charging	Iron ore, coal, and oxygen are charged simultaneously to make iron in a single step. Eliminates coking ovens.	Energy savings of up to 10%. Decreases iron ore and coal requirement by 10 to 20%.
Coal Moisture Control	Preliminary reduction in coal moisture reduces fuel consumption in coke oven.	Energy savings of up to 40,000 kcal/t of steel produced.
Coke Dry Quenching	Uses inert gas for cooling hot coke and uses the heated gas to generate steam/electricity.	Energy recovery of about 200,000 to 300,000 kcal/t of steel produced.
Scrap Preheater	Scrap is preheated to about 350°C using the heat of EAF exhaust gases.	Preheating of scrap reduces EAF power consumption by about 5 to 10%.
Blast Furnace Top-Pressure Power Generation	Uses flash turbines for converting excess pressure energy of (high top-pressure) blast furnaces into electricity.	Axial flow turbines provide higher efficiencies. Energy savings/recovery of up to 120,000 kcal/t-s.
Direct Linking of Continuous Casting and Rolling Operation	Eliminates ingot making and primary rolling processes. Reduces fuel consumption for reheating slabs, blooms, and billets.	Estimated energy savings up to 21%; up to 200,000 kcal/t-s.
Hot Direct Rolling and Hot Charge Rolling	Eliminates the reheating furnace and reduces fuel consumption in the rolling process.	Energy savings of up to 300,000 kcal/t-s.
Slag Heat Recovery	Recovers slag heat by cooling hot slag in air.	Energy savings of up to 280,000 kcal/t-s.
Improved Utilization of Rolling Mill	Combines proper workload, preventative repair, and maintenance scheduling for the reheating furnace and the rolling process.	Energy savings of about 10 to 20%.

Table 20.3. Summary of Pollution Prevention Options for Miscellaneous Operations at Integrated Steel Mills

Waste Source	Waste Description	Pollution Prevention Options
Scrubber water used to remove pickling fumes	Acidic wastewaters	Supplement spent scrubber waters with makeup acid and recycle as pickling liquor.
Surface treatment—acid pickling	Acidic wastewaters and sludge	Complement acid pickling with mechanical cleaning methods such as abrasive blasting to reduce volume of acidic wastewaters and sludge.
Surface treatment—stainless steel products acid pickling	Fluorspar (calcium fluoride) containing corrosive wastes	Recover fluorspar by selective precipitation. Pickling stream water (at pH 2) contain fluorides, which will selectively precipitate as CaF_2 on addition of slaked lime without exceeding pH 2.5. Reuse fluorspar as flux in the electric-arc furnace.
Surface treatment—water used for cleaning pickled parts	Inorganic acids and suspended toxic metals	Minimize wastewater discharge by using multistage cascaded or countercurrent dip and spray rinsing techniques.
Surface treatment—spent acid pickling liquor	Inorganic concentrated acids, suspended toxic metals and iron	Recover recyclable-quality pickling acid. Several types of acid-recovery systems, including evaporative, crystallization, and membrane technologies, are available to recover spent HCL and H_2SO_4.
Cold rolling—miscellaneous oil leaks and spills from roll conditioning and finishing shops	Tramp oil containing wastewaters	Implement spill-prevention programs.
Cold rolling—water-oil emulsion sprayed directly on the product and the rolls	Wastewater contaminated with oily emulsions	Modern cold-rolling mills employ recovery systems to recover waste oil to reduce oily wastewaters and save on oil costs.
Continuous-casting—hydraulic cylinder leakage	Aryl phosphate ester	Collect wastewaters from scale pit and treat using membrane separation and vacuum distillation to recover hydraulic fluid.[25]
Steel mill sludge from various mill operations	Mill sludge containing hazardous metal oxides and oil and grease	Dewater sludge using belt-filter or screw presses.
Cooling towers	Relatively good quality contact cooling water	Recycle water through noncontact cooling towers. Use treated water from municipal sewage treatment plants.
Kolene descaling salt baths	Highly alkaline sludge	Use Kolene sludge for neutralizing spent pickle liquor. If Kolene sludge contains high concentrations of chromium, it should be segregated and not used for neutralization.
Lime rock handling (iron making)	Lime rock dust and particulate matter	Implement a fully enclosed pneumatic handling and transfer system.
Wind erosion of slag and other material storage piles (general)	Slag particulates	Methods include reducing storage in open areas, adding vegetation, sprinkling with water, erecting wind screens, and using chemical dust suppressants.

study of a pollution prevention audit (for a metal-surface-treatment facility) and its results are summarized below.

GENERAL METHODOLOGY FOR DEVELOPING POLLUTION PREVENTION PLANS

This section explains six basic steps needed to develop pollution prevention options for a facility. To foster a better understanding of the methods used, a case study will be used as an example for developing pollution prevention alternatives. This method may have to be modified when preparing a pollution prevention audit for a different plant.

Step 1. To develop pollution prevention alternatives, it is essential to understand the various processes and operations that take place in the facility. First, prepare a list of all unit processes and operations (including relatively unimportant operations such as washing the floor for cleaning) that contribute to the generation of wastes. The level of detail in the list is a function of the level and quality of pollution prevention options the audit team wishes to develop, and applies to all the steps that follow.

Once all operations are listed, develop a block flow diagram representing the various materials and waste flows from each of the operations and their relationship to each other.

A steel tube surface-treatment facility was chosen for the hypothetical case study. For brevity, it is assumed that this facility has a sulfuric acid pickling operation only. It is designed for rinsing the pickled tubes in a continuously overflowing water tank. A conical bottom neutralization tank is used for settling the suspended matter in the spent pickling liquor. For this case study, Step 1 is summarized in Table 20.4 and Figure 20.6. The steel tubes are manually racked and dipped into the pickling tank.

Step 2. Account for process inputs and outputs (see Table 20.5) from design parameters provided by the facility's process design engineers. Take into account the cross-media mass-transfer effects as shown in Table 20.6. The total weight of material going into a process must be equal to the total weight of materials exiting a process. However, it should be noted that because of cross-media mass transfer, the weight of liquid material entering a process need not be equal to the weight of liquid material leaving a process. For example, a part of the liquid material may evaporate or may bind to solids that are generated in the process. Cross-media mass transfer

Table 20.4. Unit Processes/Operations

Unit Process/Operations	Functional Description
Pickle tank	6 m^3, steel, epoxy-lined
Continuously flowing rinse tank	500 L/min of rinse water
Fume scrubber	Packed-bed wet scrubber with internal recycle
Lime storage	Bulk quicklime storage (8 m^3)
Lime slaker	Slakes to 30% lime hydrate
Lime holding tank	4 m^3
Neutralization basin	Conical basin with rudimentary solids removal

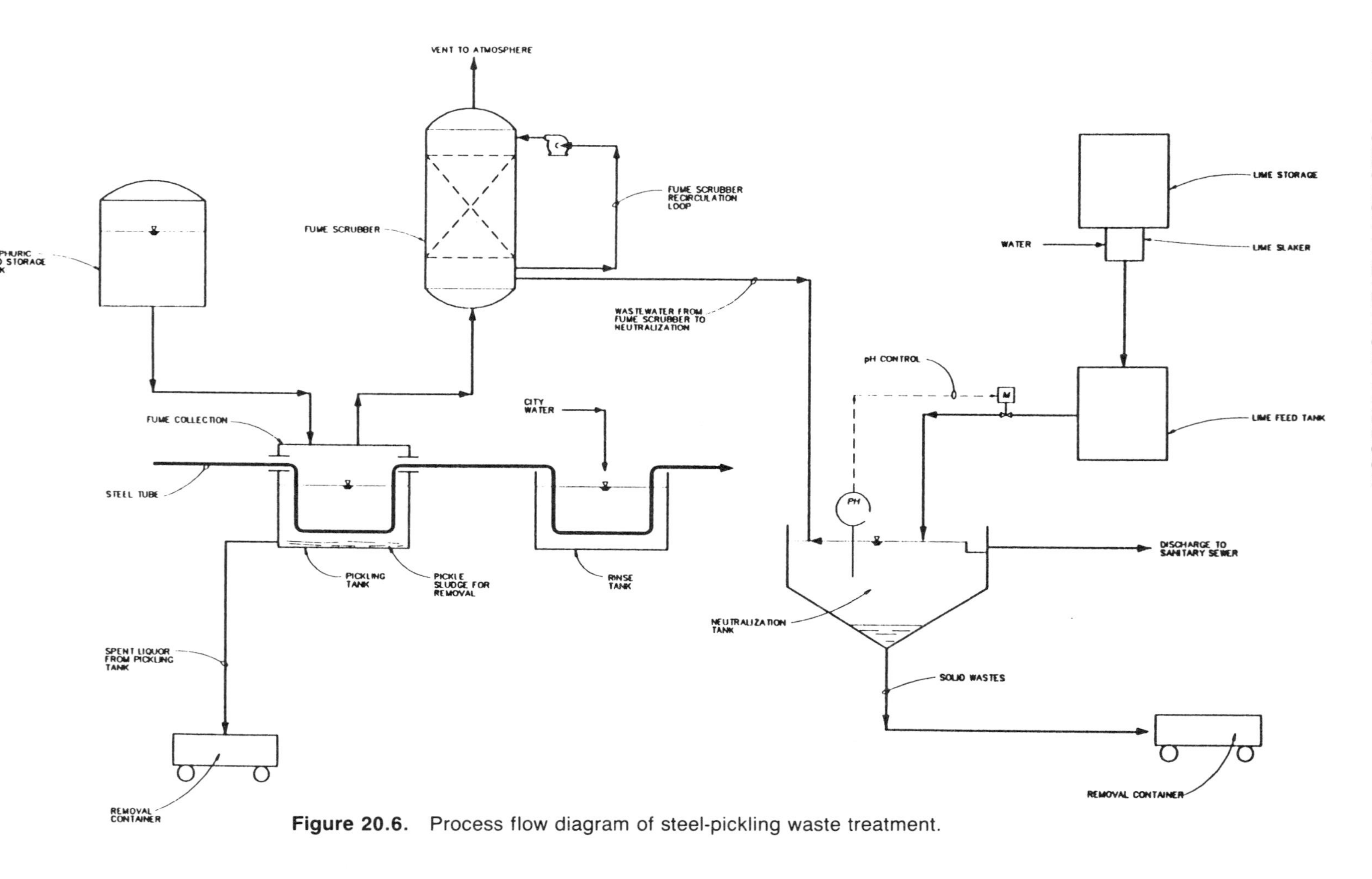

Figure 20.6. Process flow diagram of steel-pickling waste treatment.

Table 20.5. Process Inputs and Outputs of Case Study Facility

Pickle Tank Inputs (tonnes)		Rinsing and Chemical Treatment Inputs (tonnes)	
Steel	8,000	Rinse water	7,700
Acid	190	Scrubber water	300
Water (makeup)	500	Hydrated lime (closest at 30% concentration)	20
Water (steam)	500	Miscellaneous waters[a]	390
Total	9,190	Total	8,410
Outputs (tonnes)		**Outputs (tonnes)**	
Steel	7,920	Sewer discharge	8,400
Pickle liquor	700	Chemical-treatment sludge	10
Pickle-liquor sludge	160		
Evaporation	100		
Drag-out	310		
Total	**9,190**	**Total**	**8,410**

[a]Water from floor washing.

should be considered for gaseous, solid, and liquid material entering or leaving a process.

For the case study facility, it may be observed that the steel tube loses mass during the pickling process. This is because pickling is an etching operation that chemically removes the rust and oxide skin of the steel tube.

Similarly, pickle sludge is not supplied to the pickle tank. The pickle sludge will be composed of metal oxides removed from the steel tubes and water bound to the solids settled in the tank; it is generated by the pickling process.

Water entering the pickling process ends up in the pickle liquor, pickle-liquor sludge, evaporation, and drag-out. Drag-out is the residual water that clings to the steel tubes when they are removed from the pickle tank. A similar pattern in the distribution and generation of water can be observed in the case of rinsing and chemical treatment.

Step 3. Generate pollution prevention options. The options generated for each of the waste generating processes in the case study facility are shown in Table 20.7.

Step 4. Evaluate each option for technical feasibility. The technical evaluation process for the case study facility is summarized in Table 20.8.

Step 5. Evaluate each option that has passed the technical feasibility criteria for economic feasibility as described in the following section.

Step 6. Evaluate each option that passed as well as failed the economic feasibility criteria to check if any overriding intangible benefits can be obtained. This step is discussed in detail after the economic analysis section.

Table 20.6. Process Outputs Accounting for Gaseous, Solid, and Liquid Wastes in the Case Study Facility

Unit Process	Product	Waste Reused	Airborne Waste	Wastewater Disposed of in Sewer	Waste to be Disposed of as Solid or Hazardous Waste	
					Liquid Waste	Solid Waste
Pickling	Descaled rod/steel tubes	None as per existing (proposed) design	Evaporation and negligible quantity of acid fumes	Not applicable	Spent pickle liquor (700 tons)	Pickle-liquor sludge (700 tons)
Rinse	Cleaned rod/steel tubes	Same as above	Negligible evaporation	7,700 tons	None	None
Scrubber	High pH wastewater to treatment	Same as above	Negligible evaporation	300 tons	None	None
Wastewater treatment	Dischargeable wastewater and chemical-treatment sludge	Same as above	Negligible evaporation	8,400 tons	None	Chemical-treatment sludge

Table 20.7. Pollution Prevention Options for the Waste Generating Processes in the Case Study Facility

Waste Source	Waste Description	Pollution Prevention Option
Pickling operation	Spent pickle liquor and pickle sludge	**Source reduction:** Input material change: Use HCl instead of H_2SO_4 to improve recoverability of spent pickling liquor. Technology change: Change the pickling operation from a manual to a semi-manual operation (i.e., hoist-assisted) so the operator can control the movement of parts. Good operating practices: a. Hold processed parts over the tank for at least 60 seconds (the drain time depends on the specific process used) before moving them to the rinse tank. This practice reduces dragout of pickling solution. As a result, acid makeup, the quantity of rinse water, and the treatment chemicals typically required are reduced. b. Rack steel tubes so residual liquid can drain back into the tank easily and quickly as the parts are removed from the tank. Product reformulation: None at present. **Recycle/recovery:** a. Do nothing (i.e., contract disposal). b. Acid recovery. c. Neutralization system reduces the quantity of pickle liquor that needs to be hauled.
Continuously overflowing rinse tank	Acidic wastewater that needs to be neutralized	**Source reduction:** Input material: None at present. Technology change: a. Use flow restriction to throttle rinse water, feed to the rinse tank. b. Countercurrent rinsing to reduce rinsing water requirements by 60 to 80%. Good operating practice: Increase drag-in time and reduce dragout. Product reformulation: None. **Recycle/recovery:** None.

Table 20.8. Technical Feasibility of Pollution Prevention Options

Waste Source	Pollution Prevention Option and Technical Feasibility
Pickling operation	a. Use of HCl instead of H_2SO_4: This is not feasible because HCl is not compatible with the type of steel etched in the facility.
	b. Change to semi-manual operation: Using a hoist will considerably ease the burden on the worker. The work will have better control in terms of increasing drain time and decreasing dragout. This option is technically feasible.
	c. Increase drain time: It can be easily implemented with semi-manual operations and worker training. However, with manual operation it is impractical to implement the practice (a minimum 25-kg load has to be handled by one worker according to the current design).
	d. Improve design of the tube rack: This option involves minor changes in rack design and can be accomplished without difficulty. It can reduce dragout significantly.
	e. Contract disposal: This option entails renewing contracts with hauling companies that find the most economical means for disposing of the hazardous spent pickle liquor. Although no major investment costs are incurred with this alternative, management can become less dependent on hauling contractors and the spiralling disposal costs. Management will have to upgrade the waste-water treatment facility in the future. This option is technically feasible.
	f. Neutralization: For this option, the spent pickle liquor and rinse waters are neutralized with lime. The process produces a calcium sulphate/iron hydroxide sludge that requires subsequent dewatering (by filter pressing) and disposal. Neutralization is a technically feasible option.
	g. In the acid-recovery process, the spent pickle liquor is pumped to a crystallizer where the iron salts are crystallized from the spent pickle liquor by cooling and the remaining acid solution is separated and made up to pickling strength with fresh acid. The effect of the acid-recovery system is to remove the untreated H_2SO_4 for reuse instead of discharging it as a waste product. In addition, ferrous sulphate crystals are generated and can be sold as a by-product. The price obtained for the crystals is very dependent on market conditions. Apart from the acid-recovery equipment, it would be necessary to a) balance water in and out of the pickling tank by installing indirect heating, air agitation, and countercurrent rinsing systems, and b) install an improved fume exhaust system for the acid-recovery operation to minimize air emissions. This option is difficult to implement. It has been implemented successfully in similar plants and is a technically feasible option.
Continuously overflowing rinse tank	a. Use flow-restrictors: This option is extremely effective and easy to implement. Water savings of about 10 to 20% can be realized from implementing this option.
	b. Countercurrent rinsing: The countercurrent rinsing equipment is an easy and simple system to install. It will reduce consumption of rinse water by at least 60 to 80% and requires adding one or two more tanks and some plumbing.
	c. Increase drain time: Technically feasible only with semi-manual operations.

General Approach to Detailed Economic Analysis of Pollution Prevention Options

A detailed economic evaluation of options is carried out using standard measures of profitability, such as breakeven point or payback period, internal rate of return (IRR), and net present value (NPV) analysis.

Payback period assessments are carried out as a quick means of evaluating the economic viability of an investment. The payback period for a project is the amount of time it takes to recover the initial cash outlay on the project. The payback period in years on a pretax basis is computed as the annual savings in capital investments and annual operating costs. Typically, the payback period for an investment should be 4 years or less.

The results of the economic evaluation for the case study facility are summarized as follows.

Change to Semi-Manual Operation

The cost of a hoist system for this facility is estimated at $28,000. The annual savings by way of decreasing drag-out with the help of the hoist system is estimated at $4,000 (i.e., from reduced acid, chemical treatment, water and sewer costs; annual operating and maintenance costs are negligible in this case). The payback period is estimated at $28,000 ÷ $4,000/yr = 7 years. Typically, this payback is not acceptable; hence, this option is not economically feasible.

Increase Drain Time

This option needs operator training. The costs involved to train operators are considered negligible; hence, the option is economically feasible.

Improve Rack Design

No additional capital or operating costs will be incurred by redesigning table racks; hence, the option is economically feasible.

Contract Disposal, Neutralization, and Acid Recovery

Economic evaluations for these three options are given in Table 20.9. To arrive at the estimates used in Table 20.9, a thorough understanding of the engineering and economic concepts associated with the recovery processes is essential.

Countercurrent Rinsing

The countercurrent rinsing system will cost $12,000 to install. It will reduce sewer costs by $3,000 annually. Its payback period is less than 4 years, considering the savings it will contribute by way of reduced water usage. Hence, this option should be implemented unless it has adverse intangible effects.

Table 20.9. Comparison of Economics for Waste Treatment and Reduction at Case Study Facility (in thousands of dollars)

Item	Basic	Acid Recovery	Neutralization	Contract Hauling
Capital investment	8,000 tonne/yr pickling plant	170	180	–
Labor				
Operators	$25,000/man-year	25	50	25
Foreman	$30,000/man-year	14	10	0
Utilities				
Steam	11/1,000 kg	10	–	–
Process water	$0.35/m^3	(3)	–	–
Electricity	$0.04/kWh	5	–	–
Raw materials				
H_2SO_4 (93%)	$110/tonne	(9)	–	–
Ca	$75/tonne	0	10	10
Shipping, haulage, and disposal costs				
Crystals ($FeSO_47H_2O$)	$17/tonne	7	0	0
Neutralization sludge	$40/tonne	0	40	0
Pickle liquor	$129/tonne	0	0	90
Pickle-liquor sludge	$62/tonne	0	0	10
Maintenance		28	18	–
Wastewater costs				
Sewer fees	$0.38/m^3	0	3	3
pH adjustment	$1.50/m^3	0	–	–
Other regulatory costs		0	–	–
By-product credit	$27/tonne	(2)	0	0
Total annual cost		75	131	138

The relative payback periods of a neutralization system and an acid-recovery system are present in Table 20.10. The neutralization system is not cost effective in comparison to the present option of contract disposal. The acid-recovery system is cost effective, and should be recommended for implementation.

Table 20.10. Comparison of Payback Periods of Process Options for Case Study Facility

Process Option	Capital Investment	Annual Savings (operating costs)	Payback Period (years)
Neutralization system	$180,000	$7,000	26
Acid-recovery system	$170,000	$63,000	2.7

Intangible Benefits

Sometimes the benefits may be intangible, such as the creation of a more hygienic workplace with better employee morale, reduced absenteeism, and fewer work-related accidents. Similarly, the public-relations value of implementing waste reduction measures is difficult to measure and can be an overriding consideration in selecting an option.

In evaluating intangible benefits, a criteria table should be developed with lists of perceivable benefits such as reduced regulatory burden, improved health and safety, consumer-perceived improvement in product quality, and improved community relations.

One striking example for the case study facility in terms of intangible benefits is the implementation of the hoist system. Although the hoist cannot be justified on the basis of the payback period, the added convenience and comfort to the worker can improve productivity, reduce spillage and accidents, and enhance the overall aesthetics of the plant. Since it is a small capital investment, this option could be recommended for the above reasons in spite of its poor economic rating.

Conclusions

On the basis of the waste audit and cost/benefit analysis, installing the acid-recovery system will save on raw materials and nearly eliminate waste and pollution problems. Furthermore, pickling efficiency is expected to improve by approximately 25% as a result of better control of pickling-bath acidity and improved agitation.

It may be apparent from the hypothetical case study that production efficiency, waste management, and pollution are interrelated. Through hard work and a deliberate approach to waste auditing and waste reduction, and by implementing action plans for waste reduction, the following can be achieved:

- improvement in process efficiency
- savings on raw materials
- improvement in product quality through better process control
- elimination or reduction of waste and disposal problems
- improvement in the working environment
- improved relations with regulatory authorities and the public

Because of the benefits of conducting a waste audit on the pickling process, the company can proceed to conduct a similar study in other parts of the manufacturing plant.

REFERENCES

1. Parkinson, G. "Steel Making Renaissance," *Chemical Engineering*, May 1991.
2. Nemerow, N.L., and A. Dasgupta. *Industrial and Hazardous Waste Treatment*. (New York, NY: Van Nostrand Reinhold, 1991).
3. Kemmer, F.N. "Coke Industry," The Nalco Water Handbook (New York: McGraw-Hill Book Company, 1979).

4. "Development Document for Effluent Limitations Guidelines and New Source Performance Standards for the Steel Making Segment of Iron and Steel Manufacturing Point Source Category," U.S. Environmental Protection Agency, EPA-440/1-74-024-a, June 1974.
5. Kemmer, F.N. "Steel Industry," The Nalco Water Handbook (New York: McGraw-Hill Book Company, 1979).
6. "Compilation of Air Pollutant Emission Factors," Volume I, Stationary Point and Area Sources AP42, 4th Ed., U.S. Environmental Protection Agency, September 1985.
7. Prabhu, D.V., and P.F. Cilione. "1990 Clean Air Act Amendments: Technical/Economic Impact on U.S. Coke and Steel Making Operations," *Iron and Steel Engineer*, January 1992.
8. Labee, C.J., and N.L. Samways. "Developments in the Iron and Steel Industry-U.S. and Canada-1990," *Iron and Steel Engineer*, February 1991.
9. Labee, C.J., and N.L. Samways. "Developments in the Iron and Steel Industry-U.S. and Canada-1991," *Iron and Steel Engineer*, February 1991.
10. Dunlap, R.W., and F.C. McMichael. "Reducing Coke Plant Effluent," *Environmental Science and Technology*, 10(7):(July 1976).
11. Holstein, H., and H.J. Kohlmann. "Integrated Steel Plant Pollution Study for Total Recycle of Water," Office of Research and Development Industrial Environmental Research Laboratory, Research Triangle Park, NC, Contract/Grant No. 68-02-2626, U.S. Environmental Protection Agency, EPA/600/13, July 1979.
12. Grover, J.D. "100% Elimination of Disposal and Offsite Treatment of Hazardous Waste by Recycle and Waste Minimization at Geneva Steel, An Integrated Steel Mill," Waste Management Conference, College of Engineering, Utah State University, 1992.
13. Allie, G. Personal Communication, American Iron and Steel Institute.
14. Odendall, P.E., and P.E. Van Vurren. "Reuse of Wastewater in South Africa-Research and Application," *Water Reuse* (Ann Arbor, MI: Ann Arbor Science, 1982).
15. Cooper, A.W. "Blast Furnace Slag Granulation," *Iron and Steel Engineer*, July 1986.
16. Garidis, I. "Utilization of Iron and Steel Making Slags-An Overview," Helsinki University of Technology, Faculty of Process Engineering and Science, Laboratory of Metallurgy, Report TKK-V-B63, NTIS-PB92-136498, 1991.
17. Steven, W., and H. Bertling. "Environmental Protection: A Challenge for Modern Innovative Cokemaking Technology," presented at the International Iron and Steel Institute, ENCO STEEL World Conference, 1991.
18. MacRae, D.R., and R.M. Hurd. "Electric Arc Furnace Dust-Disposal, Recycle, and Recovery," Center for Metals Production, Mellon Institute, Carnegie-Mellon University, Report No. 85-2, Project No. RP-2570-1-2, May 1985.
19. Labee, C.J. "Update on Electric Arc Furnace Dust Treatment," *Iron and Steel Engineer*, May 1992.
20. Kotraba, N.L., and M.D. Lanyi. "Inclined Rotary Reduction System for Recycling Electric Arc Furnace Baghouse Dust," *Iron and Steel Engineer*, April 1991.
21. Teoh, L.L. "Environmental Improvements for Minimills: Recent Developments and Future Trends," presented at the International Iron and Steel Institute, ENCO STEEL World Conference, 1991.
22. Aichinger, H.M., G.W. Hoffmann, and M. Seger. "Effluent and Environmentally Compatible Utilization of Energy in the Steel Industry of the Federal Republic of Germany," presented at the International Iron and Steel Institute, ENCO STEEL World Conference, 1991.
23. Report prepared for confidential client, 1992.
24. Yoshida, M. "Management of Environmental Control in the Japanese Steel Industry," presented at the International Iron and Steel Institute, ENCO STEEL World Conference, 1991.
25. Hunter, R.A. "Recovery and Reuse of Aryl Phosphate Ester Hydraulic Fluid," Environmental Research, ARMCO, Inc., Middletown, OH.

CHAPTER 21

Petroleum Exploration and Refining

*Teresa Barthel and Ron Advani**

THE PETROLEUM INDUSTRY

The exploration and production sector of the petroleum business explores, develops, and extracts crude oil and gas resources from natural formations beneath the earth. Exploration involves mapping geological features and drilling test wells in areas where there is a high probability of finding reserves. Proven oil fields are then developed through a well drilling program, which leads to a producing well field. The extraction techniques used depend on the geological characteristics of the area, the type and formation of the oil and/or gas deposits, and the design of the well field.

Crude-oil refining operations involve extracting useful petroleum products (e.g., gasoline) from crude oil. Crude oil contains fractions of napthas, jet fuel, gasoline, diesel fuel, gas oils, lubrication oils, and asphalt. Refineries extract these fractions from crude oil thermally or catalytically.

SOURCES AND TYPES OF WASTE

Various wastes are generated during petroleum exploration, production, and refining. Large quantities of production wastes—low-toxicity solid waste and wastewater—are produced during exploration and production. The waste stream from refineries is smaller but is considerably more toxic, containing listed hazardous wastes and characteristic hazardous wastes. In addition, poor housekeeping practices during exploration, production, and refining can produce excessive waste.

Waste from Exploration and Production

There are more than 800,000 oil and gas production wells in the United States, distributed over 38 states.[1] Consequently, the wide variety of waste types and quantities are covered by a range of regulations. These wastes are currently exempt from

*Ms. Barthel and Mr. Advani are environmental engineers in CH2M HILL's Oakland, California, office and can be reached at (510) 251-2426.

RCRA, in part because of their unusually high volume, making it technically impractical to apply some of RCRA's requirements. Table 21.1 lists wastes generated during well drilling and oil production activities and shows their status under RCRA.

Production wastes can be grouped broadly into two classes: wastes related to drilling and well completion, including drilling mud, drill cuttings, and chemical additives, and wastes related to oil production, primarily produced water. The volume of produced water far exceeds the volume of drilling wastes. For 1985, the American Petroleum Institute (API) estimated that 21 billion barrels of produced water were generated versus 361 million barrels of drilling wastes. This amounted to 5,183 barrels (217,700 gallons) of drilling waste per well.[1]

Treatment and Disposal

Treatment and disposal of oil drilling wastes take place either on or off the drilling site. Wastes that are regulated by RCRA must be tested for hazardous waste characteristics before disposal. Wastes can be treated and disposed of on the site in reserve pits, by land spreading, or through annular disposal. Liquid wastes can be treated in closed systems or treated and discharged to surface water. Offsite treatment and disposal methods include disposal in centralized pits, reinjection, and application on commercial landfarms. Drilling muds can be reconditioned. The majority of produced water is disposed of underground through injection wells. Produced-water injection is permitted under either the U.S. EPA underground injection control program or a comparable state program that is approved by the U.S. EPA.

Strategies for Preventing Pollution

The petroleum industry has no control over the composition of the crude oil they extract. Most waste generated during production is the result of this basic lack of control over the composition of the crude oil. Lower-quality crude oil requires more processing and generates more waste than high-quality crude.

Although source reduction is the preferred strategy in overall waste management, the petroleum industry has limited opportunities to employ this strategy. Therefore, the key elements of pollution prevention in petroleum production and exploration

Table 21.1. Wastes Produced from Petroleum Exploration

Waste	RCRA Status[a]	Source
Drill cuttings and drilling fluids	Exempt	Oil Drilling
Well completion and stimulation fluids	Exempt	Oil Drilling
Waste crude oil from primary field operations	Nonexempt	Oil Drilling
Produced water and produced water treatment residues	Exempt	Oil Production
Hydrocarbon bearing soil	Exempt	Oil Drilling
Liquid hydrocarbons removed from the production stream	Exempt	Oil Production
Spent catalysts	Nonexempt	Oil Production
Waste solvents	Nonexempt	Oil Production
Waste sponge oil, filters, and other separation media	Nonexempt	Oil Production

[a]40 CFR 261.4 and *Federal Register*, March 22, 1993.

are recycling and careful handling of waste. Pollution prevention techniques used during production include the following:

- collecting used solvents, hydraulic fluids, and motor oils that are generated during drilling operations and shipping them to a petroleum recycler
- using smaller reserve pits for disposal of drilling muds and produced water
- segregating drilling muds for disposal as nonhazardous waste before they enter a hydrocarbon zone
- recycling drilling muds in a closed-loop system
- solidifying drilling wastes in reserve pits

Although implementing pollution prevention practices during production is not mandated by regulations, several factors provide some motivation to do so. These factors include the high cost of commercial disposal, the potential for future liability for environmental damages, and the difficulty of obtaining disposal permits.

Solid Waste from Refineries

In 1984, Congress passed the Hazardous and Solid Waste Amendments (HSWA), which amended RCRA in an attempt to reduce groundwater contamination by hazardous wastes. This law was promulgated to ensure that hazardous wastes are reduced through recycling, recovery, detoxification, and volume reduction. Congress sought compliance with HSWA by banning or regulating landfill disposal of specific listed hazardous wastes.

Since the passage of HSWA, the level of awareness of refinery waste management personnel has increased. Past disposal practices resulted in tedious and expensive site cleanups. To reduce waste management costs, the petroleum industry is developing waste reduction programs that limit the volume and toxicity of its wastes. Because of the often "tight" specifications placed on petroleum products and the limited process flexibility, waste reduction through recycle and reuse is preferred over often-expensive process modifications.

Listed Hazardous Wastes

The refining sector of the petroleum industry generates both listed and characteristic hazardous wastes. The U.S. EPA has identified five waste streams as listed hazardous wastes and therefore subject to HSWA. These streams are dissolved-air flotation (DAF) float (K048), slop-oil-emulsion solids (K049), heat-exchanger-bundle cleaning solids (K050), API separator sludge (K051), and leaded-gasoline-tank bottoms (K052). Metals and organic constituents form the basis of the hazardous designation for these wastes.

Nonlisted Hazardous Wastes

A major portion of refining sector hazardous wastes are not listed. These wastes include primary treatment plant sludge, secondary treatment plant sludge, biological treatment sludge, cooling tower sludge, ion exchange regenerant, fluidized catalytic cracking (FCC) catalyst, and hydrocracking catalyst. These wastes are not listed;

however, testing has shown that all of them have one or more characteristics of hazardous waste (that is, toxicity, ignitability, reactivity, and corrosivity).[1]

Additional Wastes Classified Hazardous as a Result of TCLP

In March 1990, the Toxicity Characteristic Leaching Procedure (TCLP) was adopted. The TCLP is a modification of the existing Extraction Procedure, designed to include wastes containing leachable organic compounds. The TCLP procedure measures the potential for toxic constituents to leach out of waste and contaminate groundwater. Since TCLP was adopted, 38 new organic constituents have been added to the list of toxic constituents of concern; the TCLP analysis placed many previously nonhazardous wastes into the hazardous classification.

Wastewater from Refineries

The major sources of wastewater in a petroleum refinery are cooling-tower blowdown, boiler blowdown, oily wastes, and other process wastes. The relative volume contribution of the various sources from two typical refineries is shown in Table 21.2.

Cooling-tower blowdown, which constitutes a large component of the total wastewater flow, contains relatively low concentrations of conventional contaminants such as organics, suspended solids, and heavy metals, but contains high levels of TDS. The number of cycles of concentration for the cooling tower may be limited by TDS, alkalinity, magnesium, silica hardness, aluminum, iron, or some other parameter.

Boiler blowdown is dependent on boiler feedwater quality, which is dictated by the boiler specifications. Boiler blowdown is high in TDS and minerals and relatively free of organic compounds and heavy metals.

Oily wastewater is generated from processes in which water contacts crude oil or refined products. Typical sources in a refinery include crude desalter bottoms, storage tank bottoms, foul-water stripper (FWS) bottoms, and storm water. Other process wastewaters include chemical plant effluent, ammonia recovery unit (ARU) bottoms, and sulfide-bearing wastes.

Data on the composition of wastewater from two refineries are shown in Table 21.3.

Opportunities for Pollution Prevention

Pollution prevention opportunities in the refining industry include recycling and reusing waste, segregating wastewater, modifying process equipment, and improv-

Table 21.2. Percent of Total Flow

Stream	Typical Range	Refinery A	Refinery B
Cooling Tower Blowdown	20 to 50%	18%	22%
Boiler Blowdown	2 to 15%	6%	2%
Oily Wastewater	30 to 50%	51%	50%
Other Process Wastewater	15 to 30%	25%	26%

Table 21.3. Combined Wastewater Flow Contaminants

Parameter	Refinery A	Refinery B
Biochemical Oxygen Demand	200 mg/L	250 mg/L
Chemical Oxygen Demand	200 mg/L	75 mg/L
Total Suspended Solids	175 mg/L	65 mg/L
Total Dissolved Solids	2,800 mg/L	2,350 mg/L
Aluminum		0.30 mg/L
Iron	90 mg/L	0.40 mg/L
Hardness/Alkalinity	80 mg/L	50 mg/L

ing general housekeeping. The goals of pollution prevention include reduction of waste toxicity or volume.

Recycling and Reusing Waste

Recycling and reuse consist of recovering materials that would otherwise be wasted and putting them to beneficial use. Treatment can play a significant role in product reuse. The following waste recycling and reuse opportunities are feasible for the refining industry:

- Recycle FWS bottoms as crude desalter water.
- Recycle boiler blowdown for cooling-tower make-up water.
- Soften cooling-tower side-stream to increase cycles of concentration.
- Recycle wastewater treatment plant (WWTP) effluent as cooling-tower make-up water.
- Regenerate spent catalyst.

Most of the water used in refining is for process heating or cooling. The easy availability of water and its high heat capacity have made it the favored heat-transfer medium in industries. However, recent water shortages and environmental awareness have resulted in an increased interest in conserving water used for heating and cooling. In the past, many systems used water once and returned it to the receiving stream. Currently, water is reused to a much greater extent and as a result, refineries spend millions of dollars a year for treatment chemicals to prevent system corrosion, deposition, and fouling. This section presents information on several in-plant techniques for recycling and reusing water.

Recycling Foul-Water-Stripper Bottoms as Crude Desalter Water

Ammonia and hydrogen sulfide are present in varying concentrations in crude oil stocks. NH_3 and H_2S gases are generated during cracking operations. When steam is used to strip the product streams, these contaminants are transferred to the steam and then they dissolve in the water produced when the steam condenses. Steam condensate water is collected at various points and treated in an FWS. When NH_3 and H_2S are removed from the condensate, the NH_3 can be thermally destroyed or can be recovered for the production of fertilizers, and H_2S can be converted to elemental sulfur for resale.

The FWS effluent can then be used as wash water in the crude oil desalter. The FWS bottoms may be high in phenol, and using the water for this purpose has the additional benefit of phenol removal from the water. The amount of FWS effluent that may be used in this manner is limited. Because crude oil is mixed with 5% (of crude volume) water before it is processed through the desalter and the maximum amount of recycled water used for this purpose should be limited to 3%, for a 100,000-barrel-per-day refinery, the maximum water that can be used for this purpose would be 90 gal/min.

Recycling Boiler Blowdown for Cooling-Tower Make-up Water

Boilers and cooling towers service many of the refinery process plants and, therefore, have the greatest water demand. Boilers require high-quality make-up water to prevent tubes from accumulating scale deposits. The pressure and design of a boiler determine the quality of water required for operation. Even good-quality domestic water is seldom of sufficient quality to be used as feed water for even low-pressure boilers. Blowdown water from boilers is of high quality and can be considered for a number of uses. Cascade reuse of boiler blowdown is common refinery practice and is an excellent opportunity for reducing waste. In cascade reuse, blowdown water from high-pressure boilers is fed to the low-pressure boilers, and low-pressure boiler blowdown is subsequently used for cooling-tower make-up water.

Disadvantages of cascade reuse are that hot boiler blowdown can reduce cooling tower performance, and chemical additives for boiler water may not be compatible with those used for cooling towers. These factors could limit the amount of boiler blowdown that can be used as cooling-tower make-up water.

Softening Cooling-Tower Side Stream

Improving the quality of recirculating water can increase the number of cycles of concentration at which the cooling tower can operate, reducing raw water usage and waste blowdown. Side-stream softening can remove hardness and silica from cooling-tower recirculating water and reduce the potential for scaling.

The side-stream softening process treats a portion of the circulating water to remove calcium, magnesium, and silica. High concentrations of these constituents can limit the number of cycles the towers can operate. Treating a small stream of recirculating water is more economical than treating the entire cooling-water flow.

Recycling Wastewater Effluent for Cooling-Tower Make-up Water and Other Industrial Uses

The refinery sewer system typically collects wastewaters from various individual process units. The individual streams are combined or segregated and are treated in the refinery's WWTP. WWTP effluent is discharged either to a publicly owned treatment works (POTW) or to a body of water under an NPDES permit. Refinery WWTPs typically include oil-water separation, followed by secondary biological treatment.

Refinery WWTP effluent can be of high quality and recyclable or reusable. The potential for reuse of the wastewater depends on the levels of several water quality parameters including TDS, oil and grease, organic matter, and ammonia. The WWTP water quality should be evaluated along with the cooling-tower recirculating water requirements to determine the level of additional treatment required to use WWTP effluent as cooling-tower make-up water.

Regeneration of Spent Catalysts

Catalysts are used in refining processes for desulfurizing product streams and converting heavy stocks to lighter products, such as gasoline and napthas. Spent catalysts from FCC, hydrocracking, and hydrotreating processes are usually pyrophoric and therefore designated hazardous. FCC catalyst is an inexpensive zeolite sand and is not worth regenerating. Typically, FCC catalyst is sent to landfills or used in concrete.

Hydrocracking and hydrotreating catalysts are made of valuable noble or base metals. The composition of the catalysts can be nickel, molybdenum, palladium, vanadium, tungsten, cobalt, or a mixture of these metals. Over time, the catalyst builds up a layer of coke on its surface, which inhibits its reactivity. The catalyst can then be removed from the unit and regenerated at a catalyst regeneration facility. These facilities typically use a thermal regeneration process to oxidize the coke and remove it from the catalyst. The catalyst can then be placed back into its original service, or if necessary, into a less-critical service. The coke generated from the thermal oxidation can be used in portland cement or asphalt production.

Segregating Wastewater

Hazardous wastes and wastewater requiring unique treatment should be separated from other wastewaters that require relatively limited treatment. The wastewater collection system and the WWTP are the logical places to begin segregating wastewater because much of the hazardous waste generated by refineries is generated at the WWTP. Ideally, a refinery storm-water and wastewater collection system should include four sewers: an offsite storm-water sewer, a clean-storm-water sewer, a nonpoint-source sewer, and a point-source (oily) sewer.

Offsite Storm-Water Sewer

An offsite storm-water collection system collects all offsite storm water, including clean storm water, from areas such as access roads around the process areas. Typically, this storm water should be discharged without treatment.

Clean-Storm-Water Sewer

Storm water from clean areas (i.e., roofed areas or areas with no process units) should be segregated from storm water originating from dirty areas. Clean storm water requires no treatment before discharge or reuse. The rainwater falling on access roads around and inside the process areas should be directed to a clean-storm-water sewer. Run-on from adjacent areas also should be intercepted and sent to this sewer. Roofs should be installed over process units that could contribute pollutants to storm-water run-off. Roof drains should be connected to the clean-storm-water sewer.

Nonpoint-Source Sewer

A nonpoint-source sewer collects and conveys washdown water and storm water that falls on the process areas. Effective segregation of point-source contaminants can result in relatively clean water being collected by a nonpoint-source sewer system. This water should need only minimal treatment for oil removal prior to discharge. The chemical constituents in the water and the potential for variation must be considered in determining appropriate treatment. Provisions for directing this water to the WWTP during a spill must be included in the design and operation of this system.

Other pretreated wastewater streams could be directed to this sewer for final discharge. Examples of such streams include neutralized water softener regenerant waste and treated groundwater.

Point-Source Sewer

Process wastewater containing oil requires a significant level of treatment before discharge. Oily wastewater consists of process waste that has come into contact with process fluids. A point-source sewer collects all point-source discharges from inside the process areas of the units. Wastewater from sampling stations, pump seal water, and other contaminated wastewater also should be directed to this sewer.

Opportunities for reducing waste within the actual process units include optimizing the process and minimizing the volume of washdown water used to clean up the process units. Every effort should be made to contain potentially polluting equipment and prevent spreading contaminants to the surrounding and general process areas. Areas around pumps, compressors, and other potentially polluting equipment should be curbed and roofed. Wastewater generated inside the curbed areas should be directed to a point-source sewer.

Modifying Process Equipment

Several process equipment modifications that can improve operating efficiency and reduce wastewater generation are:

- When sampling, use continuous recycling loops on sampling ports to eliminate the volume of water necessary for purging.
- Substitute pressure wands for fire hoses during unit washing operations to reduce the volume of washdown water.
- Use an automatic shutoff, rather than a manual shutoff or a holding tank, when removing water from crude, intermediate, and product storage tanks to avoid excess oil discharges to the oily sewer and to return oil to the storage tank.

Improving General Housekeeping

Excessive waste production often results from poor housekeeping practices. Leaking tanks, valves, or pumps can cause oil spills, requiring cleanup and disposal of the oil, contaminated soil, and cleanup material. Key advantages of housekeeping changes are that they can usually be implemented quickly; they require little, if any, capital investment; and they are likely to reduce waste generation substantially.

Spill Prevention and Control

Spill prevention and control programs should be developed to enforce the management of waste generated from activities that occur in process areas, including routine sampling, equipment decontamination, equipment cleaning and maintenance, and spilling.

Contained Work Areas

All areas that have potential for spills should be paved rather than covered with gravel or other granular media. All containment areas should be adequately sloped to drain to the appropriate collection system.

Designated maintenance areas should be paved with concrete and used for equipment overhaul, cleaning, and maintenance. These areas should be curbed and drain to the oily sewer. High-pressure hot-water cleaners should be provided in the maintenance area so that the paved area can be cleaned after maintenance activities are over. After cleanup, rainwater should be diverted to the clean sewer.

Concrete-paved pads should be installed for areas behind heat-exchanger banks. This will provide adequate safeguard against clean sewer contamination when heat-exchanger bundles are pulled for cleaning or maintenance.

Case Study 21.1: Zero-Discharge Wastewater Management Plan for a San Francisco Refinery

Background and Objectives. A refinery in the San Francisco Bay area processes 140,000 barrels per day of Californian and Alaskan crude oil. This complex produces 10% of the West Coast's gasoline and diesel fuels. The refinery hired CH2M HILL to conduct a study to identify wastewater streams suitable for reuse, refinery processes where the water could be reused, and wastewater treatment options available to satisfy process requirements. The ultimate goal was to develop a plan for a phased approach to zero discharge.

Data Collection and Analysis. The study began with an extensive review of background material and collection of samples representative of wastewater streams. Background and analytical data were used to identify the quality and quantity of wastewater discharged to the industrial sewers. CH2M HILL conducted a review of historical data on process operations to determine the quality and quantity of makeup water required by individual processes. This information was used to develop an overall water balance for the refinery. A table of water users and wastewater sources was developed to evaluate reuse options. Four large-volume wastewater sources were identified for potential direct reuse or were treated for reuse:

- bottoms from the foul-water stripper
- bottoms from the ammonia recovery unit
- 250 psig, 400 psig, and 600 psig boiler blowdown
- effluent from the wastewater treatment plant

Potential uses of recycled water were identified as:

- cooling-tower make-up
- boiler feed water

Evaluation and Recommendations. On the basis of the raw water and wastewater quality and quantity data and knowledge of unit processes, three potential reuse options were identified and are presented below.

Bottoms from the FWS and the ARU were neutralized before discharge into the industrial sewer. These streams are moderately basic (pHs of 10 to 12). For this option, CH2M HILL recommended that the FWS and ARU bottoms be neutralized in the existing neutralization facility and treated to remove oil and grease, high ammonia, and high phenol. Following this proposed treatment, the wastewater could be used for cooling-tower make-up water.

Cooling-tower and boiler blowdowns were discharged to the industrial sewer and treated in the WWTP. For this option, CH2M HILL recommended recycling the blowdown from the 600-psig boilers to feed the 400-psig and 250-psig boilers and using the 400- and 250-psig boiler blowdown as cooling-tower make-up water. The hot-boiler blowdown could potentially reduce cooling tower performance, and boiler-water chemical additives may not be compatible with cooling-tower additives. However, CH2M HILL determined that as long as boiler blowdown did not exceed 5% of the make-up stream for an individual cooling tower, these complications would not significantly affect cooling tower performance.

The WWTP effluent was discharged into a local water body under an NPDES permit. For this option, CH2M HILL recommended that the WWTP effluent be demineralized using reverse osmosis and be used as boiler feed. Reverse osmosis is required for removing dissolved solids and producing high-quality water suitable for use as boiler feed. The lower-quality permeate could be used for cooling-tower make-up water. This option is the most expensive, but would be required to meet the client's ultimate goal of zero discharge.

Case Study 21.2: Wastewater Management Master Plan for a Southeast Asian Refinery

A petrochemical complex in Southeast Asia refines 550,000 barrels per day and provides feedstocks to 18 downstream processes in the petrochemical complex. The complex produces olefins, ethylene, benzene, toluene, xylene, polyethylene, polypropylene, para-xylene, and styrene monomer. The complex has 12 WWTPs. Faced with increasingly stringent effluent discharge regulations, the refinery contracted with CH2M HILL to develop a wastewater management master plan. When the plan was prepared, wastewater flow reduction and treatment system alternatives were evaluated. This case study describes the highlights of the waste reduction options that were recommended.

The major water users in the complex are summarized in Table 21.4.

The largest water use is for cooling-tower make-up water, and the largest wastewater flow is the treatment plant effluent. Therefore, options to reuse WWTP effluent as cooling-tower make-up water were explored. In addition, other waste reduction options, including recycling wastewater from one process to feed another, were considered. Five options survived the first-level screening. Three of the options

Table 21.4. Major Water Users in the Petrochemical Complex

Description	Gallons/Day	Percent
Boiler Feed Water	3,168,000	20
Water for Process Units	1,700,000	10
Backwashes and Rinses	792,000	5
Potable Water	950,000	6
Cooling-Tower Make-up Water	7,850,000	48
Fire Water and Construction Water	1,840,000	11
TOTAL	16,300,000	100

involve recycle of WWTP effluent as cooling-tower make-up water; two of the three required no further treatment; the third would require extensive treatment. The remaining two options were reuse of FWS bottoms in the crude desalters and use of boiler blowdown for cooling-tower make-up water.

The flow schematics for the effluent recycle options are shown in Figures 21.1, 21.2, and 21.3.

In Option 1 the WWTP effluent would be reused without additional treatment as cooling tower make-up water. Cooling-tower blowdown would need to be treated by filtration and granular activated carbon to remove the constituents built up in the

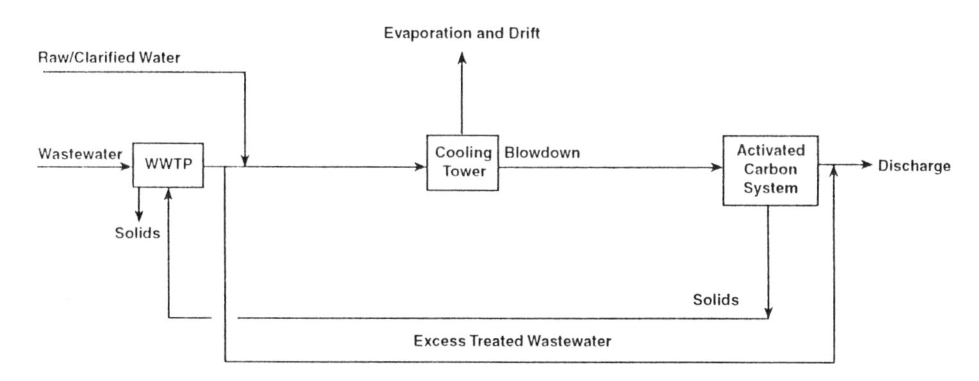

Figure 21.1. Flow schematic for effluent recycling: Option One.

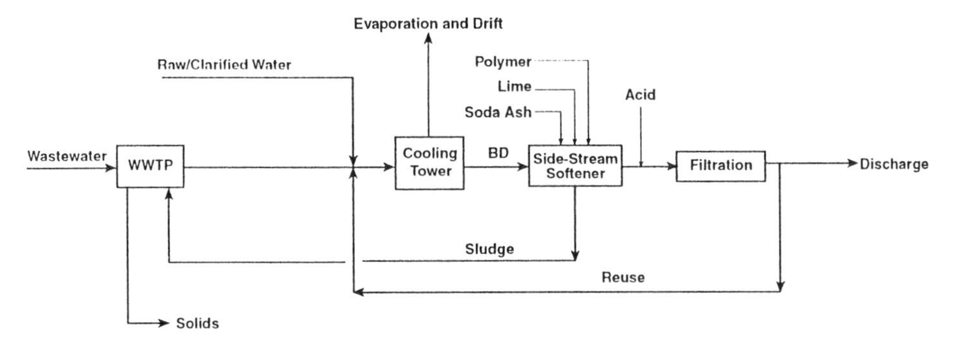

Figure 21.2. Flow schematic for effluent recycling: Option Two.

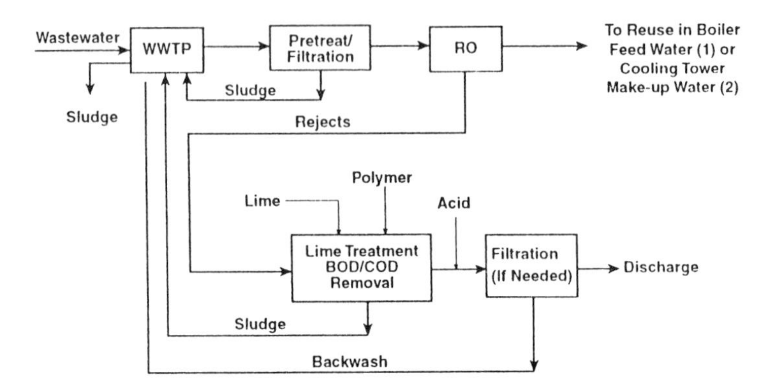

Figure 21.3. Flow schematic for effluent recycling: Option Three.

recirculating water. The amount of wastewater that can be recycled would depend on the water chemistry and the cycles of concentration achievable in the cooling tower. Extensive analysis of the ion buildup was conducted to calculate the cooling-tower cycles achievable and the limiting ions. This in turn determined the amount of wastewater effluent that could be recycled to the cooling tower and hence the water conserved. Option 2 also allows direct reuse of wastewater treatment plant effluent; however, the buildup of ions would be adjusted by side-stream softening and filtra-

Table 21.5. Comparison of Costs for Recommended Recycle/Reuse Alternatives

Option No.	Component	Water Conserved (1,000,000 gal/yr)	Unit Cost (USD/ 1,000,000 gallon)[a]
1	WWTP effluent to cooling towers, blowdown treatment using filtration and GAC	1,225	206[b,c]
2	WWTP effluent to cooling towers, blowdown treatment using filtration and side-stream softening	1,590	1,116[b]
3	WWTP effluent to cooling towers, filtration and reverse osmosis	1,265[d]	1,531[b,c]
4	Sour-water stripper as desalter feed	55	533
5	Boiler blowdown as cooling tower make-up water	47.5	444

[a]Unit cost is total 15-year present worth divided by water conserved during 15 years. Total present worth is the sum of initial capital costs and present worth of O&M costs. Present worth is calculated over 15 years assuming 5% inflation and 8% interest.
[b]Costs for Options 1, 2, and 3 do not include capital, installation, or O&M costs of piping.
[c]Costs for Option 1 do not include cost for biological treatment of blowdown from Cooling Towers 1, 2, 3, 4, 7, and NAC because it is assumed that the existing WWTP No. 2 will be converted for this purpose.
[d]Estimate of water conserved with Option 3 assumes 75% rejection rate.
[e]Costs for Option 3 do not include possible vaporization/crystallization or biological treatment of high COD effluent from side-stream softening treatment of RO rejects.

tion. The amount of wastewater that could be recycled in this option also was calculated on the basis of water chemistry and buildup of critical ions. Option 3 involves treating the WWTP effluent further by using reverse osmosis before the wastewater is recycled as cooling-tower make-up water. This option would not require separate treatment of the cooling-tower blowdown because the ion adjustment occurs before the wastewater reaches the cooling tower.

A cost analysis of the options was conducted to evaluate the unit costs for the water savings. The cost analysis presented in Table 21.5 shows a wide range of unit costs, from $800 per million gallons to $5,800 per million gallons.

Use of untreated WWTP effluent as cooling-tower make-up water is a viable option and would conserve 18% of the raw water used in the complex.

REFERENCE

1. Leemann, J., *Hazardous Waste Minimization Industrial Overviews*, Air and Waste Management Association, 1988.

CHAPTER 22

Pharmaceuticals

*Jeff Enzminger**

PHARMACEUTICAL OPERATIONS AND WASTES

Pharmaceutical companies manufacture a wide range of products using diverse processes. Products include natural substances extracted from plants or animals, chemical modifications of such substances, synthetic organic chemicals, inorganic chemicals, and substances generated by microbial fermentation. Processes include chemical synthesis, fermentation, extraction of naturally occurring materials, purification, formulation, and packaging. Many of these processes are carried out in batch operations.[1]

Fermentation

Antibiotics and steroids are among the pharmaceutical products produced by fermentation.[1] Fermentation is carried out using a carefully maintained microbial strain. Organisms used include both bacteria and fungi. A concentrated cell suspension of up to 20% of the fermenter volume is produced from a few initial cells by gradually culturing the strain in progressively larger containers. Typically, the fermenter is sterilized with steam. Reaction conditions, such as degree of mixing, temperature, oxidation/reduction potential, and pH, are carefully controlled. After the fermentation step is complete, processes such as filtration, solvent extraction, ion exchange, and precipitation are used for recovering the product from the fermentation broth.

The following materials are used in fermentation:[1]

- nutrient broths, including sugars, starches, phosphorus and nitrogen compounds, and protein
- solvents, including methylene chloride, benzene, chloroform, 1,1-dichloroethylene, trans-1,2-dichloroethylene, acetone, ethyl acetate, and methanol

*Dr. Enzminger is an environmental engineer at CH2M HILL's San Jose, California, office and may be reached at (408) 436-4909.

- salts of metals such as copper and zinc
- cleaning and sterilizing agents, including phenol

Chemical Synthesis

Most pharmaceuticals are produced by batch chemical synthesis.[2] Various reaction chemistries and downstream processing and purification steps are used, making it difficult to generalize about resulting waste streams.

Natural Extraction

Natural extraction typically involves extracting an active ingredient from a large volume of a naturally occurring mixture, such as plant material. Product purification follows initial extraction, and may include precipitation and additional extractions.

Materials used include the following:[1]

- salts of metals such as zinc and lead used for precipitation
- residues from natural raw materials
- chemicals, including phenol, used for cleaning and sterilizing
- extraction solvents, including benzene, chloroform, 1,2-dichloroethane, acetone, 1,4-dioxane, ethyl acetate, methanol, and ammonia

Formulation

Pharmaceutical formulations include solids (tablets and capsules), liquids (injectable and oral forms), and creams and ointments (petrolatum base and oil/water emulsions). Tablets are typically made by forming a core of the active ingredient in combination with fillers and binders. Coatings are often added to provide additional active ingredients, water barriers, and coloring. Capsules have a gelatin coating, or shell, formed around a preparation of the active ingredient. Preparation of liquid formulations involves mixing, filtration, addition of preservatives, and packaging. Formulations for injection (parenterals) must be sterilized by means of autoclaving or filtration. Materials used include fillers, solvent carriers, petroleum base, and cleaning and sterilizing agents.

Waste Streams

General categories of waste streams include:[2]

- fermentation broth, which is typically high in suspended solids and biochemical oxygen demand
- spent fermentation media
- filter cake
- extraction aqueous raffinate
- cleaning solutions for equipment

- scrubber blowdown
- spent solvents (extractants) and still bottoms from solvent recycling
- extracted natural material
- dust in air emissions
- solvent in air emissions

OPPORTUNITIES FOR POLLUTION PREVENTION

For existing processes and facilities, an approach to pollution prevention includes the following steps:

- Quantify and characterize liquid, solid, and vapor waste streams through sampling and analysis and mass-balance development.
- Evaluate the regulatory issues involving the various waste streams identified.
- Develop waste reduction alternatives for each of the waste streams identified and estimate costs.
- Rank pollution prevention projects with respect to benefits and costs and develop an implementation plan.

One of the greatest impediments to implementation of pollution prevention projects is a lack of knowledge of waste streams. A second major impediment in the pharmaceutical industry is that the process used in manufacturing a particular pharmaceutical is subject to U.S. Food and Drug Administration (FDA) approval. Obtaining USFDA approval is complex, costly, and time-consuming. Process modifications for an approved substance may require filing supplemental applications and demonstrating to the USFDA that the final product quality has not been adversely affected.

A pollution prevention strategy for new pharmaceutical processes includes the following steps:[3]

- Prepare a material balance to analyze information on potential waste streams.
- Develop and evaluate pollution prevention opportunities.
- Incorporate feasible opportunities.
- Update the material balance.

The U.S. EPA has established a pollution-prevention hierarchy consisting of source reduction as the preferred method, followed by recycling, treatment, and disposal. Several examples of pollution prevention projects for the pharmaceutical industry are presented below, including waste reduction in both manufacturing operations and in ancillary services, such as packaging and glassware cleaning.

Case Study 22.1: Solvent Use in Package Labeling

A pharmaceutical facility employs a flexographic process to manufacture relief plates that are used to stamp expiration dates and lot numbers on product labels and tags.[4] The first step in the process consists of covering a flexible photopolymer plate with a photographic negative of the image to be printed. The plate is then exposed to ultraviolet (UV) light. UV exposure causes the photopolymer to crosslink.

A schematic diagram of the process steps following UV exposure of the photopolymer plate is presented in Figure 22.1. The UV-exposed photopolymer plate was placed in a rotary washer that uses the combined action of brushes and 1,1,1-trichloroethane (TCA) to wash out the uncrosslinked polymer. Removal of the uncrosslinked polymer left a raised image for printing. During the process, some of the solvent was absorbed into the plate, which caused the plate to swell slightly and distorted the printing image. The plate was removed from the washer and dried in an oven at 140 to 150°F to evaporate the solvent.

TCA was recirculated between a 55-gallon drum and the washer. Typically, enough photopolymer residue accumulated in the TCA in one week to impair the quality of the finished plate, at which time the spent TCA was disposed of off the site as a hazardous waste (F002). At the time of disposal, only two-thirds of the original 55 gallons of TCA remained, one-third having evaporated. Volatile organic air emissions from plate washing and drying passed to the atmosphere through a dedicated venting system.

Approximately 30,400 pounds of TCA were used to process 345 square yards of plates each year. Using a 12-inch by 15-inch plate as the label-making production basis, TCA usage is approximately 12 pounds per plate. The objective of the

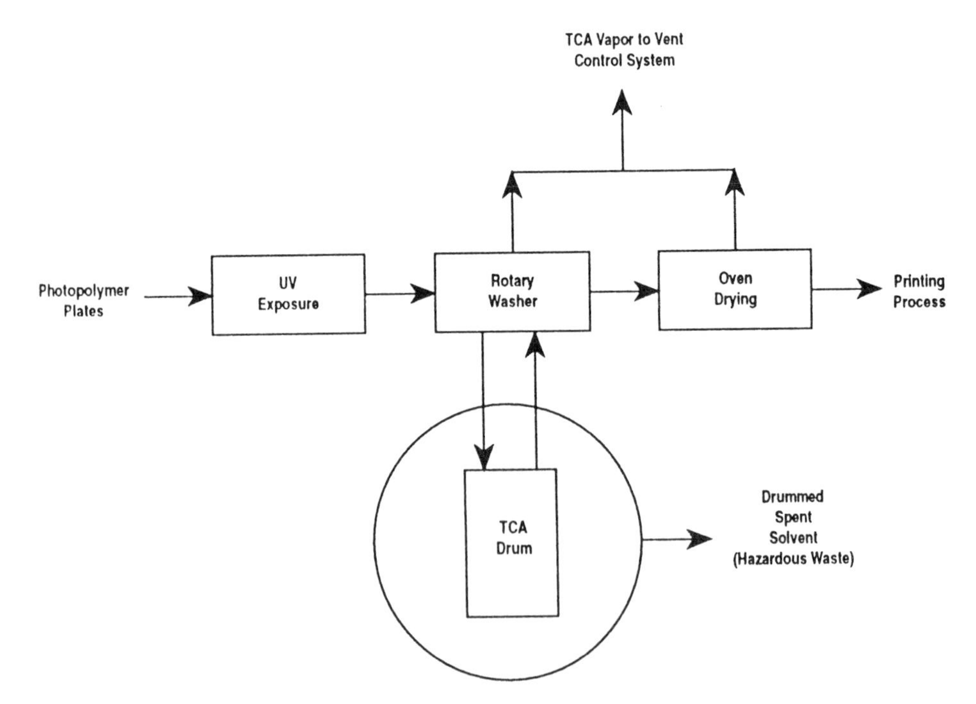

Figure 22.1. Flow diagram of the photopolymer plate process.

pollution-prevention evaluation was to develop alternatives for reducing the per-plate TCA usage.

Modified Housekeeping and Operating Procedures

Approximately two-thirds of the TCA used in the plate-making process was discharged as liquid hazardous waste for offsite disposal, and one-third was emitted to the air. The decision to discard solvent was based on plate quality. The solvent was used until the plate quality was unacceptable. Historical solvent usage was compared with photopolymer plate production. The ratio of solvent usage to plate production did not vary significantly. Therefore, the potential for modifying operation of the existing equipment to reduce TCA discharge was minimal.

Solvent losses from the apparatus are negligible when the equipment is not in use. Of the TCA loss of 200 lb per week to volatilization—when the equipment was in use—the estimate was that at least 60 lb could be accounted for at the plate-washing tank because of air exchange when the tank wash door was opened. Minimizing the number of times that the tank cover is opened was recommended. TCA vaporization in the drying oven was believed to account for the remainder of the loss of TCA from volatilization. Allowing the washed photopolymer plate to drain in the washing tank before transferring it to the drying oven also could reduce TCA losses.

The use of a vent condenser was considered for reducing air emissions and recovering usable solvent. However, the concentration of TCA in the air exhaust was too low to make a vent condenser practical.

Solvent Substitution

Four alternative solvents were identified that could eliminate the use of hazardous substances as defined in the Pollution Prevention Act. These are:

- Solvit, manufactured by W.R. Grace
- Polysafe V, manufactured by 3D, Inc.
- Optisol, manufactured by DuPont
- Nylosolv, manufactured by BASF

Polysafe V and Nylosolv were rejected because they have low flash points, and after use these solvents would become hazardous wastes. Optisol was rejected because its compatibility with the photopolymer plates was questionable. Solvit was retained for further evaluation because it is compatible with the existing photopolymer plates. Table 22.1 compares the use of Solvit with the use of TCA in the existing equipment. The major disadvantage of using Solvit was that, unlike TCA, it is considered a combustible solvent and would present safety concerns.

Solvent Reclamation

TCA and Solvit can both be reclaimed by vacuum-assisted distillation and reused for plate washing. Reclaiming Solvit would not provide pollution-prevention benefits as defined in the regulation but would reduce the cost of using Solvit. Two alternatives were identified for reclaiming plate-washing solvents: using onsite equipment or using an onsite service. Using onsite equipment would involve install-

Table 22.1. Comparison of Alternative Solvent and Solvent Reclamation Alternatives

	No Solvent Reclamation Disposal		Reclamation with Onsite Equipment		Reclamation Using an Onsite Service	
Issue	**TCA**	**Solvit**	**TCA**	**Solvit**	**TCA**	**Solvit**
Pollution Prevention	TCA usage of 50 drums per year; hazardous waste production of 33 drums per year	Eliminates use of a hazardous substance and generation of hazardous waste	TCA usage of 23 drums per year; hazardous waste production of 7 drums per year	Eliminates use of a hazardous substance and generation of hazardous waste	TCA usage of 23 drums per year; hazardous waste production of 7 drums per year	Eliminates use of a hazardous substance and generation of hazardous waste
Reliability	High	High	High	High	Depends on contractor	Depends on contractor
Operability	Easy	Combustible solvent; precautions required	Moderate skill/ attention necessary	Moderate skill/ attention necessary	Easy	Easy
Implementability	Unchanged	Combustible solvent concerns	Operator training, proper maintenance	Operator training, proper maintenance, explosion-proof environment required.	Depends on proper maintenance and operation by contractor	Depends on proper maintenance and operation by contractor
Capital Cost	$0.00	$1,500	$38,100	$52,350	$1,000	$15,250
Annual Operating Cost						
Photopolymer	$42,400	$42,400	$42,400	$42,400	$42,400	$42,400
Utilities/ Supplies	400	400	2,500	2,500	400	400
Solvent	16,600	25,100	7,700	7,200	7,700	7,200
Disposal	3,200	10,100	2,000	2,000	2,000	2,000
Service	0	0	0	0	6,400	9,600
Total	$62,600	$78,000	$54,600	$54,100	$58,900	$61,600

ing and operating a vacuum-assisted still. The onsite solvent-recovery service option would consist of regularly scheduled visits by a contractor with a mobile distillation system. The contractor would leave the still bottoms for disposal by the facility. These alternatives are compared in Table 22.1.

Organic-Solvent-Free Plate-Making Processes

The final pollution-prevention alternative evaluated was switching to an organic-solvent-free plate-making process. The three processes evaluated were:

- Printight plates with an Orbital VIII washer. Printight plates are manufactured by Toyobo Co. Ltd. of Japan. The Orbital VIII washer is manufactured by Anderson & Vreeland.

- Cosmolight plates with a Cosmolight closed washout system, manufactured by Toyobo Co. Ltd. of Japan.

- The Merigraph liquid photopolymer and water washout system, manufactured by Hercules, Inc.

The Printight photopolymer plates were determined to be incompatible with the ink solvents in use. Therefore, the Orbital VIII system was eliminated from further consideration. Table 22.2 compares the Cosmolight and Merigraph processes.

Summary and Recommendations

Table 22.3 compares the pollution-prevention alternatives. Specific recommendations included the following:

1. Housekeeping procedures should be implemented immediately. Solvent losses should be monitored to estimate the level of pollution prevention achieved.

2. The organic-solvent-free alternative was recommended over the use of Solvit because Solvit costs more and, being combustible, requires special procedures for safe storage and handling.

3. The compatibility of the Merigraph and Cosmolight photopolymers with the ink solvent should be evaluated through immersion tests to be performed by the system manufacturers. The printing process ink requirements and the potential for using inks that are compatible with either the Merigraph or Cosmolight systems also should be evaluated.

4. Onsite reprocessing of TCA should be considered if the organic-solvent-free processes cannot be implemented. This pollution-prevention alternative should reduce overall costs slightly while reducing hazardous substance usage and hazardous waste generation substantially.

Table 22.2. Comparison of Alternative Plate-Making Processes

Issue	Existing Process	Cosmolight	Merigraph
Pollution Prevention	No reduction	Eliminates use of hazardous substances and generation of hazardous waste	Eliminates use of hazardous substances and generation of hazardous waste
Reliability	Existing system	Approximately 10 systems in the U.S.; maximum operating history is approximately 1 year	Approximately 12 systems of Model 1826 in operation in the U.S.; another 300 installations of higher-capacity units
Implementability	Existing system	Unit would fit into current plate-washing room; existing UV exposure unit and drying oven would be reused	Unit would fit into current plate-washing room
Operability	Existing system	Manpower requirements similar to current system	Manpower requirements similar to current system
Capital Cost	$0.00	$33,000	$58,000
Annual Operating Cost			
Photopolymer	$42,400	$43,000	$19,600
Utilities/Supplies	400	1,400	1,100
Solvent	16,600	0	0
Disposal	3,200	0	0
Total	$62,600	$45,200	$20,700

Table 22.3. Summary of Label-Making Alternatives

	Use of TCA According to Current Practice	Use of Solvit with Solvent Reclamation Service and Improved Housekeeping Practices	Use of TCA with Solvent Reclamation Service and Improved Housekeeping Practices	Cosmolight System	Merigraph System
Pollution Prevention	Uses 50 drums of TCA per year	Eliminates use of hazardous substance and production of hazardous waste	50% reduction in hazardous substance usage	Eliminates use of hazardous substance and production of hazardous waste	Eliminates use of hazardous substance and production of hazardous waste
Reliability	Existing system	Depends on contractor	Depends on contractor	The system was recently introduced; there are 12 installations in the U.S.	There are 6 installations in the U.S.
Implementability	Existing system	Need to evaluate liability and safety issues regarding onsite reclamation service, use of a combustible solvent, and ink/plate compatibility	Need to evaluate liability and safety issues regarding onsite reclamation service	Testing required	Testing required
Operability	Existing system	Same as existing system	Same as existing system	Similar to existing system	Similar to existing system
Capital Cost	$0	$15,250	$1,000	$33,000	$58,000
Annual Operating Cost[a]	$62,600	$61,600	$58,900[b]	$45,200	$20,700

[a]Operating cost includes power, water, photopolymer plates, solvent, utilities, and solvent or still bottoms disposal.

[b]TCA reprocessing operating costs assumes that up to 12 55-gallon drums of spent solvent can be stored at the facility pending reprocessing

Case Study 22.2: Alternatives for Labeling

Company B uses sulfuric acid for removing indelible marks used to label glassware.[2] The glassware is loaded into steel baskets, placed in a sulfuric acid bath for 10 minutes, and rinsed with water. The glassware is then washed in a commercial dishwasher. The sulfuric acid bath is used several times before being sent to an offsite treatment and disposal facility at a cost of $380 per drum. A proposal was made to use adhesive labels in place of indelible marker, saving the cost of purchasing and disposing of sulfuric acid (approximately $15,000 per year).

Case Study 22.3: Solvent Recovery in Chemical Synthesis

An alternative chemical synthesis was developed for an antibiotic intermediate, which eliminated the use of methylene chloride.[5] However, the alternative synthesis generated 64 kg of spent solvent per kilogram of intermediate. Because of this solvent waste load, additional development efforts focused on improving the recovery of the solvent.

Subjecting the spent solvent streams to aqueous extractions at various pH values removed water-soluble solvents, thereby improving the recovery of the major immiscible solvents. Of the 64 kg of solvent per kg of process intermediate, 85% could be recovered, 3% to the treatment plant, and 12% was sent off the site for incineration.

Case Study 22.4: Water-Based Coatings

A company was coating tablets using several organic solvents as coating carriers.[6] A water-based solvent was developed in conjunction with new spray application equipment. The payback was less than one year on the basis of savings in solvent costs. The emission of organic solvents was reduced by 24 tons per year.

Case Study 22.5: Alternatives for Tablet Coating

A company used methylene chloride as a carrier in coating tablets. Various technologies were evaluated for reducing solvent emissions, including direct condensation, compression condensation, use of the Brayton Cycle to recover solvent, adsorption, and membrane pervaporation.[7] In each case, whether the solvent would be of sufficient purity to use in the original process was questionable, but it was expected to be saleable to solvent reclaimers.

REFERENCES

1. United States Environmental Protection Agency, *Preliminary Data Summary for the Pharmaceutical Manufacturing Point Source Category*, National Technical Information Service, PB90-126533, September 1989.
2. United States Environmental Protection Agency, *Guides to Pollution Prevention—The Pharmaceutical Industry*, EPA/625/7091/017, October 1991.
3. Dienemann, E. et al., Waste Minimization Strategies During Pharmaceutical Process Development, Second EPA Conference on Waste Minimization, San Diego, 1991.

4. Enzminger, J.D. et al., "Pollution Prevention Evaluation of a Printing-Plate Making Process at a Pharmaceutical Manufacturing Plant," Presented at the Air and Waste Management Association Annual Meeting, Denver, April 1993.
5. Venkataramani, E.S., "Waste Minimization in a Leading Ethical Pharmaceutical Company," Presented at the Waste Minimization/Pollution Prevention Workshop, New Jersey Water Pollution Control Association, New Brunswick, New Jersey, February 1990.
6. Huisingh, D. et al., *Proven Profits From Pollution Prevention—Case Study 14*, Institute for Local Self Reliance, Washington, DC.
7. Housel, G., VOC Containment Technology Alternatives. Presented at the specialty conference titled "Controlling Air Toxics in the Chemical and Pharmaceutical Industry," Philadelphia, September 1991.

CHAPTER 23

Pulp and Paper Industry

*Ulf Wallendahl**

DESCRIPTION OF PULP AND PAPER OPERATIONS AND WASTES

Pulp and paper manufacturing is a major industry in the United States. There are approximately 600 paper and paperboard mills and about 350 pulp mills throughout the country. The paper and allied products industry ranks about tenth among the national manufacturing industries, having sales totaling $115 billion and a capital investment of $15.5 billion in 1990.

Pulp and paper manufacturers process fibrous raw material to produce a large variety of products, mainly paperboard, paper, and pulp. Paperboard is used for paper boxes, product packaging, milk cartons, paper cups and plates, and posterboard. Paper products include book, magazine, and newsprint paper; copying and computer paper; writing and art paper; tissue paper; and paper bags and wrappings. Pulp is not considered an end product because the principal use of pulp is in paper and paperboard. In addition, the cellulose fibers are used as fluff pulp in products such as diapers; as dissolving pulp in the manufacture of man-made fibers, films, plastics, and chemicals; and in insulation and other products.

The raw material, the fiber, is predominantly of wood origin. Roughly one-third is recycled fiber, one-third is lumber and plywood by-product chips, and one-third is pulpwood. Other raw materials used, where available, are sugarcane bagasse, straw, grass, and rags, and (for specialty papers) natural hemp and cotton fibers. Synthetic fibers are used in small amounts where special product properties are demanded. Less than 1% of the fiber used in paper and paperboard in the United States is of nonwood origin.

Typically, production facilities are integrated, with both pulping and paper machine operations located at the site. Also common are pulp or paper mills operating next to raw-material producers, such as lumber and plywood operations, recycling operations, or converting operations. Integrating production and sharing facilities has significant benefits. Shared facilities for steam and power distribution, water and effluent treatment, and reuse of waste and by-products, and the elimination of

*Mr. Wallendahl is an environmental engineer in CH2M HILL's Seattle, Washington, office and may be reached at (206) 453-5000.

extra process steps, such as drying pulp for long-distance transport, are some of the more obvious benefits.

The pulp and paper industry has a long history of recycling. For many years, pulp and paper companies have reprocessed not only their own wastes but also their customers' (converters) waste and, in increasing amounts, consumers' waste. The industry is the largest recycler in the United States. Figure 23.1 shows that close to 40% of the post-consumer waste of paper and paperboard is collected. Recyclable paper, lumber and plywood waste, and internally generated waste streams constitute the majority of the raw material that is recycled. Processing and utilization of waste paper constitute the fastest growing sector in manufacturing. At the current pace, the growth rate exceeds 7% per year.

Basic Pulp and Paper Manufacturing Operations

The principal process steps in the pulp and paper industry are:

Pulping

The production of virgin pulp is accomplished by mechanical and chemical separation of fibers from wood chips, sawdust, or logs. The produced pulp can be screened, cleaned, and bleached to meet the product requirement where it will be used.

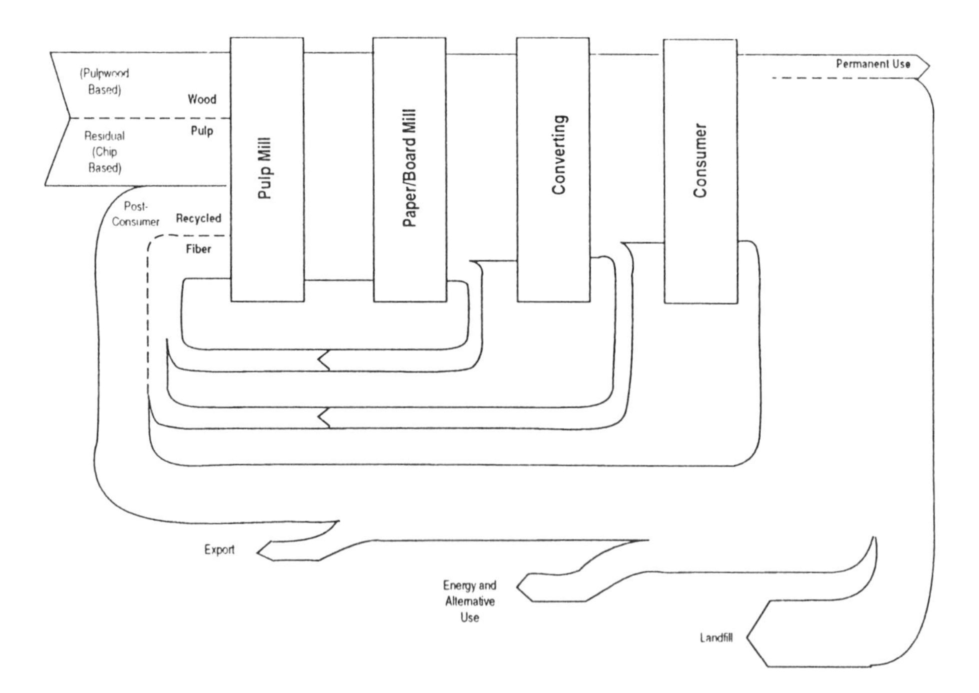

Figure 23.1. Fiber balance diagram for paper products raw materials used.

Processing Recycled Fiber

The processing of recycled fiber can be as simple as beating the return fibers in water to a pulp suspension or can also involve a more sophisticated process of screening, cleaning, deinking, and bleaching. The method of processing depends on how contaminated the raw material is and the requirements of the product it is going into.

Making Paper and Paperboard

Paper and paperboard are made on paper machines in a process consisting of conditioning, blending, and mixing of fiber and additives to a pulp suspension that is then formed into a sheet, pressed, and dried.

To support these operations, pulp and paper mills need utility services, such as process steam, electrical power, water treatment and supply, process cooling, and water and wastewater treatment.

Industry Trends

Several distinct trends in the industry will influence both the amount of waste generated and the future opportunities for dealing with waste.

Development of Paper Machinery

Paper machines have improved, and modern installations can use lower-grade raw materials. For example, newsprint can be produced entirely from mechanical pulp without chemical pulp content, as well as from 100% recycled fiber. Coated and multilayer sheets allow greater flexibility in raw-material selection.

Increase in Pulp Yield from Wood

Improvements in paper machinery and the development of new pulping processes have made it possible to obtain a higher yield from wood. The higher yield conserves wood resources and produces less by-product material.

Improved Use of Dissolved Wood Material

Recovery of dissolved wood material has increased slowly but steadily. Extended delignification and the development of new bleaching methods involving counter-current recovery will continue this trend.

Energy and Water Conservation

The industry has made significant steps toward conserving energy and water. Since 1972, the industry-wide use of purchased energy has dropped from 33 million Btu/ton to 27 million Btu/ton of pulp, paper, and board produced. Significant water reductions are achieved through better reuse methods and by separating cooling water from process water. Some grades of paper can be produced at very close to zero discharge.

TYPES OF WASTE PRODUCTS

The waste streams generated in the industry can be best classified by their origins. The main groups are materials that originated in the raw materials, such as dirt and bark with wood; nonfiber constituents in wood; contaminants in waste paper and make-up chemicals; reaction products, such as dissolved wood substance from mechanical or chemical action; fiber fragments; by-products of chemical recovery and combustion; and fiber and nonfiber process losses and discharges of used water, air, and heat.

This chapter focuses on wastes that are specific to the industry. It does not discuss waste reduction related to processes, such as treatment of mill water supply, ash from combustion of solid fuels (coal, woodwaste, etc.), and emissions from steam generation, unless specifically applicable to the industry.

Virgin Pulp

Wood-based fiber is the predominant raw material for the pulp and paper industry in the United States. Various pulps, having different properties and yields, are produced from both softwoods and hardwoods. Figure 23.2 shows main industry processes and the typical yield of pulp. The figure also shows where recovery of

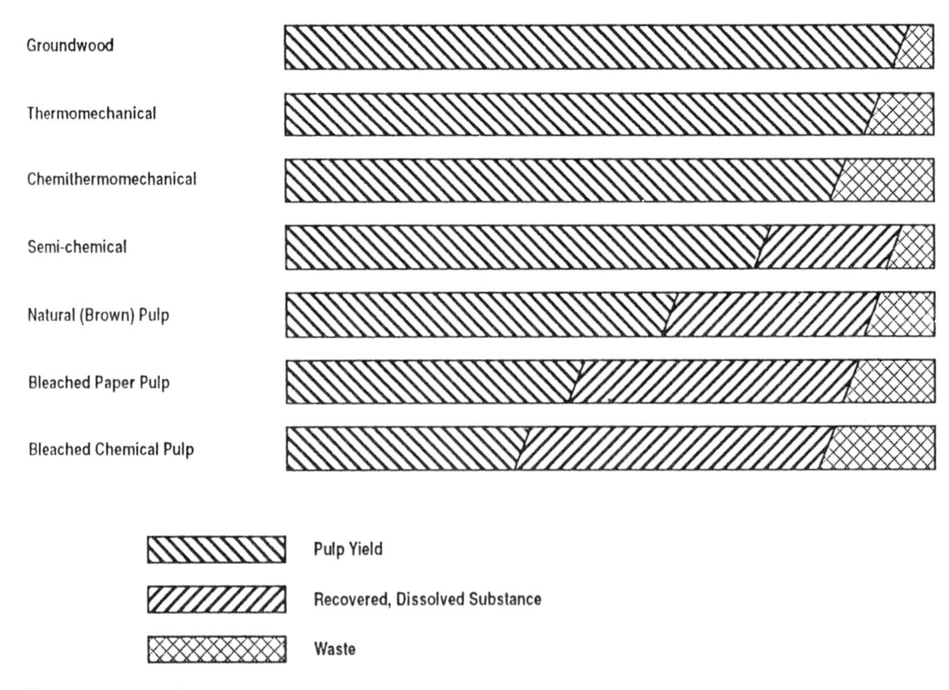

Figure 23.2. Pulping of wood; typical yields with recovered and reuse wood substance and waste (not used material).

chemicals and dissolved wood substance is practiced. Table 23.1 shows the approximate production of pulp by process.

The kraft process, which is used for about 80% of the wood pulp production, has shown a steady improvement in waste reduction and material reuse. Figure 23.3 shows the main trends in the industry and steady improvements in all areas. The biggest change in the near future will be in reduced water consumption and discharge and the move towards closed mills.

Table 23.2 illustrates the process steps involved in producing wood pulp (all the grades do not have to go through all the steps) and the type of waste most commonly generated in each step.

Solid Waste in Wood Pulping Operations

The main solid wastes produced in wood pulping operations are wood and fibrous rejects, residues, and process losses; caustic waste solids, mainly calcium-based grits, dregs, and excess lime; and yard, wood yard, and general mill solid waste.

The bark and woodroom rejects usually are processed and used in bark boilers on the site or are sold as hog fuel. Rejects from pulping operations can be minimized by recooking knots and recovering good fiber and either adding the fiber to the hog-fuel stream or selling it to pulping operations for making lesser grades. The balance of the lost fiber is collected in effluent primary clarifiers. It is then pressed and used as hog fuel. Effluent sludge from secondary treatment plants, if continuously recovered from the process, is dewatered with the primary sludge in a screw press and also is used as hog fuel.

In kraft mills, a portion of the lime recycle stream needs to be blown down to remove impurities entering the process with the natural raw materials, the wood, and the limestone. The two principal outlets for the inert materials are the dregs from green-liquor clarification and slaker grits. Both streams must be washed to remove free caustic solution for soda recovery and then dewatered for safe disposal. For washing the grits, the installation of core washers is often advisable.

The balance of solid waste from mills is mostly dirt and general trash, which is disposed of in landfills.

Table 23.1. Wood Pulp Production (United States, 1990)

Grade	Quantity (Million Tons)
Natural (unbleached) kraft	20.0
Bleached softwood kraft	13.2
Bleached hardwood kraft	11.8
Mechanical, groundwood, and thermo-mechanical	5.8
Semi-chemical	3.8
Bleached and unbleached sulfite	1.4
Special and dissolving pulp	1.2
Total wood pulp	57.2

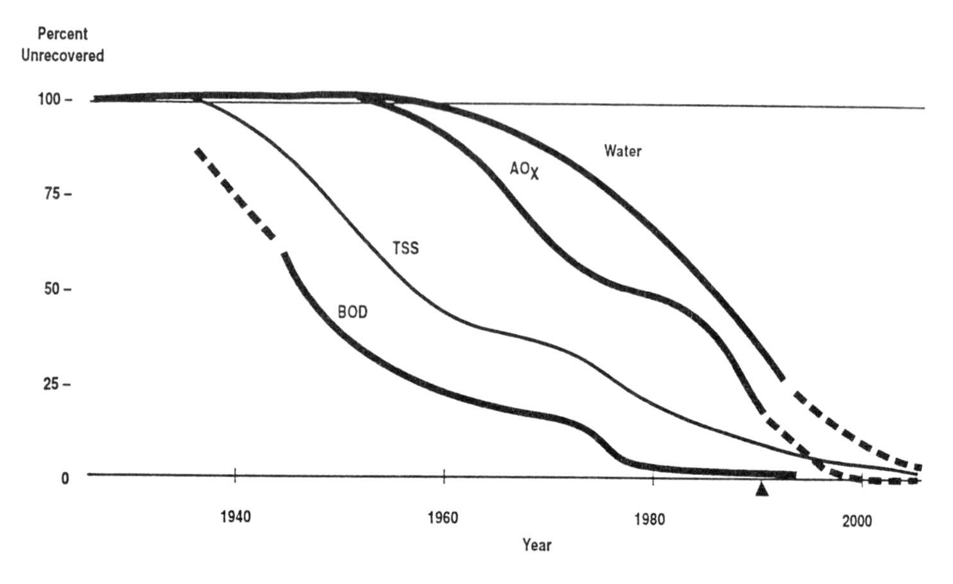

Historical technology developments:

Pre 1960	Kraft pulping with chemical recovery gaining production share Sulfite pulping with chemical recovery emerging Better pulping equipment and first recovery systems
1960s	Bleaching advancements - introduction if chlorine dioxide Improved liquor recovery Primary effluent treatment starting Installation of evaporator surface condensers
1970s	Secondary treatment Water reduction for energy conservation Switching bleaching chemistry towards chlorine dioxide
1980s	Increased use of high chlorine dioxide substitution Closing up bleach plants Separation of process and cooling waters Oxygen enforced bleaching
1990s	Extended delignification in cooking Oxygen delignification Short bleaching sequences Full substitution with chlorine dioxide Peroxide bleaching

Figure 23.3. Environmental trends in wood pulping.

Effluent from Wood Pulping Operations

The wastewater from pulping operations always has been the industry's greatest waste-handling challenge, but it also has the greatest potential for waste reduction and reuse. The main constituents in pulping wastewater originate from wood constituents, reaction by-products, and pulping chemicals and additives.

For the wastewater constituents to be captured, the pulping process must be modified to reduce water usage, reuse as much water as possible, and eliminate leakage. Modern pulping processes can be operated with much less water if process cooling and process water systems are separated, internal recycling and cleaning

Table 23.2. Pulping of Wood

Process Step	Potential Waste Products: Solid Waste	Liquid	Air
Step 1: Wood yard • Barking • Chipping • Storage	• Dirt and debris • Hog fuel	• Barking effluent • Wood acids • Extractives	• VOCs (e.g., turpentine) • Dusting
Step 2: Pulping • Cooking • Defibration • Washing • Screening	• Knots and rejects • Fiber loss	• Dissolved wood • Extractives • Volatiles from wood and reaction products • Organic and inorganic sulfur • pH and heat	• Organic and inorganic sulfur • VOC from wood and and reaction products • NCGs and odor
Step 3: Bleaching • Delignification • Brightening • Cleaning	• Cleaner rejects • Fiber loss	• Reaction products from dissolved wood, some chlorinated • Waste from bleach chemical preparation • pH and heat	• Reaction products from bleaching • Bleaching chemicals
Step 4: Pulp Dryer • Forming and pressing • Drying	• Fiber loss	• Residuals from steps 1–3 above • Heat	• Residuals from above • Heat
Step 5: Chemical Recovery • Evaporation • Combustion • Chemical regenerant	• Dregs, grits • Lime and mud • Scale and inert materials • Filtration residue	• Chemical spills • pH • Soap RFA • Liquor spills	• TRS odor • NO_x, SO_x • NCGs, VOCs • Particulates
Step 6: Utilities • Water supply • Effluent • Process cooling • Energy	• Water treatment sludge • Primary effluent sludge • Secondary effluent sludge	• TSS, COD, BOD, AO_x • Color and toxicity as result of the above • Heat and chemithermomechanical (CTWR) blowdown	• VOC odor • CTWR drift • CTWR VOC • Flue gas

systems are installed, and processes that generate less wastewater are used. For example, bleached kraft pulp mills used to discharge 100 m^3 of wastewater per ton of pulp but now can discharge 50 to 60 m^3/t, and proposed processes aim for 15 m^3/t or less.

The effluent constituents of most concern are:

- wood extracts, in particular the fatty and resin acids in fresh wood, especially softwood, because of their toxicity
- volatile organic compounds (VOCs), such as turpentine, naturally occurring in wood; volatile reaction products, such as methanol; and organic sulfur

compounds because of their contribution to VOC discharge, odor, and potential toxicity

- the balance of dissolved wood extract compounds, which contribute to color, chemical oxygen demand, and biochemical oxygen demand, and react with chlorine in the bleaching process to produce toxic compounds, such as dioxins

The wood extracts need to be recovered. In alkaline pulping (kraft), the extracts are soluble in the cooking liquor. To separate out the extracts, processing plants must have good practices for brown-stock washing and must operate a tight black-liquor system where the extracts separate out as soap. Pulp-mill soap can be transported and sold as such or can be acidified and sold as tall oil. Today's market for soap and tall oil does not favor the recovery of this material for external sale, but both products make an excellent fuel for the mills. In sulfite pulping, the pulp usually is deresinated after brown-stock washing and before bleaching. Processes can recover and use the fibers containing the extracts as fuel, although a more common practice is to recapture the material through primary clarification. In mechanical pulping, the extracts in the effluent remain a major concern. For the chemithermomechanical process (CTMP), the first installations with recovery and concepts are emerging.

The new emphasis on VOC emission controls and the conversion of mill water systems to closed cycle will increase the pressure to improve the capture and removal of VOCs in the process. In new kraft mills, condensate stripping and turpentine recovery already is the standard, and even better methods will be required as more plants are converted to closed cycle. The collected turpentine is sold as a by-product. Other collected VOCs replace fossil fuel in boilers or lime kilns. In sulfite mills, liquors are neutralized before evaporation to tie up the acids produced in the process and use them as fuel in the recovery boiler.

The balance of the dissolved wood extracts need to be minimized through improved brown-stock washing and extended delignification with countercurrent recovery before final bleaching. Figure 23.4 summarizes the trend in the industry for the bleached softwood kraft process, which shows a 60% reduction in dissolved material and an even greater reduction in dioxin and adsorbable organic halides (AOX) discharges.

Air Emissions

The pulp and paper industry has eliminated most atmospheric emissions of inorganic pulping chemicals. With efficient precipitators on the recovery boilers and lime kilns or with good lime-kiln scrubbers, particulate chemical loss has been reduced drastically.

Sulfur losses have been reduced through efficient recovery operations. The kraft mills need to practice efficient capture and incineration of both strong and weak noncondensible gases (NCGs), both to recover chemicals and to abate total reduced sulfur (TRS) and odor.

For volatile material, turpentine recovery can be improved, but in current market conditions there is little incentive for investing in recovery equipment. Methanol-type compounds can and must be captured more efficiently, not only to reduce VOC emissions but also to reduce fuel use in the plant.

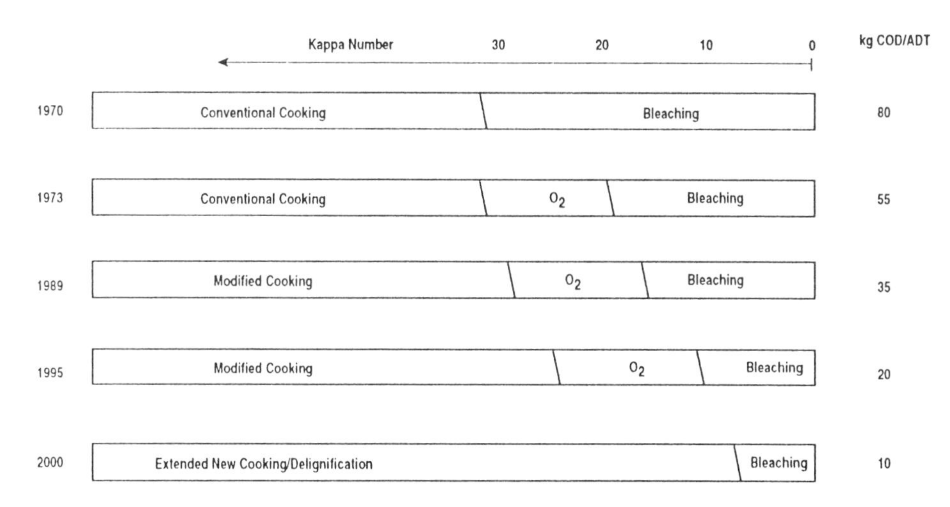

Figure 23.4. Trends in delignification and bleaching.

Energy and Water Conservation

The production of wood pulp consumes large quantities of water and energy, but the processes also present great opportunities for reuse and conservation, usually in connection with other waste reduction efforts. In the preparation of wood, dry debarking and chip screening yield a by-product that is an excellent fuel. This practice also reduces contaminants that would otherwise have to be rejected later in the process. In mechanical pulping, installing heat-recovery equipment reduces energy use while making recovery of VOCs possible. Chemical pulping can be energy self-sufficient for both electrical power and thermal (steam) energy. Modern installations can supply surplus energy for offsite sale or use.

Recycled Fiber

The recycled-fiber industry has been the fastest-growing sector in the industry, and technology developments in paper machines have broadened the grades where recycled fibers can be used. However, secondary fiber technologies still need to be developed, in particular to use lower-grade waste papers. Waste-to-energy technologies are needed for recycled paper not suitable for reuse in paper and board.

Figure 23.5 shows wastepaper recovery rates and industry targets for 1995. Waste papers commonly are graded as mixed papers, newspapers, old corrugated containers (OCC), pulp substitutes (usually printed sheets and cuttings), and deinking stock. In 1990, the utilization of wastepaper was estimated as shown in Table 23.3.

The pulping of waste paper varies with grade and type of waste and the intended use, but a typical recycling process includes pulping, contaminant removal, and conditioning or further treatment to suit the application. Table 23.4 outlines a recycled-fiber process involving deinking and lists the potential waste products.

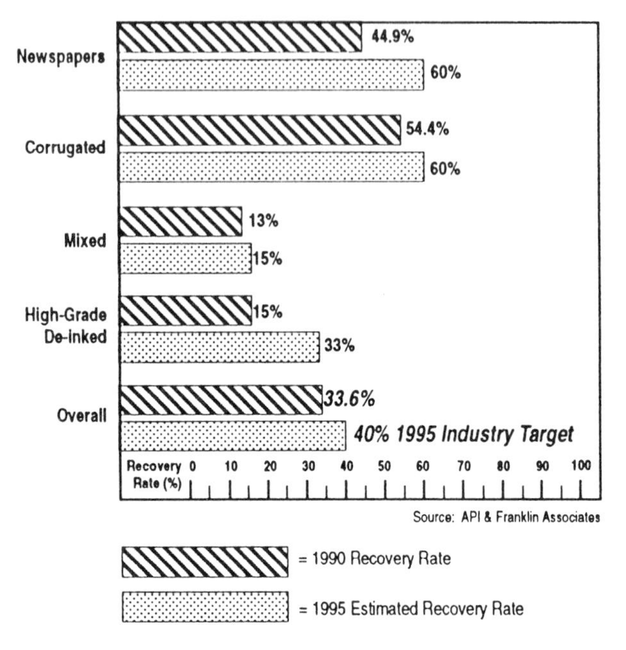

Figure 23.5. Wastepaper recovery rates.

Solid Waste from Secondary Fiber

The solid waste generated in the receiving and pulping areas can include baling wire, pallets, boxes, plastics, and miscellaneous contaminants that come in with recycled paper. Baling wire can be shredded and sold for scrap, and the other wastes can either be used as fuel or put in landfills.

Some of the paper received does not meet specifications and has to be rejected as not suitable for recycling. Rejected paper can be sold, returned, or downgraded to fuel, but it often has to be disposed of. Landfilling is an option, but the landfill has to be suitable both for the paper and contaminants. Rejects from the secondary-fiber processing include plastic, string, staples, glue, and labels. This waste stream usually is dewatered for landfilling or incineration.

The largest component of solid waste, especially for a deinking operation, is sludge. Sludge from magazine deinking contains large amounts of fillers, fibers, and

Table 23.3. Wastepaper Use in 1990

Product Using Waste Paper	Million Tons
Corrugated	10.7
Newspaper	4.1
Pulp Substitutes	2.7
Mixed Papers	2.6
Deinking	1.9
Subtotal	22.0
Exported	6.4
Total Collected	28.4

Table 23.4. Recycled-Fiber Processing with Deinking

Process Step	Potential Waste Products		
	Solid Waste	**Liquid**	**Air**
Step 1: Receiving • Storage • Conveying	• Paper • Wire		• Dust
Step 2: Pulping • Coarse screening and cleaning	• Tramp metal • Dirt, plastic, rope	• Broken fiber • Dissolved fiber • Starch and fillers	• Odor, dust • VOCs
Step 2: Deinking • Flotation or washing	• Deinking sludge • Fiber rings • Fillers, coating	• Broken and dissolved fiber compounds • Starch and fillers • Ink and solvents	• Odor • VOCs
Step 4: Cleaning and screening • Forward cleaners • Fine screens • Reverse cleaners	• Dirt sieves • Plastics	• Same	
Step 5: Washing	• Fiber loss	• Same	• Odor • VOCs
Step 6: Bleaching • Bleach tower(s) • Bleach extractors and washers	• Fiber loss	• Same plus reaction products	• Reaction products from bleaching • VOCs, odor
Step 7: Storage or pressing and shipping	• Fiber loss		
Step 8: Effluent treatment	• Sludge	• BOD, COD, color, TSS, VOCs, sludge	• VOCs, odor

ink. The principal methods for sludge handling and disposal are pressing and incineration, landfilling, reuse as filler in low grades of paper or board, and land application.

Effluent from Processing Secondary Fiber

Processing secondary fiber generates an effluent whose characteristics vary, depending on the type of wastepaper and process used. OCC operations generate an effluent similar to that of a brown paperboard mill. Effluent from deinking has less color but higher concentrations of dissolved and suspended material, the dissolved part originating from starch and fiber fragments and paper additives and the suspended material coming from fillers, coatings, tines, inks, glue, and plastic. The effluent usually is treated in primary and secondary treatment facilities and, in addition, often is treated at the mill, or at least filtered, to facilitate reuse.

Air Emissions

Air emissions from secondary-fiber operations are principally fugitive dust (from the handling and transport of the waste paper) and potentially odor and VOCs from contaminants in the waste paper. Trace amounts of sulfur in the raw material can produce reduced sulfur compounds that contribute to odor. Air emissions are minimized by good housekeeping and enclosed operations, including minimizing areas where the pulp could decay. In addition, the use of oxidants, such as peroxide, in the process can reduce or eliminate odor formation. In plants where the pulp is bleached, additional emissions could come from residual bleaching chemicals and VOCs formed as reaction by-products.

Energy and Water

Modern secondary-fiber plants can be very efficient in water use and benefit from very low heat demand to maintain temperature during processing. In well-balanced systems, water use often is determined by the need to bleed contaminants entering with the raw material. In the production of packaging grades of paper, the water use can be very low, and even in the production of deinked market pulp, water use of 5,000 gallons/ton or less is possible. Water use can be reduced further by installing internal systems for treating circulating process water.

Installing sludge-dewatering facilities and incinerators with heat recovery can make the secondary-fiber process more than self-sufficient in terms of thermal (steam) energy.

Paper and Paperboard

The paper and paperboard industry today has the capacity to produce more than 85 million tons of products annually. A characteristic feature of the industry is the paper machine, in which a continuous sheet of paper or paperboard is produced for a variety of end users. The production data for 1989 for paper machines operating in the 600 mills in the country are in Table 23.5.

In general, the manufacturing process includes the following process steps. First is

Table 23.5. Production Data for Paper and Paperboard Machines

Product	Number of Machines	Installed Capacity (tons per year)
Printing and Writing	421	23.6
Newsprint	40	6.3
Tissue	193	5.9
Packaging	176	5.4
Total Paper	840	41.2
Unbleached Kraft	67	20.9
Recycled	187	9.6
Semi-chemical	38	5.8
Solid Bleached	26	4.7
Total Paperboard	318	41.0

stock preparation, where the fibers are refined, cleaned, screened, and blended to meet the product demands and where additives and paper-making chemicals are added. A continuous sheet is then formed as a single sheet or in layers and is pressed and dried in the drying section. Sizing or a coating may then be applied to the sheet, or it may be calendared for smoothness. The product sheet usually is delivered on large rolls for subsequent finishing (cutting into sheets or made into finished products) and is packed for shipping. Table 23.6 outlines the process steps and the common potential waste products.

Solid Waste in Paper and Paperboard Production

The principal solid waste streams generated are "trim," "broke," and "rejects" and the sludge from primary and secondary wastewater treatment. Rejects systems often are multistage recovery systems for minimizing the loss of good fiber. Trim and broke are reprocessed into the process, used in different products, or used as fuel. Effluent primary sludge is minimized by using in-process "saveals" for fiber recovery and eliminating fiber-rich water discharges. Nonfibrous solids losses are mini-

Table 23.6. Processes and Wastes at Paper and Paperboard Mills

	Potential Waste Products		
Process Step	**Solid Waste**	**Waste**	**Air**
Step 1: Stock preparation • Refining • Cleaning, screening • Broke handling • Wet-end additives	• Rejects • Broke	• Dissolved wood, additives • Broken fibers	
Step 2: Forming and pressing • White water system • Forming • Pressing • Vacuum pumps	• Wet end broke	• Same	• Moisture heat • VOCs
Step 3: Drying • Steam drying • Gas dryers • Size presses	• Dryer broke	• Heat	• VOCs, dust • Heat
Step 4: Coating and screening • Coating preparation • Coating application • Coater drying	• Coater broke • Coating material	• Coating residues	• VOCs, dust
Step 5: Finishing and packaging • Rewinders • Roll handling • Packaging • Sheeters			• Dust
Step 6: Effluent treatment	• Sludge, TSS	• BOD, COD, TSS • VOCs, filler additives	• VOCs, odor

Table 23.7. Waste Reduction Measures for Wood Pulping

	Principal Impact Area				
Waste Reduction Measure	**Solid Waste**	**Effluent Load**	**Air Emission**	**Water Use**	**Energy Use**
Dry debarking	•	•		•	•
Chip-pile management	•		•		
Chip screening	•	•			•
Heat recovery from TMP (refiner) exhausts			•		•
Solids recovery/incineration of (BC)TMP effluent		•			•
Pitch recovery/removal system	•	•			
Fully enclosed brown stock gas/vent systems			•		
Improved brown stock washing		•			•
Black-liquor spill recovery systems		•			•
Closed/multistage screen room	•	•			
Condensate stripping/methanol recovery		•	•		•
Improved soap/turpentine recovery		•	•		•
Extended cooking/oxygen bleaching		•		•	
Chlorine-free bleaching with countercurrent recovery		•		•	
Lime slaker grits (rejects) washer	•				
Separate cooling water service and process water				•	•
Defined warm and hot water systems				•	•
Reuse of secondary heat				•	•

mized by improving retention of filler material and reusing process white waters wisely and efficiently.

Liquid Discharges

Liquid discharges from the process contain solids, mainly fiber, fillers, and colloidal and dissolved material. The fiber and fillers are minimized, as described above, and are reused in the sheet. For the most part, dissolved and colloidal materials are by-products of the refining of the fibers, originate in the chemical additives, or are carried over from the pulp mill. Discharges of dissolved material are minimized by washing the stock and displaced carryover from pulp mills and by practicing good water reuse and conservation strategies that reduce the volume and the concentrations of waste in the wastewater.

Air Emissions

The predominant air emission from a paper mill is the hot, humid air exhausted from dryers and the machine room. The exhaust can contain particulates, essentially fiber, and VOCs originating from paper-making additives or carryover from the pulp mill. In cases where exhaust-heat recovery is practiced, scrubbers or particulate-recovery equipment can reduce waste.

Table 23.8. Waste Reduction Measures for Secondary-Fiber Processing

Waste Reduction Measure	Principal Impact Area				
	Solid Waste	Effluent Load	Air Emission	Water Use	Energy Use
Collect baling wire for scrap metal sale.	●				
Recycle pallets and reject fiber and use as fuel.	●				●
Return rejected raw material or use as fuel.	●				●
Recover and press rejects for transport or use as fuel.	●				●
Dewater deinking sludge for transport, use in lesser paper grades, or use as fuel.	●				●
Install sludge and waste fuel boiler with heat recovery.	●				●
Practice closed-loop water reuse and countercurrent flow to minimize process water use.		●		●	
Install system for removing contaminants from internal water.	●		●		
Practice good housekeeping and odor control.					

Table 23.9. Waste Reduction Measures for Paper and Paperboard Mills

Waste Reduction Measure	Principal Impact Area				
	Solid Waste	Effluent Load	Air Emission	Water Use	Energy Use
Collect and reuse trim and broke paper.	●				
Install savealls and needed white-water storage for fiber recovery and water conservation.	●	●		●	
Install dry vacuum pumps or water-recycling system for liquid ring vacuum pumps.		●		●	●
Install heat recovery from dryer exhausts.			●		●
Maintain and upgrade machine steam and condensate systems.					●
Install technology for using waste-type raw materials.	●	●			

Energy and Water Conservation

In paper and board production, the greatest inefficiencies and material losses are associated with high water use, which correlates directly with high energy use. Installing filtration equipment for white water—fiber saveålls—and equalizing water storage make it possible to operate the white-water system without adding fresh water. Internal water cleaning systems make it possible to substitute filtered white water for clean water. Separating process cooling and clean water is often necessary to achieve balance in operations and water use. The steam and condensate system must be maintained at optimal conditions for the grades produced. Installing heat recovery for machine exhausts minimizes heat losses.

The waste reduction processes are summarized in Tables 23.7, 23.8, and 23.9.

REFERENCES

1. *Pulp & Paper, North American Fact Book*. Miller Freeman Inc. ISBN 0-87930-254-2 (1992).
2. U.S. Environmental Protection Agency. *Pollution Prevention Opportunity Assessment and Implementation Plan*, EPA 910/9-92-027, 1992.
3. U.S. Environmental Protection Agency. *Model Pollution Prevention Plan for the Kraft Segment of the Pulp and Paper Industry*, EPA 910/9-92-030, 1992.
4. U.S. Environmental Protection Agency. *Pollution Prevention Technologies for the Bleached Segment of the U.S. Pulp and Paper Industry*, EPA/600-R-93/110, 1993.
5. Erickson, D. "Closing Up the Bleach Plant, Industry and Regulatory Update." Paper presented at 1993 TAPPI Pacific Section Seminar, Seattle, WA, September 16, 1993.
6. Campbell, M.E., and W.M. Glenn. "Profit from Pollution Prevention." The Pollution Probe Foundation. ISBN 0-920668-21-6 (1982).
7. Carroll, R.C., and T.P. Gajda. "Mills Considering New Deinking Line Must Answer Environmental Questions." *Pulp and Paper* 201 (September 1990).
8. Whitford, A.D., and S.E. Frase, "Reducing Waste at Longview Fibre Company." *TAPPI Journal* 73(3):167 (March 1990).
9. Weyerhaeuser Company. *Recycling—and Beyond*, Bulletin BC-373-3-93. Tacoma, WA, 1993.
10. Rolf Ryham, Ahlstrom, Recovery Inc. "Kraft Pulping: An Environmentally Adaptable Process." Envirotech Finland '92, Montreal/Thunder Bay, December 8-10, 1992.
11. Diehn, K., and B. Zuercher. "A Waste Management Program for Paper Mill Sludge High in Ash." *TAPPI Journal* 81 (April 1990).

Glossary of Terms

AFB	air force base
AISI	American Iron and Steel Institute
ANPRM	advanced notice of proposed rule making
ANSI	American National Standards Institute
AOX	adsorbable organic halides
API	American Petroleum Institute
ARU	ammonia recovery unit
ASTM	American Society for Testing and Materials
ATF	automatic transmission fluids
BCSD	Business Council for Sustainable Development
BIF	Boiler and Industrial Furnace
BOD	biochemical oxygen demand
BOF	basic oxygen furnace
BS	British Standard
BS&W	bottom sediment and water
BSI	British Standards Institute
Btu	British thermal unit
C&D	construction and demolition
CAAA	Clean Air Act Amendments
CDQ	coke-dry-quenching
CEMAS	EC environmental management and audit system
CERCLA	Comprehensive Environmental Response, Compensation, and Liability Act
CFC	chlorofluorocarbon
cfm	cubic feet per minute
CFR	Code of Federal Regulations
CIP	clean in place
CLI	commercial leasehold improvements
CMA	Chemical Manufacturers Association
cmd	cubic meters per day
CWA	Clean Water Act
CWMS	comprehensive waste management strategy
DAF	dissolved-air flotation

DCES	direct control evacuation systems
DfE	Design for the Environment
DI	deionized
DOD	Department of Defense
DOE	Department of Energy
DRI	directly reduced iron
DSM	direct steel making
EAF	electric arc furnace
EC	European Community
ECE	Economic Commission for Europe
ED	electrodialysis
EN	electroless nickel
EP	extraction procedure
(U.S.) EPA	Environmental Protection Agency
EPCRA	Emergency Planning and Community Right-to-Know Act
ESAP	Environmental Self-Assessment Program
F.O.B.	Free on board
FCC	fluidized catalytic cracking
FGD	flue gas desulfurization
FIFRA	Federal Insecticide, Fungicide, and Rodenticide Act
FLOCS	Fast-Lube Oil-Change System
FWS	foul-water stripper
FY	fiscal year
GEMI	Global Environmental Management Initiative
gpd	gallons per day
gph	gallons per hour
gpm	gallons per minute
GSA	General Services Administration
HCFCs	hydrochlorofluorocarbons
HFCs	hydrofluorocarbons
HMT	high-mass-transfer
HSWA	Hazardous and Solid Waste Amendments
HVAC	heating, ventilation, and air conditioning
ICC	International Chamber of Commerce
IEC	International Electrotechnical Committee
INEL	Idaho National Engineering Laboratory
IRR	internal rate of return
ISO	International Standards Organization
IVD	ion vapor deposition
IX	ion exchange
LCA	life-cycle analyses
LLDPE	linear low-density polyethylene
LSA	low surface area
MEK	methyl ethyl ketone
mgd	million gallons per day
MIBK	methyl isobutyl ketone
MMA	methylmethacrylate
MSDS	material safety data sheet
NAAQS	National Ambient Air Quality Standards
NADEP	Naval Aviation Depot

NAFG	North American Fire Guardian
NAS	naval air station
NASA	National Aeronautic and Space Administration
NCG	noncondensible gas
NEMA	National Electrical Manufacturers Association
NESHAP	National Emission Standards for Hazardous Air Pollutants
NIPER	National Institute for Petroleum and Energy Research
NMP	N-methyl-2-pyrrolidone
NO_x	oxides of nitrogen
NPCA	National Paint and Coating Association
NPDES	National Pollution Discharge Elimination System
NPV	net present value
NSPS	new source performance standards
NSS	nickel sulfamate strike
NSY	naval shipyard
NVR	nonvolatile residue
O&M	operation and maintenance
OCC	old corrugated containers
ODC	ozone-depleting chemical
OHF	open hearth furnace
OSHA	Occupational Safety and Health Administration
P&W	Pratt & Whitney
PCB	polychlorinated biphenyl
PCE	perchloroethylene
PM10	respirable particulate matter
PMB	plastic-media blasting
POTW	publicly owned treatment works
ppb	parts per billion
ppm	parts per million
psi	pounds per square inch
psig	pounds per square inch gauge
PVC	polyvinylchloride
RCRA	Resource Conservation and Recovery Act
RO	reverse osmosis
SAGE	Strategic Advisory Group for the Environment
SARA	Superfund Amendments and Reauthorization Act
SCAs	supplemental coolant additives
SEA	Single European Act
SOX	Oxides of Sulfur
SRRP	Source Reduction Review Project
STEP	Strategies for Today's Environmental Partnership
SWDA	Solid Waste Disposal Act
TCA	trichloroethane
TCE	trichloroethylene
TCLP	Toxicity Characteristic Leaching Procedure
TDS	total dissolved solids
TQEM	Total Quality Environmental Management
TQM	Total Quality Management
TRI	Toxics Release Inventory
TRS	total reduced sulfur

TSCA	Toxic Substances Control Act
TSD	treatment, storage, and disposal
TSS	total suspended solids
UF	ultrafiltration
UNCED	United Nations Conference on Environment and Development
USDA	United States Department of Agriculture
USFDA	U.S. Food and Drug Administration
UV	ultraviolet
VOC	Volatile Organic Compounds
WAL	waste ammonia liquor
WAVE	Water Alliances for Voluntary Efficiency
WWTP	wastewater treatment plant

Index